PLASMIDS OF MEDICAL, ENVIRONMENTAL AND COMMERCIAL IMPORTANCE

DEVELOPMENTS IN GENETICS
Volume 1

PLASMIDS OF MEDICAL, ENVIRONMENTAL AND COMMERCIAL IMPORTANCE

Proceedings of the Symposium on Plasmids of Medical, Environmental and Commercial Importance held in Spitzingsee, F.R.G., 26-28 April, 1979.

Editors
K.N. TIMMIS
and
A. PÜHLER

1979

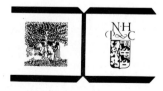

ELSEVIER/NORTH-HOLLAND BIOMEDICAL PRESS
AMSTERDAM · NEW YORK · OXFORD

ISBN for this volume: 0-444-80161-8
ISBN for the series: 0-444-80160-X

Published by:
Elsevier/North-Holland Biomedical Press
335 Jan van Galenstraat, P.O. Box 211
Amsterdam, The Netherlands

Sole distributors for the USA and Canada:
Elsevier North Holland Inc.
52 Vanderbilt Avenue
New York, N.Y. 10017

Printed in The Netherlands

PREFACE

Plasmids are autonomous genetic elements that encode a wide range of functions and that are found in many different bacteria. Recently, it has become increasingly evident that the special characteristics exhibited by bacterial strains which are of importance in medicine, agriculture, commerce and the environment are often plasmid-determined. Such characteristics include (1) a variety of virulence properties, such as exotoxins, haemolysin, adhesion antigens, resistance to antibiotics and the lytic activity of serum (animal pathogens) and tumour-formation ability (plant pathogens), (2) the ability of nitrogen-fixing Rhizobium strains to nodulate roots of legumes, (3) antibiotic production by Streptomycetes, and (4) the ability of certain strains of Pseudomonas and related organisms to detoxify a range of deleterious organic environmental pollutants, including xenobiotics.

Progress in the study of these important bacterial properties accelerated markedly during very recent years with the finding that many are plasmid-determined and the development of important technical advances for the analysis and manipulation of plasmid genes and functions. This rapid progress was accompanied by much excitement and a great expansion in the number of groups working in the field. We therefore considered the time to be ripe for a meeting devoted to applied aspects of plasmids. The excitement generated during the Symposium: "Plasmids of Medical, Environmental and Commercial Importance" that was held in Spitzingsee, Bavaria, from April 26-29, 1979, amply confirmed this conviction. The present volume contains by a series of review articles that summarize fundamental and applied aspects of plasmids. It is the first attempt to present the state of the art in this significant and rapidly growing field of the genetics, biology and manipulation of plasmids that encode bacterial properties of direct importance to Man.

The Spitzingsee Symposium was generously financed by the Bundesministerium für Forschung und Technologie. Additional support was provided by Bayer AG, Beckman Instruments, BioRad, Biotronik, Gilford Instruments GmbH, Hoechst AG, ICI Corporate Laboratory, Kontron Technik GmbH, LKB Instruments, Miles GmbH, Philips, Sandoz Forschungsinstitut, Sartorius GmbH, Schering AG, Shell Biosciences Laboratory, Smith-Kline, Springer-Verlag and Uniequip.

We wish to express our gratitude to A. Chakrabarty, N. Datta, W. Goebel, D. Helinski, and D. Hopwood who skillfully chaired the scientific sessions and initiated many stimulating discussions; to Anneliese Hoffman, Irmgard Schallehn and Joan Timmis who typed endlessly and who cheerfully and energetically assisted in the organization of the meeting, and to Gillian Ritchie of Elsevier/North Holland who contributed much effort in the production of this volume and with whom it was a pleasure to collaborate.

K.N. Timmis A. Pühler

CONTENTS

VII - GENE CLONING WITH PLASMIDS

REVIEW ARTICLES

© 1979 Elsevier/North-Holland Biomedical Press
Plasmids of Medical, Environmental and Commercial Importance
K.N. Timmis and A. Pühler, editors

PLASMID CLASSIFICATION: INCOMPATIBILITY GROUPING

NAOMI DATTA
Royal Postgraduate Medical School, Du Cane Road, London W12 0HS

INTRODUCTION

Incompatibility of two plasmids is shown by their inability to coexist stably in the same cell line. Distinguishable variants of a single plasmid are incompatible with one another and with their parent plasmid. Examples are F' plasmids[1], mini-derivatives of F or other plasmids[2,3] and cloning vehicles such as ColE1 with various deletions and/or inserted genes[3]. Plasmid pSC101 was once thought to be an exception to this rule, being derived from R6[4] yet compatible with its parent[5] but subsequently its DNA sequence was found to be different from any part of R6, and its origin was reinterpreted[6].

When two naturally-occurring plasmids are incompatible their relationship may be analogous to that of the laboratory-constructed exemplars, i.e. they may both be derived from a common ancestor. But the observed phenomenon of incompatibility does not prove such a relationship.

Naturally occurring plasmids, in all bacterial genera where they have been examined, show compatibility and incompatibility relationships. Examples are in staphylococci[7], streptococci[8], pseudomonads[9] and in the enterobacteria. Thus in any bacterial genus a classification may be made by testing pairs of plasmids, in a chosen host strain, for their ability to co-exist; incompatible plasmids are assigned to the same group, compatible ones to separate groups. Plasmids of staphylococci[7] and pseudomonads[9] have been classified in this way. This review, however, concerns the classification of those plasmids, conjugative and non-conjugative, which can be transferred to *Escherichia coli* K12.

METHODS AND DEFINITIONS

To test a pair of plasmids, introduce (by conjugation, transduction or trans formation) plasmid B to a culture already carrying plasmid A. Select for a character of plasmid B and examine transcipients for the continued presence of a character (or characters) of plasmid A. If plasmid A is eliminated from every clone, then A and B appear to be incompatible. Confirm this by testing in the other direction, introducing plasmid A to cells carrying plasmid B. When the resident is retained in all clones tested, the two plasmids are probably compatible. If the resident is eliminated from some but not all, grow a clone carrying both in non-selective conditions and then test for the continued presence of both plasmids. If the double carriage is unstable, the plasmids are considered incompatible; if stable, the culture is used as a donor with separate selection for transfer of each plasmid. Their stable and separate co-existence indicates compatibility.

It may be asked how instability is defined: obviously the stability of each plasmid in a double must be compared with its stability when replicating alone in the same host. Difficulties arise in testing compatibilities of a plasmid that is itself unstable in the chosen host; such instability is not regularly reproducible but shows itself, for unknown reasons, more in some clones than in others. This has led to mistaken classification of some plasmids (see below). But in general compatibility or incompatibility of two plasmids is clear and unambiguous, and the necessity to decide upon what degree of instability is required to define incompatibility does not arise.

CONJUGATIVE PLASMIDS

The first transferable antibiotic resistance (R) plasmids were identified in Japan and were subdivided into two classes, fi^+ (fertility inhibition) and fi^-, according to whether or not they inhibited F-mediated conjugation when present with plasmid F in the same culture[10]. Members of each class, fi^+ or fi^-, were compatible with F and with members of the other class, but within each class, pairs of plasmids were incompatible.

This then, was the beginning of a plasmid classification that has since been much extended. The schemes of classification worked out in different laboratories[11,12,13] using different plasmid collections from different bacterial genera have for the most part identified the same incompatibility (Inc) groups. Many plasmids fall neatly and unambiguously into an Inc group and members of some of the groups are common all over the world, as are the bacterial species that harbour them.

Currently recognised Inc groups of conjugative plasmids are B, C, D. E, FI, FII, FIII, FIV, H, Iα, I_2, Iγ, Iδ, Iζ, J, K, M, N, P, T, V, W, X. The characteristics and constituent members of most of these groups are listed in the recent Cold Spring Harbor book[14]. Not listed there are 1) IncD, a newly designated group[15] that includes two plasmids R711b and R778b, rather unstable in *E. coli* K12, and previously assigned to IncX and 2) IncE, that includes plasmids identified in fish pathogens[16]. Not all identified plasmids fall into these groups.

Relegated Inc groups are A, FV, L and S. The only IncA plasmid available is

RA1[17] that appeared compatible with IncC plasmids, although evidently related to them[18]. Many plasmids are incompatible with RA1 and with the IncC plasmids and have been called IncA-C[14]. Tested again this year (by N. Datta) RA1 was clearly incompatible with the prototype IncC plasmids pIP55 and pIP40a[12] and therefore the designations IncA and A-C can be omitted from the list of groups. IncFV should not have been so designated (see next section). The plasmids formerly called IncL belong to IncM[19] and those called IncS belong to IncH[20].

Correlation of Inc grouping with pilus type

Plasmids of groups FI, FII etc. were so called because they all determine F pili[21,22]. Their transfer operons show homology both physically in heteroduplex studies[23] and genetically[24]. The plasmid Fo*lac* from *Salmonella typhi* was called IncFV because its pili are receptors for filamentous F-specific phages like fd, that attach to pilus tips. Its pili, however, are now shown to be serologically unrelated to F pili and not to allow attachment of RNA phages like MS2 that absorb to the sides of F pili; its transfer genes are physically and genetically unrelated to those of F[25]. No known plasmids are incompatible with it, so it need not be given any Inc grouping. IncD plasmids determine yet other pili to which phage fd but not MS2 attaches[25].

The plasmids of Incα have serologically identified pili (I pili) whose tips are receptors for phage If1[26,27]. Other members of the "I complex"[28] also produce receptors for phage If1 and this was taken to indicate that they determined I pili. The tips of pili, however, are evidently relatively non-specific phage receptors, so members of the "I complex" do not necessarily have closely related transfer genes.

The sides of pili of IncP plasmids are receptors for phage PRR1[29,30] which only lyses bacteria carrying IncP plasmids, bacteria which may belong to a wide range of Gram negative genera. A group of other plasmid-specific phages, of which PR4[30] is an example, attaches to the tips of IncP-determined pili and also to the tips of the morphologically different pili determined by plasmids of IncW[30] and IncN[31]. D. E. Bradley (personal communication) has recently identified two new kinds of pili, determined by RA1 (IncC, see above) and by R391 (IncJ) respectively. It remains to be seen whether all plasmids in these groups determine the same pili as the prototypes.

All known IncT plasmids determine another type of pilus[32], now designated T pili[33]. No pili have been identified on bacteria-carrying the IncX plasmid, R6K[34], but R485, a plasmid incompatible with R6K[35], determines the production of long thin pili[36,37].

Only relatively few plasmids have been examined for pilus-determination, but as far as is known, all plasmids within an Inc group have pili of one type. (IncX plasmids may prove to be an exception.) Between groups, pili tend to be different but exceptions are among the plasmids that determine F pili; their transfer operons are closely related but they are separated by their compatibility properties. The same probably applies to members of the IncI groups (see below).

<u>Correlations of Inc group with specificity of surface exclusion</u>

Usually plasmids exhibit surface exclusion specifically towards others in the same Inc group. But not all conjugative plasmids show surface exclusion and when they do, it is sometimes exerted against members of compatible, yet related groups eg. the distinguishable IncF or IncI groups[38,28].

<u>Non-correlation of Inc group with plasmid-determined phenotypic characters</u>

The carriage of a particular antibiotic resistance gene shows little correlation with plasmid Inc group and this is now understood to be because the genes determining many phenotypic characters of bacteria, including those for drug resistance, are carried on transposable DNA sequences and widely disseminated among unrelated plasmids. LeMinor *et al.*[38a] have shown that metaboli and haemolysin genes are not confined to plasmids of particular Inc groups. Carriage of colicin-determining genes is more often correlated with Inc group eg. I colicins are usually, but not always, determined by plasmids of the IncI groups. Of the many other plasmid-determined characters of bacteria, more anc more are proving to be transposable, and so correlation with Inc group is unlikely.

NON-CONJUGATIVE PLASMIDS

Large non-conjugative plasmids fall into the same Inc groups as conjugative ones[39,40]. Small ones might be expected to do likewise, since when large plasmids are cut down to mini-size, they retain their Inc grouping[2]. No naturally-occurring small plasmids, however, although themselves showing incompatibility properties, have so far been found to fall into any of the groups identified for the conjugative plasmids. Most small plasmids exist as multiple copies in their host cells. Smith *et al.*[41] tested the Inc properties of non-conjugative plasmids. One large Inc group, called SSu, consisted of plasmids determining streptomycin and sulphonamide resistance, and incompatible with a laboratory recombinant ampicillin-sulphonamide (ASu) R plasmid[42]. Into this group also fell a naturally-occurring ampicillin-streptomycin sulphonamide (ASSu) R plasmid, NPT7. This Inc group, whose members are very common among bacteria of many species, has been designated IncQ[43]. The other small plasmids tested by Smith *et al.*, determining ampicillin resistance, tetracycline resistance or colicin E1, E2 or E3 were all compatible with one another. We have found five small plasmids, pSC101, ColE1, pHH509, R831a and an IncQ plasmid, R300B, all compatible with one another as well as with conjugative plasmids of all known Inc groups[44]. There are thus no designations, except IncQ, for incompatibility groups among the small non-conjugative plasmids.

ColE1 is a multicopy plasmid of which many laboratory-constructed variants exist. The variants are incompatible with one another, different pairs showing different rates of segregation[3].

Correlation of Inc group with DNA homology

Plasmids within an Inc group show much DNA homology, and plasmids of different groups, little. This is generally true of the conjugative plasmids[45,46,47] and of the SSu or IncQ nonconjugative ones[41,48]. Exceptions are of two kinds, much homology between compatible plasmids and little homology between incompatible ones. The former is seen among the F and I groups eg. FI and FII plasmids are compatible, but have long DNA sequences in common which can be explained by their having common, or closely related, transfer operons. The same is true of plasmids of groups Incα, Iδ, Iζ which also probably share transfer genes. This could be interpreted as meaning that the whole F or I transfer operon has at some time been inserted into plasmids of different groups; alternatively it could mean that the F and I plasmids, retaining their respective transfer genes, have evolved into separate incompatibility groups. Compatibility relationships between members of the various F and I Inc groups are sometimes complex and difficult to interpret (see below) which seems to favour the latter hypothesis.

IncB plasmids show considerable homology with those of IncIα[45,46]. In addition, they are incompatible with some I groups, viz Iζ[46] and Iγ[49]. Iζ plasmids have I pilus genes (at least provide receptors for phage If1) and are incompatible with Iα and IncB plasmids and were therefore called Iζ[46], but I am not sure that they should not simply be described as belonging to both groups Iα and B, simultaneously, since incompatibility grouping is a descriptive classification rather than an interpretation of phylogenetic relationships. We have suggested[46] that IncB plasmids are part of an "I complex" in view of their homologous DNA sequences and overlapping incompatibility properties. IncB plasmids, however, differ from those of the "I complex" in being transferable to, and replicating stably in Proteus strains[50], (H. Richards, unpublished).

$IncI_2$ plasmids[45,38a,51] provide an example of lack of DNA homology between presumed related plasmids, though not in this case incompatible ones. They were called I_2 because like $I\alpha(I_1)$ plasmids they determine receptors for phage If1 and constitute a separate Inc group. They have not enough DNA homology with $I\alpha$ plasmids to allow of common transfer genes (unlike $I\gamma$ or $I\delta$ plasmids) and thus probably they are related to the other IncI groups only in so far as F*olac* is related to the IncF group plasmids and IncW and IncN plasmids to those of IncP i.e. the tips of their pili act as receptors to the same phages.

True examples of incompatible plasmids showing little DNA homology are found in group IncH. Plasmid TP116 was found some years ago to be incompatible with IncH plasmids, yet its DNA did not hybridise to a significant extent with other IncH plasmids[45]. IncH plasmids, all incompatible with one another, were therefore subdivided into groups H1 and H2 on the basis of DNA hybridisation tests as well as on compatibility/incompatibility relationships with other plasmids (see below). H1 and H2 plasmids have now been identified in many bacteria[52,20,38a] and a third IncH subgroup, H3 has been found, incompatible with the others, but showing little homology with either H1 or H2[47].

Plasmid molecular weights

Within an Inc group, plasmids have molecular weights of the same order eg. IncW plasmids are 25 megadaltons or less, IncC and IncH plasmids much larger, often over 100 megadaltons[14]. (This applies to the H1 and H2 subdivisions of IncH.)

Anomalous incompatibility relationships

The results of incompatibility tests are not always reproducible eg. plasmid RA1 which once seemed compatible with pIP55 and pIP40a is now incompatible (see above). ColB-K98 was compatible with FI and FII plasmids[53] and therefore allotted to IncFIII, but it is sometimes incompatible with IncFII plasmids[54]. The IncFIV plasmid, R124 also behaves rather unpredictably in tests with other IncF plasmids[54,15].

Not infrequently incompatible plasmids appear to be arranged in a 'hierachy', in which one is more likely than the other to be eliminated, or to be lost by segregation[55,39]. Testing pairs of plasmids for compatibility in both ways (transfer B into a culture carrying A and A into a culture carrying B) detects this.

Plasmids of the H1, but not the H2, subdivision of IncH are incompatible with autonomous F. However F incompatibility was lost from an H1 plasmid along with tetracycline resistance, the plasmid retaining its IncH character[56], so incompatibility with F is not a reliable distinguishing marker between the H1 and H2 subgroups if these are defined by their DNA composition (see above). Plasmids of IncFI are subdivisible according to their ability to coexist with autonomous F and plasmid MP10, which is compatible with F and some other FI plasmids but not with those designated FI*me* (because they have been found in many strains of Middle Easten origin)[40].

These results are difficult to interpret. They suggest that not all incompatibility relations have the same underlying mechanism. A diversity of incompatibility mechanisms is also suggested by recent experimental cloning of incompatibility determinants[57,58,59].

Unreproducible results are rare and the other odd relationships, even though not understood, help in identifying wild plasmids.

CONCLUSIONS

On logical grounds and on experimental evidence it has been accepted that incompatibility is to be expected, and is seen, between two plasmids whose replication is controlled by the same regulatory mechanism[60,61,62]. Each is subject to both sets of regulatory control(s). The expression of incompatibility is different, as would be expected, with different control systems eg. more rapidly expressed with plasmids normally present as one copy per chromosome than with multicopy plasmids. Novick and Hoppensteadt[63] dismiss mutual interference with replication as the basis of incompatibility whenever both plasmids can replicate in incompatible heteroplasmid cells. But some replication of both plasmids, before they segregate into separate clones, is surely not inconsistent with the view that incompatibility results from their having the same, or related, replication control?

If all incompatibility between two plasmids was of this nature, it would be a good basis on which to classify wild plasmids since the replication genes may be considered as the very nub of the whole plasmid. An ideal classification would be one that rested upon recognition of the nub. With the current rapid advances in techniques of molecular genetics perhaps such an ideal may be achieved, although it is already evident that the genetic determinants of copy-number control and incompatibility are complex[57,58]. Meanwhile "steam-age" classification by incompatibility testing, though expensive in man hours (often woman hours) can ofter answer epidemiological or other questions about the ecology of plasmids.

REFERENCES

1. Echols, H. (1963) J. Bacteriol, 85, 262-268.
2. Timmis, K., Cabello, F. and Cohen, S. N. (1975) Proc. Nat. Acad. Sci. (USA) 72, 2242-2246.
3. Warren, G. and Sherratt, D. (1978) Molec. gen. Genet, 161, 39-47.
4. Lebek, G. (1963) Zbl. Bakt. Abt. 1, Orig. 188, 494-505.
5. Cohen, S. N. and Chang, A. C. Y. (1973) Proc. Nat. Acad. Sci. (USA) 70, 1293-1297.
6. Cohen, S. N. and Chang, A. C. Y. (1977) J. Bacteriol, 132, 734-737.
7. Novick, R. P., Cohen, S., Yamamoto, L. and Shapiro, J. A. (1977) DNA Insertion Elements, Plasmids and Episomes, Cold Spring Harbor Laboratory, USA, pp. 657-662.
8. Jacob, A. E. and Hobbs, S. J. (1974) J. Bacteriol, 117, 360-372.
9. Jacoby, G. A. (1977) DNA Insertion Elements, Plasmids and Episomes, Cold Spring Harbor Laboratory, USA, pp. 639-656.
10. Nakaya, R., Nakamura, A. and Murata, Y. (1960) Biochem. Biophys. Res. Commun, 3, 654-659.
11. Grindley, N. D. F., Grindley, J. N. and Anderson, E. S. (1972) Molec. gen. Genet, 119, 287-297.
12. Chabbert, Y.-A., Scavizzi, M. R., Witchitz, J. L., Gerbaud, G. R. and Bouanchaud, D. H. (1972) J. Bacteriol, 112, 666-675.
13. Datta, N. (1975) Microbiology - 1974, American Society for Microbiology, Washington D.C., pp. 9-15.
14. Jacob, A. E., Shapiro, J. A., Yamamoto, L., Smith, D. I., Cohen, S. N. and Berg, D. (1977) DNA Insertion Elements, Plasmids and Episomes, Cold Spring Harbor Laboratory, USA, pp. 607-638.
15. Unpublished observations, Division of Enteric Pathogens, Central Public Health Laboratory, London N.W.9. and Department of Bacteriology, Royal Postgraduate Medical School, London W.12.
16. Aoki, T., Kitao, T. and Arai, T. (1977) Plasmids, Medical and Theoretical Aspects, Springer Verlag, pp. 39-45.
17. Hedges, R. W. and Datta, N. (1971) Nature (London), 234, 220-221.
18. Datta, N. and Hedges, R. W. (1972) Ann. Inst. Past. 123, 879-883.
19. Richards, H. and Datta, N. (1979) Plasmid, 2, in press.
20. Taylor, D. E. and Grant, R. B. (1977) Antimicrob. Ag. & Chemother, 12, 431-434.
21. Meynell, E., Meynell, G. G. and Datta, N. (1968) Bact. Rev, 32, 55-83.
22. Hedges, R. W. and Datta, N. (1972) J. gen. Microbiol, 71, 403-405.
23. Sharp, P. A., Cohen, S. N. and Davidson, N. (1973) J. Mol. Biol, 75, 235-255.
24. Alfaro, G. and Willetts, N. (1972) Genet. Res, 20, 279-289.

25. Bradley, D. E. and Meynell, E. (1978) J. gen. Microbiol, 108, 141-149.

26. Meynell, G. G. and Lawn, A. M. (1968) Nature (London), 217, 1184-1186.

27. Meynell, E. (1978) Pili, International Conferences on Pili, Vienna VA 22180 USA, pp. 207-235.

28. Hedges, R. W. and Datta, N. (1973) J. gen. Microbiol, 77, 19-25.

29. Olsen, R. H. and Thomas, D. D. (1973) J. Virol, 12, 1560-1567.

30. Bradley, D. E. (1976) J. gen. Microbiol, 95, 181-185.

31. Bradley, D. E. (1979) Submitted to Plasmid.

32. To, C.-M., To, A. and Brinton, C. C. (1975) Abstracts of the ASM Annual Meeting p. 259.

33. Bradley, D. E. (1979) unpublished.

34. Kontomichalou, P., Mitani, M. and Clowes, R. C. (1970) J. Bacteriol, 104, 34-44.

35. Hedges, R. W., Datta, N., Coetzee, J. N. and Dennison, S. (1973) J. gen. Microbiol, 77, 249-259.

36. Bradley, D. E. (1978) Plasmid, 1, 376-387.

37. Bradley, D. E. (1978) Pili, International Conferences on Pili, Vienna VA 22180,

38. Willetts, N. and Maule, J. (1974) Genet. Res. Camb, 24, 81-89.

38a. LeMinor, L., Coynault, C., Chabbert, Y., Gerbaud, G. and LeMinor, S. (1976) Ann. Microbiol. (Inst. Pasteur) 127B, 31-40.

39. Datta, N. and Hedges, R. W. (1973) J. gen. Microbiol, 77, 11-17.

40. Anderson, E. S., Threlfall, E. J., Carr, J. M., McConnell, M. M. and Smith, H. R. (1977) J. Hyg. (Camb.), 79, 425-448.

41. Smith, H. R., Humphreys, G. O. and Anderson, E. S. (1974) Molec. gen. Genet, 129, 229-242.

42. Anderson, E. S., Keleman, M. V., Jones, C. M. and Pitton, J.-S. (1968) Genet. Res. (Camb.), 11, 119-124.

43. Grinter, N. J. and Barth, P. T. (1976) J. Bacteriol, 128, 394-400.

44. Barth, P. T., Richards, H. and Datta, N. (1978) J. Bacteriol, 135, 760-765.

45. Grindley, N. D. F., Humphreys, G. O. and Anderson, E. S. (1973) J. Bacteriol, 115, 387-398.

46. Falkow, S., Guerry, P., Hedges, R. W. and Datta, N. (1974) J. gen. Microbiol, 85, 65-76.

47. Roussel, A. F. and Chabbert, Y.-A. (1978) J. gen. Microbiol, 104, 269-276.

48. Barth, P. T. and Grinter, N. J. (1974) J. Bacteriol, 120, 618-630.

49. Datta, N. Unpublished observations.

50. Datta, N. and Hedges, R. W. (1972) J. gen. Microbiol, 70, 453-460.

51. McConnell, M. M., Smith, H. R., Leonardopoulos, J. and Anderson, E. S. (1979) J. infect. Dis, 139, 178-190.

52. Anderson, E. S., Humphreys, G. O. and Willshaw, G. A. (1975) J. gen. Microbiol, 91, 376-382.

53. Frydman, A. and Meynell, E. (1969) Genet. Res. (Camb.), 14, 315-332.

54. Gasson, M. J. and Willetts, N. S. (1975) J. Bacteriol, 122, 518-525.

55. MacFarren, A. C. and Clowes, R. C. (1967) J. Bacteriol, 94, 365-377.

56. Smith, H. R., Grindley, N. D. F., Humphreys, G. O. and Anderson, E. S. (1973) J. Bacteriol, 115, 623-628.

57. Timmis, K. N., Andrés, I. and Slocombe, P. M. (1978) Nature, 273, 27-32.

58. Manis, J. J. and Kline, B. C. (1978) Plasmid, 1, 492-507.

59. Crosa, J. H., Luttropp, L. K. and Falkow, S. (1978) J. mol. Biol, 124, 443-468.

60. Cabello, F., Timmis, K. and Cohen, S. N. (1976) Nature (London), 259, 285-290.

61. Uhlin, B. E. and Nordström (1975) J. Bacteriol, 124, 641-649.

62. Matsubara, K. and Otsuji, Y. (1978) Plasmid, 1, 284.

63. Novick, R. P. and Hoppensteadt, F. C. (1978) Plasmid, 1, 421-434.

© 1979 Elsevier/North-Holland Biomedical Press
Plasmids of Medical, Environmental and Commercial Importance
K.N. Timmis and A. Pühler, editors

MECHANISMS OF PLASMID INCOMPATIBILITY

KENNETH N. TIMMIS
Max-Planck-Institut für Molekulare Genetik, Berlin-Dahlem, BRD.

Plasmid Inheritance Functions

Plasmids are extrachromosomal genetic elements that encode a wide range of metabolic functions. Despite their phenotypic diversity, all plasmids have in common one essential characteristic, namely the possession of determinants that collectively constitute the minimal plasmid replicon, and that are required for controlled plasmid replication and stable inheritance in dividing bacteria*. These determinants include (a) a specific DNA sequence at which initiation of plasmid replication takes place (origin of replication), (b) a mechanism for the control of initiation of replication which regulates the cellular concentration of plasmid copies (the copy control system), and (c) in some instances, one or more essential plasmid replication genes[1,2]. Different plasmids are maintained at different copy numbers ranging from 1 to 50 plasmid copies per genome equivalent, but any given species will be maintained at a constant level in a particular host bacterium grown under defined conditions. Plasmid replication is therefore strictly regulated. Because plasmid mutants with altered copy numbers may be readily isolated[3,4,5], at least part of the replication control system is plasmid-determined.

Most low copy number plasmids seem to be as stably inherited in dividing bacteria as high copy number plasmids. This indicates that at least the former type (and perhaps also the latter) must utilize a specific mechanism which ensures that one or more plasmid molecules are distributed to each daughter bacterium at cell division. If this were not the case, i.e. if the cellular popula-

*It should be noted that plasmids may not be stable in all host bacteria to which they can be transferred. Powerful selection fores frequently generate "unnatural" plasmid-host combinations that are unstable in the absence of such pressures. "Unstable" plasmids of this type may, however, be quite stable in alternative hosts.

tion of plasmid molecules were distributed randomly (dynamically) to daughter bacteria, low copy number plasmids would be very unstable (e.g. it can be calculated that at each cell division, 1.6% of the bacteria would lose a plasmid having a copy number of 3.[24]). The mechanism that is postulated to distribute plasmid copies to daughter bacteria at cell division has been termed the partition system. A simple partition system could consist of plasmid-specific membrane attachment sites that bind plasmid molecules prior to cell division, and that are spatially distributed within the cell in such a way that at least one plasmid molecule is transmitted to each daughter bacterium[2,7]. Plasmid partition systems could be composed of both plasmid- and host-encoded components[3,8]. The partition and copy control systems, together with essential host and plasmid replication determinants, collectively may be termed plasmid inheritance functions.

Plasmid Incompatibility

It is well documented that not only single plasmids, but also that mixtures of plasmids, can be stably inherited over many generations within a single bacterial clone. Indeed, wild strains of bacteria usually contain multiple plasmid species and it has been possible to construct in the laboratory bacterial clones that stably maintain up to 8 different plasmids at their normal cellular levels[9]. In contrast to this general situation, it has been found that certain specific pairs of plasmids are incapable of stable maintenance. Two distinct plasmids that can be stably co-inherited in dividing bacteria are said to be compatible, whereas those pairs of plasmids that cannot are termed incompatible. Incompatibility may be defined as the inability of two distinct plasmids to be stably co-inherited in a single clone of dividing bacteria in the absence of continued selection pressure for both plasmid types. This definition is clearly an operational definition and does not presuppose the involvement of particular cellular mechanisms although it does exclude all of those, e.g. plasmid-specified surface exclusion, plasmid- or host-specified restriction, or any other form of plasmid-promoted reduction in host cell recipient ability*; that may prevent initial establishment of the two plasmids in a single bacterial clone. The specificity of incompatibility, and the large number of different plasmid incompatibility types, indicates that non-specific causes such as the competition of two plasmids for a limiting essential host replication function are rarely, if ever, responsible for this phenomenon. Incompatible plasmids are almost always very closely related in that they exhibit extensive polynucleotide sequence homology[10] and, for this reason, incompatibility tests have proved of great value in plasmid classification[11,12].

In dividing bacteria incompatible plasmids may segregate either symmetrically (i.e. after complete segregation, 50% of the bacteria carry only one plasmid type and 50% carry only the other) or, to varying degrees, asymmetrically (after segregation, between 50% and 100% of bacteria carry only the "stronger" plasmid species and between 0% and 50% carry only the "weaker"). Asymmetric segregation is often, though not always, caused by transient or permanent differences in the copy numbers of the two plasmids, e.g. a plasmid that is introduced by transformation or conjugation into a plasmid-carrying bacterial clone will usually be at a numerical disadvantage to the resident plasmid.

Incompatibility, then, being the mutual destabilization of two closely related plasmids, presumably results from detrimental interactions of one or more of their specific inheritance functions. All current models that deal with plasmid incompatibility assume, explicitly or implicitly, that one or more of the inheritance mechanisms cannot distinguish individual plasmid molecules, i.e. that molecules are chosen at random for replication or partition events. This assumption of a "random" step in the plasmid life cycle is crucial for generation of the imbalances in proportions of two incompatible plasmids which subsequently lead to total segregation of one from the other. The random choice of plasmid molecules for replication is of course well established.

---

*Therefore, in order to assign two plasmids to a single incompatibility group it is essential to demonstrate experimentally their unstable co-inheritance after they have been established together in the same cell.

Incompatibility Models

The Replicon Model[13] postulates that the replication of cellular replicons takes place on specific membrane attachment sites. These sites also distribute daughter replicons symmetrically to daughter bacteria. Replication occurs when sufficient plasmid-specified positively-acting initiator product accumulates but overall plasmid copy number is determined by the number of attachment sites per cell. According to this model, compatible plasmids utilize different attachment sites whereas incompatible plasmids compete for the same sites. While the involvement of the cell membrane in replicon duplication and/or segregation is generally accepted, those specific aspects of the model that deal with the control of replication and replicon partitioning are not (see refs. 7,14,15,16 for more detailed discussions of this model).

The Inhibitor Dilution Hypothesis of Pritchard et al.[15,16] proposed that control of DNA replication is effected by a trans-dominant, freely diffusible, inhibitor substance that limits the frequency of replication events. It was supposed that cell growth reduces the cellular concentration of inhibitor and eventually leads to initiation of replication. This, in turn, was thought to produce an increase in inhibitor concentration, due either to a gene-dosage effect or to a burst of inhibitor synthesis during replication, which prevents further initiations until sufficient new cell growth has taken place. Plasmid incompatibility was concluded to result from the mutual inhibition of two related plasmids that produce cross-reacting repressor molecules. Random choice of DNA molecules for replication from a population of two incompatible plasmids generates numerical disproportions in the two DNA species (Fig. 1). These disproportions are amplified during subsequent replication events and eventually lead to the production of bacteria that lack one of the original plasmid species.

Several lines of evidence support the proposal that copy number control is effected by an inhibitor of replication and is responsible for plasmid incompatibility, e.g. plasmid mutants with elevated copy numbers are readily isolated and all of these have altered incompatibility properties[3,4,5]. One type of mutant expresses weaker than wild-type incompatibility (predicted for a replication inhibitor mutation) whereas another expresses stronger

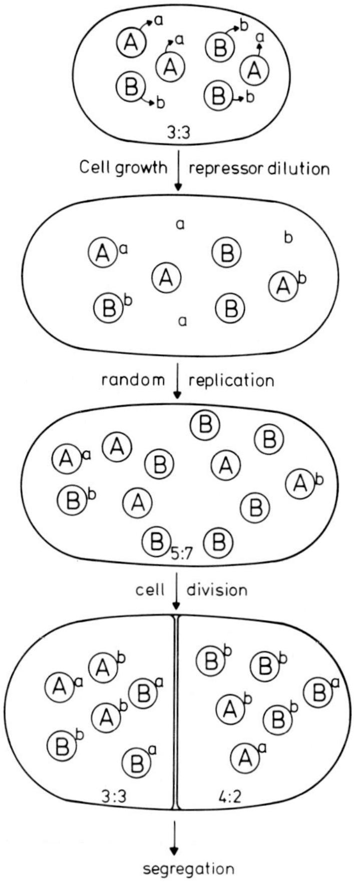

Fig. 1
Plasmid segregation due to random replication. It is assumed for simplicity that inhibitor molecules a and b specified by plasmids A and B inhibit equally well initiation of replication of homologous and heterologous replicons. After some cell growth, the total inhibitor concentration falls below a threshold level and permits the replication of one or more plasmid molecules. The molecules that replicate during each cell cycle are chosen randomly from the total population of replicons: that is, in individual cells, equivalent replication of each type of plasmid rarely occurs. Either equi- or random-partitioning of the plasmid molecules at cell division will generate daughter bacteria containing unequal proportions of the two plasmid species. These inequalities will tend to be amplified during subsequent cell generations until bacteria stably carrying only one plasmid species arise.

than wild-type incompatibility[3,5] (predicted for a replication inhibitor "target" mutation, having reduced binding affinity for inhibitor). Similarly, copy number control and incompatibility of λdv, a plasmid derived by a series of laboratory manipulations from bacteriophage λ, have been shown to be mediated by the λcro (or tof) gene product, an autoregulated repressor protein [17,18]. Furthermore, it is possible to "switch off" and "switch on" the replication of one component of a two component composite replicon, such as a ColE1-pSC101 hybrid plasmid[19], by elevating or reducing its copy number through appropriate laboratory manipulations[20]. Such manipulations also bring about parallel changes in the degree of incompatibility expressed by that component[20]. Th

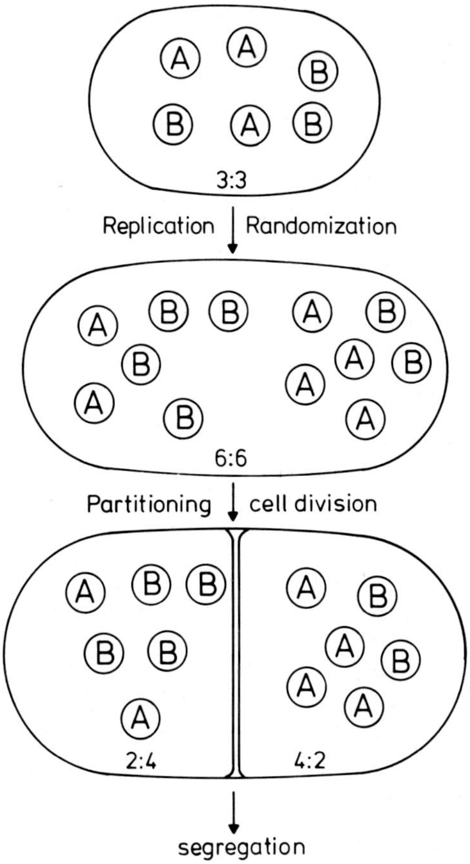

Fig. 2

Plasmid segregation due to random partitioning. It is assumed for simplicity that during the cell cycle one doubling of the molecules of each plasmid type occurs, but that the spatial distribution of the molecules within the cell is random (whether or not specific partition structures are utilized). Random partitioning of plasmid molecules frequently generates daughter bacteria having unequal proportions of the two plasmid types. These disproportions will be amplified by subsequent partition events until bacteria arise that stably carry only one plasmid species.

is consistent with the idea that the cellular concentration of an inhibitor substance determines the replication activity and level of incompatibility expressed by a plasmid replicon.

Two further models of incompatibility that were recently proposed[6,7,14], suggest that random spatial distribution of the copies of a single plasmid species occurs prior to cell division, resulting in a random population of DNA molecules being partitioned to each daughter bacterium. Two or more compatible plasmid types in a single bacterial clone will be partitioned as distinct, individual plasmid populations and will be inherited stably. In contrast, two incompatible plasmid types will be partitioned as a single plasmid population and their random assortment during cell division will generate disproportions of the two species in daughter bacteria, eventually leading to the production of bacteria containing a single plasmid type (Fig. 2). As indicated above, specific partition systems must be responsible for the stable inheritance of low copy number plasmids. It should be stressed however that, according to the above models, random partitioning of plasmids must occur whether or not a specific partition system is utilized. If a partition system is utilized, it is envisaged that compatible plasmids partition independently on different partition sites, just as in the Pritchard Model[15] they are regulated by different repressors. In contrast, incompatible plasmids must exhibit the same partition specificity. Although superficially similar to the membrane attachment sites proposed in the Replicon Model[13], the partition sites discussed here are stated to differ in two fundamental respects, namely that they are not involved in replication, or in the regulation of replication, and that they partition a random population of plasmid copies between daughter bacteria[7].

At this point, there is no evidence directly supporting these latter models of incompatibility. Nevertheless, structural and functional evidence suggests that two distinct mechanisms can be responsible for plasmid incompatibility. Firstly, the incompatibility group incFI plasmid, F, has been shown to contain two physically distinct incompatibility determinants that are located on F DNA segments having F kilobase coordinates 43-46 and 46-49[4]. All of three F copy mutants show loss of the incompatibility function located on the 43-46 kb segment[4]. Similar findings have been made with the incFII plasmid R1[25,26]. Secondly, a PstI endonuclease-generated DNA fragment carrying the R6-5 incompatibility determinant, has been cloned in the pBR322 vector plasmid[21] (the high copy number pBR322 vector is compatible with the low copy number R6-5 plasmid). This hybrid plasmid, because it lacks the R6-5 origin of replication and the essential replication gene RepA, can only replicate using the pBR322 replication system. It cannot therefore be influenced by the R6-5 replication control system. According to the Inhibitor Dilution Model, if the cloned incR6-5 region contains the copy control gene, the pBR322-incR6-5 hybrid plasmid should strongly destabilize a co-existing R6-5 plasmid, but should not itself be destabilized by R6-5. This prediction was tested by introducing the pBR322-incR6-5 hybrid by transformation into bacteria carrying a mini R6-5 plasmid, and observing the inheritance of the two plasmid types. In the majority of clones obtained, the mini R6-5 plasmid was very rapidly lost, in agreement with the prediction that the hybrid plasmid directs the synthesis of high amounts of an R6-5 negatively-acting copy control element. However, in a minority of clones, the mini R6-5 plasmid was retained and the hybrid plasmid was lost[21]. This destabilization of the hybrid plasmid is unlikely to result from the activity of the R6-5 copy control system and probably reflects the abnormal activity of a second R6-5 inheritance function, perhaps its partition system, which in some way interferes with the partitioning and/or replication of the pBR322-*inc*R6-5 hybrid. Similar findings have been observed with hybrid plasmids containing *inc* regions of the R1[22] and ColE1[23] plasmids.

A model that combines elements of all of the three foregoing classes of model was recently proposed[24] and will be described briefly because it offers explanations for several puzzling experimental findings. In essence, it accepts that replication control is effected by an inhibitor of replication and that most of the plasmid copies are distributed randomly to daughter bacteria. It differs from other models in postulating that replication of each plasmid type takes place on a single incompatibility group-specific membrane site and that high concentrations of replication inhibitor prevent productive binding to the site. The model further postulates that stable plasmid inheritance is normally assured because daughter replicons produced during the last replication event prior to cell division are distributed to the two daughter bacteria, i.e. each daughter must acquire at least one plasmid copy. In this model, incompatible plasmids are assumed to utilize the same replication/partition site and therefore, at each cell division, only one plasmid type of an incompatible pair is assured distribution to both daughter bacteria. Because this model postulates that the mechanisms for plasmid copy control, replication, and partitioning interact, it provides a neat explanation for the perplexing finding that high copy number plasmid mutants may be less stably inherited than their low copy parents[3,4], and that the plasmid determinants for two distinct mechanisms of incompatibility are thus far physically indistinguishable[5,21].

CONCLUSIONS

In conclusion it may be stated that the bulk of experimental evidence currently available strongly supports the Inhibitor Dilution Model of control of plasmid DNA replication and indicates that the copy control system is the most important component of plasmid incompatibility. Experiments in which the influence of copy control is abolished do, however, indicate the existence of a second incompatibility mechanism that may result from another plasmid inheritance function such as plasmid partitioning.

ACKNOWLEDGEMENT

The author gratefully acknowledges stimulating discussions with H. Danbara, N. Datta and T. Hashimoto-Gotoh.

REFERENCES

1. Andres, I., Slocombe, P.M., Cabello, F., Timmis, J.K., Lurz, R., Burkardt, H.J. and Timmis, K.N. (1979) Molec. Gen. Genet., 168, 1-25.
2. Timmis, K.N. Andrés, I., Slocombe, P.M. and Synenki, R.M. (1979) Cold Spring Harbor Symp. Quant. Biol., 43, in press.
3. Uhlin, B.E. and Nordstörm, K. (1975) J. Bact., 124, 641-649.
4. Manis, J.J. and Kline, B.C. (1978) Plasmid, 1, 492-507.
5. Danbara, H. and Timmis, K.N. (1979) submitted for publication.
6. Ishii, K., Hashimoto-Gotoh, T. and Matsubara, K. (1978) Plasmid, 1, 435-445.
7. Novick, R.P. and Schwesinger, M. (1976) Nature, 262, 623-626.
8. Wada, C. and Yura, T. (1971) Genetics, 69, 275-287.
9. Barth, P.T., Richards, H. and Datta, N. (1978) J. Bact., 135, 760-765.
10. Grindley, N.D.F., Humphries, G.O. and Anderson, E.S. (1973) J. Bact., 115, 387-398.
11. Chabbert, Y.A., Scavizzi, M.R., Gerband, J.L. and Bouanchaud, D.H. (1972) J. Bact., 112, 666-675.
12. Datta, N. (1975) Microbiology-1974, ed. Schlessinger, D., American Society for Microbiology, Washington, D.C. pp. 9-15.

13. Jacob, F., Brenner, S. and Cuzin, F. (1963) Cold Spring Harbor Symp. Quant. Biol., 28, 329-348.

14. Novick, R.P. and Hoppensteadt, F.C. (1978) Plasmid, 1, 421-434.

15. Pritchard, R.H., Barth, P.T. and Collins, J. (1969) Symp. Soc. Gen. Microbiol., 19, 263-297.

16. Pritchard, R.H. (1978) DNA Synthesis - Present and Future, ed. Kohiyama, M. and Molineux, I., Plenum Press, New York, pp. 1-26.

17. Berg, D.E. (1974) Virology, 62, 224-233.

18. Matsubara, K. (1976) J. Mol. Biol., 102, 427-439.

19. Timmis, K., Cabello, F. and Cohen, S.N. (1974) Proc. Nat. Acad. Sci. U.S.A., 71, 4556-4560.

20. Cabello, F., Timmis, K. and Cohen, S.N. (1976) Nature, 259, 285-290.

21. Timmis, K.N., Andrés, I. and Slocombe, P.M. (1978) Nature, 273, 27-32.

22. Molin, S. and Nordström, K. (1979) J. Bact., in press.

23. Hashimoto-Gotoh, T. and Inselberg, J. (1979) submitted for publication.

24. Hashimoto-Gotoh, T. and Ishii, K. (1979) submitted for publication.

25. Goebel, W. and Kollek, R., personal communication.

26. Danbara, H., Timmis, J.K. and Timmis, K.N. (1979) submitted for publication.

© 1979 Elsevier/North-Holland Biomedical Press
Plasmids of Medical, Environmental and Commercial Importance
K.N. Timmis and A. Pühler, editors

MOLECULAR RELATIONSHIPS BETWEEN PLASMIDS.

PAUL BRODA
Department of Molecular Biology, King's Buildings, Edinburgh University, Edinburgh, EH9 3JR, Scotland.

Plasmid incompatibility was first shown between two F' factors. This result implied that related plasmids could not co-exist. It was then suggested that incompatibility is an aspect of replication control[1]. This substantially accurate hypothesis is the subject of Timmis's article in this volume and will not be discussed further here. However, since replication is recognised as the fundamental attribute of plasmids, the finding that a workable classification could be based on the criterion of incompatibility (that is, upon a replication function) was very satisfactory. Yet classification based on only one criterion is always inadequate. In particular, it is now known that plasmid replication functions only occupy a small region of the total DNA. Also, although incompatibility tests are often straightforward, this is not always the case (see Datta's article). Moreover nothing is established about any relationship that might exist between two plasmids that are compatible.

With the discovery of transposons and the wider use of biochemical methods of studying plasmids there is interest in the extent to which classifications, according to different criteria (e.g., incompatibility, fertility inhibition[2] and drug resistance phenotypes), reflect evolutionary relationships. The most direct method of assessing relationship is through comparison of DNA sequences. In the absence of actual base sequences to compare, the principal methods that have been used are the following:

1. Gross homology. A number of DNA:DNA hybridisation procedures are available[3-5]. The limitation of this approach is that it does not reveal which parts of the molecules are homologous. This led to the use of the next method.

2. Electron microscopy of heteroduplexes. This method has, for instance, been applied to relationships of F and R factors[6,7] and to the evolution of a plasmid in laboratory stocks[8]. Very detailed pictures of relationships can be built up with the help of reference points (as are now available in the form of transposons, for instance). However, it is a complex method, especially with larger molecules, and is not easily applied to the study of large collections of plasmids.

3. A less direct approach has been to suppose that in related regions of plasmids target sites for site-specific endonucleases will be conserved. Complete cleavage of two related plasmids will yield a number of fragments of identical sizes, recognisable by electrophoresis. This was shown to be the case with a series of F-related plasmids for which the extent of homology had already been determined by the heteroduplex method[9].

4. A further test of relatedness of cleavage fragments incorporated DNA: DNA hybridisation. After transferring the fragmentation pattern onto a nitrocellulose filter (Southern transfer[10]) it may then be "probed" with labelled DNA of another plasmid species[11]. This approach is both less laborious than heteroduplex analysis, especially for larger plasmids, and more suitable for detecting small amounts of relatedness, which perhaps would not yield stable heteroduplexes. However, any relatedness is a maximum estimate, since clearly not the whole of a fragment that hybridises need be homologous with the probe DNA.

The first test[12] of whether incompatibility tests constitute a "natural" classification i.e., evolutionary relationship, was when the extent of homology of 16 plasmids of six incompatibility groups from different enterobacterial species was examined by DNA:DNA hybridisation. In general the incompatible plasmids showed more than 80% homology, and the compatible ones showed less than 10% homology. But plasmids of the B and I1 groups had as much as 25% homology whereas among four H plasmids three were closely related but the fourth had no homology with them. Therefore usually but not always incompatibility indicates relatedness, but compatible plasmids can also be related. Both points have been reiterated in the results of other studies. Thus representatives of different incompatibility groups among Pseudomonas plasmids show less homology than do ones belonging to the same group, and there was homology between plasmids of the P-9 group, which contains degradative plasmids from soil bacteria as well as resistance plasmids from clinical isolates (Table 1).

The overall extent of relatedness among plasmids has not been extensively documented. Rather, "clusters" of relatedness have been demonstrated (Table 1). Enough is now known to show that there is extensive homology between plasmids of different strains, isolated in different places. This suggests that plasmids have not arisen in a large number of entirely independent ways. However, whether this merely implies that a particular type of plasmid-forming event occurred repeatedly, or whether a group of related plasmids are of truly common origin, cannot yet be said.

Heteroduplex analysis shows in great detail how homology is distributed in segments. The best example is from studies on F, ColV, three R factors, and an Ent (for enterotoxigenic) plasmid, P307[7,13]. Such studies, in conjunction with genetic analysis, showed that homology was confined to the transfer regions, which explained why most transfer-deficient mutants of F can be complemented by the R factors R100-1 or R6-5[25,26]. But the relationships of the different components of the transfer systems of F and a series of other plasmids were found to be very complex[27].

TABLE 1

SOME GROUPS OF RELATED PLASMIDS

Host species	Plasmids	Criteria[(a)]	References
Enterobacteria	F,R1,R6,R100,ColV,EntP307	DNA	7,13
	Ent(ST+LT)	DNA	14
	Colicin factors group A	size, colicins	15,17,18
	Colicin factors group B	size, colicins	16,17,18
	Inc H1 R plasmids	DNA, inc	12,19
	SmSu R plasmids	DNA, size	20
	Lactose	β-galactosidases	21
Agrobacterium tumefaciens	Ti	DNA, endo	22,23
Pseudomonas spp.	TOL,NAH,R2,pMG18	DNA, inc	(b)
Streptococcus faecalis and S.pyogenes	R plasmids	DNA	24

(a) DNA: homology of DNA; inc: incompatibility; endo: products of endonuclease digestion.

(b) S. Bayley, D. Morris and P. Broda (unpublished data).

As mentioned earlier, plasmids can also be classified according to their transfer systems, or their phenotypes. It is evident that an evolutionary description of a plasmid will involve these as well as the replication function. This has become very clear with the realisation that many functions are carried in transposons. The best-studied of these, transposon A (which specifies a β-lactamase of wide specificity), has been shown to occur naturally in enterobacterial plasmids of at least eight incompatibility groups, which were otherwise unrelated[28]. It is also the basis of the carbenicillin resistance of RP4 in Pseudomonas aeruginosa and the acquisition of ampicillin resistance by indigenous cryptic plasmids in Haemophilus influenzae and gonococci[29,30]. The spread of such a determinant through populations of plasmids is likely to be of more interest to medical microbiologists than are any prior relationships of those host plasmids. Resistance in H.influenzae and the gonococci appeared only many years after ampicillin was first used against these infections. This suggests that in some cases at least there are only a few originating events.

Thus plasmids have probably evolved by an accretion of separate functions which may have come from entirely different sources. The selective advantage to a population in maintaining cryptic plasmids may be to provide a vehicle for such incoming functions, and in turn to facilitate their transmission between cells in the population.

REFERENCES

1. Dubnau, E. and Maas, W.K. (1968) J.Bact. 95, 531-539.

2. Meynell, E., Meynell, G.G. and Datta, N. (1968) Bact.Revs. 32, 55-83.

3. Guerry, P. and Falkow, S. (1971) J.Bact. 107, 372-374.

4. Barth, P.T. and Grinter, N.J. (1975) J.Bact. 121, 434-441.

5. Roussel, A.F. and Chabbert, Y.A. (1978) J.Gen.Micro. 104, 269-276.

6. Sharp, P.A., Hsu, M-T., Ohtsubo, E. and Davidson, N. (1972) J.Molec.Biol. 71, 471-497.

7. Sharp, P.A., Cohen, S.N. and Davidson, N. (1973) J.Molec.Biol. 75, 235-255.

8. Timmis, K.N., Cabello, F., Andrés, I., Nordheim, A., Burkhardt, H.J. and Cohen, S.N. (1978) Molec.Gen.Genet. 167, 11-19.

9. Thompson, R., Hughes, S.G. and Broda, P. (1974) Molec.Gen.Genet. 133, 141-149.

10. Southern, E. (1975) J.Molec.Biol. 98, 503-517.

11. Heinaru, A.L., Duggleby, C.J. and Broda, P. (1978) Molec.Gen.Genet. 160, 347-351.

12. Grindley, N.D.F., Humphreys, G.O. and Anderson, E.S. (1973) J.Bact. 115, 387-398.

13. Santos, D.S., Palchaudhuri, S. and Maas, W.K. (1976) J.Bact. 124, 1240-1247.

14. So, M., Crosa, J.H. and Falkow, S. (1975) J.Bact. 121, 234-238.

15. Davies, J.K. and Reeves, P. (1975) J.Bact. 123, 102-117.

16. Davies, J.K. and Reeves, P. (1975) J.Bact. 123, 96-101.

17. Hardy, K.G., Meynell, G.G., Dowman, J.E. and Spratt, B.G. (1973) Molec.Gen. Genet. 125, 217-230.

18.Hughes, V., Le Grice, S., Hughes, C. and Meynell, G.G. (1978) Molec.Gen. Genet. 159, 219-221.

19.Smith, H.R., Grindley, N.D.F., Humphreys, G.O. and Anderson, E.S. (1973) J.Bact. 115, 623-628.

20.Barth, P.T. and Grinter N.J. (1974). J.Bact. 120, 618-630.

21.Guiso, N. and Ullmann, A. (1976) J.Bact. 127, 691-697.

22.Sciacy, D., Montoya, A.L. and Chilton, M-D. (1978) Plasmid 1, 238-253.

23.Drummond, M.H. and Chilton, M-D. (1978) J.Bact. 136, 1178-1183.

24.Yagi, Y., Franke, A.E. and Clewell, D.B. (1975) Antimicrob. Agents and Chemother. 7, 871-873.

25.Ohtsubo, E., Nishimura, Y. and Hirota, Y. (1970) Genetics 64, 173-188.

26.Achtman, M., Kusecek, B. and Timmis, K. (1978). Molec.Gen.Genet. 163, 169-179.

27.Data of N. Willetts, summarised in P. Broda (1979) Plasmids. W.H. Freeman.

28.Heffron, F., Sublett, R., Hedges, R.W., Jacob, A. and Falkow, S. (1975) J.Bact. 122, 250-256.

29.Kaulfers, P.M., Laufs, R. and Jahn, G. (1978) JGM 105, 243-252.

30.Elwell, L.P., Saunders, J.R., Richmond, M.H. and Falkow, S. (1977) J.Bact. 131, 356-362.

Plasmids of Medical, Environmental and Commercial Importance
K.N. Timmis and A. Pühler, editors

PLASMID DNA REPLICATION

C. M. THOMAS and D. R. HELINSKI
Department of Biology, B-022, University of California at San Diego, La Jolla, California 92093 (USA)

INTRODUCTION

A plasmid is a replicon that is stably inherited as an extrachromosomal element. Naturally occurring plasmids specify a variety of bacterial properties including sexuality, antibiotic resistance and bacteriocin and toxin production and exhibit various modes of replication. Combined with their relatively small size they thus provide convenient model systems for studying the problem of the control of replication and maintenance of a DNA molecule in a cell. Plasmids also have been derived from other types of replicons such as the bacteriophage λ and the *E. coli* chromosome. Studies of these various plasmid systems are providing important information on their parent replicons specifically and the plasmid state in general.

Analysis of replication of a variety of plasmids clearly indicates that plasmids share much of the host DNA replication machinery. Emphasis in current research, therefore, has been directed largely at the nature of plasmid specified information required for controlled replication of a specific plasmid replicon. The aim of this review will be to try to give an overall picture of our current state of knowledge of events in the replication cycle of a plasmid with particular emphasis on recent developments. Given the limitations on space, no attempt has been made to cover all of the literature but rather examples from the various plasmid systems studied are presented. In general examples are drawn from studies with *E. coli* to allow a minimum of explanations of individual systems. For a more comprehensive coverage of plasmid replication the reader is referred also to other recent review articles[1,2,3].

STRUCTURE OF PLASMID REPLICONS

Considerable effort has been engaged in identifying regions of plasmid genomes important for replication. The replication cycle of DNA molecules includes stages of initiation, elongation, and termination. Replicative intermediates can be purified by virtue of their sedimentation velocity or their density in caesium chloride/ethidium bromide gradients and have been examined in the electron microscope. The problem of instability of these intermediates has been partially overcome by the use of psoralen derivatives to cross link DNA in order to reduce fork migration[4]. Studies on these intermediates have confirmed that initiation occurs at specific sites that have been mapped relative to partial denaturation patterns[5] or restriction endonuclease cleavage sites[6]. Most plasmids appear to have a unique initiation point (origin of replication). Elongation from this origin may be unidirectional (e.g., ColE1[7,8,9] and RK2[10]), or bidirectional (e.g., mini-F[11]). Some plasmids have been reported to have two sites at which initiation occurs (e.g., R6K[12,13] and NR1[14]). In the case of NR1 and R100 it appears that the bacterial host and the exact experimental conditions can influence whether one or more origins are observed[14,15]. Initiation from the two origins in R6K is of particular interest since it is sequentially bidirectional, i.e., predominantly in one direction at first followed by replication in the other direction[13,12].

Once initiation has occurred, movement of the replication fork continues until terminated either by meeting a replication fork proceeding in the opposite direction, returning to the origin of replication or arriving at a region of DNA that prevents or impedes further elongation. In the case of R6K a specific terminus region has been identified[13]. To complete the duplication cycle of R6K when the primary elongation fork reaches the terminus a second elongation fork starts from the origin in the opposite direction and proceeds to the terminus. At present the reasons for these different overall patterns of replication observed for different plasmid elements are not known.

Various manipulations, including *in vitro* recombinant DNA techniques and insertion of transposons, have been successful in either removing from plasmid genomes DNA that is not required for replication or interrupting essential regions. In the case of using recombinant DNA techniques, a plasmid generally is cleaved partially or to completion with a restriction endonuclease and the fragments joined to each other or to an added DNA fragment carrying a selective marker (a list of such fragments and their plasmid sources has been prepared[16,3]). A suitable bacterial strain is transformed with the resultant DNA mixture and transformants for a genetic marker (usually antibiotic resistance) of the parent plasmid or of the added fragment are selected[17,18]. This procedure is illustrated in Fig. 1. Minireplicons have been derived from a number of plasmids both by these methods and by spontaneous deletion events (Table 1) and the essential replication regions mapped relative to the parent plasmid. A feature common to most plasmids studied so far is that replication functions are clustered in a 0.5-4 kb region on the genome. An

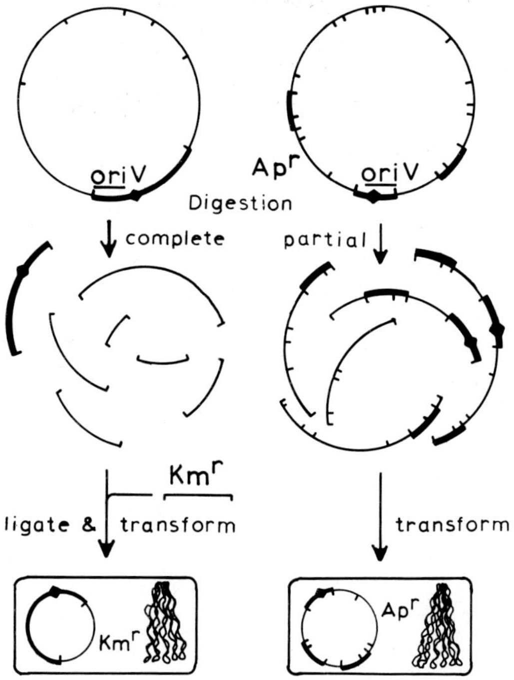

Fig. 1. Isolation of mini replicons. Plasmid DNA is digested with a restriction endonuclease either to completion or only partially. If suitable selective markers do not exist on the parent plasmid then a purified restriction fragment carrying such a marker (e.g., resistance to kanamycin (Km^r) or ampicillin (Ap^r)) may be added before ligation. After transformation of a suitable host strain with the DNA, transformants are analysed to determine how much of the parental DNA is present in the new plasmids.

exception to this is the broad host range plasmid RK2 where minireplicons consist of DNA from three well-separated regions of the parental genome[27]. The gross replicon structure of RK2 and R6K are compared in Fig. 2. The significance of clustering is not understood at present. It may represent evolutionary selection to reduce the likelihood of disruption of the replicon by recombination events. Alternatively, as in the model proposed for control of λdv replication[26,28] it may be necessary for all replication genes to lie in a single transcriptional unit.

Derivatives of RK2 and R100 contain and require for maintenance the region mapped as the replication origin by electron microscopy[27,29,15]. Similarly the minimum replicon of R6K was derived from the general region of R6K

TABLE 1

REPLICATION PROPERTIES OF VARIOUS PLASMIDS

Plasmid	Plasmid size (kb)	Copy number[a]	Mode of replication	Size of mini replicons[b] (kb)
F and F's	100-150	1-2	bidirectional	2.8[19]
R1	100	0.5-2		1.8[20]
R100(NR1, R12, R222)	100	~1	unidirectional	1.45[21]
R6-5	100	2-3	unidirectional	2.6[22]
RK2	56	4-10	unidirectional	5.4[23]
R6K	40	10-20	sequentially bi-directional to a terminus	2.0[24]
pSC101	8.7	5-10	unidirectional	-
ColE1	6.4	15-20	unidirectional	0.58[25]
λdv's	2.2-13.5	> 10	unidirectional(?)	2.2[26]

[a]Copy number per chromosome equivalent

[b]The values given represent the amount of parental plasmid DNA common to small derivatives.

containing the two origins of replication[24]. In the case of the mini-F plasmid, it was possible to obtain derivatives that were deleted for the origin of replication located by electron microscopy analyses of replicative intermediates[19]. These derivatives utilized an origin of replication located some 2 kb from the primary origin of replication. This finding suggests that more than one potential replication origin can exist within a DNA sequence in certain cases with one origin preferred in the intact plasmid and a second origin activated or utilized only upon deletion of the DNA region carrying the primary origin. Additional alterations in replication properties that have been observed for certain plasmids on reduction in size are decreased stability of the derivatives (e.g., with RK2[30] and R1drd19 [31]) and altered copy number (e.g., with R1drd19 [31] and ColE1 [32]). Despite these changes in replication properties upon deletion of segments of DNA of a replicon, the small derivatives of plasmids when compared with the parent plasmids have in general retained the replication and incompatibility properties of the parent plasmid and,

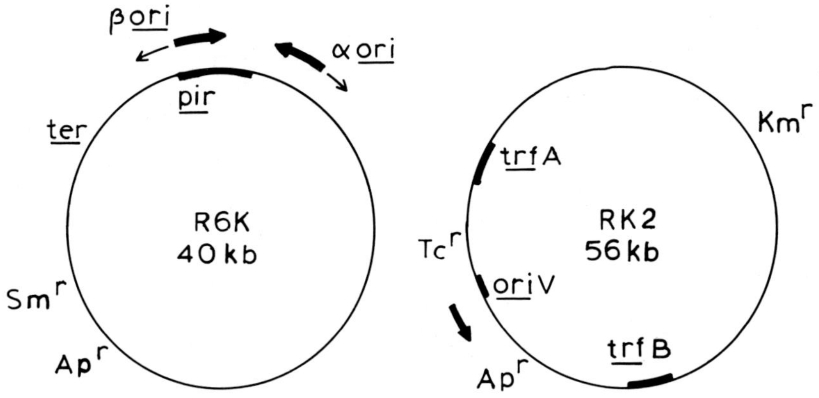

Fig. 2. Comparison of the organization of replication functions on RK2 and R6K. Sm^r, Ap^r, Tc^r and Km^r refer to genes conferring resistance to streptomycin, ampicillin, tetracycline and kanamycin, respectively. *Ori* refers to the origin of replication as determined by electron microscopy with the arrows indicating the direction of replication (for R6K the thick arrows indicate the initial direction of replication). *ter* refers to the replication terminus. The thick sections of the circle indicate regions required for replication and maintenance of mini replicons derived from R6K and RK2. For R6K *pir* refers to a *trans*-acting function involved in initiation of replication. For RK2 *trf*A and *trf*B refer to *trans*-acting replication functions.

therefore, analysis of these derivatives can provide considerable insight to the control and maintenance properties of the parent plasmid.

Considerable information on the control of plasmid replication also has been obtained from studies on DNA fragments that are not themselves capable of autonomous replication but have been derived by restriction enzyme cleavage from the essential replication region of plasmids. One important finding has been the demonstration that certain replication functions can act in *trans*. With both RK2[33] and R6K[34] it has been possible to isolate a fragment that is dependent for its autonomous replication in an *E. coli* host on other regions

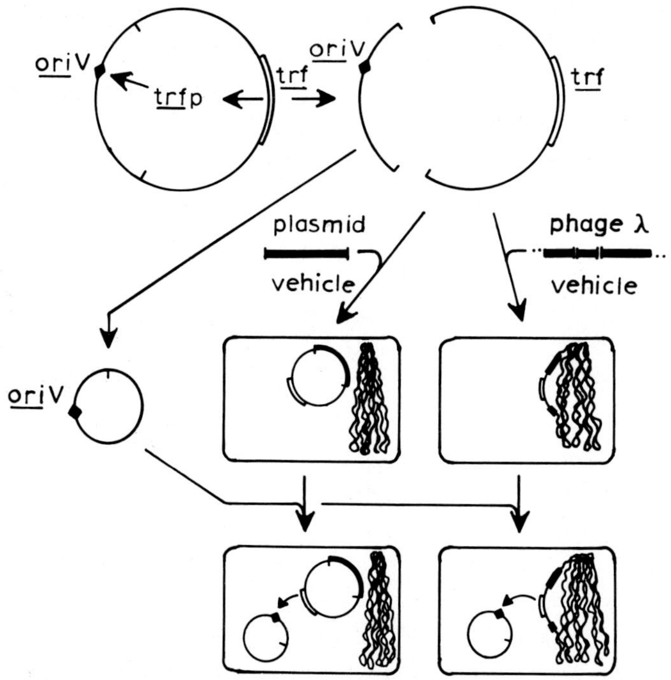

Fig. 3. Construction of helper strains capable of supplying replication functions in *trans*. A plasmid is cleaved into a segment that contains the vegetative replication origin (*ori*V) and a segment that encodes a gene or genes (*trf*≡*trans*-acting replication function) that supply products (*trf*p) required for replication. This latter segment can be cloned into either a cleaved plasmid cloning vehicle; or a cleaved λ phage cloning vehicle and strains producing *trf*p either from genes on the plasmid or on the chromosome of a lysogen can be used as a host for replication and maintenance of the defective origin containing plasmid.

of the parent plasmid that have been cloned into an independent plasmid vehicle or integrated into the host chromosome as shown in Fig. 3. This functional separation allows a more precise localization of the replication origin and *trans*-acting replication genes. In particular it has been possible to demonstrate for both R6K and RK2 that a functional replication origin is contained within a region of about 380 bp[34,23]. Similarly in the case of derivatives of bacteriophage λ small fragments (90-200 bp) have been derived from the cII gene or replication origin region of λ that will replicate autonomously when the products of gene O and P are provided in *trans* and a promoter is present for transcription through these fragments[35].

For R6K the *trans*-acting gene(s) has been localized to two contiguous *Hind* III fragments that probably code for a single protein, designated π, that has been shown to be required for the initiation of R6K replication[36]. In the case of RK2, two regions are required in *trans* and, therefore, a minimum of two *trans*-acting genes may be involved. For R1, although it has not been possible to demonstrate a clear separation of replication origin and *trans*-acting genes, it has been shown that the function that controls copy number can act in *trans* and is encoded by a small fragment (520 bp)[37].

Incompatibility is the phenomenon whereby related plasmids can not stably co-exist; upon cell division daughter cells frequently lose one or the other plasmid. In certain instances of incompatibility a relationship has been demonstrated between this phenomenon and the mechanism of copy number control[38]. In other cases arguments have been made for the involvement of a plasmid partitioning mechanism in this phenomenon[39,40]. Restriction fragments that express incompatibility but are incapable of autonomous replication have been obtained from plasmids miniF[19,40], R6-5[22], R1[37], RK2[41,42] and R6K[43]. For RK2 and R6K the small segment of DNA containing the replication origin cloned into ColE1 is strongly incompatible with parental plasmids[23,43]. For R1 copy number control, incompatibility and switch off of replication are encoded by a 520 bp fragment[37] that does not include the replication origin. An analogous situation is found for R6-5[22] while in the case of mini-F at least two non-overlapping regions of the minireplicons express incompatibility separately[40].

The replication terminus of R6K, while not essential for replication, is of interest since it must possess some structure that impedes or stops the elongation process. Fragments of R6K containing this region have been cloned into ColE1 and it has been demonstrated that although the replication fork of the replicative intermediates was halted at the terminus, replication of this hybrid plasmid was not inhibited completely[44]. In addition completion of replication appears to occur by eventual movement of the replication fork through the terminus rather than initiation in a second direction from the ColE1 replication origin. The reason for the strong preference of intact ColE1 to replicate unidirectionally is not understood.

For the ColE1 type plasmids the sequence of the entire genome of pBR345 and pBR322 is known[46,47]. The primer RNA/DNA junction at the origin has been identified[48] and other essential sequences in the approximately 580 base pair autonomously replicating region of ColE1[25] have been localized by deletion analysis and interruption of specific sites by insertion of DNA utilizing recombinant DNA techniques[25]. Functional replication origins from R6K[49], and RK2[23], and regions surrounding the replication origin of R1[50] and R100[51] that have been mapped by electron microscopy also have been sequenced. Unfortunately, the RNA/DNA junction in the case of these plasmids is not known. The sequence of the R100 replication origin contains a number of inverted repeat sequences that allow a large amount of intrastrand base pairing. The most outstanding features of the R6K and RK2 origin sequences is the existence of a series of direct repeats in units of 22-25 bp suggesting possible binding sites for regulatory proteins. The significance of putative promoter sequences in the region of the origin of replication for plasmids RK2, R6K and R100, recognized from their similarity to a 'Pribnow Box', is not known at this time.

IN VIVO STUDIES ON THE PROCESS OF REPLICATION

Biochemical studies *in vivo* have shown that replication of plasmid DNA molecules proceeds through a theta type structure where the unreplicated portion may remain supercoiled. After completion of duplication non-supercoiled covalently closed circles are produced before conversion to supercoiled DNA molecules[52,53]. Studies with ColE1 have indicated that the elongation process is discontinuous at least on one strand[54,55]. These is evidence that during the process of duplication the plasmid is associated with membrane or folded chromosome[56,57,58,59,60,61]. In the case of the ColE1 plasmid isolated from *E.coli* mini-cells cellular membrane was found associated with plasmid DNA in the region of the origin of replication. Evidence also has been obtained for the association of replicative intermediates of R6K with the folded chromosome[59,60]. Recently it was reported for NR1 that plasmid DNA synthesized in the presence of chloramphenicol cannot be isolated by ethidium bromide/caesium chloride density equilibrium centrifugation or velocity sedimentation, although it is recovered when protein synthesis is allowed to proceed[61]. These various observations point to the possible importance of highly organized cell structure-plasmid DNA complexes in either or both regulated replication or segregation of a plasmid element.

The whole of the plasmid replication cycle has been estimated to occupy a relatively short period of time; 80 to 100 sec for CloDF13[62], about 2 min for ColE1[56] and less than four minutes for R1drd19[63] and F'lac[63]. Nevertheless, these time estimates are greater than expected from the rate of fork movement calculated for the *E. coli* chromosome. Plasmids studied to date, both high and low copy number, have been shown by density shift experiments, to be selected randomly for replication[63,64,65,66]. In addition, in exponentially growing cultures carrying plasmids R1 or F'lac, it has been demonstrated that plasmid replication is equally likely at any stage of the cell cycle[63].

The requirement for RNA synthesis for replication has been clearly demonstrated for ColE1 since replication continues in the presence of chloramphenicol and rifampicin inhibits this replication in the presence of the protein synthesis inhibition[67]. The sensitivity to rifampicin may be the result of either or both prevention of transcriptional activation or prevention of synthesis of specific primer RNA. Recent experiments with NR1 have indicated that inhibition of replication of this plasmid by rifampicin can be explained entirely through its effect on protein synthesis[61].

REQUIREMENT OF PROTEINS AND PROTEIN SYNTHESIS FOR PLASMID REPLICATION

With the exception of ColE1, ColK, CloDF13, P15A, RSF1030 and other similar plasmids, replication appears to require continued protein synthesis. This may indicate that essential replication proteins for most plasmids are either unstable or used stoichiometrically. Plasmids share many of the host chromosome replication functions. For R1 it has been demonstrated that this plasmid may compete with the chromosome for polIII under limiting conditions[68]. ColE1 replication has been shown to be reduced to 20% of its normal rate in a *dna*E mutation[69] while CloDF13 cannot replicate in a *dna*Z mutant (both loci being involved in elongation of the *E. coli* chromosome)[70]. In general unless the rate of plasmid DNA replication is differentially affected when compared with chromosomal DNA replication in *dna* temperature sensitive mutants, it is difficult to draw a meaningful conclusion since there is evidence that plasmid replication may be coupled to chromosomal DNA replication under certain conditions. For instance in *dna*A mutants on shift to the restrictive temperature replication of F'lac or NR1 cannot continue after chromosome replication has stopped even though the plasmid/chromosome cointegrate involving these plasmids can continue to replicate[71,72]. On the other hand, the behaviour of plasmid NR1 in *dna*A and *dna*C mutants, both defective for chromosome initiation, is different. Plasmid DNA replication halts immediately on shift to the restrictive temperature in *dna*C mutants but continues until chromosome replication stops in *dna*A mutants[73]. This is consistent with the independence of NR1 from the *dna*A gene product and the requirement for a *dna*C gene product. While a similar relationship to *dna*A and *dna*C has been observed for certain other plasmids, it is not always the case. For example, pSC101 has been reported to be dependent on the *dna*A product[74].

Elongation defective *dna*B mutants will maintain plasmids ColE1 and CloDF13, however, a *dna*B252 mutant, defective in initiation of chromosome replication, will not support the replication of ColE1. The dependence of ColE1 on the *dna*B gene product has been confirmed by *in vitro* studies[75]. Certain plasmids are capable of suppressing chromosomal *dna*B mutations[76] and it has been suggested that these plasmids specify a *dna*B protein analogue similar to that found for bacteriophage P1. The study of plasmid replication in host chromosomal replication temperature sensitive mutations has been extended to gram positive organisms. Recent results with pUB110 a plasmid from *S. aureus* introduced into *B. subtilis* indicate that pUB110 is dependent on the equivalent of polIII but independent of polI and several other *dna* loci[77].

Searches for conditional lethal mutations that affect plasmid but not chromosomal replication have yielded both chromosomal and plasmid mutations. The chromosomal mutations can be classified by their specificity for particular plasmids or groups of plasmids[78,79]. While ColE1 specific host mutants include strains with a temperature sensitive polA enzyme, the biochemical nature of the other mutations is unknown. The sensitivity of these mutant bacteria to deoxycholate has led to the suggestion that a membrane lesion was involved[78].

In general the existence of a plasmid encoded temperature-sensitive mutation indicates that a plasmid specifies a protein required for its own replication. However, despite the existence of such mutations for ColE1[78,80], it has been demonstrated that ColE1 replication *in vitro*[81] and the replication of ColE1 derivatives *in vivo*[82,83] do not require a positively acting plasmid encoded protein. It is conceivable that these mutations affect a physiological property of ColE1 other than replication(e.g., mobilizability) and this mutational change indirectly interferes with the ColE1 replication complex. Alternatively, derivatives of ColE1 are replicated *in vivo* by a somewhat different pathway than intact ColE1. In the case of plasmids other than the ColE1-type there is increasing genetic evidence for a role of a plasmid specified protein(s) in plasmid replication. Complementation analysis of the conditional lethal mutants of these plasmids has been hampered by the problem of incompatibility but now in cases where a *trans* system has been developed complementation analysis of these mutants promises to yield a considerable amount of information.

PLASMID REPLICATION *IN VITRO*

Ultimately, a detailed understanding of the biochemistry of plasmid DNA replication will depend on the development of *in vitro* replication systems for the various plasmid types. So far such systems have been developed for three types of plasmids: ColE1[84], CloDF13[85] and RSF1030[86]; R6K[87]; and a small high copy number derivative of RI[88]. Replication of the ColE1 type plasmids *in vitro* does not require a plasmid encoded protein and can be achieved in extracts prepared from plasmid negative bacteria[81]. Synthesis of an RNA primer for initiation appears to require RNA polymerase and may also involve processing of the RNA by an enzyme such as RNaseH[89] or RNaseIII[19]. Evidence has been presented for a two stage synthesis of ColE1[2]. Initially, an RNA primer is used for elongation by polI to form a D loop type intermediate containing a 6-7S copy of the 3'--5' strand. Completion of replication is dependent upon other enzymes including the products of the *dna*E, *dna*B, *dna*C, *dna*G, and *dna*Z genes[2]. The requirement for polI *in vitro* is absolute, consistent with the *in vivo* dependence on this enzyme. Replication of ColE1 *in vivo* at 20% the normal rate at the restrictive temperature in *dna*E temperature sensitive mutants may be explained by the leakiness of these mutants[69]. Elongation *in vitro* proceeds by a discontinuous mechanism and although joining of Okazaki fragments is resistant to NMN it is dependent on DNA ligase as judged by results in a temperature sensitive ligase mutation[90].

Replication of plasmid Rsc11 (mini RI factor) *in vitro* has been demonstrated recently[88]. Interestingly, it has not been possible to show *in vitro* replication of the parent plasmid which is stringently regulated *in vivo*[88]. In contrast, both R6K and its derivatives will replicate *in vitro*[36]. Both R1 and the relaxed plasmid R6K do not require DNA polymerase I for replication[91,44]. In the case of R6K replication *in vitro*, a requirement for a plasmid encoded protein, designated π, that acts at the stage of initiation has been demonstrated[36].

CONTROL OF PLASMID DNA REPLICATION

It is generally accepted that the frequency of initiation of replication of a replicon is important in the regulation of copy number. The replicon model[92] proposed that a replicon contains a replicator region that requires a positive effector (initiator) to allow replication. Control of the production of this positive effector, as for example by an autoregulatory process,[93] could serve as a means of modulating the frequency of initiation. A frequently discussed alternative model involves synthesis of a repressor protein shortly after initiation of replication which prevents a second round of initiation until the cell volume has increased sufficiently for the repressor concentration to fall below a critical level[94]. Support for this model has come from an interpretation of studies carried out with the joint replicon between pSC101 and ColE1, where initiation of replication for the normally low copy number plasmid of the pair is completely shut off in the hybrid[95]. The detailed genetic studies on the regulation of the plasmid state of λdv, the deletion derivative of bacteriophage λ, strongly supports a negative control model for plasmid λdv replication[26]. In this system a single operon, including a repressor *tof* and positively acting initiator genes O and P, is autoregulated by the *tof* repressor protein[26].

Analysis of replication controlling elements through the study of replication mutants has allowed considerable progress with several plasmid systems. For some plasmids (e.g., R6K[96]) only chromosomal mutations have been found to give rise to a higher plasmid copy number while in other cases (e.g., R1drd19[97], NR1[98] and CloDF13[99]) plasmid mutations have been obtained that give rise to a higher copy number. In the case of F'lac and miniF both chromosomal and plasmid copy number mutations have been isolated[100,40]. In most of these cases, where copy number is affected at all growth temperatures, it is not known whether the increased copy number is due to a decrease in the concentration of a negatively acting factor or to an increase in a positively acting factor. The copy number mutations encoded by plasmid R1 have been most extensively studied and have yielded important information on the regulation of replication of this plasmid. The non-conditional mutations obtained for R1 fall into two groups, those where the effect is recessive in the presence of a wild type plasmid, and those where the effect on the mutant is *cis* dominant[38]. This suggests that copy number mutations can result from alterations in either a regulatory protein or its recognition site. In addition, conditional copy mutants, both temperature dependent and amber mutants have been isolated[101]. Mutants that show uncontrolled plasmid DNA replication at high temperature and that are thus lethal for the host can be isolated by a second mutation in the temperature dependent mutants[102]. That copy number is depressed by the wild type gene(s) acting in *trans* suggests that plasmid R1 replication is normally under negative control[37]. Recent studies on rates of plasmid DNA replication when conditions are altered so that R1 plasmid copy number is increased or decreased suggest that copy number control does not titrate the number of origins but acts by determining the rate of initiation[103].

Models for the control of replication of both R6K[34] and RK2[23] have been proposed, based partly on sequence data from the replication origins of these plasmids. An important finding with plasmid R6K was that when a replication origin containing segment of R6K replicates in strains containing the required *trans*-acting plasmid gene (specifying the π initiation protein) at various dosage levels (e.g., as a single copy in the chromosome or in a high copy number ColE1 plasmid) the copy number of the origin plasmid remained at the level expected for R6K[34]. If the plasmid encoded π protein is regulatory then it is conceivable that an autoregulatory process maintains this protein at a constant concentration which in turn provides for a constant rate of initiation of plasmid DNA replication. The region between the origin of replication in R6K and the start of the π gene contains 7 tandem repeats of 22 base pairs[49]. In RK2 there are five direct repeats with the same spacing (22-24 base pairs) but only 15 conserved base pairs[23]. It has been proposed that the repeats in R6K function as a binding site for autoregulation of synthesis of the π protein[34]. In the case of RK2 binding of a hypothetical repressor protein to the repeats has been considered to regulate synthesis of an RNA transcript that is required either for priming DNA synthesis or transcriptional activation. Also for RK2, based again on the independence between copy number of replication origin plasmid and dosage level of *trans*-acting genes it was suggested that if one of the *trans*-acting replication functions serves as a regulatory repressor then its synthesis may be autoregulated so as to give a constant level of repression of initiation[23]. This proposal differs from both the repressor dilution model of Pritchard[94] and the autoregulation model of Sompayrac and Maaloe[93].

The fact that chromosomal DNA mutations give rise to higher copy numbers of plasmids R6K and F'lac indicates that even if the supply of plasmid encoded initiator is regulatory other factors may normally also limit the rate of initiation. A general model to account for rate of plasmid initiation should therefore take into account the possibility that there are host specified components that are present at a constant but limiting concentration with regard to the initiation of replication. Transcriptional activation, required for the initiation of replication at the bacteriophage λ origin[104,104a], may be an example of this type of general control of replication.

In view of the relatively short time required for the replication of a plasmid molecule during cell growth and division, plasmid genome size is not expected to influence plasmid copy number. However, in the case of ColE1, the copy number of the plasmid is roughly inversely proportional to plasmid size[105]. While the possibility that elongation time is a determinant of copy number has not been rigorously excluded in this case, it is not unlikely that genome size influences initiation rate by an as yet to be determined mechanism. In the case of certain mini-derivatives of R1drd19 the copy number is elevated[31], despite the evidence for the preservation in these derivatives of the gene(s) postulated to control copy number[106]. Similarly, the loss of approximately one-half of ColE1 DNA in the formation of miniColE1 results in a five-fold increase in copy number of this plasmid[32].

It seems likely that the precise mechanisms that determine the rate of initiation will vary from plasmid to plasmid. A comparison of these systems is yielding fundamental knowledge of biological control mechanisms. While considerably more fundamental information must be obtained yet from *in vivo* studies on plasmid replication the ultimate understanding of replication control will undoubtedly require *in vitro* studies on the various factors that influence the rate of initiation.

CONCLUSION

Recent developments in the study of plasmid DNA replication have substantially increased our knowledge of the biochemistry of DNA replication, its control *in vivo* and the molecular structure of the genetic units involved in these processes. The considerable effort currently made at a number of levels promises to yield a detailed understanding of the genetic and biochemical factors involved in the regulation of replication of the major types of plasmid systems. The significance of this information extends beyond a basic understanding of a major class of genetic elements in bacteria since plasmids are proving to be important model systems for the study of inheritance of stably maintained replicons characteristic of all organisms.

REFERENCES

1. Rowbury, R. J. (1977) Biophys. Molec. Biol., 31, 271-317.
2. Staudenbauer, W. L. (1978) Curr. Top. Microbiol. Immunol., 83, 93-156.
3. Timmis, K. N., Cohen, S. N. and Cabello, F. C. (1978) Prog. Mol. Subcell. Biol., 6, 1-58.
4. Bolivar, F., Betlach, M. C., Heynecker, H. L., Shine, J., Rodriguez, R. L. and Boyer, H. W. (1977) Proc. Natl. Acad. Sci. USA, 74, 5265-5269.
5. Schnös, M. and Inman, R. B. (1970) J. Mol. Biol., 51, 61-73.

6. Nathans, D. and Smith, H. O. (1974) Ann. Rev. Biochem., 44, 273-293.

7. Lovett, M. A., Katz, L. and Helinski, D. R. (1974) Nature, 251, 337-340.

8. Tomizawa, J., Sakakibara, Y. and Kakefuda, T. (1974) Proc. Natl. Acad. Sci. USA, 71, 2260-2264.

9. Inselburg, J. (1974) Proc. Natl. Acad. Sci. USA, 71, 2256-2259.

10. Meyer, R. and Helinski, D. R. (1977) Biochim. Biophys. Acta, 478, 109-113.

11. Eichenlaub, R., Figurski, D. and Helinski, D. R. (1977) Proc. Natl. Acad. Sci. USA, 74, 1138-1141.

12. Crosa, J. H., Luttropp, L. K., Heffron, F. and Falkow, S. (1975) Molec. Gen. Genet., 140, 39-50.

13. Lovett, M. A., Sparks, R. B. and Helinski, D. R. (1975) Proc. Natl. Acad. Sci. USA, 72, 2905-2909.

14. Perlman, D. and Rownd, R. H. (1976) Nature, 259, 281-284.

15. Silver, L., Chandler, M., Boy de la Tour, E. and Caro, L. (1977) J. Bacteriol., 131, 929-942.

16. Kahn, M., Kolter, R., Thomas, C., Figurski, D., Meyer, R., Remaut, E. and Helinski, D. R. (1979) Methods in Enzymology, in press.

17. Timmis, K., Cabello, F. and Cohen, S. N. (1975) Proc. Natl. Acad. Sci. USA, 72, 2242-2246.

18. Lovett, M. A. and Helinski, D. R. (1976) J. Bacteriol., 127, 982-987.

19. Kahn, M. L., Figurski, D., Ito, L. and Helinski. D. R. (1978) Cold Spring Harbor Symp. Quant. Biol., 43, in press.

20. Molin, S., Stougaard, P., Uhlin, B. E., Gustafsson, P. and Nordström, K. (1979) J. Bacteriol., in press.

21. Taylor, D. P. and Cohen, S. N. (1979) J. Bacteriol., 137, 92-104.

22. Andrés, I., Slocombe, P. M., Cabello, F., Timmis, J. K., Lurz, R., Burkhardt, H. J. and Timmis, K. N. (1979) Molec. Gen. Genet., 168, 1-25.

23. Thomas, C. M., Stalker, D., Guiney, D. and Helinski, D. R. (1979) This volume.

24. Kolter, R. and Helinski, D. R. (1978) Plasmid, 1, 571-580.

25. Backman, K., Betlach, M., Boyer, H. W. and Yanofsky, S. (1978) Cold Spring Harbor Symp. Quant. Biol., 43, in press.

26. Matsubara, K. (1976) J. Mol. Biol., 102, 427-439.

27. Thomas, C. M. and Helinski, D. R., submitted.

28. Berg, D. E. (1974) Virology, 62, 224-233.

29. Ohtsubo, F., Feingold, J., Ohtsubo, H., Mickel, S. and Bauer, W. (1977) Plasmid, 1, 8-18.

30. Thomas, C. and Ditta, G., unpublished.

31. Luibrand, G., Blohm, D., Mayer, H. and Goebel, W. (1977) Molec. Gen. Genet., 152, 43-51.

32. Hershfield, V., Boyer, H. W., Chow, L. and Helinski, D. R. (1976) J. Bacteriol., 126, 447-453.

33. Figurski, D. and Helinski, D. R. (1979) Proc. Natl. Acad. Sci. USA, in press.

34. Kolter, R., Inuzuka, M. and Helinski, D. R. (1978) Cell, 15, 1199-1208.

35. Hobom, G., Lusky, M., Grosschell, R. and Scherer, G. (1978) Cold Spring Harbor Symp. Quant. Biol., 43, in press.

36. Inuzuka, M. and Helinski, D. R. (1978) Proc. Natl. Acad. Sci. USA, 75, 5381-5385.

37. Molin, S. and Nordström, K., unpublished.

38. Uhlin, B. E. and Nordström, K. (1975) J. Bacteriol., 124, 641-649.

39. Timmis, K. N., Andrés, I. and Slocombe, P. M. (1978) Nature, 273, 27-32.

40. Mannis, J. J. and Kline, B. C. (1978) Plasmid, 1, 492-507.

41. Meyer, R., Figurski, D. and Helinski, D. R., submitted.

42. Sakanyan, V. A., Yakubov, L. Z., Alikhanian, S. I. and Stepanov, A. I. (1978) Molec. Gen. Genet., 165, 331-341.

43. Kolter, R., unpublished.

44. Kolter, R. and Helinski, D. R. (1978) J. Mol. Biol., 124, 425-441.

46. Bolivar, F., Betlach, M. C., Shine, J., Heynecker, H. L., Rodriguez, R. L., Tait, R. C., Yanofsky, S. and Boyer, H. W., in preparation.

47. Sutcliffe, G. (1978) Ph.D. Thesis, Harvard Univ.

48. Tomizawa, J., Ohmori, H. and Bird, R. E. (1977) Proc. Natl. Acad. Sci. USA, 74, 1865-1869.

49. Stalker, D. M., Kolter, R. and Helinski, D. R. (1979) Proc. Natl. Acad. Sci. USA, 76, 1150-1154.

50. Oertel, W., Kollek, R., Beck, E. and Goebel, W. (1979) Molec. Gen. Genet., in press.

51. Rosen, J., Ohtsubo, H. and Ohtsubo, E. (1979) Molec. Gen. Genet., in press.

52. Crosa, J. H., Luttropp, L. K. and Falkow, S. (1976) Nature, 261, 516-519.

53. Timmis, K., Cabello, F. and Cohen, S. N. (1976) Nature, 261, 512-516.

54. Inselberg, J. and Oka, A. (1976) J. Bacteriol., 123, 739-742.

55. Katz, L., Williams, P. H., Sato, S., Leavitt, R. W. and Helinski, D. R. (1977) Biochemistry, 16, 1677-1683.

56. Sherratt, D. J. and Helinski, D. R. (1973) Eur. J. Biochem., 37, 95-99.

57. Kline, B. C. and Miller, J. R. (1975) J. Bacteriol., 121, 165-172.

58. Sparks, R. B. and Helinski, D. R. (1979) Nature, 277, 572-575.

59. Wlodarczyk, M. and Kline, B. C. (1976) Biochem. Biophys. Res. Comm., 73, 286-292.

60. Archibold, E. R., Clark, C. W. and Sheehy, R. J. (1978) J. Bacteriol., 135, 476-482.

61. Womble, D. D. and Rownd, R. H. (1979) Plasmid, 2, 79-94.

62. Veltkamp, E. and Nijkamp, H. J. J. (1976) Biochim. Biophys. Acta, 425, 356-367.

63. Gustafsson, P., Nordström, K. and Perram, J. W. (1978) Plasmid, 1, 187-203.

64. Bazaral, M. and Helinski, D. R. (1970) Biochemistry, 9, 399-406.

65. Rownd, R. (1969) J. Mol. Biol., 44, 387-402.

66. Gustafsson, P. and Nordström, K. (1975) J. Bacteriol., 123, 443-448.

67. Clewell, D. B., Evanchik, B., Cranston, J. W. (1972) Nature New Biol., 273, 29-31.

68. Nordström, U. M., Engberg, B. and Nordström, K. (1974) Molec. Gen. Genet., 135, 185-190.

69. Collins, J., Williams, P. and Helinski, D. R. (1975) Molec. Gen. Genet., 136, 273-289.

70. Veltkamp, E. (1976) Ph.D. Thesis, Amsterdam.

71. Nishimura, Y., Caro, L., Berg, C. M. and Hirota, Y. (1971) J. Mol. Biol., 55, 441-456.

72. Chandler, M., Silver, L. and Caro, L. (1977) J. Bacteriol., 131, 421-430.

73. Womble, D. D. and Rownd, R. H. (1979) Plasmid, 2, 95-108.

74. Hasunuma, K. and Sekiguchi, M. (1977) Molec. Gen. Genet., 154, 225-230.

75. Staudenbauer, W. L., Lanka, E. and Schuster, H. (1978) Molec. Gen. Genet., 162, 243-249.

76. Wang, P. Y. and Iyer, V. N. (1977) Plasmid, 1, 19-33.

77. Shivakumar, A. G. and Dubnau, D. (1978) Plasmid, 1, 405-416.

78. Kingsbury, D. T. and Helinski, D. R. (1973) Genetics, 74, 17-31.

79. Novick, R. P. (1967) Proc. 5th Intern. Cong. Chemother. Vienna, pp. 269-273.

80. Collins, J., Yanofsky, S. and Helinski, D. R. (1978) Molec. Gen. Genet., 167, 21-28.

81. Tomizawa, J., Sakakibara, Y., and Kakefuda, T. (1975) Proc. Natl. Acad. Sci. USA, 72, 1050-1054.

82. Donoghue, D. J. and Sharp, P. A. (1978) J. Bacteriol., 133, 1287-1294.

83. Kahn, M. and Helinski, D. R. (1978) Proc. Natl. Acad. Sci. USA, 75, 2200-2204.

84. Sakakibara, Y. and Tomizawa, J. (1974) Proc. Natl. Acad. Sci. USA, 71, 802-806.

85. Staudenbauer, W. L. (1976) Molec. Gen. Genet., 145, 273-280.

86. Staudenbauer, W. L. (1977) Molec. Gen. Genet., 156, 27-34.

87. Inuzuka, M. and Helinski, D. R. (1978) Biochemistry, 17, 2567-2573.

88. Bezanson, G. S. and Goebel, W. (1979) Molec. Gen. Genet., 170, 49-56.

89. Itoh, T. and Tomizawa, J. (1978) Cold Spring Harbor Symp. Quant. Biol., 43, in press.

90. Sakakibara, Y. (1978) J. Mol. Biol., 124, 373-389.

91. Mayer, H., Luibrand, G. and Goebel, W. (1977) Molec. Gen. Genet., 152, 142-152.

92. Jacob, F., Brenner, S. and Cuzin, F. (1963) Cold Spring Harbor Symp. Quant. Biol., 43, in press.

93. Sompayrac, L. and Maaløe, O. (1973) Nature New Biol., 241, 133-135.

94. Pritchard, R. H., Barth, P. T. and Collins, J. (1969) Symp. Soc. Gen. Microbiol, XIX, 263.

95. Cabello, F., Timmis, K. and Cohen, S. N. (1976) Nature, 259, 285-290.

96. Macrina, F. L., Weatherly, G. G. and Curtiss, R. (1974) J. Bacteriol., 120, 1387-1400.

97. Nordström, K., Ingram, L. C. and Lundbäck, A. (1972) J. Bacteriol., 110, 562-569.

98. Morris, C. F., Hashimoto, H., Mickel, S. and Rownd, R. (1974) J. Bacteriol., 118, 855-866.

99. Kool, A. J. and Nijkamp, H. J. J. (1974) J. Bacteriol., 120, 569-578.

100. Cress, D. E. and Kline, B. C. (1976) J. Bacteriol., 125, 635-642.

101. Gustafsson, P. and Nordström, K. (1978) Plasmid, 1, 134-144.

102. Uhlin, B. E. and Nordström, K. (1978) Molec. Gen. Genet., 165, 167-179.

103. Gustafsson, P. and Nordström, K., unpublished.

104. Dove, W., Inokuchi, H. and Stevens, W. (1971) in: The Bacteriophage Lambda. (ed. A. D. Hershey) Cold Spring Harbor, pp. 747-771.

104a. Furth, M. E., Blattner, F. R., McLeester, C. and Dove, W. F. (1977) Science, 198, 1046-1051.

105. Warren, G. and Sherratt, D. (1978) Molec. Gen. Genet., 161, 39-47.

106. Molin, S., Stougaard, P., Uhlin, B. E., Gustafsson, P. and Nordström, K. (1979) J. Bacteriology, in press.

Plasmids of Medical, Environmental and Commercial Importance
K.N. Timmis and A. Pühler, editors

CONJUGATION IN BACTERIA

PAUL BRODA
Department of Molecular Biology, King's Buildings, Edinburgh University, Edinburgh EH9 3JR, Scotland.

Until recently work on bacterial gene transfer was mainly concerned with elucidating the mechanisms in the classical systems of transformation, transduction and conjugation, usually using the organisms in which they were originally discovered, and with making use of such systems for genetic analysis. It has now also become possible to ask about the extent at least of conjugation in natural populations. It is important to assess the role of gene exchange in bacterial evolution, especially as a response to the changes in the environment brought about by Man.

All known conjugation systems are plasmid-mediated; that is, transfer is a function of plasmids as a group. Conjugative plasmids belonging to about 20 incompatibility groups have been described from the enterobacteria alone. Other organisms that naturally carry conjugative systems will be mentioned later. Except in laboratory situations (e.g., Hfr strains in E.coli K12) transfer is primarily of plasmid-borne rather than chromosomally-borne functions. Nevertheless, the discovery and improvement of conjugative systems that effect chromosome 'mobilisation' is an extremely important trend in molecular genetics[1] because genetic exchange is essential for genetic analysis.

The F mating system. Conjugation can be divided into three stages: collision leading to effective pair formation; the actual transfer of DNA; and the processing of the DNA in the recipient cell. The best-studied systems are those of F and related plasmids in E.coli. The following description of the transfer process as it occurs in these plasmids should illustrate what questions can be asked about other conjugative systems.

As expected from a collision-dependent process, F mating yields more progeny at higher cell concentrations. But at lower concentrations (such as will normally obtain in natural populations) the pairing process, once collision has occurred, is much more efficient[2]. It does not require active energy metabolism[3]. Appendages termed sex pili are involved in the process, though whether they function as conduction tubes or to bring mating cells together is not known[4-7]. F is unusual among pilus-producing plasmids in that its synthesis of pili is unregulated. Only a minority of cells of strains carrying other 'F-like' plasmids carry sex pili at any one time (see later). There are both DNA and RNA phages that absorb to these pili.

F and its relatives also specify a cell-envelope function termed surface exclusion, which prevents matings between two male cells[8]. But the great majority of females in a population can act as recipients[9]. Mutant recipient strains unable to accept donor DNA are generally but not always defective in cell envelope functions[10,11].

Contact formation probably generates a signal for the synthesis of activation of the enzymes that effect DNA transfer. One necessary function seems to be the nicking of the plasmid to give the linear structure that is transferred. Such nicking occurs at a specific site on F, so that F is always transferred with a defined polarity. Only one strand of F is transferred; therefore Hfr strains transferring from opposite directions transfer complementary strands[12,13]. Two R factors, one related to F and the other not, also transfer a specific strand[14]. It seems that DNA replication is not essential for such transfer, although it generally occurs at the same time, presumably replacing the transferred strand[15].

The first covalently closed circular plasmid DNA to be detectable in the recipient is associated with the membrane[16]. How it is circularised is not known, but the process does not require the host recA function[17]. If, as is possible, F is transferred by a rolling circle mechanism[18], it might be that cohesive ends, like those of phage λ, are generated to allow recircularisation.

The genetics of F transfer. The genetic study of the F transfer system was initiated using mutants of F'gal[19,20] and F'lac[21]. By means of complementation tests about 20 transfer genes have been defined[22-24]. Deletion mapping[25], heteroduplex analysis[26] and the use of chimaeras[22,27,28] and λ transducing phages[24] have allowed the correlation of the genetic and physical maps. The transfer region is about twice as large as that needed to code for 20 average-sized polypeptides. This fact, and the demonstration of further proteins in *in vitro* systems for the expression of this DNA[28], suggest that even 20 genes is an underestimate. Why the transfer function is so complex when, for instance, the pilus is made up of a single protein[24], is unclear. It is a striking fact that the smallest known conjugative plasmid has a size of about 26 kilobases[30].

As noted earlier, pilus production is 'repressed' in F-like plasmids apart from F itself. However, when R100 (for example) and F are present in the same cell, F fertility is also repressed ('fertility inhibition'). This is because transfer is controlled at two levels: most of the transfer genes are carried on a single operon, the transcription of which requires the presence of the traJ product[31,32]. The expression of traJ is in turn repressed by the co-operative action of the products of two further genes, finO and finP[32]. The reason that F fertility is normally de-repressed is that although F has an finP gene it does not produce a finO product. Fertility inhibition occurs when the finO product of R100 acts together with the F finP product. However, some plasmids apparently inhibit F fertility by a quite different mechanism[33].

Plasmid spread through bacterial populations. 'Epidemic spread' of plasmids through recipient populations was first studied with ColI in E.coli. Although only about 0.02% of the donor cells were originally competent donors, under optimum conditions most of the recipient cells that had received plasmids were competent[34,35]. However, their growth in the absence of further recipient cells led to competence declining to its original low level within seven generations. Transfer is probably initiated from a few cells in which the transfer system escapes from repression[36,37] and re-transfer from new progeny cells probably occurs before repression can take effect. The two main constraints on epidemic spread are first, that donors can only transfer about once per generation, and second, that 2-3 generations are needed before the descendants of newly infected cells become proficient donors[37,38]. Plasmid spread is possible because the fertility inhibition system takes still longer to come into effect.

Such systems for the control of transfer are therefore biologically very efficient. Because of repression, most cells in a population do not make pili and other transfer products. This saves energy and means that they will not be infected by phage that attach to pili. But because there are always a few cells that are competent donors, the transfer process can be started once a recipient population is encountered, and it is not prevented by the finOP system.

Mobilisation by plasmids. In the enterobacteria large plasmids can generally mediate their own transfer. Smaller ones cannot, but are often mobilisable, by unknown mechanisms, by self-transmissible plasmids. Presumably there is a strong selective advantage for a non-conjugative plasmid in being mobilisable by a wide range of conjugative plasmids. Normally, mobilisation will depend at least in part upon the presence of a target for nicking[39]. But there is a case where homology (in the form of a transposon carried by both plasmids) resulted in efficient mobilisation[40]. Such mobilisation may result from the physical joining of the two plasmids by recombination.

Chromosome mobilisation too may depend upon a covalent association with the 'sex factor'. In most cases this seems to be transient. The principal exception is the Hfr donor in E.coli K12[41]. Here, F has become stably integrated, by reciprocal recombination between homologous sequences on F and the chromosome[26,42]; different Hfr strains have F integrated at different chromosomal sites. In each, transfer proceeds in one direction from that site at a rate which, if maintained, would result in the whole chromosome being transferred in about 100'[43]. However, spontaneous interruption (at a rate that is proportional to time) means that only in about 0.01% of mating pairs is the whole chromosome transferred.

Other conjugation systems. Very little is known of other conjugation systems in E.coli. Some plasmids also specify pili; these pili differ from each other and from F pili. In some cases pilus phages have also been described[44].

Many Pseudomonas plasmids also are conjugative. Those of the IncP-1 group are transmissible to E.coli[45-47]. This implies that as well as being transferred, they can replicate in their new host and that their drug resistances are expressed. In general it is unknown whether the apparent absence of transfer between strains is really due to an inability to transfer, rather than to a failure in replication or gene expression. Thus although a TOL (toluene-degrading) plasmid from Pseudomonas putida is transmissible to E.coli this was not at first realised, because the progeny cells could not grow on toluene or m-toluate[48]. The transfer of the plasmid to the recipient strain could be demonstrated by 'labelling' it with a carbenicillin-resistance transposon, which was expressed in E.coli.

The IncP-1 plasmids (e.g., RP1) are transmissible not only to E.coli but to a very wide range of other Gram-negative bacteria. Although they themselves only mobilise the chromosome with low efficiency, derivatives have been isolated that are much more effective in this respect[1,49-51]. Such plasmids can be used as tools for genetic analysis and in vitro genetic engineering that will prove to be of very general importance.

There are also conjugative systems in the streptococci that are still in the early stages of study[54-58]. But no case has yet been described for any of the staphylococci. Two conjugative plasmids are also known in Streptomyces coelicolor[59,60]. Study of the mechanisms of conjugation in this important group is difficult because gene transfer occurs during mycelial growth.

The impact of conjugation. Commensal organisms in the gut that carry R factors, such as *E.coli* strains, can form a reservoir from which an incoming pathogen may acquire drug resistance. In the case of the *Salmonella typhimurium* type 29 epidemic in Britain around 1965, additional resistances were acquired at different times[61]; in the cases of the 1972-3 *S.typhi* epidemics in Mexico, India and Vietnam, the R plasmids probably came from the commensal flora[62]. In each case conjugation was probably involved. It is also possible to demonstrate the formation of recombinant strains *in vivo* after drug therapy[63,64] but it is difficult to show this in the absence of selection (when enterobacteria only constitute about 1% of the total gut flora). In one study over 200 days elapsed before R transfer from the predominant strain could be demonstrated[65]. On the other hand, transfer between strains in the rumen of the sheep occurs efficiently after 24 h starvation[66].

Barriers to conjugation. As discussed earlier, transfer can be prevented by absence of effective pair formation or by surface exclusion. External conditions (e.g., pH, temperature and oxygen tension[67]) will affect this process. Once inside the recipient, the incoming DNA must be replicated (which requires making efficient use of host functions, and being replicated in a controlled manner) or integrated into a resident plasmid or into the chromosome. Except with transposons, this will presumably require some DNA:DNA homology. Many of these obstacles can now be overcome by the use of methods for *in vitro* and *in vivo*[68,69] genetic engineering, transformation of plasmids[70] and cell fusion[71-74].

It is clear that plasmids have been a major agent in gene exchange and therefore the evolution of bacteria. However, the mechanisms making for genetic isolation between bacterial populations must also have been successful, since both bacteria and plasmids exist in clearly defined groups.

REFERENCES.

1. Holloway, B.W. (1979) Plasmid 2, 1-19.
2. Collins, J.F. and Broda, P. (1975) Nature 258, 722-723.
3. Stallions, D.R. and Curtiss, R. (1972) J.Bact. 111, 294-295.
4. Brinton, C.C. (1971) Crit.Revs.Microbiol. 1, 105-160.
5. Curtiss, R. (1969) Ann.Rev.Microbiol. 23, 69-136.

6. Marvin, D. and Hohn, B. (1969) Bact.Revs. 33, 172-209.

7. Novotny, C.P. and Fives-Taylor, P. (1978) J.Bact.133, 459-464.

8. LeBlanc, D.J. and Falkow, S. (1973) J.Molec.Biol. 74, 689-701.

9. Cullum, J. and Broda, P. (1979). J.Bact. 137, 281-284.

10. Achtman, M., Schwuchow, S., Helmuth, R., Morelli, G. and Manning, P.A. (1978) Molec.Gen.Genet. 164, 171-183.

11. Falkinham, J.O. and Curtiss, R. (1976) J.Bact. 126, 1194-1206.

12. Rupp, W.D. and Ihler, G. (1968) Cold Spring Harbor Symposium 33, 647-650.

13. Ohki, M. and Tomizawa, J.-I. (1968) Cold Spring Harbor Symposium 33, 651-657.

14. Vapnek, D., Lipman, M.B. and Rupp, W.D. (1971) J.Bact. 108, 508-514.

15. Sarathy, P.V. and Siddiqi, O. (1973) J.Molec.Biol. 78, 443-451.

16. Hershfield, V., LeBlanc, D.J. and Falkow, S. (1973) J.Bact.115, 1208-1211.

17. Clark, A.J. and Margulies, A.D. (1965) Proc.Natl.Acad.Sci., Wash. 53, 451-459.

18. Kingsman, A. and Willetts, N. (1978) J.Molec.Biol. 122, 287-300.

19. Ohtsubo, E., Nishimura, Y. and Hirota, Y. (1970) Genet. 64, 173-188.

20. Ohtsubo, E. (1970) Genet. 64, 189-197.

21. Willetts, N. (1972) Ann.Rev.Genet. 6, 257-268.

22. Achtman, M., Skurray, R.A., Thompson, R., Helmuth, R., Hall, S., Beutin, L. and Clark, A.J. (1978) J.Bact. 133, 1383-1392.

23. Miki, T., Horiuchi, T. and Willetts, N.S. (1978) Plasmid 1, 316-323.

24. Willetts, N. and McIntire, S. (1978) J.Molec.Biol. 126, 525-549.

25. Ippen-Ihler, K., Achtman, M. and Willetts, N. (1972) J.Bact.110, 857-863.

26. Sharp, P.A., Hsu, M.-T., Ohtsubo, E. and Davidson, N. (1972) J.Molec.Biol. 71, 471-497.

27. Skurray, R.A., Nagaishi, H. and Clark, A.J. (1976) Proc.Nat.Acad.Sci., Wash. 73, 64-68.

28. Kennedy, N., Beutin, L., Achtman, M., Skurray, R., Rahmsdorf, U. and Herrlich, P. (1977). Nature 270, 580-585.

29. Minkley, E.G., Polen, S., Brinton, C.C. and Ippen-Ihler, K. (1976) J.Molec.Biol. 108, 111-121.

30. Crosa, J.H., Luttropp, L.K., Heffron, F. and Falkow, S. (1975) Molec.Gen. Genet. 140, 39-50.

31. Finnegan, D.J. and Willetts, N.S. (1973) Molec.Gen.Genet. 127, 307-316.

32. Willetts, N. (1977) J.Molec.Biol. 112, 141-148.

33. Gasson, M. and Willetts, N. (1976) Molec.Gen.Genet. 149, 329-333.

34. Ozeki, H., Stocker, B.A.D. and Smith, S.M. (1962) J.Gen.Micro. 28, 671-687.

35. Stocker, B.A.D., Smith, S.M. and Ozeki, H. (1963) J.Gen.Micro. 30, 201-221.

36. Willetts, N. (1974) Molec.Gen.Genet. 129, 123-130.

37. Cullum, J., Collins, J.F. and Broda, P. (1978) Plasmid 1, 545-556.

38. Cullum, J., Collins, J.F. and Broda, P. (1978) Plasmid 1, 536-544.

39. Inselburg, J. (1977) J.Bact. 132, 332-340.

40. Olsen, R.H. (1978) J. Bact. 133, 210-216.

41. Hayes, W. (1968) The Genetics of Bacteria and their Viruses. Second Edition. Blackwell, Oxford.

42. Cullum, J. and Broda, P. (1979). Plasmid 2 (in press).

43. Broda, P. and Collins, J.F. (1974) J.Bact. 117, 747-752.

44. Willetts, N. (1977) in: R factors. S. Mitsuhashi, Ed. University of Tokyo Press.

45. Chandler, P.M. and Krishnapillai, V. (1974) Genet.Res. 23, 239-250.

46. Bradley, D.E. (1977) in: Microbiology 1977. D. Schlessinger, Ed. American Society for Microbiology.

47. Olsen, R.H., Siak, J.-S. and Shipley, P.L. (1977) in: Microbiology 1977. D. Schlessinger, Ed. American Society for Microbiology.

48. Benson, S. and Shapiro, J. (1978) J.Bact. 135, 278-280.

49. Haas, D. and Holloway, B.W. (1978) Molec.Gen.Genet. 158, 229-237.

50. Barth, P.T. (1979) Plasmid 2, 130-136.

51. Beringer, J.E., Hoggan, S.A. and Johnston, A.W.B. (1978) J.Gen.Micro. 104, 201-207.

52. Petit, A., Tempé, J., Kerr, A., Holsters, M., van Montagu, M. and Schell, J. (1978) Nature 271, 570-571.

53. Klapwijk, P.M., Scheulderman, T. and R.A. Schilperoort. (1978) J.Bact. 136, 775-785.

54. Jacob, A.E. and Hobbs, S.J. (1974) J.Bact. 117, 360-372.

55. Dunny, G.M. and Clewell, D.B. (1975) J.Bact. 124, 784-790.

56. LeBlanc, D.J., Hawley, R.J., Lee, L.N. and St. Martin, E.J. (1978) Proc.Natl.Acad.Sci., Wash. 75, 3484-3487.

57. Dunny, G.M., Brown, B.L. and Clewell, D.B. (1978) Proc.Natl.Acad.Sci., Wash. 75, 3479-3483.

58. Hershfield, V. (1979) Plasmid 2, 137-139.

59. Hopwood, D.A. and Wright, H.M. 1976. J.Gen.Micro. 95, 107-120.

60. Bibb, M.J., Freeman, R.F. and Hopwood, D.A. (1977) Molec.Gen.Genet. 154, 155-166.

61. Anderson, E.S. (1968) Ann.Revs.Micro. 22, 131-180.

62. Anderson, E.S. (1975) J.Hygiene 74, 289-299.

63. Anderson, J.D., Gillespie, W.A. and Richmond, M.H. (1973) J.Med.Micro. 6, 461-473.

64. Anderson, J.D., Ingram, L.C., Richmond, M.H. and Wiedemann, B. (1973) J.Med.Micro. 6, 475-486.

65. Petrocheilou, V., Grinsted, J. and Richmond, M.H. (1976) Antimicrob. Agents Chemother. 10, 753-761.

66. Smith, M.G. (1977) J.Hygiene 79, 259-268.

67. Burman, L. (1977) J.Bact. 131, 69-75.

68. Kleckner, N., Roth, J. and Botstein, D. (1977) J.Molec.Biol. 116, 125-159.

69. Hooykaas, P.J.J., Klapwijk, P.M., Nuti, M.P., Schilperoort, R.A. and Rorsch, A. (1977) J.Gen.Micro. 98, 477-484.

70. Ehrlich, S.D. (1978) Proc.Natl.Acad.Sci., Wash. 75, 1433-1436.

71. Hopwood, D.A. and Wright, H.M. (1978) Mol.Gen.Genet. 162, 307-317.

72. Hopwood, D.A. and Wright, H.M. (1979) J.Gen.Micro. 111, 137-143.

73. Fodor, K. and Alföldi, L. (1979) Molec.Gen.Genet. 168, 55-60.

74. Bibb, M.J., Ward, J.M. and Hopwood, D.A. (1978) Nature 274, 398-400.

Plasmids of Medical, Environmental and Commercial Importance
K.N. Timmis and A. Pühler, editors

PLASMIDS OF MEDICAL IMPORTANCE

FELIPE CABELLO[1] AND KENNETH N. TIMMIS[2]
[1]New York Medical College, Valhalla, N.Y., USA and
[2]Max-Planck-Institut für Molekulare Genetik, Berlin-Dahlem, BRD

INTRODUCTION

Multiple antibiotic resistance plasmids were discovered in Japan just 20 years ago[1]. Since that time, an impressive volume of epidemiological data has accumulated which demonstrates unequivocally that the widespread and increasing occurrence of such "R factors" or R plasmids in bacteria progressively reduces or at least complicates effective antibiotic treatment of human and animal bacterial infections. More recently, a number of properties that directly contribute to the pathogenic potential of bacteria have been shown to be plasmid specified[2]. Plasmids that code for several distinct toxins[3,4,5,6] haemolysins[7,8] and antigens frequently associated with virulent bacteria[9,10] have been described, as have others that specify virulence functions of as yet undetermined activity[6]. Except for toxins, however, little is known about the precise role and relative importance of these and other bacterial virulence factors in determining the pathogenic potential of bacteria and in the development of pathological conditions. It became clear during the early recombinant DNA debates that even in the case of *E. coli*, the most extensively characterized bacterium, very little is known about bacterial mechanisms involved in the production of disease in man and animals. Nevertheless it is apparent that recent advances in molecular genetics and cell biology, in particular the development of recombinant DNA techniques, have provided powerful methods for studies of molecular mechanisms of pathogenicity[11]. Progress in this area will not only provide important information on parasite-host interactions but, in some instances, may also suggest new approaches in clinical medicine and public health practices for the prevention and/or treatment of infections (e.g. the isolation of *Vibrio cholerae tox* mutants and the development of live oral cholera vaccines[12]).

In this short article we will review the recent literature on plasmid-encoded functions that have been shown or are thought to contribute to the pathogenicity of disease-producing bacteria.

E. COLI AND DIARRHOEA

The well-characterized enteric pathogens, Salmonellae, Shigellae, and some viruses and protozoa, account for less than half of the known cases of infectious diarrhoea. In recent years, however, an ever increasing volume of evidence has implicated bacteria, in particular *Escherichia coli*, that are considered to be members of the normal gut flora as important agents of enteric disease. Pathogenic *E. coli* strains have been shown to provoke diarrhoea by three distinct processes: a dysentery-like mechanism in which the organism invades the intestinal mucosa of the small bowel and elicits an acute inflammatory response; a cholera-like mechanism in which the organism multiples solely within the lumen of the small bowel and exerts its effect by producing an enterotoxin that stimulates the secretion of water and electrolytes; and a third, as yet uncharacterized, mechanism[2,13,14].

The pioneering work of H.W. Smith and coworkers revealed that a collection of *E. coli* strains which were enteropathogenic for piglets all were able to produce enterotoxin, a surface protein antigen designated K88, and haemolysin. Genetic experiments demonstrated that the determinants of all of these characteristics were transmissible and molecular studies subsequently confirmed that the K88 antigen and enterotoxin (Ent) genes are located on plasmids[15]. The genes for K88 and enterotoxin production were found to be obligatory for enteropathogenicity of the bacteria[10]. The target tissue of enterotoxin is the small bowel epithelium, a tissue that is ordinarily maintained relatively free of bacteria. Ent^+ $K88^-$ bacteria are unable to colonize the small bowel and do not produce diarrhoea. Synthesis of K88 antigen leads to the production of surface pili that enable bacteria to adhere specifically to intestinal epithelial cells (and erythrocytes) and hence to colonize the small bowel. Extensive proliferation of Ent^+ $K88^+$ bacteria in this organ with concomittant release of enterotoxin directly at its target tissue produces acute diarrhoea[2].

Two distinct types of enterotoxin have been identified. One of these, designated ST, is heat stable, of low molecular weight, and poorly antigenic[16] whereas the other, designated LT, is heat labile, of high molecular weight, and a good antigen[17]. LT stimulates the production of adenyl cyclase in the mucosal cells of the small intestine[18] and its mode of action resembles that of Vibrio cholerae enterotoxin. It is readily assayed by its induction of morphological alterations in tissue cultures of adrenal cells[19]. On the other hand, ST stimulates the production of guanylyl cyclase and is better assayed by the infant mouse test[20]. Toxigenic strains producing either ST or LT, or LT plus ST have been isolated[21]. Whereas Ent plasmids encoding ST are of differing size and G+C content, those encoding both ST and LT are homogeneous in size and G+C content[4]. The location of ST genes on plasmids of diverse origin is consistent with the recent finding that such toxin genes can transpose from one plasmid to another[22]. It is anticipated that molecular and functional analyses of enterotoxins and enterotoxin genes will be facilitated by recent experiments in which DNA fragments carrying the genetic determinants of ST and LT were cloned into ColE1-type vector plasmids[23-25]. Bacteria carrying these high copy number hybrid plasmids should be convenient sources of the toxins and toxin determinants (see Dallas et al, this volume).

Enterotoxigenic E. coli isolated from patients with diarrhoea not only synthesize ST and/or LT but also produce a cell surface antigen called colonization factor antigen, or CFA. CFA has a pilus-like structure, mediates bacterial adhesion to epithelial cells of the small intestine[26] and is analogous in function to the K88 antigen of porcine enteropathogenic E. coli strains. Furthermore, bacteria producing CFA cause mannose-resistant haemagglutination of human group A erythrocytes[27]. In the infant rabbit model antibodies against CFA are protective against infection by enterotoxigenic E. coli[23]. Epidemiological studies revealed the presence of CFA in 86% of E. coli strains isolated from adult diarrhoeal cases but in only 8% of enterotoxigenic E. coli from healthy individuals[28]. Double blind studies with human volunteers demonstrated that CFA synthesis was obligatory for the production of disease by enterotoxigenic E. coli and that a four-fold increase of antibody titres against CFA, bacterial O antigen, and enterotoxin was elicited during infection. Bacteria carrying CFA were excreted in the faeces of volunteers for a longer period of time than were bacteria not carrying CFA[29]. More recently, a new colonization factor antigen, CFAII, was found which, like CFAI, is a surface antigen that mediates adhesion to the intestinal mucosa. CFAII differs, however, from CFAI in its erythrocyte-agglutinating properties and antigenic structure[30]. CFAI is known to be encoded by a 60 mD plasmid. Although a CFAII-encoding plasmid has not thus far been identified and characterized, the high spontaneous loss of CFAII-production ability during storage of bacteria strongly suggests a plasmid location for the CFAII genetic determinant. Ent and CFA plasmids are therefore seen to be functionally related in enteropathogenic bacteria and to play crucial roles in the virulence of such strains.

Attention should also be drawn to the relationships between the Ent/CFA plasmids, on the one hand, and R and colicin plasmids, on the other. Recently, a plasmid carrying both antibiotic resistance and enterotoxin genes was isolated[31]. It is to be expected that this type of genetic combination will appear more frequently in future in view of the powerful selection pressures created by extensive non-therapeutic use of antibiotics and the ability of these genes to transpose from one genetic element to another[22,32]. Selection pressures for this type of genetic combination are most severe in underdeveloped countries, where sanitary facilities are primitive and antibiotics used indiscriminately, and are likely to increase the prevalence of enterotoxic E. coli.

It is well established that ColV confers upon E. coli an increased ability to survive in the alimentary tract of animals and human beings and that significant transfer of this plasmid can occur among intestinal bacteria in the human gut[33,34]. The increased persistence of ColV-carrying bacteria results partly from the displacement of competing E. coli bacteria that are sensitive to the lethal activity of colicin V but also perhaps from additional functions specified by the ColV plasmid. Recently, a ColIb plasmid that encodes a CFA was described[35]. In this case the selective advantage conferred by ColIb upon host bacteria in the intestinal environment is expected to promote the transfer of CFA to other gut strains. ColIb-CFA plasmid-carrying bacteria were also found to carry R plasmids. Presumably recombination of R and ColIb-CFA plasmids may eventually take place, thus conferring additional selective advantages upon host bacteria in those individuals exposed to antibiotics[35]. Eventual introduction of an enterotoxin plasmid into such bacteria could then allow further evolution of their pathogenic potential[22].

Exterotoxin plasmids are not found with the same frequency in all *E. coli* O-serotypes and, thus far, strains carrying both ST and LT plasmids have been shown to belong to only four groups. The biological basis of this serotype preference has not yet been elucidated but could result from plasmid incompatibility reactions or strain-dependent differences in the ability to (a) act as recipients for ST-LT plasmids donated by conjugation[36] (b) stably maintain ST-LT plasmids, or (c) persist in the gut. O-antigen structure might influence bacterial persistance by modifying bacterial resistance to host defence factors, colicins or bacteriophages, or the ability of bacteria to adhere to the mucosal epithelium.

E. COLI EXTRAINTESTINAL INFECTIONS

In addition to intestinal disease, *E. coli* is able to produce several other types of infection ranging in severity from asymptomatic bacteriuria to bacteraemia and meningitis.

E. coli infections of the urinary tract. The majority of urinary tract infections are caused by strains of *E. coli*. One school of thought holds that bacteria isolated from such infections necessarily possess specific virulence characteristics that enable invasion of the urinary tract. Another school, on the other hand, believes that such strains do not exhibit special properties and that they are derived from predominant bacterial types present in the gut at the time of infection[37]. Recent evidence indicates that *E. coli* strains which cause urinary tract infections commonly possess some or all of a constellation of specific characteristics and thereby supports the former view.

Properties of *E. coli* strains isolated from urinary infections which are known or thought to be plasmid-encoded include the production of colicin V and haemolysin and the ability to agglutinate red blood cells[38,39]. ColV is known to increase the persistence of host strains in the intestine and to elevate the resistance levels of such bacteria to the lethal activity of serum[6]; presumably these functions could also contribute to the invasive ability of urinary tract pathogens. Recently, ColV has been shown to encode an iron-uptake system that is of crucial importance for the provision of sufficient iron for the growth of bacteria in body tissues and fluids (see Williams and George, this volume).

The haemagglutination ability of some E. coli strains is due to the presence of specific pili on the bacterial surface that promote bacterial attachment to human uroepithelial cells[40]. This attachment is presumably required for initial bacterial colonization of, and subsequent persistance in, the urinary tract and is reminiscent of the CFA-promoted adhesion of enteropathogenic E. coli to intestinal epithelial cells. Whether haemolysin production influences the pathogenic potential of E. coli is not presently known but it seems probable that the cytotoxic activity of haemolysin can, in general, be significant for the development of infections[41,42]. Other properties that are commonly found in urinary tract E. coli strains are the possession of particular O- and K-antigens and the ability to ferment dulcitol[43]. Although it is believed that K antigens increase the ability of E. coli to resist phagocytosis, it is not presently known whether K antigen and/or these other properties do contribute to the virulence of urinary tract pathogens.

E. coli bacteraemias. Bacteraemia and septic shock can result from E. coli infections of any anatomical system, e.g. urinary, biliary, and respiratory tracts, etc. Such complications usually occur in individuals with an underlying disease[44] and are often fatal. E. coli isolated from blood cultures frequently produce colicin V and haemolysin (ref. 38 and F. Cabello, unpublished). It is tempting to speculate that the increased abilities to scavenge iron and to resist the lethal effects of serum that are conferred upon E. coli by the ColV plasmid may be important in bacterial invasion of, and persistance in, the blood stream[8]. It is also noteworthy that colicin V and endotoxin have been shown to act synergistically to increase vascular permeability[45]. The relevance of this finding to the pathology of septic shock, where endotoxin plays a crucial role, deserves further investigation. Bacterial resistance to the lethal activity of serum is also known to be coded by some antibiotic resistance plasmids[46,47,48]. If serum resistance is important for the initiation of bacteraemia, administration of antibiotics may select just those bacteria with invasive properties.

Recently a toxin that is determined by a plasmid designated Vir was shown to be elaborated by a strain of E. coli that produced bacteraemia in a lamb. The Vir plasmid is transmissible, belongs to the fi^+ group of plasmids, and is able to mobilize R plasmids[6]. The mode of action of the toxin, and its contribution to the virulence of the producing strain, has thus far not been determined.

E. coli meningitis. Studies by the Cooperative Neonatal Meningitis group (N.I.H.) demonstrated that 84% of E. coli strains which produce meningitis elaborate K1 antigen. Such K1-producing strains were found to be more pathogenic than control E. coli strains for mice, and disease production in this experimental model was prevented by K1-specific antibody[49]. Morbidity and mortality in patients with meningitis produced by K1 type E. coli strains were higher than in those with meningitis produced by non-K1 types of E. coli and were related to the amount of K1-antigen in the spinal fluid[50]. While these investigations establish the importance of K1-antigen in the ability of E. coli to invade and produce meningitis, additional factors are clearly essential. A variable proportion (20-40%) of healthy individuals (neonates, children and adults) were found to carry K1-producing E. coli strains. Acquisition of such strains is by vertical transmission[51]. In one nursery, wide fluctuations in the proportions of healthy carriers of K1-producing E. coli strains were observed. These fluctuations suggest a requirement of such strains for a colonization factor[52]. Perhaps other factors, such as those determined by ColV, are additionally required by invasive K1 types of E. coli; recently it was found that 44% of E. coli isolates that produce K1 antigen carry the ColV plasmid (F. Cabello, unpublished data) suggesting a relationship between K1 and a ColV function in such bacteria.

E. coli strains carrying K1 antigen are agglutinated by antiserum prepared against meningococcus group B cells[53]. The purification and analysis of polysaccharides from both cell types showed that they were indistinguishable by immunological methods and composed of sialic acid[54]. Genetic studies suggested that the gene coding for K1 antigen is located on the chromosome of E. coli although the frequency of conjugal transfer of this marker was low[55]. On the other hand, approximately 1-3% of K1 positive strains spontaneously lose the K antigen upon subculture (J.B. Robbins, personal communication; F. Cabello, unpublished data). The fact that the K1 antigen of some *E. coli* strains is lost rather frequently, and that two different bacterial species carry this antigenic structure, indicates that the K1 genetic determinant may be located on a plasmid or a transposon[56].

Other *E. coli* antigens have also been shown to be identical with surface antigens of different bacterial species, e.g. those of *E. coli* K92 and group C *Neisseria meningitidis*[57]; *E. coli* K100 and *Haemophilus influenza* type B[58]. The existence of cell surface antigens that are common to different bacterial genera could reflect selective advantages conferred by these structures upon bacteria that colonise similar ecological niches or, in the case of invasive pathogens, that must resist identical host defense factors. It is to be expected that there are other, as yet undetected, common properties of these unrelated pairs of bacteria which are important for their invasiveness. The identification and characterization of such properties may lead us to the development of methods for the prevention of disease formation by this class of invasive Gram negative bacteria.

STAPHYLOCOCCUS AUREUS PATHOGENICITY AND PLASMIDS

Some coagulase-positive strains of *S. aureus* are able to produce enteric disease by elaboration of an enterotoxin and it is known that 10% of all cases of food poisoning are caused by such strains[59]. Early genetic studies suggested that the production of one of the serological types of toxin, enterotoxin B, is coded by a plasmid because treatment of toxin-producing bacteria with acridine orange cured this trait[60]. Subsequently a plasmid coding for enterotoxin B was identified. Similarly, exfoliative toxin, a toxin which causes the scalded skin syndrome in infants, was shown to be plasmid encoded[61]. Bacteriocin synthesis by *S. aureus* is also plasmid-specified but thus far no correlation has been observed between the ability to produce bacteriocin and pathogenicity[62]. On the other hand, a bacteriocin-like substance would appear to be responsible, at least in some cases, for the interference which has been observed between different *S. aureus* strains, and which has been successfully used to stop the epidemic spread of *S. aureus* infection in nursery settings[63].

It is well known that most antibiotic resistance determinants in *S. aureus* are located on plasmids and that antibiotic resistant strains are able to spread epidemically[64,65]. It is not known, however, if these plasmids encode other functions that increase the infectivity and invasive ability of host bacteria.

STREPTOCOCCUS PATHOGENICITY AND PLASMIDS

The genus *Streptococcus* contains several strains that are pathogenic for Man, e.g. *S. pyogenes* (group A); *S. faecalis*, *S. agalactiae* (group B) and *S. pneumoniae*. In the last five years plasmids have been found in several of these species, although plasmid properties specifically related to pathogenicity have not been extensively investigated. Plasmids coding for antibiotic resistance have been found in a high proportion of *S. faecalis* strains[66,67]. Because *Streptococcus* is able to produce invasive diseases, such as bacterial endocarditis, in humans and because bacterial resistance to the lethal activities of serum is thought to be an important factor in pathogenicity, it will be instructive to determine whether there is a link between R plasmid carriage and serum resistance, as has been shown for *E. coli*[47,48]. It will also be interesting to investigate the relationships between plasmids encoding multiple antibiotic resistances, which recently appeared in *S. pneumoniae*, and the pathogenicity and epidemiology of this strain[69,70].

The group A β-haemolytic streptococci can be divided into approximately 61 types on the basis of their M proteins. M protein is a pilus-like protein surface antigen that is antiphagocytic and responsible for bacterial virulence. Epidemiological and genetical evidence support the view that the genes for M protein, and a frequently-associated extracellular lipoproteinase called "serum opacity factor" are plasmid-located but the physical isolation of such plasmids has not yet been reported[71,72,73]. Detailed characterization of the M protein genetic determinants may lead to a better understanding of the special pathogenic capacity of so-called nephritogenic strains, which belong to a limited number of M types and which are able to produce potentially fatal diseases, such as poststreptococcal glomerulonephritis[74].

Haemolysin and bacteriocin are known to be plasmid-specified in *S. faecalis*[75]. Although haemolysin is classically described as a determinant of pathogenicity, little is known about its real importance in disease formation. Finally, it should be noted that genetic experiments indicate the ability to produce extracellular glucan in some cariogenic strains of S. faecalis, is plasmid-specified[76].

CONCLUSIONS

From the foregoing it can be seen that plasmids contribute significantly to the overall pathogenic potential of disease-forming bacteria. In some instances, conclusive evidence has shown that plasmid-encoded functions play a crucial role in pathogenesis. Probably the best examples of this are the enterotoxins whose genetic determinants have been physically mapped to specific plasmid segments and whose central role in pathogenesis has been clearly demonstrated in both tissue culture and animal model systems, and in human volunteers. In other instances, however, only indirect albeit highly suggestive evidence exists for the involvement of plasmid functions in pathogenicity. Included in this latter category are two different classes of function.

The first class contains those bacterial properties that are known, or thought to contribute to the pathogenicity of bacteria and which, because of their genetic instability, are suspected of being plasmid-encoded. We have given several examples of such properties including streptococcal M-protein and E. coli K1 antigen. However, one other example of this class should be mentioned here, namely the IgA1 proteases which specifically cleave IgA1 and thereby destroy its antibody activity. Because IgA antibodies mediate important immunity reactions on internal surfaces of the body, the ability to destroy them may represent an important pathogenicity property of disease-forming bacteria. IgA1 cleavage activity was originally detected in an E. coli strain and was spontaneously lost during subculture of the clone. The genetic determinant of this property may well, therefore, be plasmid located. Subsequently IgA1 cleavage ability was found in other species of bacteria including pathogenic Neisseria[77,78]. The recent discovery that Clostridium perfringens, which produces a range of different toxins, harbours several plasmids[79] indicates that important virulence factors in this type of pathogen may be plasmid-determined.

The second class of properties contains a variety of plasmid-coded functions that are commonly expressed by pathogenic organisms but whose role in pathogenesis is quite speculative. Examples include haemolysins, colicins, the fermentation of dulcitol raffinose and sucrose, and the production of H_2S[80,81].

It is clear that use of current techniques for the molecular analysis and genetic manipulation of genomes will very quickly identify the location of pathogenicity genes and permit their molecular dissection (e.g. of M-proteins and serum opacity factor) and functional analysis. On the other hand, the development of tissue culture, organ culture and whole animal models that are suitable for investigating the precise role of bacterial properties that are known to be, or suspected of being, involved in bacterial pathogenesis will take considerably longer. It is to be expected, however, that the development and use of such model systems in conjunction with molecular methods will during the next years produce important advances for our understanding of mechanisms of pathogenicity at the molecular level, and for the development of new approaches to combat infections.

Finally, we wish to emphasize the ability of plasmids to undergo rapid evolution in response to prevailing selection pressures (see also Farrell and Chakrabarty, this volume). It has been amply demonstrated that gene transfer among populations of bacteria, by conjugation, transduction and perhaps transformation, and among populations of plasmids, by the translocation of DNA segments from replicon to replicon, allows the formation of a large number of different plasmids containing all possible combinations of genes that can contribute to the ability of pathogenic bacteria to produce disease. Of particular relevance here is the recent finding that the conjugal transfer proficiency of a tetracycline resistance plasmid can be induced by tetracycline[82]. The potential for plasmids to undergo rapid evolution and, under appropriate selection conditions, to transmit themselves with high efficiency among bacterial populations is expected to have a marked impact upon the epidemiology of a number of bacterial diseases in future. It may well have been important for the development of the epidemics of *Salmonella vienna* and *S. typhi* that suddenly appeared after 1970 in France and Mexico, respectively, and that were correlated with R plasmid acquisition by the causative agents[2].

REFERENCES

1. Akiba, T., Koyama, K., Ishiki, Y., Kimura, S. and Fukushima, T. (1960) Japan J. Microbiol. 4, 219.

2. Falkow, S. (1975) Infectious Multiple Drug Resistance, Pion, London.

3. Smith, H.W. and Halls, S. (1968) J. Gen. Microbiol. 52, 319.

4. Gyles, C., So, M. and Falkow, S. (1974) J. Infect. Dis. 130, 40.

5. Shalita, Z., Hertman, T. and Sarid, S. (1977) J. Bacteriol. 129, 317.

6. Smith, H.W. (1974) J. Gen. Microbiol. 83, 95.

7. Smith, H.W. and Halls, S. (1967) J. Gen. Microbiol. 47, 153.

8. Goebel, W. and Schrempf, H. (1971) J. Bacteriol. 106, 311.

9. Ørskow, I. and Ørskow, F. (1966) J. Bacteriol. 91, 69.

10. Smith, H.W. and Linggood, M. (1971) J. Med. Microbiol. 4, 467.

11. So, M. and Falkow, S. (1977) Recombinant Molecules Impact on Science and Society — 10th Miles International Symposium, Ed. Beers,R. and Basset,C., Raven Press, New York, p. 107.

12. Mekalanos, J.J., Collier, R.J. and Romig, W.R. (1978) Proc. Nat. Acad. Sci. U.S.A. 75, 941.

13. Dupont, H.L., Formal, S.B., Hornick, R.B., Snyder, M.J., Libonati, J.P., Sheahan, D.G., LaBrec, E.H. and Kalas, J.P. (1971) New Engl. J. Med., 285, 1.

14. Levine, M.M., Bergquist, E.J., Nalin, D.R., Waterman, D.H., Hornick, R.B., Young, C.R., Sotman, S. and Rowe, B. (1978) Lancet, 11, 1119.

15. Bak, A.L., Christiansen, G., Christiansen, C., Stenderup, A. and Ørskow, F. (1972) J. Gen. Microbiol. 73, 373.

16. Jacks, T.M. and Wu, B.J. (1974) Infect. Immun. 9, 342.

17. Gyles, C.L. (1971) Ann. N. Y. Acad. Sci. 176, 314.

18. Evans, D.J. Jr., Chen, L.C., Curlin, G.T. and Evans, D.G. (1972) Nature New Biol. 236, 137.

19. Donta, S.T. (1974) J. Infect. Dis. 129, 284.

20. Dean, A.G., Ching, Y.A., Williams, R.G. and Harden, L.B. (1972) J. Infect. Dis. 125, 407.

21. Levine, M.M., Caplan, E.S., Waterman, D., Cash, R.A. Hornick, R.B. and Snyder, M.J. (1977) Infect. Immun. 17, 78.

22. So, M., Heffron, F. and McCarthy, B.J. (1979) Nature 277, 453.

23. So, M., Boyer, H.W., Betlach, M. and Falkow, S. (1976) J. Bacteriol. 128, 463.

24. So, M., Dallas, W.S. and Falkow, S. (1978) Infect. Immun. 21, 405.

25. Dallas, W.S. and Falkow, S. (1979) Nature 277, 406.

26. Evans, D.G., Silver, R.P., Evans, D.J., Chase, D.G. and

Gorbach, S. (1975) Infect. Immun. 12, 656.

27. Evans, D.G., Evans, D.J. and Tjoa, W. (1977) Infect. Immun. 18, 330.
28. Evans, D.G., Evans, D.J., Tjoa, W. and Dupont, H. (1978) Infect. Immun. 19, 727.
29. Evans, D.G., Satterwhite, T.K., Evans, D.J. and Dupont, H.L. (1978) Infect. Immun. 19, 883.
30. Evans, D.G. and Evans, D.J. (1978) Infect. Immun. 21, 638.
31. Gyles, C.L., Palchaudhuri, S. and Maas, W.K. (1977) Science 19, 198.
32. Cohen, S.N. (1976) Nature 263, 731.
33. Smith, H.W. and Huggins, M.B. (1976) J. Gen. Microbiol. 92, 335.
34. Williams, P.H. (1977) FEMS Lett. 2, 91.
35. Williams, P.H., Sedgwick, M.I., Evans, N., Turner, P.J., George, R.H. and McNeish, A.S. (1978) Infect. Immun. 22, 393.
36. Evans, D.J., Evans, D.G., Dupont, H.L., Ørskow, F. and Ørskow, I. (1977) Infect. Immun. 17, 105.
37. Turck, M. and Petersdorf, R.G. (1962) J. Clin. Invest. 41, 1760.
38. Minshew, B.H., Jorgensen, J., Counts, G.W. and Falkow, S. (1968) Infect. Immun. 20, 50.
39. Cooke, E.M. and Ewins, S.P. (1975) J. Med. Microbiol. 8, 107.
40. Svanborg Eden, C. and Hansson, H.A. (1978) Infect. Immun. 21, 229.
41. Smith, H.W. (1963) J. Path. Bact. 85, 197.
42. Fried, F.A., Vermeulen, C.W., Ginsburg, M.J. and Cone, C.M. (1971) J. Virol. 106, 351.
43. Guze, L.B., Montgomerie, J.Z., Potter, C.S. and Kalmanson, G.M. (1973) Yale J. Biol. Med. 46, 203.
44. Young, L.S., Martin, W.J., Meyer, R.D., Weinstein, R.J. and Anderson, E.T. (1977) Ann. Int. Med. 84, 456.
45. Ozanne, G., Mathieu, L.G. and Baril, J.D. (1977) Infect. Immun. 17, 497.
46. Olling, S. (1977) Scand. J. Infect. Dis. S10, 7.
47. Reynard, A.M., Beck, M.E., and Cunningham, R.K. (1978) Infect. Immun. 19, 861.
48. Taylor, P.W. and Hughes, C. (1978) Infect. Immun. 22, 10.
49. Robbins, J.B., McCracken, G.H., Jr., Gotschlich, E.C., Ørskow, F., Ørskow, I. and Hanson, L.A. (1974) New Engl. J. Med. 290, 1216.
50. McCracken, G.H., Sarff, L.D., Glode, M.D., Mize, G., Schiffer, M.S., Robbins, J.B., Gotschlich, E.C., Ørskow, I. and Ørskow, F. (1974) Lancet II, 246.

51. Sarff, L.D., McCracken, G.H., Schiffer, M.S., Glode, M.P., Robbins, J.B., Ørskow, I. and Ørskow, F. (1975) Lancet I, 1099.

52. Schiffer, M.S., Oliveira, E., Glode, M.P., McCracken, G.H., Sarff, L.M. and Robbins, J.D. (1976) Pediat. Res. 10, 82.

53. Grados, D. and Ewing, W.H. (1970) J. Infect. Dis. 122, 100.

54. Kasper, D.L., Winkelhake, J.L., Zollinger, W.D., Braudt, B.L. and Artenstein, M.S. (1973) J. Immunol. 110, 262.

55. Ørskow, I., Sharma, V. and Ørskow, F. (1976) Acta Path. Microbiol. Scand. Sect.B, 84, 125.

56. Elwell, L.P., Roberts, M., Mayer, L.W. and Falkow, S. (1977) Antimicrob. Ag. Chem. 11, 528.

57. Glode, M.P., Sutton, A., Moxon, E.R. and Robbins, J.B. (1977) J. Infect. Dis. 135, 94.

58. Ginsburg, C.M., McCracken, G.H., Schneerson, R., Robbins, J.B. and Parke, J.C. (1978) Infect. Immun. 22, 339.

59. Hornick, R. (1977) Gastroenterocolitis Syndromes, in: Infectious Diseases, Ed. Hoeprich,P.H., Harper and Row, New York, p. 540.

60. Dornbush, K., Hallander, H.O. and Löfquist, F. (1969) J. Bacteriol. 98, 351.

61. Warren, R., Rogolsky, M., Wiley, B.B. and Glasgow, L.A. (1975) J. Bacteriol. 122, 99.

62. Dajani, A.S. and Taube, Z. (1974) Antimicrob. Ag. Chem. 5, 594.

63. Shinefield, H.R., Ribble, J.C., Boris, M. and Einchenwald, H.F. (1963) Amer. J. Dis. Child. 105, 148.

64. Novick, R.P. (1969) Bacteriol. Rev. 33, 210.

65. Nahmias, A.J. and Shulman, J.A. (1972) Epidemiologic Aspects and Control Methods, in: The Staphylococci, Ed. Cohen,J.O., Wiley and Sons Inc. New York, p.483.

66. Clewell, D.B., Yagi, Y., Dunny, G.M. and Schultz, S.K. (1974) J. Bacteriol. 117, 283.

67. Van Embden, J.D.A., Engel, H.W.B. and Van Klingeren, B. (1977) Antimicrob. Ag. Chem. 11, 925.

68. Durack, D.T. and Beeson, P.B. (1977) Infect. Immun. 16, 213.

69. Jacobs, M.R., Koornhop, H.J., Robins-Browne, R.M., Stevenson, C.M., Veermack, Z.E., Freiman, T., Bennie Miller, G., Witcomb, M.A., Isaacson, M., Ward, J.I. and Austrian, R. (1978) New Engl. J. Med. 299, 735.

70. Smith, M.D. and Guild, W.R. (1979) J. Bacteriol. 137, 735.

71. Widdowson, J.P., Maxted, W.R., Grant, D.L. and Pinney, A.M. (1971) J. Gen. Microbiol. 65, 69.

72. Maxted, W.R. and Valkenburg, H.P. (1969) J. Med. Microbiol. 2, 199.

73. Cleary, P.P., Johnson, Z. and Wannamaker, L. (1975) Infect. Immun. 12, 109.

74. Stollerman, G.H. (1971) Circulation, XLIII, 915.

75. Jacob, A.E., Douglas, G.J. and Hobbs, S.J. (1975) J. Bacteriol. 121, 863.

76. Oliver, D.R., Brown, B.L. and Clewell, D.B. (1977) J. Bacteriol. 130, 759.

77. Mehta, S.K., Plaut, A.G., Calvanico, N.J. and Tomasi, T.B. (1973) J. Immunol. 111, 1274.

78. Mulks, M.H. and Plaut, A.G. (1978) New Engl. J. Med. 299, 973.

79. Duncan, C.L., Rokos, E.A., Christenson, C.M. and Rood, J.I. (1978) Microbiology-1978, Ed. Schlessinger,D., American Society for Microbiology, Washington, D.C., p. 246.

80. Smith, H.W. and Parsell, Z. (1975) J. Gen. Microbiol. 87, 129.

81. Lesher, L.J. and Jones, W.H. (1978) J. Clin. Microbiol. 8, 344.

82. Privitera, G., Sebald, M. and Fayolle, F. (1979) Nature 278, 657.

© 1979 Elsevier/North-Holland Biomedical Press
Plasmids of Medical, Environmental and Commercial Importance
K.N. Timmis and A. Pühler, editors

THE PLASMIDS OF AGROBACTERIUM TUMEFACIENS

MARC VAN MONTAGU and JEFF SCHELL

Laboratories for Histology and Genetics, State University Gent, K.L. Ledeganck-straat, 35, B-9000 Gent (Belgium).

INTRODUCTION

Plant pathologists have long been interested in the gram-negative bacteria, Agrobacterium, because most strains classified as Agrobacterium tumefaciens have the capacity to induce tumors (crown gall's) on most dicotyledonous plants[1]. These tumors are self-proliferating and graftable[2].

Recently there has been a renewed interest in Agrobacterium because progress was made in the understanding of the mechanism of tumor induction . This renewed attention was stimulated by the observation that all tumor inducing Agrobacterium strains harboured a rather large plasmid (approximately 120 Mdal)[3]. Removing the Ti-plasmid from virulent strains resulted in loss of oncogenicity and reintroduction of the Ti-plasmid yielded once again oncogenic Agrobacteria strains [4,5]. It could therefore be concluded that functions important to oncogenicity were plasmid encoded.

The idea of an extra-chromosomal nature of the "Tumor Inducing Principle"[6] and hence the search for a plasmid, was triggered by the observation that in a taxonomic classification of the different Agrobacterium strains, oncogenic and non-oncogenic representatives were found in all subgroups and clusters considered[7,8]. Thus it appeared possible that the capacity for tumor formation had spread throughout the genus as an extra-chromosomal element.

Once the idea of a plasmid as the causative agent had been accepted, the study of the mechanism of tumor-induction was initiated. By analogy with what was already known for oncogenic viruses, the integration into plant DNA of the plasmid or of a segment of it was very appealing[9]. This possibility was also strengthened by the observation that crown gall tumors contain a high concentration of unusual amino acids[10,11]. These compounds have been called opines[12] (Table 1). It had been shown that the synthesis in crown gall cells of a given opine was specified by the bacteria that caused the tumor and not by the plant species on which the tumor developed[13]. The importance of these observations

TABLE 1

STRUCTURES OF THE OPINES OF THE IMINO ACID TYPE ISOLATED[a] THUS FAR

$$R_1\text{-CO-COOH} + R_2\text{-CHNH}_2\text{-COOH} \xrightarrow[\text{catalysed by an "opine synthase"}]{\text{reductive condensation}} \begin{array}{l} \quad\;\; \text{H} \\ R_1 - \text{C} - \text{COOH} \\ \quad\;\; \text{NH} \\ R_2 - \text{C} - \text{COOH} \\ \quad\;\; \text{H} \end{array}$$

		octopine series	nopaline series
R_1-CO-COOH		pyruvic acid	α ketoglutaric acid
R_2-CHNH$_2$-COOH:	arginine	octopine[29]	nopaline[10]
	ornithine	octopinic acid[29]	nopalinic acid[32] or ornaline
	lysine	lysopine[30]	lysaline[33]
	histidine	histopine[31]	

[a] An octopine[34,35] and a nopaline[35,36] synthase have been purified. They catalyse the *in vitro* synthesis of the opines belonging to respectively the octopine series and the nopaline series.

was not generally accepted[14] because of doubts about whether or not opines were present in untransformed plant cells.

However when further work had confirmed that opines were found exclusively in crown gall tissue[15,16], it became logical to postulate that one or more genes localized on the Ti-plasmid would be directly involved in the biosynthesis of opines in the plant. The simplest explanation for this could be that the plasmid or a segment of it would be stably present and expressed in the tumor cells. Recently cot-analysis[17] and Southern blotting[18,19a] showed that this was indeed the case. Results from both the Seattle group (Mary-Dell Chilton, E. Nester, M. Gordon and co-workers) and our own group show that a sequence of the Ti-plasmid referred to as T-DNA (for transforming or transferred DNA) is present in the DNA of transformed plant cells. By transposon insertion mutagenesis we were able to demonstrate [18,19b,19c] that genes involved in opine biosynthesis are part of this T-DNA.

An important aspect of the interest in Ti-plasmids stems from the observation that this plasmid has a potential as a host-vector system for plants. Evidence pointing to this possibility was obtained by analysing the DNA of crown gall tumor cells induced by a Ti-plasmid carrying an insertion in the T-region[18]. It was shown that the 9.5 Mdal long drug-resistant transposon Tn7[20] was present along with the T-DNA in these tumors.

The main currents of our present Ti-plasmid research will be briefly reviewed hereafter.

THE MULTIPLE PLASMIDS OF AGROBACTERIUM TUMEFACIENS STRAINS

Supercoiled plasmid DNAs can be separated from linear DNAs by CsCl density centrifugation in the presence of intercalating dyes such as ethidium bromide[21]. If care is taken to avoid nicking of the plasmids and trapping in the mass of bacterial chromosomal DNA, it becomes possible to reproducibly isolate plasmids of 100 to 150 Mdal on a preparative scale[3,22,23].

Using this approach, Ti-plasmids are routinely prepared for physical studies such as restriction mapping, homology determinations and cloning.

By a systematic screening of various Agrobacterium strains, it became evident that many commonly used strains harboured several plasmids in the 130 Mdal size class range[16,25]. Often length measurements by electron microscopy allowed straightforward demonstration of the heterogeneity of the plasmids populations. In some cases, however, plasmids isolated from the same strain had very similar lengths making a convincing demonstration of the existence of two different plasmids on the basis of length measurements impossible. This was the case with strain B6. The presence of two plasmids pTi-B6 (120 Mdal) and pAT-B6

(113 Mdal) was established here by gelelectrophoresis of restriction enzyme digests of the plasmid mix and by a comparison of the sum of the lengths of the restriction fragments with the length measured by electron microscopy.

In other instances, extensive homology was observed between the different plasmids isolated from a given strain. For example in K14, a strain containing three plasmids[19d]: pTi-K14A of 132 Mdal, pAT-K14B of 130 Mdal and pAT-K14C of 158 Mdal, it was found that heteroduplex molecules between the A and the B plasmids were homologous for up to 60 %. It is not clear why such a combination of homologous plasmids does not continuously undergo recombination resulting either in a heterogeneous plasmid population or in a fusion plasmid carrying all the genetic information now distributed over the different plasmids.

Although a variety of abnormal structures, possibly plasmid recombinants, where observed they never accounted for more than one or two percent of the plasmid population.

Recently it was shown that large plasmids present in Rhizobium strains can be separated by gel electrophoresis[26]. Using this approach it was possible to demonstrate the presence of a very large (300 Mdal) plasmid[19f,26] in most Agrobacterium strains. This large plasmid DNA was prepared and its lenght was measured under the electron microscope[19g]. Furthermore Southern type hybridizations indicated that although all the large plasmids of different strains are of the same size class, they only have a limited homology with one another[19g]. It turns out that at least in the case of pAT C58C these large plasmids are transferrable by conjugation to other Agrobacterium strains [19f].

Little is known about the functions encoded by the non-Ti-plasmids of the different Agrobacterium strains. Occasionally a marker has been identified, such as phage exclusion (phage S18 by pAT-K14B)[19b] or agrocin K84 production (a marker present on pAT-K84 and pAT-396)[19a,28]. Hence these plasmids are frequently refered to as cryptic plasmids. Further analyses, however, revealed different cases where such cryptic plasmids were actually Ti-plasmids with deletions of part or all of the T-DNA and of the opine-catabolism encoding DNA.

The functions found to be encoded by the Ti-plasmids are summarized in Table 2, the abbreviations commonly used to describe phenotypes are indicated.

THE Ti-PLASMIDS ARE CATABOLIC PLASMIDS

The "opines" are not esssential for the development or maintenance of crown gall tumors. This became clear when Agrobacterium strains were isolated harbouring mutant nopaline Ti-plasmids, which induced tumors that did not contain the opines found in the galls induced by the "wild type" strains. So far we have not noticed any difference between the tumors induced by the pTi Onc^+ Nos^- and those induced by the pTi Onc^+ Nos^+ strains. It is rather unlikely that the biosynthesis of such high concentrations of opines in the transformed plant cells would be fortuitous. It can be argued[18] that the establishment of opine synthesis in plant cells is one of the goals of the crown gall formation. Since Agrobacterium strains have the unique capacity to catabolise these opines, by way of the Ti-plasmid encoded enzymes[37], the availability of the opines creates a singular ecological niche for the bacteria. The Ti-plasmids can thus be considered as a particular class of catabolic plasmids. In general bacteria colonize a territory when there is a natural compound that they alone are able to degrade. Frequently the particular enzymes needed for the degradative pathways are plasmid encoded. Hence these plasmids received the name of degradative or catabolic plasmids. Well documented are the Pseudomonas plasmids which encode for enzymes able to catabolize various hydrocarbons[38]. In the case of Agrobacterium, the compounds catabolized are not "found" in nature, but rather Agrobacterium, via its Ti-plasmid, obliges the plant to synthesize the compounds which only they are able to utilize. Crown gall cells release compounds into the rhizosphere, and bacteria able to utilize these compounds have a selective advantage over other soil bacteria. This unsuspected method of naturally occurring genetic engineering, has been called "genetic colonization"[18]. Agrobacterium strains harbouring a Ti-plasmid achieve this by inserting new genetic information (the T-DNA) into the plant.

Until recently only the imino acid type opines have been considered. Since it has been shown that a sugar derivative, called agropine[39], is also found in crown galls induced by Ti-octopine strains, and is catabolized only by such Agrobacterium strains, one may expect the possible existence of more, as yet undiscovered, "opines".

It should be stressed that the T-DNA is a rather large DNA segment of about 15 Mdal. Transposon insertion and deletion mutagenesis taught us that possibly only half of it is essential for tumor induction. There is therefore enough genetic information available to code for additional opine synthesizing enzymes.

THE REGIONS OF HOMOLOGY BETWEEN OCTOPINE AND NOPALINE Ti-PLASMIDS

Measurement of DNA reassociation kinetics leads to the conclusion that the extent of the homology between the various types of Ti-plasmids was limited to 10 % [40]. Since they all have in common the capacity for tumor induction, it was worthwhile to analyse the distribution of the regions of homology along the

TABLE 2

LOCALIZATION OF SOME Ti-PLASMID ENCODED FUNCTIONS BY TRANSPOSON INSERTION MUTAGENESIS

Phenotypes		Approximate map position in Mdal pTiC58	pTiAch5
Ape	Exclusion of phage AP1[19b,19h,52]	70	9 - 11
Agr^S	Sensitivity to the pAT-K84 encoded agrocin[28,53] Only Agrobacterium strains harbouring a nopaline plasmid show this phenotype,the other strains are Agr^R.	85 - 87	—
Arc	Arginine catabolism[54] Is part of a complex, inducible pathway. Deletion mutants constitutive for arginine catabolism (Arc^C) have been isolated[19f,19h].	6 - 7	12
Agc	Agropine catabolism, an octopine tumor specific sugar derivative[39,19h].	—	50 - 51
Noc	Nopaline catabolism. An inducible function, possibly under both negative and positive control. Mutants can be grouped in several classes (according to the aspect of colony growth on nopaline as sole nitrogen or sole carbon source). The necessary enzymes are encoded by nopaline Ti-plasmids. Nopaline is an inducer of its own catabolic pathway. Noc constitutive mutants are able to catabolize octopine[12].	6 - 7 and 11 - 12	—
Nos	Nopaline synthesis in crown gall tumors. This function is encoded by nopaline Ti plasmids only[37].	0 - 1	—
Occ	Octopine catabolism. An inducible pathway involving many functions. Octopine is an inducer. Nopaline is not catabolized by the octopine Ti-plasmid encoded enzymes[12,62].	—	24 - 26

Table 2 (cont'd)

Tra	Functions responsible for the conjugative phenotype of the plasmid. The transfer genes are repressed in the wild type plasmid. Conditions for induction of transfer have been determined and Tra constitutive mutants have been isolated[48].	70-80 14-19	15-20
Onc	Oncogenicity or the capacity for induction of crown galls on at least tobacco, peas and sunflower.	129-132 115 106-108 99-101 74	115-117 97-102 85-90 63-64
Onch	Host range effect on oncogenicity. Tumors are formed on Kalanchoe and/or potatoes, but not on tobacco, peas and sunflower.	128	9 105
Orc	Catabolisme of ornithine. It can be distinguished from Arc since $Arc^{+}Orc^{-}$ mutants were found.		28

plasmid genome. The first segment of substantial homology that was identified [41,42], was part of the T-DNA. A more detailed analysis was undertaken for the determination of the homology between a standard octopine Ti-plasmid Ach5 (or B6-806) and a standard nopaline Ti-plasmid C58. Upon hybridization of the labeled octopine Ti-plasmid DNA to nitrocellulose filters containing, via Southern blotting, the unlabeled DNA fragments obtained by digestion of the nopaline Ti-plasmid, it was possible to distinguish several regions of extensive homology[43,19e]. A more detailed physical map of the regions of homology was obtained by EM heteroduplex analyses[19d]. It was concluded that the 30 % homology is restricted to four major stretches of which two are distributed in the same relative order as compared to a common reference point and of which two occur in an inversed order. Three of these common regions contain several non-homologous segments, mostly substitution loops, dispersed throughout the homologous DNA. In fact the homology within the common DNA is also not complete. Increasing the stringency of the heteroduplex formation revealed that sequences which are only partially homologous are distributed over many sites of the common DNA. Taken together the data are consistent with the concept of a limited number of essential functions which remained relatively constant through plasmid evolution. Rather than implying a common ancestral plasmid one could as well consider the independent assembly of building blocks, in analogy with the evolution of R factors. The map position of the regions of homology together with a first identification of Ti-encoded functions, is indicated in Figure 1. It is striking that many of the functions essential for oncogenicity map in regions of "conserved" homology.

THE PHYSICAL MAPS OF THE TI PLASMIDS

The physical arrangements of the SmaI and the HpaI digest fragments of the octopine Ti plasmid B_6-806 have been determined[24]. For this a "Southern blot" of a HpaI digest was hybridised to the labelled DNA fragments obtained by extraction of gel bands of a SmaI digest. Recleavage of isolated fragments and further cross hybridisations allowed the arrangement of a nearly complete map.

For the octopine Ti-plasmids pTiAch5 and pTiB_6 a map was obtained by molecular cloning[19k] in pBR322 of large DNA fragments derived from partial HindIII digests of these plasmids. Restriction enzyme digestion of the different clones and Southern blot hybridisation, established the map order. Since there are more then 70 HindIII sites in these plasmids, overlapping fragments for all the regions were not obtained. Cloning of some large EcoRI and Bam-I fragments completed the gaps. A map for the restriction endonucleases HindIII, HpaI, SmaI, KpnI, EcoRI, XbaI and BamI was then constructed[19k]. The octopine plasmids

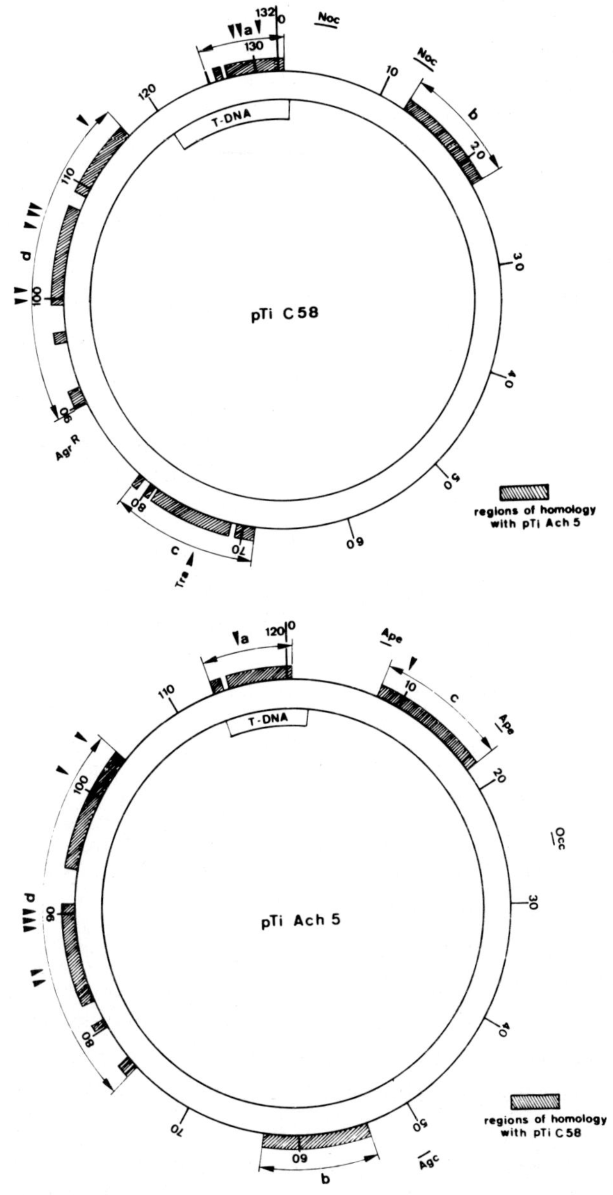

Fig. 1. Functional organization of an octopine (pTiAch5) and a nopaline (pTiC58) plasmid. The length of the plasmids (in Mdal) is indicated along the outer circle and is used as map coordinates. A SmaI restriction site in a segment common to both plasmids is chosen as zero point. The sequences common to both types of Ti-plasmid are indicated by a heavy line and are labeled a to d. The sites of transposon insertions resulting in Onc$^-$ phenotypes are indicated by an arrow.

pTiAch5 and $pTiB_6$ were virtually identical. Only one or two cleavage sites varied with the restriction endonucleases tested.

In an analogous way we constructed[44] a circular map of the nopaline Ti-plasmid pTiC58 using the same restriction enzymes. Comparison with a preliminary map[19c] of the nopaline Ti-plasmid pTiT37 shows that there are areas which are more fully conserved than others which, although homologous by EM heteroduplex analysis, show a distinctly different endonuclease cleavage pattern.

The determination of these restriction enzyme cleavage maps were a prerequisite for efficiently establishing a functional map of the Ti-plasmids. As explained further the location on this map of the different plasmid encoded functions was determined by insertion and deletion mutagenesis. The mapping of a given insertion or deletion can best be performed by the determination of the restriction fragment(s) which disappear(s) or change(s) mobility.

The length of the pTiAch5 and the pTiC58 as measured under the electron microscope[19d] and as obtained from the sum of the fragment lengths from a single restriction endonuclease digest[19e,k] correspond sufficiently well to accept a genome size of 120 Mdal for pTiAch5 and of 132 Mdal for pTiC58. A common Sma I site, located in the T-DNA was chosen as the zero coordinates for both plasmids.

All mapping data given in this article refer to distances in Mdal from this common restriction site.

TRANSPOSON INSERTION MUTAGENESIS

Undoubtedly the most efficient way to obtain a mutant plasmid is the insertion of an antibiotic resistance transposon[45]. No selection for a mutant phenotype has to be devised, but rather a straightforward selection for a plasmid carrying the transposon is sufficient. The plasmid to be mutagenized is introduced into the bacterial strain containing a drug transposon in the chromosome or in a non transferrable plasmid. The strain obtained is used as donor in a conjugation with an appropriately marked acceptor strain. The selection of those acceptor strains which acquired the antibiotic resistance typical for the drug transposon is equivalent to the selection of a strain containing a plasmid mutated by transposon insertion. Indeed, the only way, under these circumstances, by which a transposon can be transferred from the donor to the acceptor strain is by hooking up to the tranferrable plasmid.

If the plasmid under study is not conjugative, it is still possible to obtain tranfer either by cotransfer with an easily curable transferrable plasmid or by formation of a cointegrate between the plasmid and some R-type plasmids[46,19f].

Thanks to the ease of the selection, a large number of Ti-plasmids carrying a transposon (pTi::Tn) were obtained. Since a transposon is most likely to be inserted within a gene, it is clear that most of these plasmids will be mutant plasmids. Such mutations are polar, with the result that the silencing of genes is not necessarely due to an intragenic insertion. A comparison of reversion frequencies can allow a distinction between gene inactivation by polarity or inactivation by the formation of an altered gene product[19c,19j,19h]. Thousands of these mutant plasmids were screened by detecting the loss of a phenotype. The frequencies with which mutants in the different loci were obtained varied between 0.1 % and 10 %[47,19c,19h].

A further advantage of transposon insertion mutagenesis is the fact that the point at which the transposon is inserted becomes a starting point for the formation of deletions. Starting with a given pTi::Tn mutant plasmid, it was possible, by selection or screening for the loss of one or more plasmid encoded phenotypes, to obtain deletions (sometimes including 70 % of the plasmid[19j] (Table 3).

TABLE 3

LOCALIZATION OF SOME Ti-PLASMID ENCODED FUNCTIONS BY DELETION MAPPING

Approximate coordinates of the deletions in Mdal		Mutant phenotype
pTiC58	Δ 99 - 13	Onc^- Noc^- Orc^- Apr^-
	Δ 64 - 131	Onc^- Agr^R Ape^- tra^-
	Δ 35 - 110	Onc^- Agr^R Ape^- tra^-
	Δ 4 - 16	Noc^- Orc^C
	Δ 3 - 16	Noc^- Orc^-
	Δ 3 - 12	Noc^- Arc^C
pTiAch5	Δ 105 - 17	Onc^- Ape^-
	Δ 85 - 118	Onc^-

The "viability" of these deletion plasmids suggests that the origin of replication must be in the region 16 to 35.

First we used E. coli as host for the transposon insertion mutagenesis of the Ti-plasmids, since the choice of transposons available and the information on the efficiency of transposition was already extensive for this organism. Cointegrated plasmids, formed by the broad hostrange plasmid RP4 and a Ti-plasmid, were constructed [46,19f].

A first set of Tn7[20] insertion mutants were isolated in the octopine plasmid pTi Ach5[47]. Once it became possible to conjugate the Ti-plasmid with high efficiency between different Agrobacterium strain[48] a more direct approach was used. The transposon of choice was introduced into the chromosome of an appropriately marked Agrobacterium strain (mostly C58C1). There upon, the Ti-plasmid to be mutagenised was conjugated into this strain and the pTi containing strain so obtained was subsequently conjugated with a marked Agrobacterium as acceptor. To introduce a transposon into Agrobacterium a suicide plasmid vector was developed[49,19b]. This was based on the observation that plasmids carrying a phage Mu insert are unstable in Agrobacterium[50]. If such a plasmid carries a transposon, in addition to Mu, a copy of this transposon can be inserted in the host DNA before the loss of the incoming plasmid. This method has also been applied successfully for Rhizobium strains[51]. By making the mutant plasmids with the method described above it became possible to obtain a large number of independent isolates. Indeed by spotting one drop of a conjugation mix of the Agrobacterium strains on a solid medium, a multitude of crosses can readily be performed so that one can afford to keep only one isolate from each mating. For each of these mutant plasmids the location of the insert was subsequently determined.

In principle the location of the site of insertion can be determined by isolating the mutant plasmid on a preparative scale and analysing the fragmentation pattern after digestion with different restriction endonucleases. Alternatively, heteroduplexes formed with a reference plasmid which carries two mapped insertions, can be analysed in the electron microscope. The most efficient way however was a Southern blot hybridisation. A filter containing the fragments obtained from a restriction enzyme digest of the total Agrobacterium DNA (chromosomal plus plasmid DNA) was hybridised with probes, labelled by nick translation, which were consecutively : the pure Ti-plasmid, the drug transposon DNA and a cloned Ti-DNA segment. The latter was used as a confirmation in the identification of the restriction fragment in which the insertion occurred. With this method only a few micrograms of DNA, which can be obtained from a 10 ml culture, have to be prepared. This substantially increases the number of mutant colonies that can be analysed.

THE TI PLASMID ENCODED FUNCTIONS

By comparing the properties of Agrobacterium strains with and without the Ti plasmid it was possible to define Ti plasmid encoded functions. A preparation of mutant plasmids was subsequently screened for the loss of each of these functions. As described above, the use of transposon insertions for mutagenesis allowed a straightforeward physical mapping of mutant sites. These studies revealed that while some of the known functions could be mapped to single sites, other functions are more complex since many independent loci are involved. Table 2 summarizes those functions for which preliminary mapping data have been obtained.

We can consider that functionally the Ti plasmid is constructed out of a number of building blocks. Although little information is available on the number of different genes required for obtaining a particular phenotype, some generalizations can be advanced.

Oncogenicity functions. As indicated in fig 1 the Onc$^-$ mutations are distributed over the entire Ti plasmid. These findings underline that there is a diversity of functions involved in determining consecutively the contact between the bacteria and the plant, the conjugal transfer of the Ti plasmid into a plant receptor structure, the transfer of Ti plasmid DNA to the plant nucleus and the integration of the T-DNA into the plant chromosome and finally the expression of functions that interfere with the plant cell metabolism so as to create the tumor cell phenotype.

By definition the Onc$^-$ mutants which map in the T-DNA identify those genes involved in the maintenance of the tumorous growth of the plant cells; in nopaline plasmids these genes span roughly 5Mdal. It is stricking to note that many of the known oncogenicity functions map in this region which also corresponds to the major area of homology between nopaline and octopine Ti-plasmids. To date there is no biochemical information about the products of this region. It is however well established that the Onc region of the T-DNA is actively transcribed in the plant cell[55,19e].

Interestingly some onc mutants (both within and out of the T-DNA) suggest a host specific interaction; for example, some mutants affect functions which are dispensible for tumor formation on Kalanchoe but not on other plants. There is also no information about the products encoded by Onc regions outside of the T-DNA except that several of these DNA fragments cloned in E. coli synthesize proteins in minicells.

Conjugal transfer. With the observation that Ti-plasmids are autotransferable plasmids, as can be expected for most large plasmids, one could predict the existence of an elaborate set of genes necessary for this activity. In the case of the already well studied F-plasmid of E. coli about one third (20 Mdal) is occupied by transfer genes[56] . For the F-plasmid most of these gene products are involved in the formation, stabilization and, later on, active disaggregation of the mating aggregates between the donor and acceptor bacteria. Another set of genes is then responsible for initiating the transfer of DNA. It has been suggested that in the case of the Ti plasmids the part of the Tra genes essential for bacterial conjugation would also be involved in the transfer of the Ti-DNA from the bacteria to the plant[57] . This correlation was based on the outspoken thermosensitivity (32°C is the upper temperature limit) of both tumor induction and bacterial conjugation. However, most of the Tra^-plasmid mutants obtained by deletion or transposition insertion mutagenesis[19] were still Onc^+ so that these mutations impaired only bacterial conjugation. Recently, some insertion mutants were obtained which were simultaneously Onc^- Tra^-, but further analyses is needed to determine the exact function of the gene(s) inactivated here.

The preliminary mapping data available for pTiC58 show that there are Tra genes located in at least two diametrically opposed areas of the plasmid. Some mutations in the 15-19 Mdal regions (fig 1) are not fully Tra^-but show a 10^5 fold reduction of the efficiency of transfer.

It is particularly interesting that the wildtype Ti - plasmids studied thus far are all repressed for the expression of the transfer activity. Often octopine or nopaline, the substrates for the catabolic functions encoded by the Ti plasmid, will induce the Tra operon. Constitutive mutants have been isolated as point mutants but not by transposon insertion mutagenesis. Since all transposon insertions are polar, insertion into a Tra controlling gene would always inactivate more distal Tra genes and produce a negative phenotype. Thus, the Tra repressor may be the first gene of an operon containing some genes essential for transfer.

The fact that some opines are inducers of the Tra functions should be understood in the perspective that Ti plasmids are catabolic plasmids. Under conditions favorable for crown gall development the tumor cells release cellular components[58] and probably opines , so that an ecologically advantageous environment could be created for Agrobacterium strains which harbour a Ti-plasmid. It is therefore beneficial for Agrobacterium to spread the Ti plasmid as efficiently as possible to other invading strains at the crown gall site. This "in planta" conjugation was actually the first type of genetic exchange observed for Agrobacterium[59,60].

Opine biosynthesis. More than half of the T-DNA is not essential for tumor induction, nevertheless the same extent of the Ti-plasmid is inserted reproducibly in the plant DNA. The constancy of this length probably reflects the mechanism of DNA insertion. It is tempting to postulate that most of this DNA encodes for opine biosynthetic functions and hence provides a selective advantage to Agrobacteria containing opine catabolic functions.

Until now only opines of the imino acid type have been studied in detail (table 1). It is not definitively proven that the Ti-plasmid codes directly for all or part of an opine synthase since no DNA or protein sequence is a yet available. It is conceivable that the plasmid encodes an enzyme which modifies the specificity of a host enzyme or even induces the synthesis of latent host genes. The direct involvement of the T-DNA is evident since Onc^+ Nos^- Agrobacterium mutants induce tumors which no longer contain nopaline and the mutated DNA is found to be integrated in the plant genome[19a].

There has been some recent progress with the in vitro biosynthesis of imino acid type opines. A single enzyme catalyzes the synthesis of the pyruvic acid derivatives (octopine, octopinic acid, lysopine, hisopine) in octopine tumors, and another enzyme catalyzes the synthesis of the α-ketoglutarate derivatives (nopaline and nopalinic acid) in nopaline tumors[33,34,35,36]. If only one or two plasmid encoded proteins are required for the synthesis of these imino acids, we are challenged to determine whether or not T-DNA encodes genetic information for other, as yet undiscovered opines biosynthetic functions. For octopine Ti-plasmids a new class of opines, sugar derivatives called agropine, have been identified[39]. However, no further information is available on the exact structure of agropine nor on the number of biosynthetic steps involved in its biosynthesis.

Opine catabolism. The functions involved in providing Agrobacterium with the capacity to use octopine or nopaline as sole nitrogen or sole carbon source can be analyzed due to the availability of a large set of insertion and deletion mutants in these functions[19c,19j]. Furthermore, these regions of the Ti-plasmid have been isolated and propagated by molecular cloning techniques. Cloned DNA fragments have been used to study the expression of these regions in E. coli minicells[19n] or in Agrobacterium[19j,19i,19h]. Preliminary results point to the existence of a rather complex operon structure possibly containing both negative and positive controlling elements. Apart from a permease[61] and the octopine or nopaline dehydrogenase this encodes for the enzymes involved in the catabolism of arginine and ornithine[54]. For the octopine plasmid pTiB6 there is clearly some common element in the regulation of octopine catabolism and conjugal transfer[12,62]. In the case of the nopaline plasmid pTiC58 the nopaline catabolic region could span more than 8 Mdal[19j]. If all of this DNA codes for proteins involved in the efficient catabolism of nopaline, the system is indeed complex.

Mutants defective in the catabolism of agropine map in an area clearly distinct form octopine catabolism. These results may begin to identify a stretch of DNA encoding enzymes involved in the catabolism of some agropine related sugars. Also if it is correct that the T-DNA region encodes for the synthesis of as yet undiscovered "opines", one can expect that the relatively extensive regions of the Ti plasmid where no functions have as yet been mapped will encode for the catabolism of these opines.

Agrocin sensitivity. It is well established that bacteria can defend themselves against closely related strains by producing bacteriocins. In most of the cases studied, these toxic compounds are proteins and their action is limited towards those strains which have a receptor for the bacteriocin. The strain producing the bacteriocin is itself immune to the effects of the toxin. These properties were also observed for the bacteriocin produced by *Agrobacterium* strains K84 and 396. Agrocin is however not a protein but a nucleotide analogue[63,19m]. The nopaline Ti-plasmids encode for a membrane associated receptor protein which directs the uptake of agrocin (Murphy and Roberts, personal communication). By using ^{32}P-labelled agrocin it was easily demonstrated that only *Agrobacterium* strains containing nopaline Ti-plasmids will take up agrocin. The functions controlling agrocin sensitivity may play an important role in the uptake of phosphorylated compounds in general. Agrocin sensitivity is conserved in all nopaline Ti-plasmids even pTi396. This plasmid was isolated from the agrocin producing strain 396, a strain able to induce nopaline containing crown galls. The mechanism of action of agrocin is unknown but it is very efficient since an agrocin producing strain effectively protected fruit trees against crown gall formation[64,65].

Phage Exclusion. Bacteriophage AP-1 cannot complete a single growth cycle when the host bacerium harbours a nopaline or an octopine Ti-plasmid[19b,52]. The host is killed but no viable phage progeny is formed.

This Ti-plasmid encoded property can only be tested in bacteria which have an adsorption site for phage AP-1. The nature of the block in phage propagation is unknown. However it is well documented that many phages mutually interfere with each others development upon double infection. Exhaustive research has shown that these dramatic inhibitions of phage development are often the results of subtle gene dosages differences. By analogy one may suppose that the phage exclusion phenomenon is of minor importance to our knowledge of the Ti-plasmid molecular biology.

Miscellaneous functions. Among the less well documented Ti-encoded functions is the capacity of a C-58 Agrobacterium to release the plant hormone transzeatin[66,71]. It is not known if the T-DNA participates directly by encoding for enzymes involved in transzeatin biosynthesis, or in some indirect way such as by degradation of tRNAs or by modifying membranes resulting in the release of purine derivatives. Transzeatin secretion is likely to be of fundamental importance in crown gall formation since plant cells respond significantly to micromolar concentrations.

Some other functions may be identified by studying the expression of cloned fragments of Ti-plasmid. By testing the different fragments for protein production in either E. coli or Agrobacterium strains one obtains a list of proteins encoded by different areas of the Ti-plasmid. By using overlapping fragments or subclones of a particular fragment, one can position rather precisely the protein on the Ti map. Since circular DNA molecules can be efficiently mutagenized[67,19], it is possible to isolate mutants for a given protein. However, it will require a great effort to identify the functions of these different proteins.

THE EXTENT AND THE LOCALISATION OF THE T-DNA

When it became obvious that the opine synthesizing capacity, expressed by the transformed plant cell, was actually a trait determined by the Agrobacterium Ti-plasmid, it was evident to consider the possibility that a segment of the Ti-plasmid became part of the transformed plant genome. A first indication was obtained by solution DNA hybridisation studies. Some restriction fragments of the octopine plasmid pTiB6806 did indeed show an accelerated rate of association after addition of DNA from a particular tobacco crown gall tumor[17]. A much clearer picture was obtained when Southern gel blotting hybridisations were performed with the crown gall DNA. Total tumor DNA was digested with a variety of restriction enzymes, and used to drive the hybridization of nick translated cloned fragments of the Ti-plasmid[18,19a]. Some generalizations may be advanced from this approach. In the case of all tumors induced by the nopaline plasmids pTiT37 and pTiC58 the T-DNA corresponds to one large 15.6 Mdal contineous segment of the Ti-plasmid. The T-DNA is not integrated in plastid DNA (M.D. Chilton, personal communication) but is located in the nucleus[19a]. Several copies of the T-DNA are present as seen in fig 2 where border fragments

COMPOSITE FRAGMENTS REPRESENTING THE BORDERS BETWEEN T-DNA AND PLANT DNA

DRIVER : W38-C58 CROWN GALL

HINDIII DIGEST

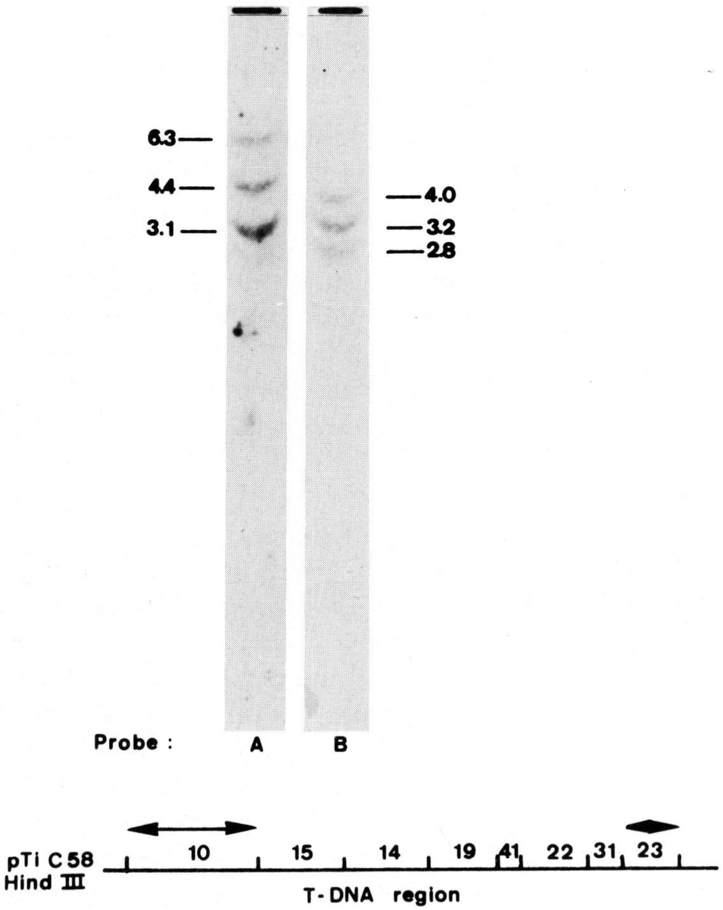

Fig 2. A "genomic blot" of Wisconsin 38 tobacco crown gall DNA, digested with HindIII restriction enonuclease was hybridised to the nick-translated T-DNA border fragments H-10 (lane A) and H-23 (lane B). The autoradiogram demonstrates the existence of at least three border fragments.

of the Ti-plasmid were used as probes in hybridizations to southern blots of tumor DNA. In this tumor line one can observe that there are at least three different right borders and three different left borders. The molecular weights of these fragments suggest that they are composed of T-DNA linked to some other DNA, probably plant DNA. The attempts to detect hybridization with clones of Ti-plasmid other than the border fragments were negative.

The border fragments of the T-DNA from tobacco teratoma T37 crown gall were cloned using bacteriophage lambda as the initial cloning vector. Subsequently, a DNA segment containing the non T-DNA border and as little T-DNA as possible was subcloned and used as a probe to obtain further information on the nature of the plant DNA integration site for T-DNA. It turns out that the T-DNA is integrated in a plant sequence which is repeated. This sequence is species specific since no hybridization was observed to the DNA from Petunia, a plant not so distantly related to tobacco. The biological role of this repeated DNA in tobacco is unknown. One can speculate that the specificity of a repeated sequence may have a functional role for example in synapse formation during meiosis. The availability of this plant sequence as a cloned probe will allow its properties to be tested and may lead to substantial new information necessary for the contruction of models of the organization of the plant genome. It will be particularly interesting to see if this DNA repeat is responsible for the formation of multiple copies of the T-DNA in the plant genome. To be sure many more tumor lines should be examined before this aspect of crown gall formation is completely resolved.

Whereas we routinely see multiple border fragments in tumors induced by nopaline Ti-plasmids, octopine Ti-plasmid induced tumors seem to have a single integration site. Fig 3 represents a typical result where only a single border fragment is observed using a probe for the left border of the T-DNA in crown galls induced by pTiAch5. Analogous results were obtained when the right hand border was looked for. For three plant species studied, Arabidopsis, Petunia and Tobacco, only one type of T-DNA border fragments was observed. The high intensity of the hybridisation signal obtained with Arabidopsis is due to the relatively small genome size of this plant.

This simple picture is probably not valid for all octopine crown galls[68] (M.D. Chilton and M. Gordon, personal communication) since the extensively analysed crown gall E-9 could contain up to twenty copies of T-DNA fragments[17].

The left end of octopine-type T-DNA in different tumour lines

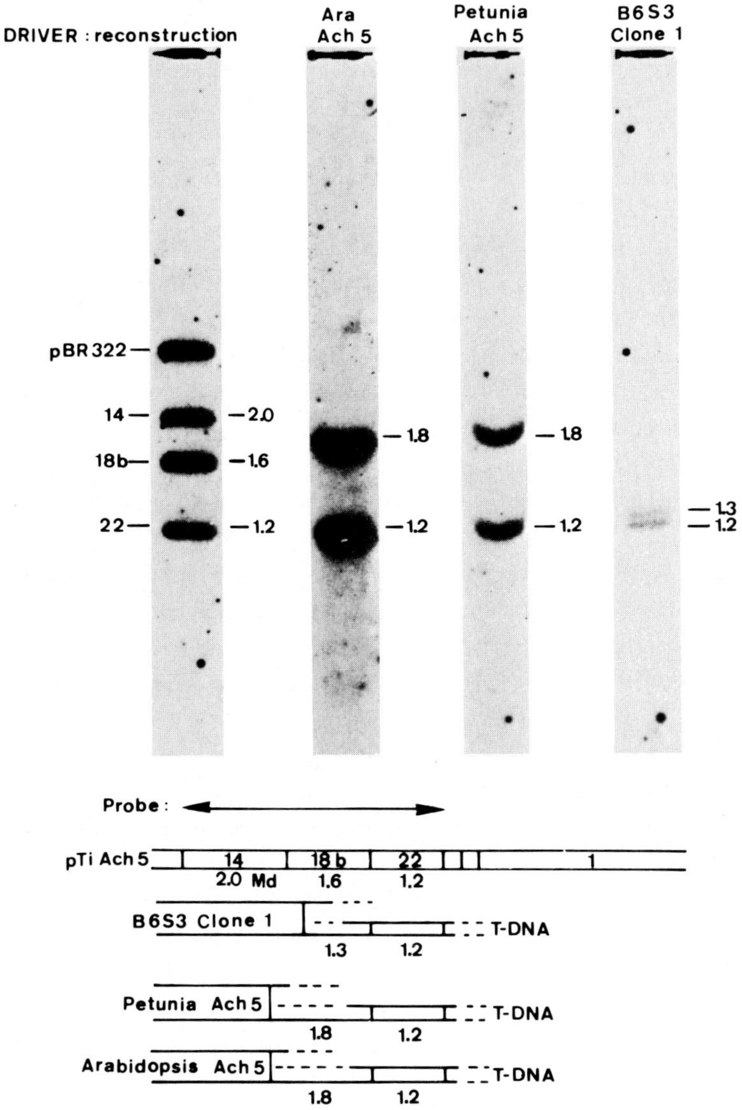

Fig 3. A "genomic blot" of respectively Arabidopsis[69], Petunia and cloned Tobacco crown gall DNA, digested with HindIII restriction endonuclease and hybridised to a probe for the left end of the octopine T-DNA.
The double headed arrow indicates the extent of the probe on the TiAch5 HindIII map. The Ti-plasmid B6S3 used for crown gall induction in Tobacco is completely homologeous to pTiAch5, as determined by hybridisation in the electron microscope. The HindIII restriction map for the T-DNA area is identical for pTiB6S3 and pTiAch5. The autoradiogram demonstrates the existence of single border fragments.

POSSIBLE USE OF THE TI-PLASMID FOR GENETIC ENGINEERING IN PLANTS

It is well established that new plants can be regenerated from a single somatic cell of several plant species. This totipotency of plant cells is gratifying for considerations of future attempts for genetic manipulations of plant cells. One may hope that new traits which are introduced will remain expressed in the plant progeny. For example, new plants have been regenerated from crown gall tumor cells and these regenerates retain the nopaline biosynthetis capacity[70]. For a T37 crown gall teratoma, Southern blotting experiments have now shown[19a] (and M.D. Chilton personal communication) that exactly the same segment of the T-DNA which is inserted in the original teratoma stays present in different tissues of the regenerated plant (Stem, leaves and flower petals). On the basis of these results it was obviously important to attempt to link a deliberately chosen DNA fragment to the T-DNA to investigate if this DNA fragment is stably inserted into the plant DNA. This result has been obtained since we have recently demonstrated that the antibiotic resistance transposon Tn7 inserted in the T-DNA (fragment H-23, fig 2) of a nopaline Ti-plasmid[19c] can be transferred upon tumor formationlinked to the whole of the T-DNA[18]. These promising results will now be followed by experiments where non transposon DNA and possibly plant genes are inserted into the T-DNA by in vivo recombination. It is a real challenge to put this natural genetic engineer, the Ti-plasmid at work to introduce new traits into plants.

ACKNOWLEDGMENTS

The authors wish to thank their collaborators[19] for providing the information presented in this article.
This work was supported by grants from the Kankerfonds van de A.S.L.K. and from the Fonds voor Wetenschappelijk Geneeskundig Onderzoek (n° 3.0052.78).

REFERENCES

1. Braun, A.C. (1978) Biochim. Biophys. Acta, 516, 167-191

2. Braun, A.C., and White, P.R. (1963) Phytopathology, 33, 85-100.

3. Zaenen, I., N. Van Larebeke, H. Teuchy, M. Van Montagu, and J. Schell (1974) J. Mol. Biol., 86, 109-127.

4. Van Larebeke, N., C. Genetello, J. Schell, R.A. Schilperoort, A.K. Hermans, J.P., Hernalsteens, and M. Van Montagu (1975) Nature, 255, 742-743.

5. Watson, B., T.C. Currier, M.P. Gordon, M.D. Chilton, and E.W. Nester (1975) J. Bacteriol., 123, 255-264.

6. Braun, A.C., and Mandle, R.J. (1948) Growth, 12, 255-269.

7. Keane, P.J., Kerr, A. and New, P.B. (1970) Austr. J. Biol. Sci., 23, 585-595.

8. Kersters, K., De Ley, J., Sneath, P.H.A., and Sackin, M. (1973) J. Gen. Microbiol., 78, 227-239

9. Schell, J., and Van Montagu, M. (1977) Brookhaven Symp. Biol., 29, 36-49.

10. Ménagé, A., and Morel, G. (1965) C.R. Soc. Biol. (Paris), 159, 561-562.

11. Goldmann, A., Thomas, D.W., and Morel, G. (1969) C.R. Acad. Sci. (Paris), 268, 852-854.

12. Petit, A., and Tempé, J. (1978) Molec. gen. Genet., 167, 147-155.

13. Petit, A., S. Delhaye, J. Tempé, and G. Morel (1970). Physio. vég., 8, 205-213.

14. Lippincott, J.A., and Lippincott, B.B. (1976) Physiol. Plant Pathol., 4, 356-388.

15. Kerr, A., and Roberts, W.P. (1976) Physiol. Plant Path., 9., 205-211.

16. Montoya, A., Chilton, M.D., Gordon, M.P., Sciaky, D., and Nester, E.W.(1977) J. Bacteriol., 129, 101-107.

17. Chilton, M.D., H.J. Drummond, D.J. Merlo, D. Sciaky, A.L. Montoya, M.P. Gordon, and E.W. Nester (1977) Cell, 11, 263-271.

18. Schell, J., M. Van Montagu, M. De Beuckeleer, M. De Block, A. Depicker, M. De Wilde, G. Engler, C. Genetello, J.P. Hernalsteens, M. Holsters, J. Seurinck, B. Silva, F. Van Vliet, and R. Villarroel (1979) Proc. R. Soc. Lond. B 204, 251-266.

19. Unpublished work of this laboratorium:
a De Beuckeleer, M. *et al.*, in preparation.
b Hernalsteens, J.P. *et al.*, in preparation.
c Holsters, M. *et al.*, in preparation.
d Engler, G., *et al.*, in preparation.
e Depicker, A., *et al.*, in preparation.
f De Block, M., *et al.*, in preparation.

g Villarroel, R., *et al.*, in preparation.
h De Grève, H., *et al.*, in preparation.
i Leemans, J., *et al.*, in preparation.
j Van Haute, E., *et al.*, in preparation.
k De Vos, G., *et al.*, in preparation.
l Seurinck, J., *et al.*, in preparation.
m Messens, E., *et al.*, in preparation.
n Dhaese, P., *et al.*, in preparation.

20. Barth, P.T., Datta, N., Hedges, R.W., and Grinter, N.J. (1976) J. Bacteriol., 125, 800-810.

21. Bauer, W., and Vinograd J. (1968) J. Mol. Biol., 33, 141-171.

22. Currier, T.C., and Nester, E.W. (1976) Analytical Biochemistry, 76, 431-441.

23. Van Larebeke, N., Genetello, C., Hernalsteens, J.P., Depicker, A., Zaenen, I., Messens, E., Van Montagu, M., and Schell, J. (1977) Molec. gen. Genet., 152, 119-124.

24. Chilton, M.-D., Montoya, A.L., Merlo, D.J., Drummond, M.H., Nutter, R., Gordon, M.P., and Nester, E.W. (1978) Plasmid, 1, 254-269.

25. Schell, J., Van Montagu, M., Depicker, A., De Waele, D., Engler, G., Genetello, C., Hernalsteens, J.P., Holsters, M., Messens, E., Silva, A., Van den Elsacker, S., Van Larebeke, N., and Zaenen, I. (1977) In: "Nucleic Acids and Protein Synthesis in Plants", L. Bogorad and J.H. Weil (Eds), Plenum Press, New York, 329-342.

26. Casse, F., Boucher, C., Julliot, J.S., Michel, M., and Dénarié (1979) J. Gen. Microbiol., in press.

27. Chilton, M.-D., McPherson, J., Saiki, R.K., Thomashow, M.F., Nutter, R.C., Gelvin, S.B., Montoya, A.L., Merlo, D.J., Yang, F.-M., Garfinkel, D.J., Nester, E.W., and Gordon, M.P. (1979) In: "Emergent Techniques for the Genetic Improvement of crops", I. Rubinstein (Ed.), University of Minnesota Press, in press.

28. Ellis, J.G., Kerr, A., Van Montagu, M., and Schell, J. (1979) Physiol. Plant Path., in press.

29. Ménagé, A., and Morel, G. (1964) C. R. Acad. Sci. (Paris), 259, 4795-4796.

30. Lioret, C. (1957) C. R. Acad. Sci. (Paris), 244, 2171-2174.

31. Kemp, J.D. (1977) Biochem. Biophys. Res. Comm., 74, 862-868.

32. Firmin, J.L., and Fenwick, R.G. (1977) Phytochem., 16, 761-762.

33. Kemp, J.D., Hack, E., Sutton, D.W., and El Wakio, M. (1978) Proc. IVth Int. Conf. Plant. Bact. - Angers, 183-188.

34. Hack, E., and Kemp, J.D. (1977) Biochem. Biophys. Res. Comm., 78, 785-791.

35. Otten, L.A., Vreugtenhil, D., and Schilperoort, R.A. (1977) Biochim. Biophys. Acta, 485, 268-277.

36. Kemp, J.D., Sutton, D.W., and Hack, E. (1979) Biochemistry, in press.

37. Bomhoff, G., Klapwijk, P.M., Kester, H.C.M., Schilperoort, R.A., Hernalsteens, J.P., and Schell, J. (1976) Mol. gen Genet., 145, 177-181.

38. Chakrabarty, A.M. (1976) Plasmids in Pseudomonas. Ann. Rev. Genet. 10, 7-30.

39. Firmin, J.L., and Fenwick, G.R. (1978) Nature 276, 842-844.

40. Currier, T.C., and Nester E.W. (1976) J. Bacteriol., 126, 157-165.

41. Depicker, A., Van Montagu, M., and Schell, J. (1978). Nature, 275, 150-153

42. Chilton, M.-D., Drummond, M.H., Merlo, D.J., and Sciaky, D. (1978) Nature, 275, 147-149.

43. Drummond, M.H., and Chilton, M.-D. (1978) J. Bacteriol., 136, 1178-1183.

44. Depicker, A., De Wilde, M., De Vos, G., De Vos, R., Van Montagu, M. and Schell, J. (1979) submitted for publication.

45. Brevet, J., Kopecko, D.J., Nisen, P., and Cohen, S.N. (1977) In" DNA, insertion elements, plasmids, and episomes", A.I. Bukhari, J.A. Shapiro, and Adhya, S.L. (Eds), New York, Cold Spring Harbor Laboratory, 169-178.

46. Holsters, M., Silva, B., Genetello, C., Engler, G., Van Vliet, F., De Block, M., Villarroel, R., Van Montagu, M. and Schell, J. (1978) Plasmid, 1, 456-467.

47. Hernalsteens, J.P., De Greve, H., Van Montagu, M. and Schell, J. (1978) Plasmid, 1, 218-225.

48. Petit, A., Tempe, J., Kerr, A., Holsters, M., Van Montagu, M. and Schell, J. (1978) Nature, 271, 570-572.

49. Hernalsteens, J.P., Holsters, M., Silva, A., Van Vliet, F., Villarroel, R., Engler, G., Van Montagu, M. and Schell, J. (1978) Archives internationales de Physiologie et de Biochimie, 86, 432-433.

50. Van Vliet, F., Silva, B., Van Montagu, M. and Schell J. (1978) Plasmid, 1, 446-455.

51. Beringer, J.E., Beynon, A.V., Buchanan-Wollaston and Johnston, A.W.B. (1978) Nature, 276, 633-634.

52. Van Larebeke, N., Genetello, C., Hernalsteens, J.P., Depicker, A., Zaenen, I., Messens, E., Van Montagu, M., and Schell, J. (1977) Molec. gen. Genet., 152, 119-124.

53. Engler, G., Holsters, M., Van Montagu, M., Schell, J., Hernalsteens, J.P., and Schilperoort, R.A. (1975) Molec. gen. Genet., 138, 345-349.

54. Ellis, J., Kerr, A., Tempé, J., and Petit, A. (1979) Molec. gen. Genet., 173, 263-269.

55. Gurley, W.B., Kemp, J.D., Albert, M.J., Sutton, D.W., and Callis, J. (1979) Proc. Nat. Acad. Sci. US, in press.

56. Achtman, M. and Skurray, R. (1977) In: "Microbial Interactions", Receptors and Recognition, Series B, Volume 3, J.L. Reissig (Ed.), Chapman and Hall, London, 235-279.

57. Tempé, J., Petit, A., Holsters, M., Van Montagu, M. and Schell, J. (1977) Proc. Natl. Acad. Sct. USA, 74, 2848-2849.

58. Lipetz, J., and Galston, A.W. (1959) Amer. J. Bot., 46, 193-196.

59. Kerr, A. (1969) Nature, 223, 1175-1176.

60. Kerr, A. (1971) Physiol. Pl. Pathol., 1, 241-246.

61 Klapwijk, P.M., Oudshoorn, M. and Schilperoort, R.A. (1977) J. Gen. Microbiol., 102, 1-11.

62 Klapwijk, P.M., Scheuldermon, T. and Schilperoort, R.A. (1978) J. Bact., 136, 775-785.

63. Roberts, W.P., Tate, M. and Kerr, A. (1977) Nature, 265, 379-381.

64. New, P.B. and Kerr, A. (1972) J. Appl. Bact., 35, 279-287.

65. Bazzi, C. and Mazzucchi, U. (1978) Proc. IVth Intern. Conf. on Plant Pathogenic Bacteria, 1, 181.

66. Claeys, M., Messens, E., Van Montagu, M. and Schell, J. (1978) Fresenius Z. Anal. Chem., 290, 125-126.

67. Heffron, F., So, M. and Mc Carthy B.J. (1978) Proc. Nat. Acad. Sci. USA, 75, 6012-6016.

68. Nester, E.W. and Montaya, A. (1979) A.S.M. News, 45, 283-287.

69. Aerts, M., Jacobs, M., Hernalsteens, J.P., Van Montagu, M. and Schell, J. (1979) Plant Sci. Lett., in press.

70. Turgeon, R., Wood, M.N. and Brown A.C. (1976) Proc. Nat. Acad. Sci. USA, 73, 3562-3564.

71. Kaiss-Chapmann, R.W. and Morris, R.O. (1977) Biochem. Biophys. Res. Comm., 76, 453-459.

© 1979 Elsevier/North-Holland Biomedical Press
Plasmids of Medical, Environmental and Commercial Importance
K.N. Timmis and A. Pühler, editors

DEGRADATIVE PLASMIDS: MOLECULAR NATURE AND MODE OF EVOLUTION

ROBERTA FARRELL
Biochemistry Department, University of Illinois, Urbana, IL 60801 (U.S.A.) and
A.M. CHAKRABARTY*
General Electric Research and Development Center, Schenectady, New York 12301 (U.S.A.)

INTRODUCTION

Degradative plasmids represent a group of naturally-occurring plasmids, first characterized and subsequently extensively studied in the laboratory of I.C. Gunsalus, that code for the dissimilation of complex organic compounds. Although basically demonstrated in *Pseudomonas putida* and related species, it is likely that they occur in other bacterial genera as well, and endow on their hosts the ability to utilize rather uncommon organic compounds as their carbon and energy source. Because many such compounds, particularly naturally-occurring hydrophobic and synthetic chlorinated hydrocarbons are toxic to the microorganisms, the presence and expression of such plasmids allows the host cells to quickly reduce the toxic concentration of the substrate hydrocarbons. Several review articles on degradative plasmids are available[1,2]. In this short article, we will discuss the molecular nature and genetic relatedness of various degradative plasmids, and genetic mechanisms governing their expression and evolution in different microorganisms.

Characteristics and regulation of plasmid-specified degradative pathways

The degradative plasmids govern the metabolism of a diverse group of aliphatic compounds (octane, decane, etc)[3], aromatic and polynuclear aromatic hydrocarbons (xylenes, toluene, naphthalene, etc)[4,5], products of their oxidative metabolism (salicylate, benzoate)[6,7], terpenes (camphor)[8], alkaloids (nicotine)[9], and chlorinated hydrocarbons such as 2,4-D (2,4-dichlorophenoxyacetic acid)[10] or pCB (p-chlorobiphenyl)[11]. Degradative plasmids may encode a complete degradative pathway (such as xylenes or toluene) or partial degradative steps such as naphthalene to salicylate (A.M. Boronin, personal comm.), 2,4-D to 2,4-dichlorophenol and pCB to p-chlorobenzoate. Other microorganisms may possess plasmids that code for the rest of the pathway,

* Present address: Department of Microbiology, University of Illinois Medical Center. 835 South Wolcott St., Chicago, Illinois 60612, U.S.A.

i.e., salicylate to pyruvate and acetate; chlorobenzoate to Cl^- and CO_2[12]. Some plasmids code for the conversion of the substrate organic compound to metabolites whose degradation may commonly be specified by chromosomal genes of the hosts. Thus, plasmids such as CAM or OCT code for the degradation of camphor and octane to, respectively, isobutyric acid and aliphatic aldehydes or acids, which can be utilized by most pseudomonads through chromosomally-coded enzymes. In bacteria unable to utilize these acid and aldehyde compounds, the presence and expression of CAM and OCT leads to an accumulation of intermediates and no appreciable cell growth. A list of some typical degradative plasmids is given in Table 1.

TABLE 1

PROPERTIES OF SOME TYPICAL DEGRADATIVE PLASMIDS

Plasmid	Degradative Pathway	Transmissibility[a]	Size ($\times 10^6$)	Reference
CAM	Camphor	+	>100[b]	2
OCT	_n_-Octane	-	>100	33
SAL	Salicylate	+	55,48,42	21,22,34
NAH	Naphthalene	+	46	21,22
TOL	Xylene/toluene	+	76	34,22
XYL-K	Xylene/toluene	+	90	34
NIC	Nicotine/nicotinate	+	N.D.[c]	9
pJP1	2,4-Dichlorophenoxy-acetic acid	+	58	10
pAC21	_p_-chlorobiphenyl	+	65	11

a Transmissible plasmids are denoted as +, non-transmissible ones as -.

b P2 plasmids are very large (>100 million); exact size unknown

c N.D., not determined

The enzymes, involved in the plasmid-specified degradative pathways, are invariably inducible, and contain genetic regulatory units corresponding to operons. There are at least two regulatory units in the xylene degradative pathway specified by the XYL plasmid, where either _p_-xylene or _p_-methylbenzyl alcohol induces the enzymes for their conversion to _p_-toluic acid, which induces the enzymes necessary for its own catabolism[13]. Similar regulation has also been described in the TOL-specified xylene degradative pathway, where the product of a regulatory gene, by interaction with the inducers (xylenes), is believed to regulate under positive control the synthesis of the entire pathway enzymes. The product of a second regulatory gene, which recognizes only toluate or benzoate is believed to control positively the synthesis of toluate or benzoate catabolic enzymes[14]. In the oxidation of aliphatic hydrocarbons, alkanes and their analogues exhibit some degree of specificity in inducing the plasmid-determined enzymes, and the range of substrates is determined principally by inducer specificity rather than by the substrates[15]. Genetic analyses demonstrate the clustering of structural genes coding for the hydroxylase components, but separate location for regulatory genes[16]. In the camphor pathway, the genes specifying the hydroxylase proteins and the alcohol dehydrogenase are clustered and coordinately regulated. A second group of clustered genes involved in the metabolism of the acidic product XI, are regulated separately[8].

There is an interesting difference in the regulation of chromosomally coded degradative enzymes from those coded by the plasmids. Chromosomally-coded enzymes involved in the degradation of common intermediates such as aliphatic alcohols in alkane degradation or isobutyric acid in camphor degradation are induced by their substrates. In contrast, such enzymes in the plasmid-determined pathways, are induced not by the substrates for the enzymes, but by the primary substrate of the plasmid-coded pathway, i.e., alkanes or camphor. It appears that during evolution of the degradative plasmids, transfer of structural genes occurs without a concommitant transfer of their regulatory genes, so that the genes encoding enzymes for the pathway are regulated as a single unit. The lack of a chromosomal genetic clustering in Pseudomonas may thus facilitate the ultimate partial clustering of various structural genes on plasmids under a single regulatory unit.

Molecular nature and plasmid inter-relationships

The degradative plasmids specify a large number of catabolic pathways, many of which have common sequential enzymatic steps. Genetic analyses of independently-isolated strains have demonstrated the plasmid nature of the genes encoding such degradative pathways[17,18]. Many of these degradative pathways appear to be coded by plasmids which differ in transfer properties and molecular sizes. The extent of genetic and molecular relatedness among plasmids that specify either a common degradative pathway or aromatic degradation in general, has therefore been studied by a number of investigators. It has been demonstrated that many degradative and antibiotic resistance plasmids undergo dissociation and extensive deletions when transferred to P. aeruginosa[19,20]. In studying genetic homology it is therefore important to isolate and characterize the degradative plasmid DNA from their natural hosts, which are usually P. putida or related strains, rather than from P. aeruginosa type of hosts.

Farrell et al[21] have reported extensive homology among three degradative plasmids SAL, NAH and TOL from a comparison of the mobility of various fragments of these plasmids generated by EcoRI, HinDIII and BamHI. SAL and NAH appear to be essentially identical except for a larger fragment present in SAL (Fig. 1). The TOL plasmid appears to have some bands, identical in mobility on an agarose gel, with those of NAH and SAL, although their extent of homology is not clear.

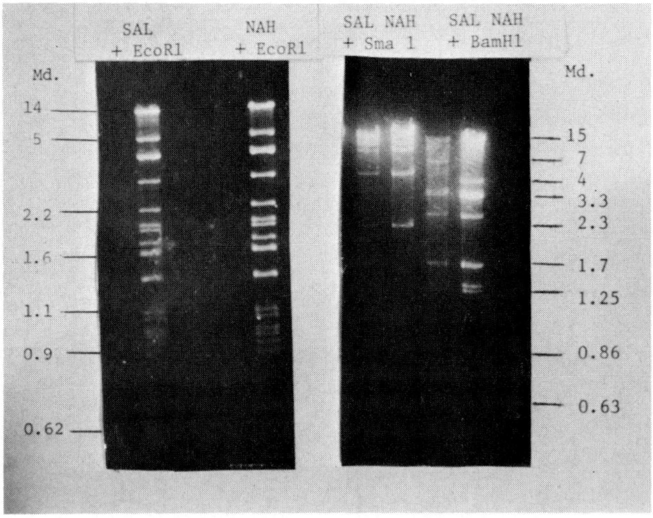

Fig. 1. Agarose gel electrophoresis of SAL (P. putida) and NAH (P. putida) DNA digested with EcoRI and SmaI, and BamHI

Heinaru, Duggleby and Broda[22] have shown from restriction hybridization studies that the endonuclease-generated fragments of the SAL plasmid, as isolated from P. aeruginosa, hybridize to all of the fragments of the NAH plasmid, isolated from P. putida, except for three bands. In the converse experiment, all of the labelled NAH plasmid fragments hybridize to the SAL fragments. They have shown that each plasmid has five fragments that do not correspond in size to any from the other plasmid when both were digested with EcoRI. Thus they inferred that although the two plasmids are closely related, the generation of SAL plasmid, which is 4 Mdal smaller than the NAH, is not due to a simple deletion of the NAH genes coding for the bioconversion of naphthalene to salicylate.

In order to get a better understanding of the relationships between SAL and NAH, Farrell and Gunsalus[23] have ordered the restriction fragments, obtained by the digestion of SAL and NAH plasmids isolated from P. putida, with SmaI. The maps appear identical between the two plasmids except for two 1.7 Mdal fragments joined in tandem in SAL, which are absent in NAH. The endonuclease SmaI produces 7 fragments with SAL and 5 with NAH, and the labelled SAL DNA hybridizes to all bands of SmaI and EcoRI-generated fragments of the NAH plasmid. The SAL plasmid may therefore have been derived by the insertion of a small DNA fragment on the NAH plasmid.

The discrepancy in the molecular sizes and the number and mobility of restriction endonuclease-generated fragments of NAH and SAL in the above two studies may be due to the fact that whereas SAL was isolated from P. aeruginosa by Heinaru et al[22], Farrell and Gunsalus[23] isolated SAL from the same host background (P. putida) as NAH. They have further demonstrated that restriction endonuclease digest pattern of SAL isolated from P. aeruginosa AC165, the same strain used by Heinaru et al[22], is different from that of SAL isolated from P. putida. Since many degradative plasmids are occasionally transferred to P. aeruginosa and then transferred back to P. putida, it is perhaps not too surprising that different molecular sizes are being reported for the same plasmid in different laboratories[22-24].

The homology of a number of plasmids specifying naphthalene degradation, isolated from P. putida and unclassified fluorescent pseudomonads, have been compared by hybridization and restriction endonuclease digestion pattern. Farrell and Gunsalus[23] have compared the NAH plasmid, isolated from two wild type strains PpG7 and PpG63 and an unclassified pseudomonad studied also by Heinaru et al[22]. Only one 46 Mdal plasmid species has been isolated from PpG7, whereas at least two, 46 Mdal and 8.2 Mdal plasmids, have been isolated from

PpG63 strain. Homology, as determined by hybridization, has been demonstrated between the 46 Mdal PpG63 NAH plasmid and the 46 Mdal PpG7 NAH plasmid as well as the SAL plasmid isolated from P. putida. By electron microscopy, an identical 'lollipop' structure usually indicative of an insertion sequence or transposon, has been observed on the two 46 Mdal NAH plasmids[23,35]. The unclassified fluorescent pseudomonad has also yielded one large plasmid, nearly 46 Mdal, which by hybridization and restriction endonuclease finger print analysis, appears to be identical to the NAH plasmid of strain PpG7. It appears, therefore, that the naphthalene degradative phenotype has arisen in several different strains, presumably by natural transfer of a single plasmid species[22,23].

Heinaru, Duggleby and Broda have demonstrated genetic homology between SAL and other degradative plasmids in that some sequences of various degradative plasmids such as TOL or OCT have homology with parts of the SAL plasmid[22]. There also appears to be extensive homology between TOL and NAH and the antibiotic resistance plasmids R2 and pMG18 (P. Broda, personal communications). Farrell and Gunsalus[23] have observed that 8 and 9 EcoRI-generated TOL and XYL-K fragments respectively hybridize to SAL plasmid (P. putida) accounting for about 25 Mdal of the two plasmids having some homology to SAL. Similarly TOL plasmid hybridizes to at least 10 bands of the EcoRI-generated NAH and the same 10 bands of the EcoRI-treated SAL plasmid. The two xylene degradative plasmids TOL and XYL-K demonstrate considerable homology as revealed by hybridization of 9 EcoRI-generated fragments (equivalent to about 45 Mdal) of XYL-K with the TOL plasmid DNA.

Evolution of degradative plasmids

The occurrence of dissimilatory pathways for compounds such as octane, xylene, toluene, naphthalene, camphor etc. coded for by plasmids raises the interesting question whether such plasmids evolve, in analogy with antibiotic resistance, as primary means of detoxification. Many of the hydrocarbons, natural or synthetic, for which degradative plasmids have been reported, are toxic to microorganisms, and plasmids may have evolved as rapid means of detoxification under conditions of high local hydrocarbon concentration. Plasmids encoding partial bioconversion steps, such as naphthalene to salicylate or pCB to p-chlorobenzoate, may simply help in such detoxification processes by converting the persistent, insoluble, compounds to less persistent, soluble, and perhaps less toxic forms. The involvement of insertion sequence (IS) elements in the transposability of antibiotic resistance genes, and in the dissociation (and ready recombination) of drug-resistance and transfer genes is believed to have played a major part in the evolution and rapid spread of antibiotic resistance among microorganisms[25,26]. Similar to antibiotic-resistance plasmids, degradative plasmids often undergo extensive deletion and dissociation into non-transmissible forms, particularly in P. aeruginosa[19]. In some cases, the non-transmissible degradative gene segments can be transposed onto antibiotic resistance plasmids or various sex factor plasmids[19]. The same genetic mechanisms, governing association-dissociation and transposability of antibiotic resistance genes may also be involved in the evolution of degradative plasmids.

Evolution of hydrocarbon degradative functions in Enterobacteriaceae

The transposition of hydrocarbon degradative gene segments such as TOL* or SAL* onto broad host range antibiotic resistance plasmids such as RP4 or R702 has enabled us to introduce and stably maintain these gene segments into a large number of gram negative bacteria including members of the facultatively anaerobic enteric bacteria. In general enteric bacteria are not known to be capable of hydrocarbon oxidation, and it was of interest to see if this inability is due to the absence of appropriate degradative gene segments in such bacteria. Several members of enteric bacteria were found to be incapable of utilizing toluene even when they harbor RP4-TOL (pAC8). Assay of the toluene degradative enzymes in E. coli harboring pAC8 has demonstrated the presence of all the enzymes at a reduced level[27]. It is not clear whether the lack of growth of E. coli with toluene or m-toluate is solely due to a reduced level of such enzymes or there are additional problems of substrate recognition and uptake by the cells.

If members of Enterobacteriaceae are incapable of functionally expressing hydrocarbon degradative genes, one may wonder how enteric bacteria such as K. pneumoniae or S. marcescens have acquired the ability to utilize as sole carbon source a chlorinated hydrocarbon such as pCB[11]. It should be stressed that anaerobic and facultatively anaerobic microorganisms comprise the major microflora in the semi-anaerobic environment of river bottom sediments where toxic liquid hydrocarbons such as PCBs tend to accumulate. The release for the last three decades of massive amounts of PCBs estimated to be about 500,000 pounds in the Hudson river sediments near the region of isolation of these strains must have exerted strong selective pressure on such microorganisms to detoxify these compounds. It appears that in response to such pressure the Klebsiella and Serratia strains have not only evolved a pCB plasmid, but also the genes that allow them to functionally express pCB, TOL and presumably other hydrocarbon degradative plasmids as well[11].

In order to determine if the genes that allow Klebsiella and Serratia species to functionally express pCB and TOL plasmid genes, evolved as part of the pCB plasmid, or separately on the chromosome or other plasmids, we have introduced pAC8, selecting for Cb^r or Km^r, into the pCB^+ parent and the pCB^- segregant of the Klebsiella and Serratia species. Examination of the Tol^+ character in such Cb^rKm^r colonies demonstrates that while pCB^+ parents harboring pAC8 can produce mutants capable of utilizing toluene, the pCB^- segregants are incapable of producing such Tol^+ mutants. Thus the pCB degradative plasmid (pAC21) and not the host chromosome harbors gene(s) needed for expression of hydrocarbon degradative functions. This is an interesting finding since it would have been useless for the enteric bacteria in the Hudson river sediments to evolve a pCB degradative plasmid only to find that they cannot express it.

Evolution of aromatic degradative functions

The mechanism(s) by which pAC21 plasmid endows on the host bacterium the ability to express toluene and pCB degradative genes is unknown at present. It appears, however, that functional expression of hydrocarbon degradation may involve the presence of specific membrane proteins for inducer and substrate recognition. Fennewald, Benson and Shapiro[16] have characterized alkane-negative (Alk^-) Pseudomonas mutants, whose Alk^- property is not due to absence of enzyme activity but rather from inaccessibility of enzymes to substrate. They interpret their observations in terms of Alk^- mutants having defects in membrane structure or biogenesis which block expression of the plasmid-determined Alk^+ phenotype. Scott and Finnerty[36] have also demonstrated the presence of intracytoplasmic hydrocarbon inclusions with specific membrane structures in Acinetobacter species HO1-N grown with hydrocarbons. In order to determine if membrane proteins, analogous to permeases, are involved in the binding and transport of hydrocarbons or their oxidative aromatic metabolites inside the cells, we have examined various laboratory strains of Klebsiella pneumoniae and Serratia marcescens for their ability to utilize hydrocarbons and aromatic compounds. While none of the strains can utilize any hydrocarbon, and none of the two Serratia strains examined has been found to utilize benzoate or other simple aromatics, a strain of K. pneumoniae KPl has been found to utilize benzoate, p-hydroxybenzoate and shikimate by the ortho pathway. Deletions spanning the his, nif and shu genetic regions have previously been isolated in this strain[28]. Interestingly, one such deletion ΔUN107 has not only lost its ability to utilize shikimate but also that of p-hydroxybenzoate (Pob) and benzoate (Ben). Introduction of the plasmid RP41 (pRD1)[29] restores the ability of this strain to fix nitrogen, grow in absence of histidine and utilize shikimate, benzoate and p-hydroxybenzoate. The plasmid pRD1, however, can not complement structural gene mutations in the benzoate and p-hydroxybenzoate pathway. Thus the deletion spans the gene(s) necessary for uptake or substrate recognition (regulatory) but not structural genes of these pathways. Introduction of pAC8 to the wild type $Ben^{+}Pob^{+}$ or various $Ben^{-}Pob^{+}$ or $Pob^{-}Ben^{+}$ mutants of K. pneumoniae KP1 allows the cells to produce rare colonies, (at a frequency of 10^{-8} per cell per generation) that can slowly utilize toluene or m-toluate. In contrast we have never observed production of mutant colonies in $pAC8^{+}$ ΔUN107 that can utilize m-toluate or toluene. Interestingly, although ΔUN107 is totally incapable of growing with benzoate or p-hydroxybenzoate, it demonstrates, in extracts of cells grown with glutamate and benzoate, the presence of all the ortho pathway enzymes at a level of 33% of the wild type. The inability of this deletion strain to grow with benzoate or p-hydroxybenzoate is, therefore, not due to absence of enzymes, but presumably due to inaccessibility of these enzymes to the substrates. The deletion may therefore span those genes which code for the respective permeases. Alternatively, there might be a single purpose permease responsible for uptake of common aromatic intermediates, which is missing in this deletion mutant. Low level of non-specific entry of benzoate will allow inefficient induction of the ortho pathway enzymes as observed in the extract. Another possibility is that the permease is present, but the deletion spans the gene coding for a membrane protein which interacts with the substrate to effect its entry inside the cell for subsequent interaction with the cytoplasmic enzymes. The enzymes can still be induced because the induction mechanism (inducer-membrane protein interaction) is different from substrate-membrane protein interaction. This substrate interacting protein may be coded by the chromosomal genes of Pseudomonas, but absent in enteric bacteria. Since degradative plasmids do not appear to carry such genes, the difference between Pseudomonas degradative plasmid and pAC21 type of plasmids present in enteric bacteria may thus be the presence of genes in the latter type of plasmid coding for the substrate recognition protein. In presence of this gene and the plasmid-borne inducer-recognition and structural genes, the enteric bacteria can produce promotor or permease mutants capable of effective induction and uptake of various hydrocarbon substrates for their subsequent metabolism by the cytoplasmic enzymes.

Mode of evolution of degradative plasmids

The genes for a biosynthetic pathway are believed to have evolved by retro-evolution, the last step evolving first, followed by gene duplication and divergence[30]. In contrast, genes for catabolic steps in a degradative pathway may evolve 'piecemeal' in different microorganisms, followed by their assembly in the form of a non-transmissible plasmid by transposition and recombination[1,31]. Evolution of individual steps, particularly in the breakdown of synthetic hydrocarbons, may occur by mutations in a gene catalyzing a similar step[18], and such mutations may or may not alter the original substrate specificity and therefore may or may not involve gene duplication[32]. The non-transmissible degradative plasmids can then recombine with transfer plasmids to form transmissible degradative plasmids (Fig. 2).

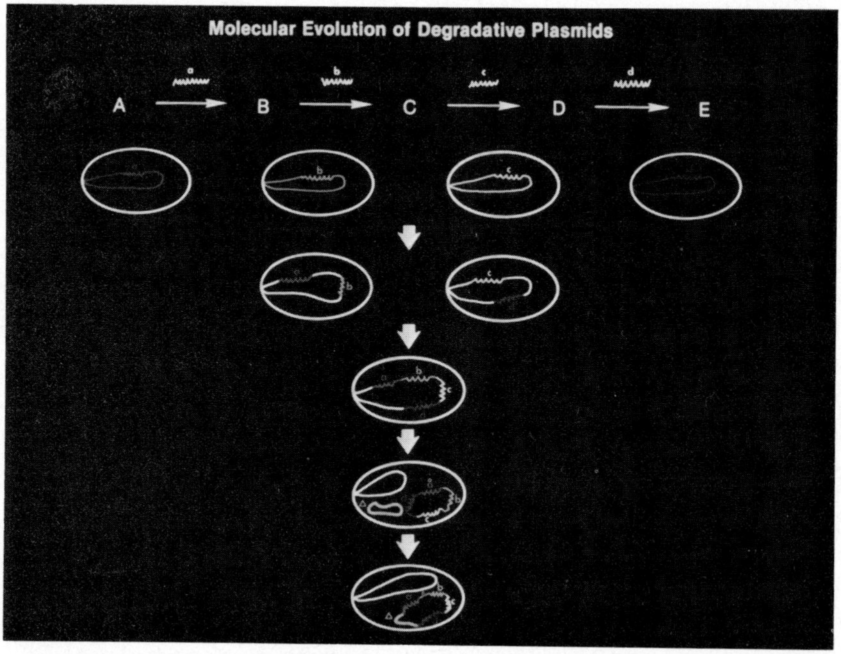

Fig.2. The molecular evolution of degradative plasmids. The primary substrate A is converted to the end product E by a series of steps a, b, c, and d. Some individual steps may require the participation of a number of enzymes. Although genes controlling the steps a, b, c, and d are shown to be chromosomal in different microorganisms, they could also develop as parts of plasmids. Such genetic segments may then be transferred by recombination or transposition from one bacterium to another, producing strains harboring several steps of the pathway. Once all the genes of the pathway are assembled, they could be transferred as a discreet unit by recombination or be transposed onto a replicon; such a replicon may later recombine with a transfer plasmid to form a transmissible degradative plasmid.

The degradative plasmids show certain characteristics that tend to support this mode of evolution. If different genes catalyzing separate steps evolve in different bacteria, and are finally assembled together, one would expect to find pure or mixed cultures that will 'cometabolize' a complex compound without obtaining appreciable energy out of such breakdown. Such cultures are widely known to occur in nature[31]. Also, if genes encoding separate steps of a pathway are assembled on a plasmid under the control of a single regulatory unit, it is likely that some steps, although they catalyze breakdown of a higher homologue, may in fact be under the control of regulatory units encoding degradation of a lower homologue, i.e., the enzymes governing the dissimilation of higher homologues will be product induced, as is often found in case of degradative plasmids such as CAM. NAH, etc. The induction of chromosomal aliphatic alcohol dehydrogenase or isobutyrate degradative enzymes by their respective substrates (aliphatic alcohol and isobutyrate) but the induction of some of the same enzymes in the plasmid-determined pathway by the primary substrate of the plasmid (alkane or camphor) and not by the substrates of these enzymes, may indicate the transfer of individual structural genes, but not the regulatory genes, in the evolution of the plasmid-determined pathway. Plasmids such as TOL also exhibit internal density heterogeneity as revealed by considerable skewing of melting temperature profiles of the sheared plasmid DNA. This may indicate the presence of genetic segments transferred from various microorganisms having varying densities on TOL. The large sizes of these plasmids may also indicate the presence of extraneous DNA that was transferred with individual genes catalyzing the sequential steps. Since genes for all the sequential steps in the degradative pathway must be present on the plasmid, extensive deletions leading to a shortening of the plasmid size but still leaving all necessary enzymatic steps can occur only rarely. A close look at the presence of insertion sequence type of elements, and the anatomy of the DNA molecules, will throw considerable light on the mode of evolution of degradative plasmids.

ACKNOWLEDGEMENTS

The work described here has been supported in part by a grant from the National Science Foundation (PCM77-25450) to AMC and PCM77-04677 to ICG. We are grateful to Dr. I.C. Gunsalus for allowing us to quote some of the unpublished results from his laboratory.

REFERENCES

1. Wheelis, M.L. (1975) Ann. Rev. Microbiol. 29, 505-524.
2. Chakrabarty, A.M. (1976) Ann. Rev. Genet. 10, 7-30.
3. Chakrabarty, A.M., Chou, G. and Gunsalus, I.C. (1973) Proc. Nat. Acad. Sci. USA, 70, 1137-1140.
4. Williams, P.A. and Murray, K. (1974) J. Bacteriol. 120, 416-423.
5. Dunn, N.W. and Gunsalus, I.C. (1973) J. Bacteriol. 114, 974-979.
6. Chakrabarty, A.M. (1972) J. Bacteriol. 112, 815-823.
7. Nakazawa, T. and Yakata, T. (1977) J. Bacteriol. 129, 39-46.
8. Rheinwald, J.G., Chakrabarty, A.M. and Gunsalus, I.C. (1973) Proc. Nat. Acad. Sci. USA, 70, 885-889.
9. Thacker, R., Rørvig, O., Kahlon, P. and Gunsalus, I.C. (1978) J. Bacteriol. 135, 289-290.
10. Fisher, P.R., Appleton, J. and Pemberton, J.M. (1978) J. Bacteriol. 135, 798-804.
11. Kamp, P.F. and Chakrabarty, A.M. (1979) This volume.
12. Reineke, W. and Knackmuss, H.-J. (1979) Nature 277, 385-386.
13. Friello, D.A., Mylroie, J.R., Gibson, D.T., Rogers, J.E. and Chakrabarty, A.M. (1976) J. Bacteriol. 127, 1217-1224.
14. Williams, P.A. and Worsey, M.J. (1978) Microbiology-1978, (D. Schlessinger, ed.) American Society for Microbiology, Washington, D.C. pp. 167-169.
15. Fennewald, M. and Shapiro, J. (1977) J. Bacteriol. 132, 622-627.
16. Fennewald, M., Benson, S. and Shapiro, J. (1978) Microbiology-1978, (D. Schlessinger, ed.). American Society for Microbiology, Washington, D.C. pp. 170-173.
17. Duggleby, C.J., Bayley, S.A., Worsey, M.J., Williams, P.A. and Broda, P. (1977) J. Bacteriol. 130, 1274-1280.
18. Pemberton, J.M. and Don, R.H. (1979) This volume.
19. Chakrabarty, A.M., Friello, D.A. and Bopp, L.H. (1978) Proc. Nat. Acad. Sci. USA, 75, 3109-3112.
20. Nagahari, K. (1978) J. Bacteriol. 136, 312-317.
21. Farrell, R., Gunsalus, I.C., Crawford, I.P., Johnston, J.B. and Ito, J. (1978) Biochem. Biophys. Res. Commun. 82, 411-416.
22. Heinaru, A.L., Duggleby, C.J. and Broda, P. (1978) Molec. Gen. Genet. 160, 347-351.
23. Farrell, R. and Gunsalus, I.C. (1979) Biochem. Biophys. Res. Commun. (in press).
24. Johnston, J.B. and Gunsalus, I.C. (1977) Biochem. Biophys. Res. Commun. 75, 13-19.
25. Cohen, S.N. (1976) Nature 263, 731-738.

26. Starlinger, P. and Saedler, H. (1976) Curr. Topic Microbiol. Immunol. 75, 111-152.

27. Ribbons, D.W., Wigmore, G.J. and Chakrabarty, A.M. (1979) Soc. Gen. Microbiol. Quart. 6, 24-25.

28. Shanmugam, K.T., Loo, A.S. and Valentine, R.C. (1974) Biochem. Biophys. Acta 338, 545-553.

29. Dixon, R., Cannon, F. and Kondorosi, A. (1976) Nature 260, 268-271.

30. Horowitz, N.H. (1965) Evolving genes and proteins (V. Bryson and H.J. Vogel, eds). Academic Press, New York.

31. Chakrabarty, A.M. (1978) ASM News 44, 687-690.

32. Hall, B.G (1978) Genetics 89, 453-465.

33. Fennewald, M., Prevatt, W., Meyer, R. and Shapiro, J. (1978) Plasmid 1, 164-173.

34. Mylroie, J.R., Friello, D.A. and Chakrabarty, A.M. (1977) Plasmids: Medical and Theoretical Aspects (S. Mitsuhashi, L. Rosival and V. Kremery, eds). Avicenum Press, Prague. pp. 395-402.

35. Agabalyan, A.S., Zakharyan, R.A., Akapyan, S.M., Bakunts, K.A., Israelyan, Y.A., and Azaryan, N.G. (1978) Mikrobiologiya 47, 97-100.

36. Scott, C.C.L. and Finnerty, W.R. (1976) J. Bacteriol. 127, 481-489.

RESEARCH ARTICLES

I – PLASMIDS OF MEDICAL IMPORTANCE: GENES AND PRODUCTS

Plasmids of Medical, Environmental and Commercial Importance
K.N. Timmis and A. Pühler, editors

THE CHARACTERIZATION OF AN ESCHERICHIA COLI PLASMID DETERMINANT THAT ENCODES FOR THE PRODUCTION OF A HEAT-LABILE ENTEROTOXIN

WALTER S. DALLAS, STEVE MOSELEY, AND STANLEY FALKOW
Department of Microbiology & Immunology, School of Medicine, University of Washington, Seattle, Washington 98195

INTRODUCTION

Certain strains of Escherichia coli are the etiologic agents for non-invasive gastroenteritis in man and young animals[1,2]. The disease is cholera-like in nature and the causative agents have been shown to elaborate one or two enterotoxins[3]. These E. coli also produce an adhesin that enables them to colonize the upper small bowel[4], a site not normally inhabited by E. coli[5]. In many of these enterotoxigenic E. coli, the genetic determinants for toxigenicity and adherence have been shown to be part of plasmids[6]. Smith et al.[6] used the transmissible nature of these plasmids to study the contribution of each virulence factor to the pathogenesis of the disease in piglets. They concluded that both virulence factors were necessary for the manifestation of the disease. They also reported that the introduction of the two virulence factors into wild-type E. coli sometimes created a fully pathogenic strain.

Enterotoxigenic E. coli have been shown to make two distinct enterotoxins, ST and LT[3]. These toxins are discernable by several characteristics including biological activity, immunogenicity, and relative heat stability[2]. Burgess et al.[7] have reported that there are two distinct ST's that can be discerned by biological assays. STa elicits fluid accumulation in suckling mice while STb is only active in weaned piglets and rabbit ligated loops. Alderete et al.[8] have reported STa to be a low molecular mass protein (5,100 daltons). LT is a larger protein[2] that shares several characteristics with the toxin made by Vibrio cholerae. Both toxins have been shown to have an ADP-ribosylating activity (M. Gill personal communication) and they both stimulate adenyl cyclase in eukaryotic cells[9,10]. LT and cholera toxin also show partial immunological cross-reactivity and have similar membrane receptors[11,12].

Smith et al.[13] first used the term Ent to describe enterotoxin encoding plasmids. Ent plasmids can encode LT only, ST only, and both LT and ST[2]. So et al.[14] showed that LT + ST plasmids constituted a homogeneous group of extrachromosomal elements that had similar mole fraction guanine + cytosine, molecular mass (about 60 Mdal), and DNA sequence homology. On the other hand, ST plasmids were found to constitute a disparate group of plasmids. Following the successful isolation of the ST genetic determinant in our laboratory[15], So et al.[16] reported that the ST genetic determinant was part of a transposon that was flanked by inverted and repeated IS1 elements.

We have been interested in exploring the relatedness of the LT gene(s) from enterotoxigenic E. coli strains isolated from humans and animals. In our work we studied an LT gene(s) that was cloned from an Ent plasmid isolated from an E. coli strain pathogenic for piglets[17]. During the course of our work, we determined that LT was a multimeric toxin. The 25,500 dalton subunit was shown to have an adenyl cyclase stimulating activity and the 11,500 dalton subunit an adsorption activity. The cistrons that encoded these proteins were located on a genetic map of the LT DNA region. We used, as a hybridization probe, a DNA fragment that contained all of the 11,500 dalton protein cistron and approximately 10% of the 25,500 dalton protein cistron to study the sequence homology among LT genes of different origins. Our results indicated that LT genes have similar structures but they are not identical.

MATERIALS AND METHODS

Bacterial strains and plasmids. The E. coli sublines used in this study have been described in detail previously[17,18]. The plasmids have also been described[17,19].

Recombinant DNA methods. Agarose gel electrophoresis, restriction endonuclease cleavage reactions, and polynucleotide ligase conditions have been described by So et al.[17]. The experiments reported here were performed under P2-EK1 conditions as specified in the NIH Guidelines for Recombinant DNA Research.

Protein synthesis in minicells. E. coli minicells were isolated from E. coli DS410 and the plasmid-specified proteins were labelled as described by Dougan et al.[18]. Labelled minicells were analyzed by the electrophoretic method of Studier[20]. Staphylococcus-antibiody-antigen reaction conditions have been described previously[21].

Adenyl cyclase assays. Adenyl cyclase stimulating activity was determined by D. M. Gill using a method that has been described[9].

Hybridization reactions. DNA was labelled by nick translation as described by Rigby et al.[22]. Colony filter hybridizations were performed as described by Grunstein[23]. Filter blot hybridization reactions were performed as described by Southern[24] and modified by Botchan et al.[25].

RESULTS

The LT genetic determinant was initially isolated on a 5.8 Mdal DNA fragment[17] and was subsequently joined to the carrier plasmid, pBR313. Normally the LT gene(s) is found in enterotoxigenic E. coli in one or two copies per chromosome equivalent. The cloned LT DNA fragment was found to be present at approximately 20 copies per chromosome equivalent. Laboratory E. coli strains harboring the cloned toxin gene(s) were found to be hypertoxin producers presumably as a result of gene amplification. A more precise location of the LT genetic determinant within the cloned DNA was established by introducing deletions into the plasmid. Specific DNA fragments were deleted from the plasmid by making partial plasmid digests with a restriction enzyme followed by rejoining the DNA fragments with ligase (Figure 1). We were able to isolate

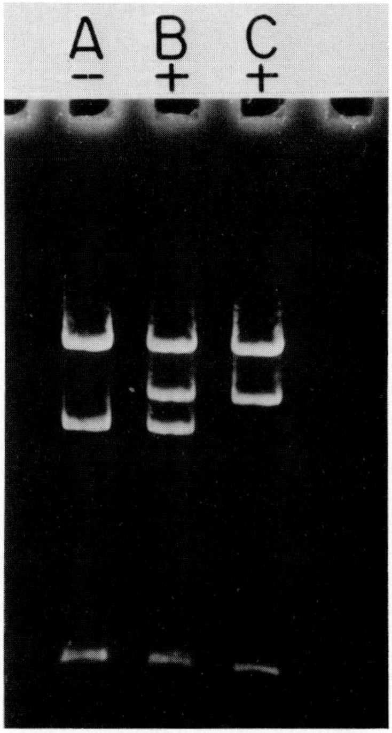

Fig. 1. HindIII cleavage of EWD022 and HindIII generated deletion plasmids. Plasmid DNA was purified, cut with HindIII, and electrophoresed through a vertical 1% agarose gel: (A) EWD306, (B) EWD022, (C) EWD300. The + or - indicates if the plasmid encoded functional LT as measured in the Y-1 adrenal

cell tissue culture assay[26]. Both EWD300 and EWD306 were derived from EWD022 by deleting a HindIII DNA fragment.

deletions which spanned the entire cloned DNA region. The effect of each deletion on LT expression was determined by using a tissue culture assay for LT[26]. In this system, Y-1 adrenal cells change shape from cubic to round in the presence of a functional LT. Using this type of analysis, we identified a 1.2 Mdal region that contained the LT genetic determinant (Figure 2). Deletions

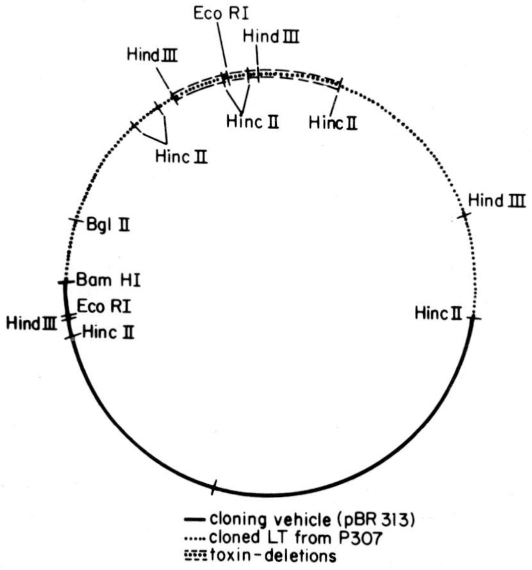

Fig. 2. Schematic diagram of restriction enzyme recognition sites in EWD022. Deletions that extended into the region bounded by the HindIII and HincII sites resulted in the loss of the LT^+ phenotype.

in this region resulted in the loss of the LT^+ phenotype. The 1.2 Mdal region had a coding capacity for about 600 amino acids. This can be considered as an upper size limit for an LT of single subunit composition.

The protein products encoded by the LT DNA region were identified using minicells as an in vivo protein synthesizing system[27]. EWD299 is a deletion plasmid that contained only the 1.2 Mdal LT DNA region. Analysis of this plasmid in minicells revealed that twelve proteins were encoded by the plasmid (Figure 3). The LT proteins were identified in minicell extracts by using an anti-serum prepared against a crude LT preparation. Since the amount of protein

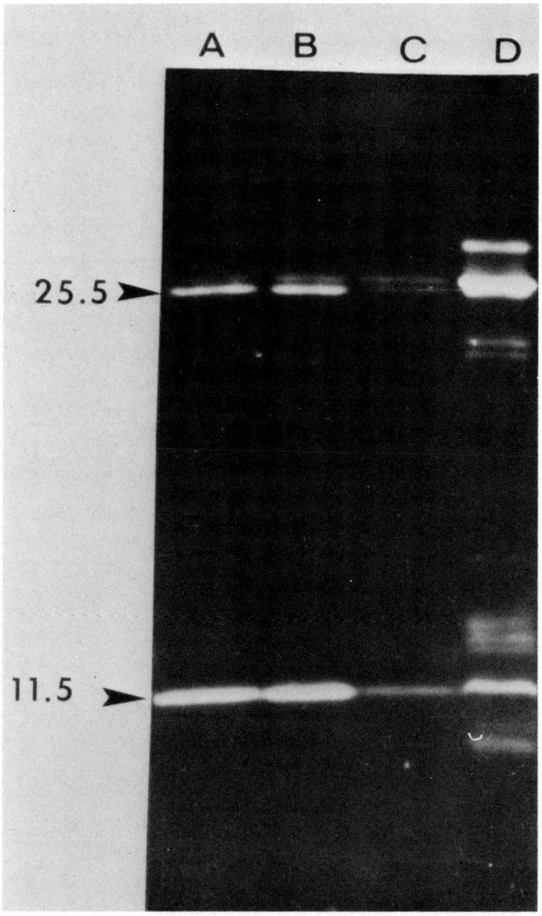

Fig. 3. Autoradiograph of a polyacrylamide gel containing total EWD299 encoded proteins and proteins that react with anti-sera. A cytoplasmic-periplasmic fraction from minicells containing EWD299 was mixed with anti-sera and formalized S. aureus. The cells were pelleted through 1M sucrose, boiled in final sample buffer, and electrophoresed through a 10-15% linear gradient SDS-polyacrylamide gel: (A) anti-cholera toxin serum, (B) anti-cholera toxin subunit B serum, (C) anti-LT serum, (D) total EWD299 encoded proteins. Two proteins, 25.5K and 11.5K, were found to react specifically with all three anti-sera.

made in minicells is too small to be separated from other antigens by immunoprecipitation, Staphylococcus aureus Cowan strain I was used to isolate antigen-antibody complexes[28]. Tertiary complexes of S. aureus-Ab-Ag are formed and these aggregates can be separated from unreacted antigens in a minicell lysate by pelleting the reaction mixture through 1M sucrose. A protein 26K in mass was found to bind non-specifically to the S. aureus cells. Two other proteins 25.5K and 11.5K in mass were complexed in the presence of anti-LT serum. Both of these proteins were also complexed in the presence of anti-cholera toxin subunit B serum. These results indicated that LT was composed of two distinct subunits, at least one of which immunologically cross-reacted with cholera toxin subunit B.

LT has been shown to have two separate activities: an adsorption activity for certain eukaryotic cell membranes and an adenyl cyclase stimulating (ACS) activity. ACS activity can be measured independently in pigeon erythrocyte lysates (PEL)[10]. Most of the ACS activity in extracts from strains harboring EWD299 migrated in SDS-polyacrylamide gels at a molecular mass of 25.5K (Figure 4). A smaller peak of activity was identified at a mass of 21.5K. We

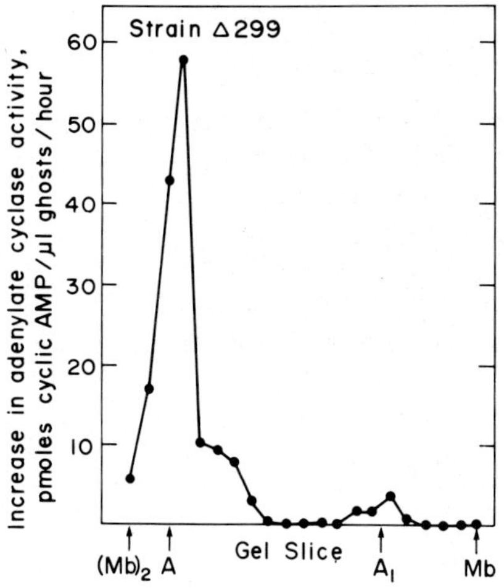

Fig. 4. Adenyl cyclase stimulating(ACS)activity encoded by EWD299. Extracts were prepared from strains carrying EWD299 and fractionated by 10-15% linear SDS-gradient PAGE. The gels were sliced into fractions, the proteins in each gel slice were eluted, and the preparations were assayed for ACS activity. A and A_1 indicate the migration distance of cholera toxin subunit A and A_1, respectively[29]. Mb is the monomer form of myoglobin and Mb_2 is the dimer form. This experiment was performed in collaboration with D.M. Gill and a more complete description of this collaborative work will appear elsewhere.

now have evidence that indicated that the smaller molecular mass activity was a breakdown product of the 25.5K protein (data not shown). The tissue culture assay for LT required both the adsorption activity and the ACS activity. Therefore, strains that had ACS activity but were not positive in tissue culture would be deficient in adsorption activity. Analysis of the deletion plasmids revealed that the 11.5K protein perfectly correlated with adsorption activity. These results indicated that LT was a multimeric toxin composed of a 25.5K protein that had ACS activity and an 11.5K protein that had an adsorption activity. We also have evidence which indicated that the holotoxin was composed of one 25.5K subunit and five 11.5K subunits.

We have been able to locate the relative positions of the two LT cistrons on a genetic map of the LT DNA region (data not shown). A 0.5 Mdal HindIII DNA fragment was identified and shown to encompass all of the 11.5K protein cistron and approximately 10% of the 25.5K protein cistron. This DNA fragment was used as a nucleic acid hybridization probe to determine the relatedness of LT genes of diverse origins. Initially, we wanted to determine if Ent plasmids could be detected by the colony filter blot hybridization technique[23]. Our results demonstrated that LT genes present at a concentration of one or two copies per chromosomal equivalent could be detected by colony filter hybridization (Figure 5). Furthermore, all LT^+ E. coli isolates tested were found to hybridize to the probe. These preliminary data indicated that LT genes share significant sequence homology. The colony hybridization method may prove useful for studying the epidemiology of LT enterotoxigenic E. coli.

To more precisely determine the degree of homology among LT genes, we chose four Ent plasmids to study in detail. Two of the plasmids were isolated from E. coli pathogenic for humans and the other two were isolated from porcine strains. One of the porcine Ent plasmids was P307, the plasmid from which the hybridization probe was isolated. The other porcine Ent plasmid was pCG86[30]. Plasmid DNA was purified from each strain, cut with HindIII, and hybridized to the HindIII probe by the method of Southern[24]. Only one DNA fragment in each of the four samples hybridized to the probe and in all cases the fragment was 0.5 Mdal in size (data not shown). This result indicated that most, if not all, of the HindIII DNA portion of the LT DNA region was conserved among the Ent plasmids. We performed the same type of analysis using six different restriction enzymes. In this way, we studied the internal structure of the common HindIII fragment. For example, the HindIII probe was found to contain a single SmaI site. Therefore, cleavage of P307 with SmaI (followed by blotting and hybridization) resulted in the appearance of two DNA bands that hybridized to the probe (Figure 6). Using enzymes that cleaved within the hybridization

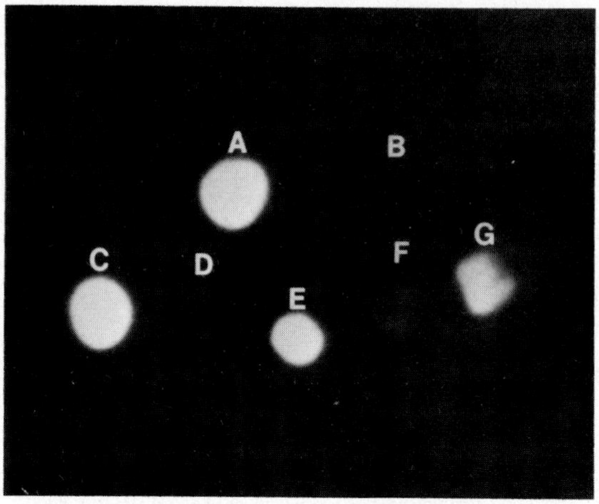

Fig. 5. Colony filter hybridization of LT^+ and LT^- E. coli. The 0.5 Mdal HindIII fragment was isolated and labelled with ^{32}P by nick translation and hybridized to colonies that were lysed on nitrocellulose filters: (A) P307, the Ent plasmid from which the LT genetic determinant was cloned; (B) C600, an LT^- laboratory strain; (C) H10407, an LT^+ strain pathogenic for humans; (D) a spontaneous LT^- derivative of H10407; (E), (F), (G) LT^+ strains pathogenic for humans.

probe, we could determine if these sites were conserved in all the Ent plasmids. Similarly by using enzymes that did not cleave within the hybridization probe, we could determine if new restriction enzyme sites were present in the LT DNA region of the Ent plasmids (Figure 6). From this analysis, we determined that the human Ent plasmids apparently demonstrated a small degree of structural divergence when compared to each other and to the hybridization probe. One human Ent plasmid was found not to have an SmaI site within the common HindIII fragment (Figure 6) and the other human Ent plasmid did not have an EcoRI site within the common HindIII fragment (data not shown). However, an enzyme that produced two DNA fragments of equal size both of which hybridized to the probe would appear as a single fragment in this type of analysis. Initially, the two porcine Ent plasmids appeared to have LT genetic determinants with very different structures. With five of the enzymes, pCG86 was found to have one extra DNA fragment that hybridized to the probe when compared to P307. However, all but the "extra" DNA fragment were found to be the same molecular mass as the corresponding restricted P307 DNA fragment(s). This unexpected observation

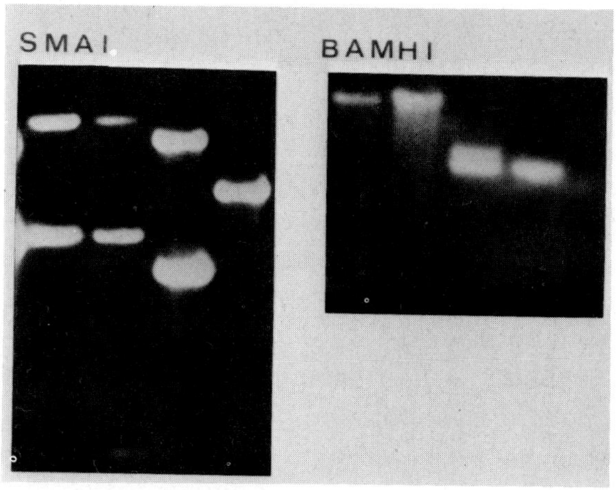

Fig. 6. Autoradiographs of Southern blots of restricted Ent plasmids. The DNA in the first autoradiograph was cut with SmaI. The order of the samples is (1) P307, (2) pCG86, (3) and (4) human Ent plasmids. The DNA in the second autoradiograph was cut with BamHI. The order of the samples is (1) and (2) human Ent plasmids, (3) pCG86, (4) P307.

can best be accomodated by a model in which all or part of the LT DNA region is duplicated in pCG86. We have other evidence that the LT DNA duplication probably occurs as an inverted and repeated sequence. Confirmation of this hypothesis awaits heteroduplex analysis. The similarity or divergence of the sequences adjacent to the LT DNA region is indicated by the sizes of DNA fragments that hybridized to the probe. A comparison of the sizes of the hybridized fragments of P307 and pCG86 indicated that the DNA sequences adjacent to the LT DNA region in each of these plasmids are the same (Figure 6). In contrast, the human Ent plasmids differed markedly from each other and from the porcine Ent plasmids with respect to the sequences adjacent to the LT DNA region. By analyzing more human and porcine Ent plasmids, we will be able to determine if indeed the porcine Ent plasmids constitute a homogeneous group and the human Ent plasmids a heterogeneous group as the preliminary data indicated.

Acknowledgements

The work reported in this paper was supported by the National Institutes of Allergy and Infectious Diseases grant AI10885-07 and contract DADA17-72-C-2149 from the Army Research and Development Command. W.D. and S.M. were supported by a Biological Infection Training Grant 1 T32 AI07149-01 from the National Institutes of Health.

References

1. Moon, H. W. (1974) Adv. Vet. Sci. Compar. Med., 18, 179-211.
2. Sack, R. B. (1975) Ann. Rev. Micro., 29, 333-353.
3. Gyles, C. L. and Barnum, D. A. (1969) J. Infect. Dis., 120, 419-426.
4. Smith, H. W. and Jones, J. E. T. (1963) J. Pathol. Bacteriol., 86, 387-412.
5. Gorbach, S. L. (1971) Gastroenterology 60, 1110-1129.
6. Smith, H. W. and Linggood, M. A. (1971) J. Med. Microbiol.,4, 467-485.
7. Burgess, M. N., Bywater, R. J., Cowley, C. M., Mullan, N. A., and Newsome, P. M. (1978) Infect. Immun., 21, 526-531.
8. Alderete, J. F. and Robertson, D. C. (1978) Infect. Immun., 19, 1021-1030.
9. Gill, D. M. and King, C. A. (1975) Biochemistry, 15, 1242-1248.
10. Gill, D. M., Evans, D. J., Jr., and Evans, D. G. (1976) J. Infect. Dis. Suppl., 133, S103-S107.
11. Clements, J. D. and Finkelstein, R. A. (1978) Infect. Immun., 22, 709-713.
12. Holmgren, J. (1974) Infect. Immun., 8, 851-859.
13. Smith, H. W. and Halls, S. (1968) J. Gen. Microbiol., 52, 319-334.
14. So, M., Crosa, J. H., and Falkow, S. (1975) J. Bacteriol., 121, 234-238.
15. So, M., Boyer, H. W., Betlach., M., and Falkow, S. (1976) J. Bacteriol., 128, 463-472.
16. So, M., Heffron, F. and McCarthy, B. J. (1979) Nature, 277, 453-456.
17. So, M., Dallas, W. S. and Falkow, S. (1978) Infect. Immun., 21, 405-411.
18. Dougan, G. and Sherratt, D. (1977) Molec. Gen. Genet., 151, 151-160.
19. Bolivar, F., Rodriguez, R., Betlach, M., and Boyer, H. W. (1977) Gene, 2, 95-113.
20. Studier, F. W. (1973) J. Mol. Biol. 79, 237-248.
21. Dallas, W. S. and Falkow, S. (1979) Nature, 277, 406-407.
22. Rigby, P. W. J., Dieckmann, M., Rhodes, C., and Berg, P. (1977) J. Mol. Biol., 113, 237-251.
23. Gruenstein, M. S. and Hogness, P. (1975) Proc. Natl. Acad. Sci. USA, 72, 3961-3965.
24. Southern, E. M. (1975) J. Mol. Biol. , 98, 503-518.
25. Botchan, M., Topp, W., and Sambrook, J. (1976) Cell, 9, 269-287.
26. Donta, S. T., Moon, H. W., and Whipp, S. C. (1974) Science, 183, 334-335.
27. Frazer, A. C., and Curtiss, R. III, (1975) Curr. Topics Microbiol. Immunol., 69, 1-84.
28. Kessler, S. W. (1975) J. Immunol., 115, 1617-1624.
29. Finkelstein, R. A. (1976) in Mechanisms in bacterial toxinology. John Wiley & Sons, New York pp 53-84.
30. Gyles, C. L., Palchaudhuri, S., and Maas, W. K. (1977) Science, 198, 198-199.

Plasmids of Medical, Environmental and Commercial Importance
K.N. Timmis and A. Pühler, editors

PLASMID CISTRONS CONTROLLING SYNTHESIS AND EXCRETION OF THE EXOTOXIN α-HAEMOLYSIN OF ESCHERICHIA COLI

W. GOEBEL, A. NOEGEL, U. RDEST and W. SPRINGER
Institut für Genetik und Mikrobiologie der Universität Würzburg,
Röntgenring 11, D-8700 Würzburg

INTRODUCTION

Haemolysis, i. e. the ability to cause lysis of erythrocytes is a rather wide-spread phenomenon among bacteria and other microorganisms. The molecular basis of this property may be different from organism to organism. Extra-cellular enzymes such as proteases, lipases and other hydrolytic enzymes have been shown to be responsible for for haemolysis in some bacteria[1-5] as well as in fungi[6]. In other bacteria "nonenzymatic" proteins which disrupt the membrane of erythrocytes, have been identified as the causative agents for haemolysis[7-10]. All these agents are termed haemolysins. They are often secreted by the cell or are bound to the cellular surface. Haemolysis may be associated with the pathogenic trait exerted by such organisms. In pathogenic *E. coli* strains, causing more or less severe diarrhoea especially in newborn animals and humans, the capability of haemolysin production is frequently encountered[11-12]. Three types of haemolysins termed α, β and , have been identified in *E. coli*[13-15]. The distinction of these three types is mainly based on rather superficial properties, such as more or less free diffusion into the surroundings or different appearance of the lysis zones on blood agar. The α-haemolysin of *E. coli*, which has been shwon in some cases to be genetically determined by transmissible plasmids[15-20], is secreted by the producing cells and can be isolated from the supernatant. Despite considerable effort, little has been known so far about the nature of this agent[21]. Here we would like to report recent data obtained in our laboratory which show that the genetic information required for the production and secretion of active α-haemolysin is completely plasmid-borne and determined by at least three clustered cistrons.

MATERIALS AND METHODS

The applied material and methods will be published in detail elsewhere (Noegel et al., manuscript submitted).

RESULTS

The haemolytic plasmid of E. coli PM152 and mutants defective in synthesis or excretion of α-haemolysin

The haemolysin-determinant of the α-haemolytic E. coli strain PM152 is carried by a transmissible plasmid which belongs to the incompatibility group I_2 and has a molecular weight of 41×10^6 dalton. A physical map constructed from EcoRI cleavage products of this plasmid is shown in Fig. 1.

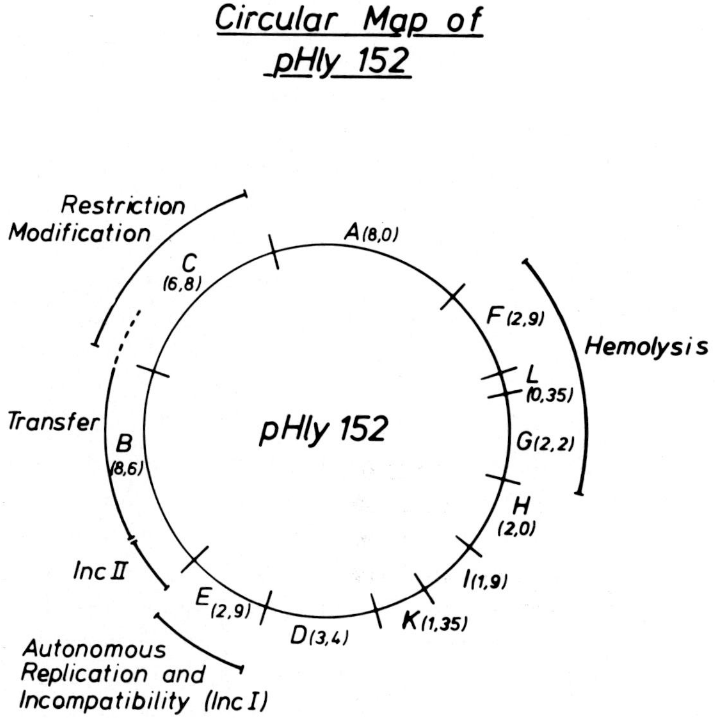

Fig. 1 Physical and functional map of the haemolytic plasmid pHly152. The numbers given represent the molecular weights of the EcoRI fragments in 10^6 dalton.

Two pools of active α-haemolysin can be detected in logarithmically growing E. coli cells harbouring this plasmid. Haemolysin of pool I exists in a free (i. e. not cell-bound) extracellular form, whereas that of pool II is found inside the cell, either in the periplasmic space or bound to the cytoplasmic membrane (Springer, unpublished results). We shall refer in the following to α-haemolysin in these two pools as "external haemolysin" or Hly_{ex} and "internal haemolysin" or Hly_{in}.

Mutagenesis with nitrosoguanidine leads to the isolation of plasmid-specific mutants which are unable to secrete α-haemolysin. The pool of Hly_{in}, however, is not affected in these of mutants. The mutations are either constitutive or temperature-sensitive. The other type of mutants can no longer synthesize α-haemolysin, i. e. neither Hly_{in} nor Hly_{ex} can be detected. Again, constitutive as well as temperature-sensitive mutants of this type were isolated.

To further characterize the genetic loci underlying these defects, mutations affecting the synthesis or excretion of α-haemolysin were induced by transposition of the ampicillin transposon Tn3 to the haemolytic plasmid pHly152 using a procedure described previously[22]. Again both types of mutants (Hly^{-}_{ex}/Hly^{-}_{in} and Hly^{-}_{ex}/Hly^{+}_{in}) were obtained (Table 1).

CHARACTERIZATION OF HEMOLYSIS-NEGATIVE MUTANTS OBTAINED BY TN3-INSERTIONS

MUTANT	RESTRICTION FRAGMENTS CARRYING THE TN3 INSERTION IN FRAGMENTS			PROPERTY AFFECTED BY THE MUTATION	
	ECORI	PSTI	HINDIII	HLY_{IN}	HLY_{EX}
pHLY 152-T5	G	F	C	-	-
pHLY 152-T10	G	F	C	-	-
pHLY 152-T12	G	F	C	-	-
pHLY 152-T21	G	F	C	-	-
pHLY 152-T8	G	J	C	+	-
HLY 152-T22	G	J	C	+	-
pHLY 152-T14	G	D	NT	+	-
HLY 152-T4	F	C	E	-	-
HLY 152-T18	F	F	C	-	-
HLY 152-T19	F	F	C	-	-
HLY 152-T15	F	C	NT	+	+

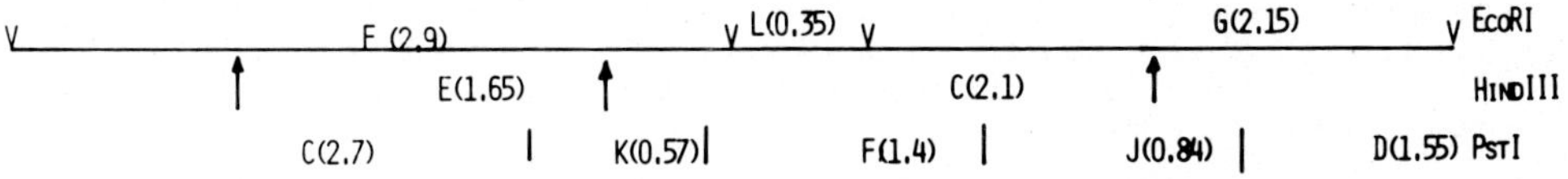

Table 1 Hly_{in} (+) or (-) indicates that mutant produces either active haemolysin which accumulates in the periplasmic space (+) or no active haemolysin (-). Hly_{ex} (-) indicates that no active haemolysin appears in an extracellular form. The numbers given in the physical map of the Hly-region of plasmid pHly152 drawn underneath the table represent molecular weights in 10^6 x dalton. (NT) = not tested.

Cleavage of the mutant plasmids with EcoRI indicates that the Tn3 insertions in mutants of type I are located on EcoRI fragments F or G, whereas those of type II are always located on EcoRI fragment G. The EcoRI-fragments F and G are separated by the small EcoRI-fragment L (Fig. 1). By additional cleavage with PstI of the isolated EcoRI fragments containing the Tn3 insertions the precise location of Tn3 within these fragments could be determined. The analyses are summarized in Fig. 2. From these data it is obvious that a

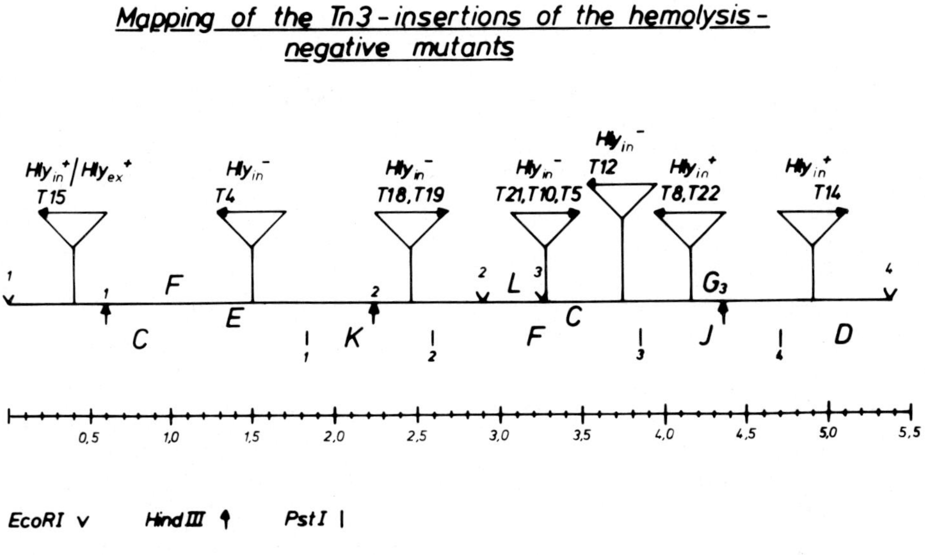

Fig. 2 Mapping of the Tn3-insertions leading to a defect in the synthesis of haemolysin (Hly^-_{in}) or in the excretion of intracellular haemolysin (Hly^+_{in}). The sites of the Tn3 insertions of mutants T5, T10 and T21 and of mutants T18 and T19 can not be distinguished by this method. The scale below the map is given in 10^6 dalton.

rather large stretch of the plasmid DNA comprising at least $3.4x10^6$ dalton is necessary for determining the production of extracellular α-haemolysin.

Cloning of α-haemolysin synthesis and excretion functions and complementation or recombination of haemolysis-negative insertion mutants by the recombinant plasmids

To obtain more detailed information about the genetic apparatus and the gene products required for production and excretion of α-haemolysin, the *Eco*RI- and *Hind*III-fragments of pHly152 were cloned on the vector plasmid pACYC184[23]. As expected from the above data none of the obtained recombinant plasmids carrying single *Eco*RI- or *Hind*III-fragments inserted into pACYC184 could express biologically active extracellular α-haemolysin. They were, however, able to complement some of the above described Tn3 insertion mutants to full haemolytic activity, i. e. production and secretion of haemolysin (Table 2).

Mutants	Tn3 insertion in	Hly-Phenotype of mutant		Complementation to production of Hly_{EX} by							
		Hly_{IN}	Hly_{EX}	pAN 104	pAN 104-1	pAN 250	pAN 250-1	pAN 202	pAN 202-1	pAN 215	pAN 215-1
pHly 152-T4	EcoRI F	-	-	+	+	-	-	+	+	-	-
pHly 152-T8		-	-	-	-	-	-	-	-	-	-
pHly 152-T19		-	-	-	-	-	-	-	-	-	-
pHly 152-T5	EcoRI G	-	-	-	-	+	-	-	-	+	-
pHly 152-T10		-	-	-	-	-	-	-	-	+	-
pHly 152-T21		-	-	-	-	-	-	-	-	+	-
pHly 152-T8	EcoRI G	+	-	-	-	+	-	-	-	-	-
pHly 152-T22		+	-	-	-	+	-	-	-	-	-
pHly 152-T14		+	-	-	-	+	-	-	-	-	-

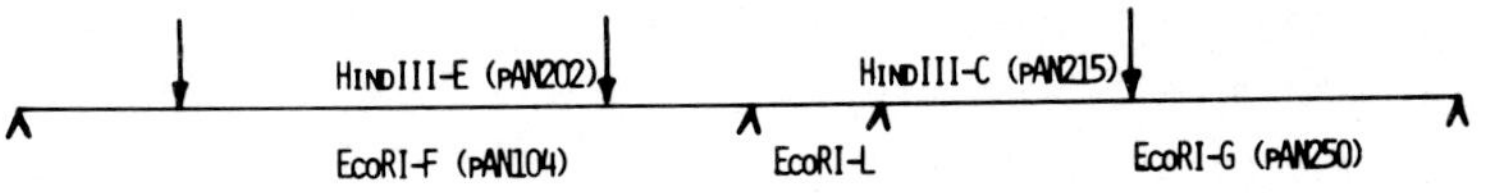

Table 2 The description of the recombinant plasmids used in the complementation studies is given in the physical map of the Hly-region drawn underneath the Table. Complementation was carried out in C600 *recA*, r^-, m^-. The recombinant plasmids were transformed into C600 *recA* carrying the indicated Tn3 insertion mutant Hly plasmid. R indicates recombination between the resident recombinant plasmid and the Tn3 mutant Hly plasmid introduced by conjugation, which leads to haemolysis-positive transconjugants in a C600 *rec*$^+$ strain and with decreased frequency in a *recA* strain.

Transformation of plasmid pAN250 into mutants defective in the secretion of Hly_{in} resulted in all cases in transformants capable of secreting active α-haemolysin. This indicates that the EcoRI fragment G carries gene(s) required for the secretion of α-haemolysin. Plasmid pAN215 having the HindIII fragment C inserted into pACYC184 could not complement either of the "transport" mutants, indicating that the genes required for this process are located on the right half of EcoRI fragment G in the map given in Fig. 2. Since the location of the Tn3 insertions leading to a defect in the secretion of Hly_{in} in the mutants T8, T22 and T14 are 0.7×10^6 dalton apart, it cannot be excluded that more than one gene is involved in the secretion of active intracellular α-haemolysin (Hly_{in}).

The hybrid plasmid pAN104 having the EcoRI fragment F inserted into pACYC184 was able to complement the Tn3 insertion mutant T4 which is Hly^-_{in}/Hly^-_{ex} (Table 2). The same complementation was obtained when the recombinant plasmid pAN202 with the cloned fragment HindIII-E, comprising the middle part of the EcoRI fragment F is transformed into this mutant (Table 2). The observed haemolytic after complementation corresponded to about 25 % of the wild type activity.

None of the other mutations caused by Tn3 insertions into the EcoRI fragment F, i. e. T18 and T19, or the EcoRI fragment G, i. e. T5, T10, T21 and T12 could be complemented by pAN104, pAN202, pAN215 or pAN250 when the latter recombinant plasmids were transformed into a recA strain harbouring the corresponding insertion mutant plasmid (Table 2). The Tn3 insertions of these mutants are located on the right half of the EcoRI fragment F and the left half of EcoRI fragment G, respectively (Fig. 2). This suggests that none of the fragments in the recombinant plasmids used for the complementation tests carries the whole information for the gene product(s) which is (are) defective in these Tn3 mutants. However, when the mutant plasmids of T5, T10, T21 or T12 were transferred by conjugation in a rec^+ strain carrying already pAN215 as resident plasmid, two types of transconjugants could be selected. Type I (about 30 % of the obtained transconjugants) is Hly^+_{in}/Hly^+_{ex} and carries a recombined large plasmid whereas type II (about 70 % of the transconjugants) is still unable to produce α-haemolysin and contains both plasmids as separate units. This indicates that the haemolysis-positive phenotype is caused by recombination between both plasmids rather than by complementation. The frequency of recombination between pAN215 and these Tn3 mutant Hly plasmids (T5, T10, T21 and T12) is only sligthly reduced in a recA background suggesting some kind of illegitimate recombination between these plasmids (Noegel and

Goebel, unpublished). Only very few Hly$^+$ transconjugants (<1%) carrying recombined plasmids were obtained, when these Tn3 mutant plasmids were transferred by conjugation into a rec$^+$ or recA strain harbouring pAN250. The frequency of recombination leading to haemolysis-positive transconjugants is also extremely low (less than 0.1 % of the obtained transconjugants), when the Tn3 mutant plasmid T18 and T19 were transferred in a rec$^+$ strain harbouring pAN215. These data indicate that the Tn3 insertions of mutants T18, T19, T5, T10, T21 and T12 (Fig. 2) which are all Hly^-_{in}/Hly^-_{ex} may inactivate a single rather large cistron.

The complementation and recombination results thus suggest that at least three cistrons are involved in the synthesis and secretion of α-haemolysin (Fig. 3). Cistron C which is inactivated by the Tn3 insertion of mutant T4

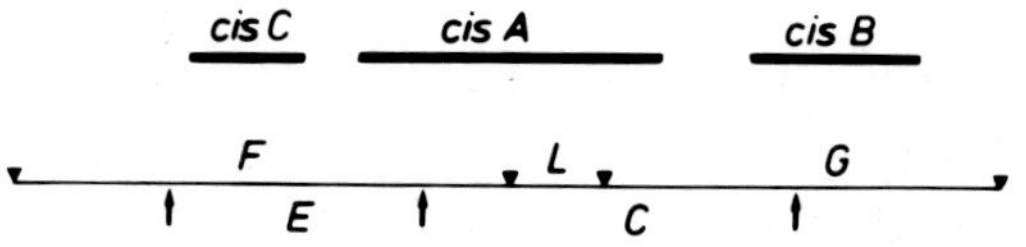

Fig. 3 Preliminary map of the location of the cistrons involved in the synthesis of active α-haemolysin and its excretion. The map is based on the complementation and recombination results obtained with the recombinant plasmids and the Tn3 insertion mutants.

can be complemented by fragment HindIII-E (pAN202). Cistron B which is inactivated in the secretion mutants T8, T14 and T22 is located on the left half of fragment EcoRI-G. The other Tn3 insertions seem to inactivate at least one other cistron (cistron A), which is not carried as an entire unit by either of the cloned fragments. The defects caused by these Tn3 insertions can be only restored by recombination.

Identification of gene products involved in the production of biologically active α-haemolysin and its secretion

To obtain information on the nature of the gene products required for the production and secretion of active α-haemolysin, we have analysed the proteins expressed in minicells of *E. coli*[24] harbouring the complete haemolysin plasmid pHly152 and the recombinant plasmids carrying the EcoRI fragments F and G and the HindIII fragments C and E. Again, we shall describe here only the major points of interest. A more detailed analysis will be published somewhere else (Rdest, Springer, Noegel and Goebel, in preparation). Fig. 4 shows the complete set of ^{35}S-labeled proteins coded by pHly152. Protein A (mw 58.000 dalton) is

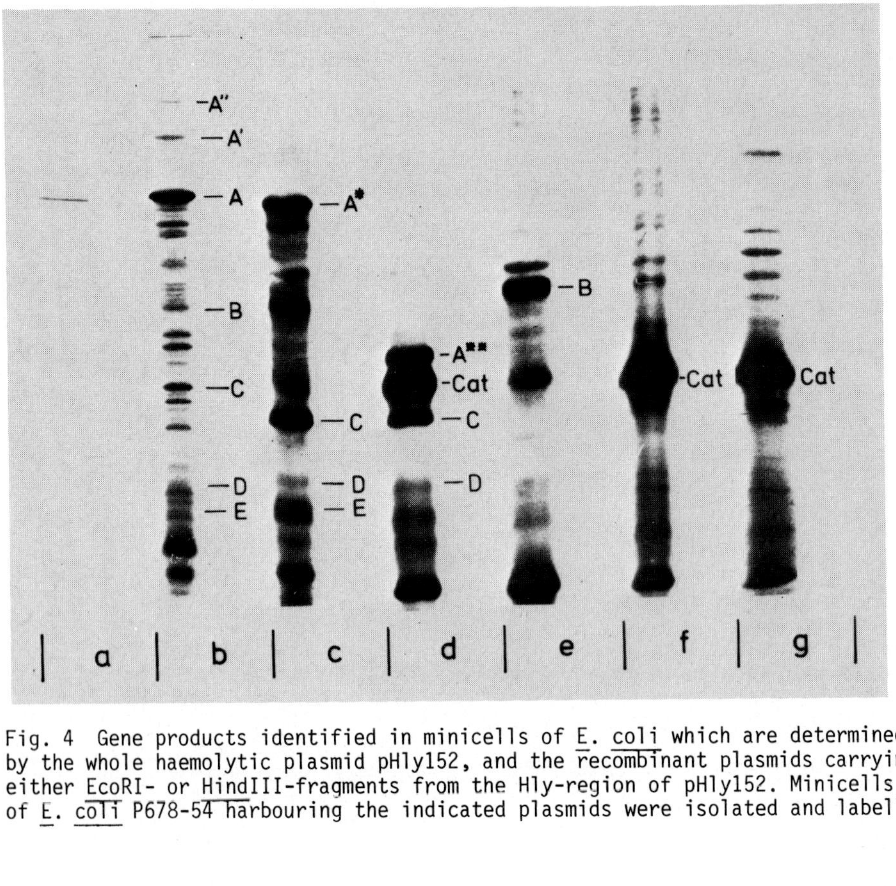

Fig. 4 Gene products identified in minicells of *E. coli* which are determined by the whole haemolytic plasmid pHly152, and the recombinant plasmids carrying either EcoRI- or HindIII-fragments from the Hly-region of pHly152. Minicells of *E. coli* P678-54 harbouring the indicated plasmids were isolated and labelled

with ^{35}S-methionine as described[30]. The labelled proteins were separated on SDS-containing polyacrylamide gradient gels (10 to 20 %) and visualized by autoradiography[31]. (a) purified ^{35}S-labelled extracellular haemolysins, proteins determined by (b) pHly152, (c) pAN104, (d) pAN202, (e) pAN250, (f) pAN215 and (g) vector plasmid pACYC184 as control. (cat) is chloramphenicol acetylase.

presumably α-haemolysin as demonstrated by comparison with purified active extracellular haemolysin (Fig. 4a) (Springer, Rdest and Goebel, in preparation). The larger proteins A' (mw 75.000 d) and A" (mw 90.000 d) carry also the α-haemolysin antigen determinant as demonstrated with specific antibody against Hly_{ex} (data not shown) and may represent precursors of α-haemolysin, which are processed to protein A. Hybrid plasmid pAN104 (Fig. 4c) expresses four major proteins all of which are also determined by the complete haemolysin plasmid Hly152. (1) Protein A^+ (mw 56.000 d) reacts with α-haemolysin antibody indicating that it may represent α-haemolysin, which is however smaller by about 2.000-3.000 dalton than protein A. This suggests that the EcoRI site 2 cleaves off part of the C-terminal sequence of the haemolysin protein. (2) Protein C (mw 18.000 d) which is also expressed by pAN202 (HindIII fragment E inserted into pACYC184) is required for the activity of α-haemolysin since it is missing in the Tn3 mutant T4 and can be complemented by the recombinant plasmids pAN104 and pAN202. (3) Protein D is expressed by minicells carrying pAN104 or pAN202, indicating that it may also be involved in a step leading to active haemolysin. (4) Protein E (mw 10.000 d) which is expressed by pAN104 and the complete haemolytic plasmid pHly152 but not by the hybrid plasmid pAN202 (Fig. 4d) is determined by a gene located left of the HindIII site 1 and is therefore not involved in the synthesis of active haemolysin. The recombinant plasmid pAN202 expresses in addition to proteins C and D another protein, which bands above cat (Fig. 4e). This proteins seems to represent a N-terminal fragment of haemolysin (protein A) (data not shown) which indicates that the start of the gene for haemolysin is located on the HindIII fragment E. The recombinant plasmid pAN215, carrying the HindIII fragment C inserted into pACYC184 does not seem to express specific proteins besides those determined by the vector (Fig. 4f and g), which further suggests that this fragment carries the C-terminal part of the haemolysin gene (cistron A), but no additional cistron. The recombinant plasmid pAN250 (carrying the EcoRI fragment G) expresses one major protein in minicells (Fig. 4e) which is termed protein B (mw 33.000 d). It is determined by pAN250 but not by pAN215 indicating that the cistron for this product is located on the right half of EcoRI fragment G (Fig. 2). This protein seems to be absent or altered in mutants which are unable to secrete haemolysin ($Hly^{+}_{in}/Hly^{-}_{ex}$). Preliminary results indicate that it may be located in the outer membrane (Rdest, unpublished).

DISCUSSION

The data described indicate that haemolysis exhibited by some *E. coli* strains is a rather complex process requiring at least three clustered cistrons which comprise about 3.4×10^{6} dalton of a transmissible plasmid in the *E. coli* strain PM152. Similar to the plasmid-determined heat-stable (ST) and heat-labile (LT) enterotoxins of *E. coli*[25-27], α-haemolysin represents an exotoxin. The genetic determinants for ST and LT which were recently isolated by the gene cloning technique, seem to be, however, considerably smaller[28, 29]. There is no evidence for a specific excretion function for either ST or LT enterotoxins, although more than one polypeptide seems to be coded for by the LT determinant. The production of active α-haemolysin requires at least two cistrons (A and C). One of them (cistron A) determines a gene product which appears to be the precursor of the protein identified as extracellular α-haemolysin.

The gene product of the other cistron C is required for the production of active haemolysin and may be involved in the conversion of an inactive haemolysin precursor into its active form. Since the presumed precursor protein is considerably larger than the mature extracellular haemolysin, the gene product of cistron C may function as a specific protease converting the haemolysin precursor protein into the active form. The active intracellular haemolysin appears to accumulate first in the periplasmic space (Springer, unpublished). Its secretion through the outer membrane is controlled by at least one cistron (cistron B). It seems to determine a protein which is located in the outer membrane of haemolytic *E. coli* cells.

ACKNOWLEDGEMENTS

We wish to thank E. Siebenhaar for skillful technical assistance. This work was supported by grants from the Deutsche Forschungsgemeinschaft (SFB 105 - A11 and A12).

REFERENCES

1. Wiseman, G.M.and Caird, J.D., Can. J. Microbiol. 13, 369 (1967).
2. Hauschild, A.H., Lecroisey, A. and Alouf, J.E. (1973) Can. J. Microbiol. 19, 881 (1973).
3. Lochmann, O., V'Ymola, F. and Chaloupeck'y, V. (1975) J. Hyg. Epidemiol. Microbiol. Immunol. (Praha) 19, 61.

4. Sakurai, J., Matsuzaki, A., Takeda, Y., and Miwatani, T. (1974) Infect. Immun. 9, 777

5. Fujita, M., Koshimura. S. (1975) J. Exp. Med. (JPN) 45, 457

6. Seeger, R., Burkhardt, M., Haupt, M. and Feulner, L. (1976) Arch. Pharmacol. 293, 163

7. Freer, J.H., Arbuthnott, J.P. and Bernheimer, A.W. (1968) J. Bact. 95, 1153

8. Rennie, R.P., Freer, J.H. and Arbuthnott, J.P. (1974) J. Med. Microbiol. 7, 189

9. Fackrell, H.B. and Wiseman, G.M. (1976) J. Gen. Microbiol. 92, 1

10. Bernheimer, A.W. (1974) Biochim. Biophys. Acta 344, 27

11. Ratiner, I.U.A., Kanare'ikina, S.K., Bondarenko, V.M. and Golubera, I.V. (1976) ZH Mikrobiol. Epidemiol. Immunbiol. 117-21

12. Smith, H.W.(1969) In Bacterial Episomes and Plasmids, 213-226, J. and A. Churchill Ltd. London

13. Lovell, R. and Rees, T.A. (1960) Nature 188, 755

14. Snyder, J.S. and Koch, N.A. (1966) J. Bact. 91, 763

15. Walton, J.B. and Smith, D.H. (1969) J. Bact. 98, 304

16. Smith, H.W. and Halls, S., (1967) J. Gen. Microbiol. 47, 153

17. Goebel, W., Royer-Pokora, B., Lindenmaier, W. and Bujard, H. (1974) J. Bact. 18, 964

18. Royer-Pokora, B. and Goebel, W. (1976) Molec. gen. Genet. 144, 177

19. Monti-Bragadin, L., Samer, L., Rottini, G.D. and Pani, B., (1975) J. Gen. Microbiol. 86, 367

20. Le Minor, S. and Le Coueffic, E. (1975) Ann. Microbiol. (Paris) 126,313

21. Short, E.C.J.R. and Kurtz, H.J. (1971) Infect. Immun. 3, 678

22. Goebel, W., Lindenmaier, W., Pfeifer, F., Schrempf, H. and Schelle, B. (1978) Molec. gen. Genet. 157, 119

23. Chang, A.C.Y. and Cohen, S.N. (1978) J. Bact. 134, 1141

24. Adler, H.J., Fischer, W.D., Cohen, A. and Hardigree, A.A. (1967) Proc. natn. Acad. Sci. 57, 321

25. Smith, H.W. and Halls, S., (1968) J. Gen. Microbiol. 52, 319

26. Gyles, C.L., So, M. and Falkow, S. (1974) J Infect. Dis. 130, 40

27. Gyles, C.L., Palchaudhuri, S. and Maas, W.K. (1977) Science 198, 198

28. So, M., Boyer, H.W., Betlach, M. and Falkow, S. (1976) J. Bact. 128, 463

29. Dallas, W.S., Dougan, D. and Falkow, S. (1978) Genetic Engineering, H.W. Boyer and S. Nicosia eds., Elsevier/North Holland, Biomedical Press

30. Levy, S.B. (1974) J. Bact. 120, 1451

31. Bonner, W.M. and Laskey, R.A. (1974) Eur. J. Biochem. 46, 83

Plasmids of Medical, Environmental and Commercial Importance
K.N. Timmis and A. Pühler, editors

PLASMIDS AND THE SERUM RESISTANCE OF ENTEROBACTERIA.

PETER W. TAYLOR, COLIN HUGHES AND MARTYN ROBINSON
Microbial Surfaces Research Group, Department of Microbiology, The University, Leeds LS2 9JT (England), and Sandoz Forschungsinstitut, Brunner Strasse 59, A-1235 Wien (Austria)

SERUM BACTERICIDAL REACTION. A wide range of both smooth and rough Gram-negative bacteria are susceptible to the bactericidal action of human and animal sera; this lethal effect is mediated by the activated components of either the classical[1], or alternative[2,3] complement pathways. The classical pathway is activated by the interaction of antibody, usually of the IgM class[4,5] and bacterial surface antigens, although recent evidence suggests that under certain conditions the classical pathway may be activated directly by the Lipid A region of lipopolysaccharide in the absence of antibody[6,7]. The alternative pathway is activated by the polysaccharide region of lipopolysaccharide[6]. Both pathways lead to the generation of a functional complex, the membrane attack unit, involving components C5b, C6, C7, C8 and C9; this complex is responsible for the poorly understood membrane alterations that culminate in the death of the bacterial cell.

Although almost all rough strains prove to be very sensitive to the bactericidal action of serum, many smooth strains are serum resistant, and resistant strains may be isolated from a wide variety of human and animal infections [8, 9, 10, 11]. A number of studies indicate that resistance to serum may be an important determinant of bacterial virulence in certain infections. For example, it has been repeatedly observed that strains isolated from cases of bacteraemia tend to be more serum resistant than strains from other sources [8, 12, 13], and bacteraemia caused by sensitive *Escherichia coli* strains is less often associated with shock and death than bacteraemia due to resistant strains [14]. Only resistant strains were able to consistently cause experimental *E.coli* endocarditis in rabbits following intracardiac catheterisation [15]. In urinary tract infections, an increased incidence of serum resistant strains is associated with both kidney involvement [16] and the ability to produce symptomatic infection [17]. In an experimental rat model, serum sensitive strains were able to produce kidney infection only after the animals had been depleted of complement using cobra venom factor [18]. These observations are consistent with the hypothesis that complement-mediated host immune mechanisms do play a role in the biology of bacteraemia and of renal infection.

BASIS OF SERUM RESISTANCE. As the primary target for complement action almost certainly involves lipid-containing structures in the outer membrane[19], studies of serum resistance have generally been concerned with cell surface components that might impede the attachment or subsequent activity of the serum factors responsible for the bactericidal effect. A number of investigators have provided good evidence that the 0-specific side chain moiety of lipopolysaccharide plays a central role in the determination of serum resistance. Mutations leading to reduction or loss of the 0-side chain result in a marked increase in serum sensitivity [20,21,22]. Feingold [23] has shown that phenotypic modification from serum resistance to sensitivity as a result of growth in the presence of sublethal concentrations of diphenylamine is accompanied by a striking reduction in the amount of 0-side chain material associated with lipopolysaccharide from an enteropathogenic strain of E.coli. Mutants of Salmonella deficient in UDP-galactose-4-epimerase form incomplete lipopolysaccharides which lack both 0-side chains and that part of the core distal to the point of the biosynthetic lesion, unless supplied with exogenous galactose [24]. Dlabac [25] found that, when grown in galactose-free medium, these mutants were sensitive to serum; when galactose was supplied to growing cells the cells became increasingly serum resistant.

However, many strains carrying a full complement of lipopolysaccharide 0-side chains may be susceptible to the bactericidal action of serum, although such strains are not usually promptly killed but exhibit a "delayed sensitive" response [26] (Figure 1). Inheritance by a promptly sensitive rough E.coli mutant of the rfb locus, determining 0-side chaim biosynthesis, from a serum resistant, K-negative, smooth 08 Hfr donor resulted in smooth recombinants that displayed the delayed serum killing response, suggesting that 0-side chains are responsible for the delay in killing but that other factors determine complete resistance [29]. A serum resistant mutant (E.coli I7) derived from delayed sensitive strain LP729 produced greater amounts of an envelope polypeptide (m.w. 46,000) than the parent strain; a rough mutant obtained from I7 produced comparable amounts of the polypeptide but lacked 0-side chains and was promptly sensitive [30]. It was therefore suggested that the polypeptide was involved in the determination of resistance but that it was only functional when superimposed upon a full complement of lipopolysaccharide 0-side chains. Similar results were found when serum sensitive strains of Neisseria gonorrhoeae were transformed to resistance; transformed cells acquired a new principle outer

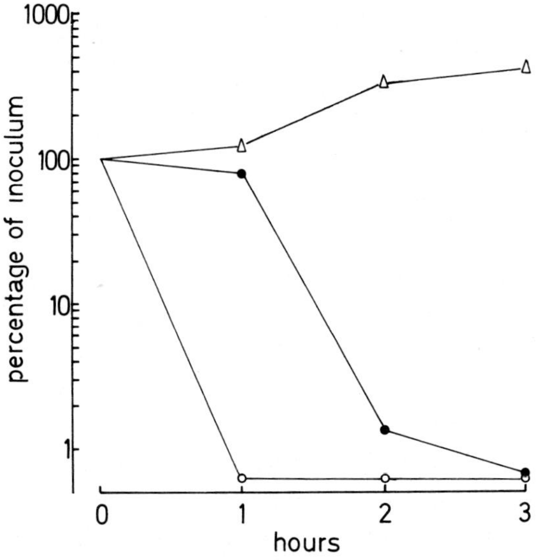

Fig. 1. Examples of promptly serum sensitive, delayed serum sensitive and serum resistant enterobacterial strains. Serum bactericidal assays were performed by the method of Taylor "et al". [27]. Symbols 0, Citrobacter C14; E.coli C5; Δ, E.coli C42. For description of strains, see Taylor and Hughes [28].

membrane protein [31].

Some workers have suggested that acidic polysaccharide K-antigens are responsible for the serum resistance of Gram negative bacteria [32] but an increasingly large body of evidence [28, 29, 33, 34, 35], appear to indicate a minimal role for these polymers.

Role of plasmids in the determination of serum resistance. The earliest report of an altered serum response following the acquisition of foreign DNA appears to be that of Muschel and associates [36, 37], who found increased serum resistance after lysogenisation of E.coli K12 with λ phage. In 1976, Reynard and Beck [38] reported that the two well characterised antibiotic-resistance plasmids R1 and NR1 (R100) were both able to confer relatively high degrees of resistance against commercial, reconstituted rabbit serum on a number of E.coli K12 strains. This ability appeared to be restricted to F-like plasmids; plasmids from four out of 16 clinical isolates examined were able to protect E.coli K12 strain J6-2N against the bactericidal action of rabbit serum [39]. Other workers, however, have been unable to confirm that either R1 or NR1 are able to protect K12 strains against killing by human and fresh rabbit serum [28,40]. There are no differences between the complement systems of human and rabbit serum that would appear to explain this ambiguity and Reynard and co-workers have themselves found that F-like plasmids do not provide K12 strains with protection against human serum [39].

Fietta "et al". [40] have recently reported that eight of 26 plasmids examined conferred relative serum resistance to E.coli K12 strains; plasmids able to do this belonged to a variety of incompatability groups. A plasmid was said to have conferred resistance if the amount of serum needed to reduce the viable count to 1% after 30 min was greater for R^+ progeny than for R^- parents. The amounts of serum needed to effect this degree of killing were extremely small (3% of the total reaction mixture) and it is possible that at these concentrations some serum components essential for complement-mediated serum killing were present in limiting amounts.

In our laboratories we have examined the effect of plasmid carriage on the serum sensitivity of enterobacteria isolated from polluted river water [28]. The 17 strains, carrying antibiotic resistance (R) and colicin (Col) determinants, gave a variety of responses to human serum; six strains were promptly sensitive, seven produced the delayed sensitive response and four were serum resistant. Attempts were made to cure all strains of plasmid-determined characters. Strains C8 (delayed sensitive) and P21 (promptly sensitive) were cured of all R and Col markers with acridine orange.

The serum sensitivities of nine cured colonies derived from E.coli C8 were measured; all showed the delayed serum killing response but were slightly more serum sensitive than the parent strain. One cured colonial form ($C8^-$) was selected for statistical evaluation, and a spontaneous Nal^R mutant was obtained. The serum responses of $C8^-$ and $C8^-Nal^R$ were identical. Replicate experiments confirmed that the loss of plasmid determinants from E.coli C8 let to a small but statistically significant increase in serum sensitivity. To assess the extent to which plasmids could modify the response of $C8^-Nal^R$ to human serum, this strain was mated with five river and four laboratory strains and progeny inheriting appropriate R markers were selected (Figures 2 and 3).

Inheritance by $C8\text{-}Nal^R$ of R-Utrecht and of the R and Col determinants from serum resistant river isolate E.coli C25 resulted in no significant change in serum response, whereas inheritance of markers from E.coli river isolates C11, C15, C42 and P43 led to comparable, small but statistically significant increases in serum resistance, although one of the donor strains was promptly sensitive and two were serum resistant. $C8^-$ Nal^R progeny inheriting plasmids

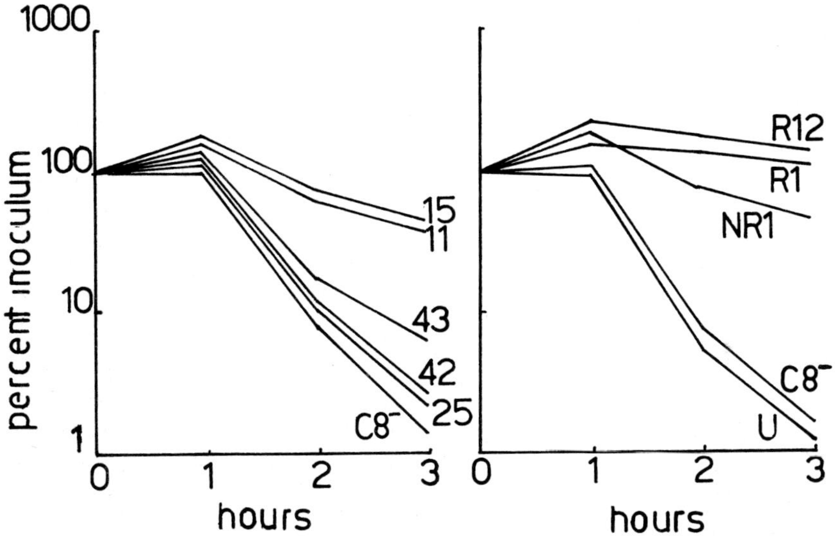

Figures 2 and 3: Response to human serum of C8⁻Nal^R and of C8⁻Nal^R progeny carrying plasmids from river isolates C15, C11, P43, C25 and C42 and from E. coli strains M941 (R-Utrecht), SFI585 (NR1), SFI587 (R1) and SFI609 (R12). See Taylor and Hughes [28] for details. Each point represents the mean of at least six determinations.

R1 and NR1 were considerably more resistant than the R⁻ parent (Figure 3). R12, a round of replication mutant of plasmid NR1 [41], conferred much higher levels of serum resistance on C8⁻Nal^R than the normal NR1, suggesting that the effect might be related to the number of plasmid copies per cell.

Loss of R determinants by the promptly sensitive strain P21 was not accompanied by any change in serum reactivity and inheritance of plasmids R1 and NR1 had no effect on the serum sensitivity of P21⁻Nal^R. These findings suggest that lipopolysaccharide O-side chains, the cell surface components considered responsible for the delay in serum killing, are essential for the expression of plasmid factors that modify sensitivity to serum.

In order to clarify the relationships between the various factors that might contribute towards serum resistance, strains were constructed carrying combinations of the O8 antigen, the K27 antigen and various plasmids. The recipient used was E.coli F470 [42], a K-negative rfb mutant that carries no extrachromosomal DNA and is promptly killed by serum; Hfr donor strains F639 (derived from E.coli F445 by Dr G. Schmidt, Freiburg), F459 [42] and F445 (Hfr 45) [43] facilitated the transfer of the his-linked K27 alone, the his-linked 08 antigen alone and both the 08 and K27 antigens simultaneously (Table 1.)

TABLE 1.

Analysis of his^+ recombinants selected from crosses between rough recipient F470 and various Hfr donor strains.

Hfr donor	Donor serotype	Number of recombinants analysed	Serological types of recombinants				
			$08:K^-$	08:K27	08:K27i	0^-:K27	Rough
F639	0^-:K27	120	0	0	0	115	5
F459	$08:K^-$	120	106	10	0	0	4
F445	08:K27	120	0	3	112	0	5

Of 120 his^+ recombinants analysed from the cross using F639 as donor, 115 (96%) inherited the K27 antigen. All nine $K27^+$ recombinants examined were as serum sensitive as E. coli F470. In fact, no colonial survivors were detected with any recombinant at any sampling time interval; $his^+:K27^-$ recombinants were also completely sensitive. Plasmids R1 and NR1 were transferred to strain F470 and a number of $0^-:K27^+$ recombinants; all progeny were completely serum sensitive.

Recombinants inheriting the 08-side chain (F459 x F470) gave the delayed sensitive response to human serum. Results obtained with $08:K^-$ recombinants 1, 6 and 10 are shown in Figure 4. Ten recombinants were found to produce the K27 antigen in addition to the 08 antigen; both donor and recipient strains were derived by mutation from E.coli E56b ($08:K27:H^-$) and loss of K antigen by both strains presumably resulted from point mutations at different chromosomal sites [44]. The response to serum of one of these recombinants, Rec 20, was found to be identical to those obtained with the $08:K^-$ recombinants (Figure 4).

The majority of his^+ recombinants from the cross between strain F470 and 08:K27 donor strain F445 agglutinated in both 08 antiserum and K27 antiserum, indicating that, although they had inherited loci determining the biosynthesis of both antigens, these hybrids had not inherited trp-linked genes essential for the production of enough antigen to confer 0-inagglutinability. These

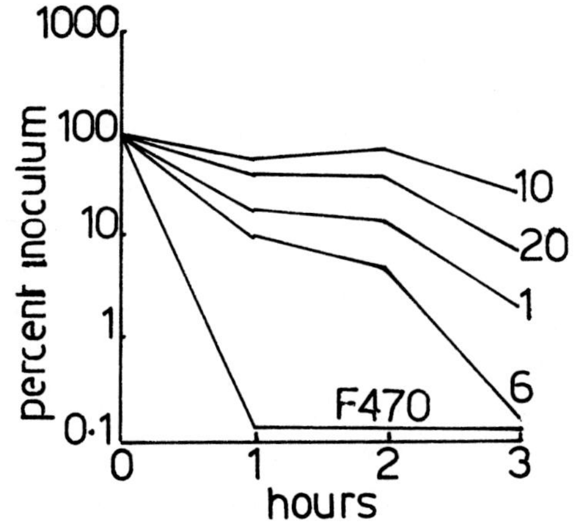

Figure 4. Response to human serum of E.coli F470 and his+ recombinants from the cross F459 x F470

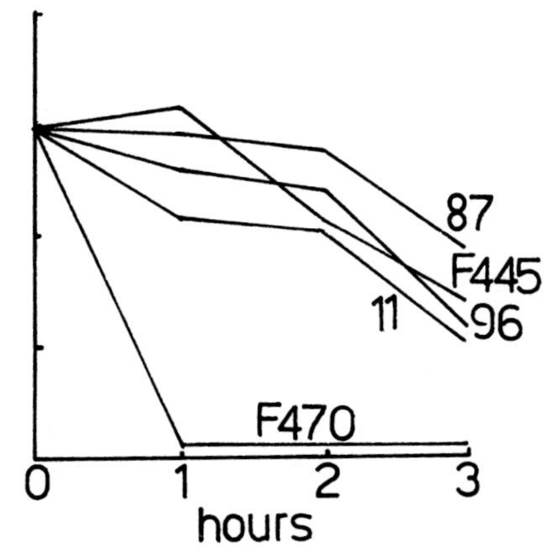

Figure 5. Response to human serum of E.coli F470, F445 and his+ recombinants from the cross F445 x F450

intermediate hybrids have been termed K27i forms by Schmidt and coworkers [43]. Three recombinants, 11, 87 and 96 were inagglutinable in 08 antiserum unless suspensions were heated at 115°C for 2h, and had thus inherited trp-linked genes determining complete formation of the K27 antigen. The reactivity in serum of these three recombinants was identical to those found with smooth recombinants from the cross F459 x F470 (Figure 5), demonstrating the inability of this antigen to confer any degree of serum resistance to strain F470, even in the presence of 0-specific side chains.

Plasmids were then transferred to a number of the smooth recombinants; plasmids belonging to incompatability groups W, P, Ia and N had no effect on the serum reactivity of any of the recombinants shown in Figures 4 and 5., Previous results had suggested that inheritance of plasmids R1 and NR1 by these recombinants should result in progeny with increased serum resistance. However, acquisition of either R1 or NR1 by the recombinants led to progeny showing increased sensitivity; one recombinant, Rec 10 (Figure 4) has been studied in some detail and Rec10 progeny carrying either plasmid were completely sensitive, even though agglutination characteristics in 0 and K antisera remained unaffected. The possibility that acquisition of plasmids R1 or NR1 is accompanied by cell surface changes leading to complete sensitivity is currently being investigated.

Martyn Robinson is currently Glaxo Research Fellow in the Department of Bacteriology, University of Bristol.

REFERENCES

1. Inoue, K. "et al." (1968) Biken J., 11, 203
2. Frank, M.J. "et al.". (1973) J. Infect. Dis., 128, S176.
3. Traub, W.H. and Kleber, I. (1976). Infect. Immun. 13, 1343.
4. Chernokhvostova, E.V. "et al." (1968) Zh. Mikrobiol. (Moskva) 45, 82.
5. Wiedermann, D. "et al". (1970) J. Infect. Dis., 121, 74.
6. Morrison, D.C. and Kline, L.F. (1977). J. Immunol. 118, 362.
7. Loos, M. "et al". (1978) Infect. Immun. 22, 5.
8. Roantree, R.J. and Rantz, L.A. (1960) J. Clin. Invest. 39, 72.
9. Fierer, J. "et al". (1972) J. Immunol. 109, 1156.
10. Larsson, P. and Olling, S. (1977). Med. Microbiol. Immunol. 163, 77
11. Carroll, E.J. and Jasper, D.E. (1977) Amer. J. Vet. Res. 38, 2019.
12. Vosti, K.L. and Randall, E. (1970) Amer. J. Med. Sci. 259, 114.
13. Simberkoff, M.L. "et al". (1976) J. Lab. Clin. Med. 87, 206.
14. McCabe, W.R. "et al". (1978) J. Infect. Dis. 138, 33.
15. Durack, D.T. and Beeson, P.B. (1977). Infect. Immun. 16, 213.
16. Gower, P.E. "et al". (1972) Clin. Sci. 43, 13.
17. Olling, S. "et al". (1973) Infection, 1, 24.
18. Miller, T.E. "et al". (1978). Clin. Exp. Immunol. 33, 115.
19. Melching, L. and Vas, S,I. (1971) Infect. Immun. 3, 107.
20. Nelson, B.W. and Roantree, R.J. (1967). J. Gen. Microbiol. 48, 179.
21. Muschel, L.H. and Larsen, L.J. (1970). Proc.Soc. Exp. Biol. Med. 133, 345.
22. Dlabac, V. (1968). Folia Microbiol, Praha 13, 439.
23. Feingold, D.S. (1969). J. Infect. Dis. 120, 437.
24, Osborn, M.J. and Rothfield, L.I. (1971). Microbial Toxins IV. Bacterial Endotoxins, Academic Press, London, pp 331-350.
25. Dlabac, V. (1969). Colloques internationaux du centre national de la recherche scientifique No. 174, pp 297-305.
26. Taylor, P.W. (1974). J. Clin. Pathol. 27, 626.
27. Taylor, P.W. "et al". (1972). Med. Lab. Technol. 29, 272.
28. Taylor, P.W. and Hughes, C. (1978). Infect. Immun. 22, 10.
29. Taylor, P.W. (1975). J. Gen. Microbiol. 89, 57.
30. Taylor, P.W. and Parton, R. (1977). J. Med. Microbiol. 10, 225.
31. Hildebrandt, J.F. "et al". (1978). Infect. Immun. 20, 267.
32. Glynn, A.A. and Howard, C.J. (1970). Immunology, 18, 331.
33. Bjorksten, B. "et al". (1976). J. Pediatr. 89, 892.
34. Olling, S. (1977). Scand. J. Infect. Dis. Suppl. 10, 1.
35. Taylor, P.W. (1976). J. Med. Microbiol. 9, 405.

36. Muschel, L.H. and Schmoker, K. (1966). J. Bacteriol. 92, 967.
37. Muschel, L.H. "et al". (1968). J. Bacteriol. 96, 1912.
38. Reynard, A.M. and Beck, M.E. (1976). Infect. Immun. 14, 848.
39. Reynard, A.M. "et al". (1978). Infect. Immun. 19, 861.
40. Fietta, A. "et al". (1977). Infect. Immun. 18, 278.
41. Morris, C. "et al". (1974). J. Bacteriol. 118, 855.
42. Schmidt, G. "et al". (1969). Eur. J. Biochem. 10, 501.
43. Schmidt, G. "et al". (1977). J. Gen. Microbiol. 100, 355.
44. Olson, A.C. "et al". (1969). Eur. J. Biochem. 11, 376.

© 1979 Elsevier/North-Holland Biomedical Press
Plasmids of Medical, Environmental and Commercial Importance
K.N. Timmis and A. Pühler, editors

PLASMID GENE THAT SPECIFIES RESISTANCE TO THE BACTERICIDAL ACTIVITY OF SERUM

KENNETH N. TIMMIS, ALBRECHT MOLL AND HIROFUMI DANBARA
Max-Planck-Institut für Molekulare Genetik, Berlin-Dahlem, BRD.

INTRODUCTION

It is well known that serum has bactericidal activity for some Gram-negative bacteria and it seems likely that this activity may represent an important host defence mechanism against infection. Current evidence suggests that the bactericidal activity of serum results primarily from the action of the complement system on the bacterial surface, followed by lysozyme attack of the mucopeptide layer. The complement system may be activated via the classical[1] or alternate[2,3] pathways. Although non-pathogenic bacteria and non-invasive pathogens are usually sensitive to the lethal activity of serum, invasive pathogens are frequently highly resistant[4,5,6]. These findings and studies with experimental animal models[7] indicate that the serum resistance of invasive bacteria is important for their pathogenicity[4,8].

Considerable interest has developed in recent years in the mechanisms whereby bacteria may be resistant to serum, although the experiments published so far are somewhat inconclusive[5,9]. Clearly, systematic analysis of the mechanism(s) of serum resistance requires detailed study of the bacterial determinant(s) responsible for this property, followed by a biochemical and physiological investigation of the gene product(s).

Recent reports that some plasmids substantially increase serum resistance levels of their host bacteria[10,11,12] indicated to us that the mechanism of at least one type of serum resistance would be readily amenable to analysis by the powerful genetic and molecular approaches, including gene cloning techniques, that have been developed for the analysis of plasmid functions[13]. We report here preliminary experiments on the analysis of a serum resistance determinant that is encoded by the FII incompatibility group conjugative antibiotic resistance plasmid R6-5.

MATERIALS AND METHODS

Most reagents, bacterial strains, and experimental methods have been described in detail elsewhere[14,15]. E. coli 59 is a wild-type strain that is relatively sensitive to the lethal activity of serum and which was isolated from the faeces of a healthy child[16,17]. E. coli 59 rif is a spontaneous mutant of E. coli 59 that is resistant to 100 µg/ml of rifampicin. CR34nal is a spontaneous mutant of E. coli K-12 CR34[18] that is resistant to 50 µg/ml of nalidixic acid.

RESULTS

Colorimetric serum resistance assay. In order to make a genetic analysis of serum resistance, it is necessary to obtain a series of serum-sensitive mutants from an appropriate serum-resistant host. This in turn, requires the availability of a rapid screening method for testing the serum resistance levels of bacteria. The most reliable method that has been described requires the construction of survival curves of bacteria exposed to serum. This method is extremely slow and laborious and quite unsuitable for screening large numbers of clones. We have therefore developed a rapid colorimetric serum sensitivity assay that is quantitative and that enables up to 100 strains of bacteria per day to be tested. The test has the further advantage that it can be scaled down in volume to a qualitative microtest, which can be performed in microtiter plates, allowing up to several thousand clones per day to be screened. The colorimetric assay is described in detail elsewhere[16], but in essence exponentially-growing bacteria are added to peptone-glucose broth containing the pH indicator bromothymol blue and serum. If the bacteria are killed, the medium remains green whereas if they survive and grow, they produce acid from the glucose and the medium turns yellow (Fig.1A). The degree of survival of the bacteria is reflected by the rapidity of acid production/ colour change which is monitored spectrophotometrically at OD_{500} (Fig.1B). In the micro assay, bacterial clones to be tested are transferred from agar plates to wells of a microtiter plate containing L-broth. When the bacteria reach exponential phase of growth, they are transferred with a multipoint inoculator to two microtiter plates containing peptone-glucose-bromothymol blue broth, with and without serum. The microtiter plates are incubated at 42°C

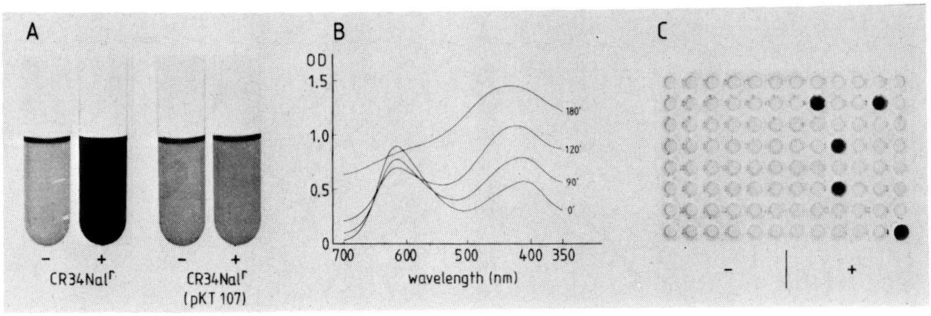

Fig. 1. Colorimetric serum resistance test. A. Survival of CR34Nal and CR34Nal(pKT107) bacteria in the presence (+) and absence (-) of 3% rabbit serum. B. Changes in absorption spectra during incubation of a culture of CR34Nal in peptone-glucose-bromothymol blue broth without serum. O' is the time of inoculation of the culture. C. Microcolorimetric test: serum resistance of different clones of CR34Nal containing mutagenized pKT107 plasmids.

for about 3 hours and the serum sensitivity/resistance of each clone is determined by comparing the colour changes in the wells of the plus and minus serum plates (Fig.1C). The micro assay is particularly useful for the preliminary screening of a large number of different strains, or of a large number of derivatives from a mutated strain.

Cloning and mapping of R6-5 serum resistance determinant. Plasmid R6-5 is a large (100 kb), low copy number, conjugative, incompatibility group FII plasmid that specifies resistance against chloramphenicol, kanamycin, fusidic acid, streptomycin, spectinomycin, sulfonamide and mercury salts (Fig.2). Its genetical and physical structure have been extensively investigated and its EcoRI and HindIII fragments have been cloned on the high copy number vector plasmids ColEl and pML21, respectively[14]. The R6-5 plasmid protects the very sensitive E. coli K-12 strain CR34 against low levels of serum and the moderately sensitive wild E. coli strain 59 against higher levels[17]. In order to locate the approximate position of the R6-5 serum resistance determinant, we measured the sensitivity/resistance of clones of CR34 containing individual R6-5 hybrid plasmids to 3% rabbit serum. Two hybrid plasmids were found to specify serum resistance. One plasmid carried R6-5 EcoRI fragment E-7, and the other carried R6-5 Hind fragment H-1. From the restriction endonuclease cleavage map of R6-5 (Fig.2) it can be seen that

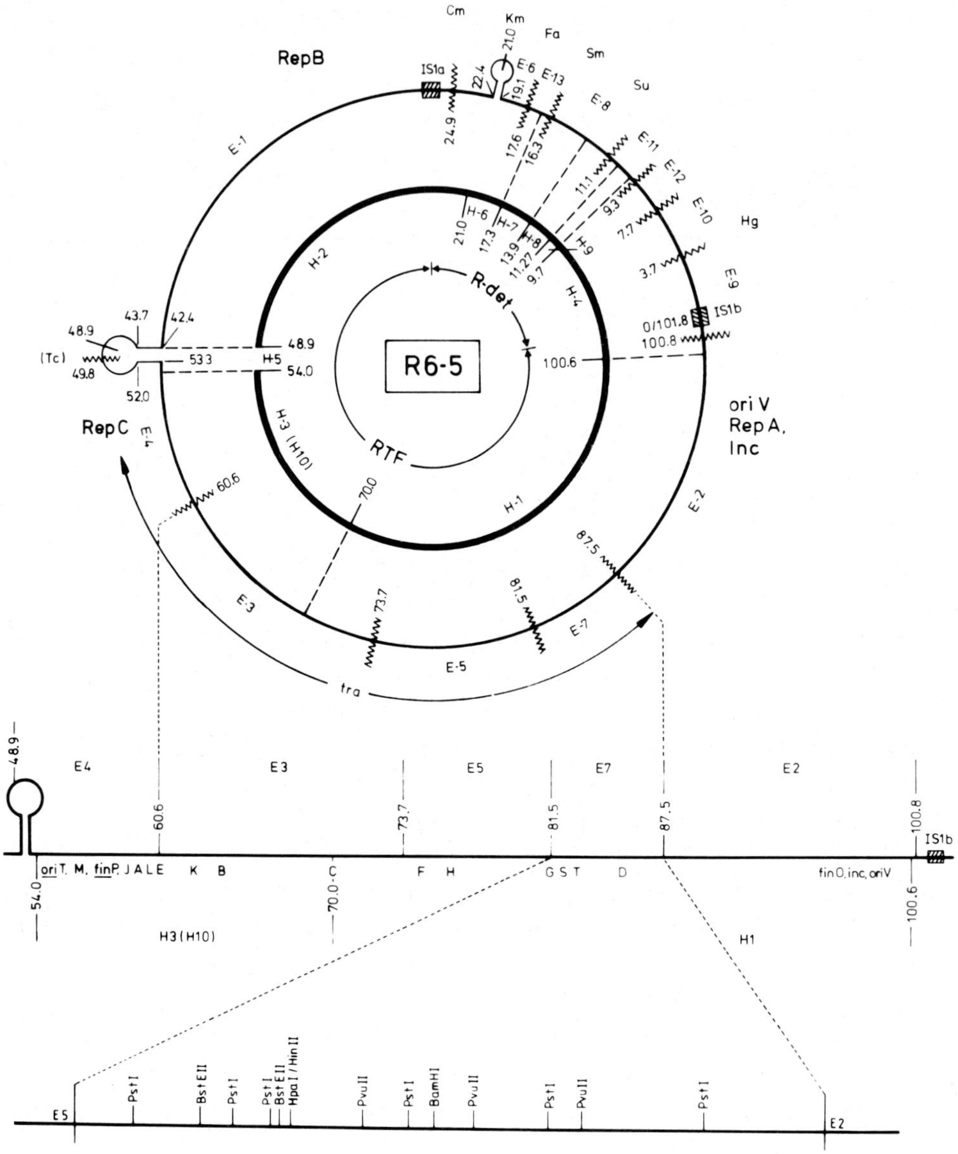

Fig. 2. Physical and genetical map of plasmid R6-5. The wavy and straight cross lines in the upper part of the figures indicate EcoRI and HindIII cleavage sites, respectively.

E-7 is entirely contained within the large H-1 fragment. The serum resistance determinant is therefore located on a 6kb DNA segment having termini with R6-5 coordinates 81.5 kb and 87.5 kb. This segment of DNA contains the traS and traT (surface exclusion), and traD genes of the transfer region of R6-5, i.e. the region responsible for its conjugal proficiency[19].

The ColE1 vector plasmid specifies only colicin E1-immunity as a selection function and is inconvenient for a variety of genetic manipulations. In order to have the E-7 fragment on a more easily selected vector plasmid, it was cloned into the tetracycline resistance pACYC184 vector plasmid[20]. One pACYC184/E-7 hybrid plasmid, designated pKT107, was selected for further studies and shown by the colorimetric serum resistance test (Fig.1A) and by the viable count method to confer upon host bacteria resistance to serum. This hybrid plasmid was subsequently analysed by cleavage with a variety of restriction endonucleases and a detailed restriction map of the E-7 fragment obtained (Fig.2).

Generation of serum-sensitive deletion mutant derivatives of the pKT107 plasmid. Although there are no PstI sites in the pACYC184 vector plasmid there are 6 in the R6-5 E-7 fragment (Fig.2). Partial cleavage of pKT107 plasmid DNA with PstI, followed by ligation and transformation into CR34Nal bacteria, was therefore used for the isolation of deletion derivatives of this plasmid that lack one or more PstI fragments. A total of 392 transformant clones were screened with the microcolorimetric serum sensitivity test described above and 22 were found to be serum-sensitive. Plasmid DNA was prepared from these clones, treated with PstI endonuclease, and analyzed by agarose gel electrophoresis (Fig.3). All serum-sensitive PstI deletion mutants of pKT107 were found to lack one of the two adjacent PstI fragments, P-4 and P-6, indicating that the serum resistance determinant is contained on these two fragments.

Relationship of the R6-5 serum resistance determinant to identified tra functions. Because complement and lysozyme attack the bacterial cell surface, it is to be expected that plasmid-induced serum resistance results from structural modification of the cell surface. Many of the tra proteins are known to be incorporated into the cell envelope. Surface exclusion, a phenomenon

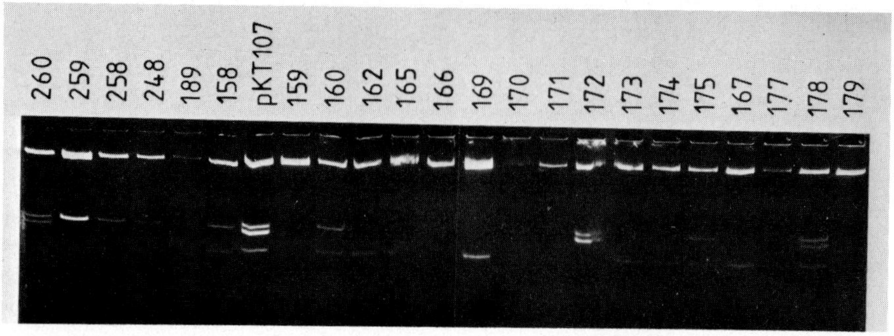

Fig. 3. Serum sensitive PstI- deletion derivatives of plasmid pKT107.

that results from the expression of the traS and traT genes, prevents plasmid-carrying bacteria from accepting from other bacteria by conjugation the same plasmid or a species closely related to that which is already carried. As its designation implies, this is a cell surface phenomenon and results, at least in part, from the large amount of the traT protein that is incorporated into the bacterial envelope. Localization of the serum resistance determinant in a region that contains the traS and traT genes, therefore suggested that serum resistance and surface exclusion might result from the same genetic determinants. Definitive identification of the serum resistance determinant as a tra gene requires, however, the genetic analysis of a number of serum sensitive point mutants of the pKT107 plasmid.

pKT107 DNA was treated in vitro with the mutagen hydroxylamine and transformed into CR34Nal bacteria. 166 transformants were subsequently screened with the micro colorimetric test for serum resistance and 22 clones were tentatively identified as serum sensitive. By means of the quantitative serum resistance assay, the 22 clones were subsequently confirmed as carrying plasmid derivatives mutated in the serum resistance determinant. Because hydroxylamine tends to produce point mutations in DNA, it was expected that the majority of the hydroxylamine-induced serum sensitive pKT107 derivatives would contain mutations only in the serum resistance gene. Examination of such derivatives for loss of specific conjugation functions should therefore demonstrate conclusively whether or not the serum resistance determinant is a known tra gene. The genetic analysis of these pKT107 mutant plasmids is currently in progress.

We have also examined the ability to confer resistance to serum upon host bacteria of a series of mini-R100 plasmid derivatives that carry insertions in traS, traT and traD (kindly provided by P.Manning); only the traT insertion mutant of R100 failed to express serum resistance. This result therefore tentatively identifies the serum resistance gene as traT, although analysis of the large number of pKT107 mutants will be required for a firm conclusion to be drawn.

DISCUSSION

Bacterial resistance to the lethal activity of serum is thought to contribute to the pathogenicity of some invasive bacteria. The finding that serum resistance can be encoded by antibiotic resistance plasmids is consistent with the increasing incidence of antibiotic resistance plasmids in pathogenic bacteria and the tendency for genetic determinants of a variety of pathogenicity functions to be plasmid-borne[23]. Nevertheless, at this stage it would be premature to conclude that all forms of serum resistance are plasmid-encoded, or that the serum resistances encoded by different plasmids will be found to result from identical genes and mechanisms. Furthermore, experiments by Taylor[5,9] suggest that additional specific structural components of the cell envelope are necessary for full expression of plasmid-mediated serum resistance. A detailed study of serum resistances determined by different plasmids, and analysis of the interaction of the serum resistance gene products with cell surface components and complement fractions, will be necessary for a basic understanding of this bacterial property.

Plasmid R6-5 has been shown to increase the levels of resistance of host bacteria towards the lethal activities of serum. Physical mapping of the serum resistance determinant of R6-5 and analysis of the serum resistance properties of a series of mini R100 insertion mutants suggest that the serum resistance determinant consists partly or solely of the surface exclusion gene, traT. TraT protein is synthesized in large quantities in plasmid-carrying bacteria and is incorporated into the outer membrane where it constitutes a major component.

The finding that R6-5-specified serum resistance is due to a conjugation protein is not surprising. The conjugal transfer of plasmids between bacteria is a cell surface phenomenon. It involves the transportation of single-stranded plasmid DNA molecules through the cell wall and membrane of donor and recipient bacteria and requires the interaction of specific surface structures of both partners. Conjugation requires the expression of at least 15 plasmid genes and the products of most of these are located in the bacterial membranes[24]. It was therefore to be anticipated, that one or more of the *tra* gene products would produce profound changes in the structure and properties of the cell envelope. It is improbable, however, that the elevated serum resistance of bacteria carrying conjugative plasmids is a fortuitous consequence of their fertility functions and it is more likely that one or more specific *tra* function has evolved to provide this advantageous property of plasmid-carrying pathogens. Precisely how advantageous the R6-5-specified serum resistance is will require pathogenicity studies in animal models.

The rapid colorimetric serum resistance assay described here will enable the large scale screening of bacterial isolates from clinical specimens for their serum resistance/sensitivity and facilitate future investigations and epidemiological studies of this property. The high copy number pACYC184 hybrid plasmid carrying the R6-5 serum resistance determinant will be a useful DNA hybridization probe to screen plasmids isolated from serum resistant pathogens for the presence of this determinant. Such molecular epidemiological experiments will reveal how prevalent this type of resistance is in such bacteria.

ACKNOWLEDGEMENTS

We thank P.Manning, P.Taylor, F.Cabello, R.Stephan and B.Hansen for helpful advice and stimulating discussions. H.Danbara is a postdoctoral fellow of the Alexander von Humboldt-Stiftung.

REFERENCES

1. Inoue, K., Yonemasu, K., Takamizawa, A. and Amano, T. (1968) Biken J. 11, 203-206.
2. Frank, M.M., May, J.E. and Kane, M.A. (1973) J. Inf. Dis. 128, S176-S181.
3. Traub, W.H. and Kleber, I. (1976) Infect. Immun. 13, 1343-1346.
4. Olling, S. (1977) Scand. J. Infect. Dis. Suppl. 10, 1-40.
5. Taylor, P.W. (1976) J. Med. Microbiol. 9, 405-421.
6. Simberkoff, M.S., Ricupero, I. and Rahal, Jr. J.J. (1976) J. Lab. Clin. Med. 87, 206-217.
7. Archer, G. and Fekety, F.R. (1976) J. Infect. Dis. 134, 1-7.
8. Schoolnik, G.K., Buchanan, T.M. and Holmes, K.K. (1976) J. Clin. Invest. 58, 1163-1173.
9. Taylor, P.W. (1975) J. Gen. Microbiol. 89, 57-66.
10. Reynard, A.M. and Beck, M.E. (1976) Infect. Immun. 14, 848-850.
11. Fietta, A., Romero, E. and Siccardi, A.G. (1977) Infect. Immun. 18, 278-282.
12. Taylor, P.W. and Hughes, C. (1978) Infect. Immun. 22, 10-17.
13. Timmis, K.N., Cohen, S.N. and Cabello, F. (1978) Prog. Mol. Subcell. Biol. 6, 1-58.
14. Timmis, K.N., Cabello, F. and Cohen, S.N. (1978) Mol. Gen. Genet. 162, 121-137.
15. Andrés, I., Slocombe, P.M., Cabello, F., Timmis, J.K., Lurz,R., Burkardt, H.J. and Timmis, K.N. (1979) Mol. Gen. Genet. 168, 1-25.
16. Moll, A., Cabello, F. and Timmis, K.N. (1979) submitted for publication.
17. Moll, A. and Timmis, K.N. (1979) submitted for publication.
18. Bachmann, B.J. (1972) Bacteriol. Revs. 36, 525-557.
19. Achtman, M., Kusecek, B. and Timmis, K.N. (1978) Mol. Gen. Genet. 163, 169-179.
20. Chang, A.C.Y. and Cohen, S.N. (1978) J. Bacteriol. 134, 1141-1156.
21. Kennedy, N., Beutin, L., Achtman, M., Skurray, R., Rahmsdorf, U. and Herrlich, P. (1977) Nature 270, 580-585.
22. Willetts, N.S. and McIntire, S. (1978) J. Mol. Biol. 126, 525-549.
23. Cabello, F. and Timmis, K.N. (1979) this volume.
24. Manning, P.A. and Achtman, M. (1979) in: Bacterial Outer Membranes, Inouye, M., ed., Wiley and Sons, in press.

Plasmids of Medical, Environmental and Commercial Importance
K.N. Timmis and A. Pühler, editors

DETERMINANTS OF PATHOGENICITY OF E. COLI K1

FELIPE C. CABELLO

New York Medical College, Valhalla, NY 10595 U.S.A.

INTRODUCTION

The factors that make *E. coli*, usually a inhabitant of our intestinal tract, invasive and able to produce extraintestinal infection are not well known[1]. Several phenotypical traits have been associated to the ability of some strains to invade, namely: synthesis of Col V, hemolysin production, hemagglutination, K1 antigen, dulcitol fermentation[2,3,4]. These properties can be coded by genes located in plasmids or in the chromosome[1]. The fact that these traits are carried more frequently by *E. coli* strains isolated from disease than from *E. coli* found in normal intestinal flora makes the theory, that pathogenic *E. coli* have special properties a likely one[2].

Neonatal meningitis is an acute bacterial infection usually produced by coliform bacilli and group B streptococci. Epidemiological studies demonstrated that 77% of the strains isolated from the spinal fluid of neonates had K1 antigen. Further studies indicated that these strains, with differences among them, were more virulent for the mice that non K1 *E. coli* and that antibodies against K1 antigen were protective. It was also found that K1 antigens were structurally identical to the meningococcal group B polysaccharide[3,5]. Epidemiological data also showed that in healthy newborns the carrier rate, for *E. coli* K1, fluctuates between 7% to 38% in the same nursery[6]. The mortality of neonatal *E. coli* meningitis is close to 35% and the sequelae among survivors is around 50%, indicating that new forms of prevention and/or treatment are needed[6].

The existance of fluctuations in the LD50 for mice, of different *E. coli* K1 strains isolated from spinal fluid and the variation in the rate of healthy newborns that carry *E. coli* K1 strains may indicate that, other bacterial factors such as Col V could be influencing the basic pathogenic potential of *E. coli* K1. Moreover, the fact that two different bacterial species carry the same antigenic determinant suggests that these genes could be located in plasmids and/or transposons[7].

MATERIAL AND METHODS

E. coli K1 strains were from the collection of H.W. Smith, J. Pitt, and G.H. McCracken. Meningococcal group B antiserum was a gift of J.B. Robbins. K1 antigen specific phages were a gift of R. Rowe. Presence of K1 antigen was detected by immunoprecipitation in minimal agar as described by Pitt[8], and using K1 specific phages as described by Gross[9]. Colicin V production, hemolytic activity, and mannose resistant hemagglutination was determined as described[10,11]. Col V negatives variants were obtained by treatment with SDS[12] K1 antigen negatives variants were obtained isolating K1 bacteriophage resistant mutants, and the absence of K1 antigen confirmed by immunoprecipitation[8]. LD50 were determined inoculating intraperitoneally adult male mice with different bacterial concentrations mixed with 1% gastric hog mucine[13]. Serum resistance was measured as described by Pitt[8] Resistance to phagocytosis was determined using mice peritoneal macrophages[14].

RESULTS AND DISCUSSION

Properties related to pathogenicity.

To investigate the presence of other properties, besides K1 antigen, that could increase the pathogenic potential of *E. coli* K1, we tested *E. coli* K1 strains isolated from blood, spinal blood and stools for production of colicin V, hemolytic activity and hemagglutination. As it can be seen in Table 1, *E. coli* K1 strains carry these properties with high frequency but, no marked differences with respect to the frequency of these characters among strains isolated from different sources could be detected. Then, it seems that the ability of some E. coli K1 to invade the blood and the central nervous system is not influenced by the presence of these traits. Because some of these properties are plasmid coded and the strains studied are not fresh isolates, before reaching a final conclusion regarding this aspect, more fresh isolates will be studied.

TABLE 1

INCIDENCE OF COLICIN V PRODUCTION, HEMOLYSIN AND HEMAGGLUTINATING ACTIVITY AMONG E. COLI K1 FROM DIFFERENT SOURCES

	PERCENTAGE OF STRAINS POSITIVE FOR		
Source of Strains	Colicin V	hemolysin	hemagglutination
Stools	50	20	50
Blood	66	11	55
Spinal Fluid	36	9	30

Contribution of Col V to E. coli K1 pathogenicity.

To ascertain the contribution of Col V to the pathogenicity of E. coli K1 strains, the LD50 for the mice of one E. coli K1 isolate and its K1 negative and Col V negative variants was determined. As shown in Table 2, the lost of K1 antigen and Col V is accompanied by a decrease in pathogenicity. This decrease is almost total when K1 antigen is lost, and Col V by itself does not seem to increase the pathogenicity of these strains. Nonetheless the lost of the Col V character on a K1 positive background is enough to decrease the pathogenicity of this strain for the mice. This finding points out that in these *E. coli* K1 strains other traits, such as Col V or hemolysin may modulate their basic pathogenicity. We are now studying the influence of hemolytic and hemagglutinating activity on the pathogenicity of *E. coli* K1 strains.

TABLE 2

LD50 FOR THE MICE OF ONE *E. COLI* K1 AND ITS $K1^-$ AND COL V^- DERIVATIVES

Strains	LD50 (bacteria/mouse
$K1^+$ Col V^+	3.5×10^3
$K1^+$ Col V^-	9×10^4
$K1^-$ Col V^+	$>2 \times 10^6$
$K1^-$ Col V^-	$>2 \times 10^6$

The Nature of the Col V Mediated Pathogenicity Increase

Non-toxigenic, extracellular bacteria are usually eliminated by host defenses through bacteriolysis and/or phagocytosis. To measure the influence of the Col V character upon *E. coli* K1 ability to survive host defenses, we determined the resistance to serum and phagocytosis of the Col V and K1 negatives variants described above. It can be seen in Table 3 that the absence of Col V does not decrease the resistance to phagocytosis or serum of this *E. coli* K1 strain, as the absence of K1 antigen does.

TABLE 3

RESISTANCE TO SERUM AND PHAGOCYTOSIS OF ONE E. COLI K1 STRAIN AND ITS K1- AND COL V- DERIVATIVES

Strains	Serum resistance[a] surviving bacteria (%)	Resistance to phagocytosis[b] Surviving bacteria (%)
$K1^+$ Col V^+	100	100
$K1^+$ Col V^-	100	100
$K1^-$ Col V^+	8	10

[a] Results of four experiments with concentrations of serum of 40% and 80%.
[b] Results of three experiments.

It seems clear that Col V is able to increase to basic pathogenic potential of E. coliK1 for the mice and that this increase is not mediated by an increase in the resistance to serum or phagocytosis but could be mediated by an increased uptake of iron as shown by Williams and George (this volume). More research is needed to determine if the presence of other characters such as hemagglutinating and hemolytic activity modify the pathogenic potential of E. coli K1. Experiments to demonstrate that the K1 genes could be carried by a plasmid and/or a transposon have not been successful until know.

In summary it seems clear that other bacterial factors, besides K1 antigen, influence the pathogenic potential of E. coli K1 strains, work is in progress to see if these factors are relevant to the different ability of E. coli K1 strains to colonize and invade. This knowledge will be useful to understand E. coli K1 epidemiology.

ACKNOWLEDGEMENTS

This research was supported by a Biomedical Research Support Grant to New York Medical College. I gratefully acknowledge the help of Dr. I. Rappaport with the phagocytosis assay.

REFERENCES

1. So, M. and Falkow, S. (1977) Recombinant Molecules: Impact on Science and Society. Raven Press, N. York pp 91-105.

2. Minshew, B., Jorgensen, J., Swanstrum, M., Grootes-Reuvecamp, G.A. and Falkow, S. (1978) J. Infect. Dis. 137, 648.

3. Robbins, J.B., McCracken, H.M., Gotschlich, E.C., Ørskov, F., Ørskov, J. and Hanson, L.A. (1974) New Engl. J. Med. 290, 216.

4. Guze, L.B., Montgomerie, J.Z., Potter, C.S. and Kalmason, G.M. (1973), Yale J. Biol. Med., 46, 203.

5. Sarff, L.D., McCracken, G.H., Schiffer, M.S., Glode, M.P., Robbins, J.B., Ørskov, I. and Ørskov, F. (1975) Lancet, I, 7916.

6. Schiffer, M.S., Oliveira, E., Glode, M.P., McCracken, G.H., Sarff, L.M. and Robbins, J.B. (1976) Pediat. Res., 10, 82.

7. Elwell, L.P., Saunders, J.R. Richmond, M.H. and Falkow, S. (1977) J. Bact. 131, 356.

8. Pitt, J. (1978) Infec. Immun. 22, 219.

9. Gross, R.J. Cheasty, T. and Rowe, B. (1977) J. Clin. Microbiol., 6, 548.

10. Minshew, B.H., Jorgensen, J., Counts, G.W., and Falkow, S. (1978) Infect. Immun. 20, 50.

11. Evans, D.G. Evans, D.J. and Tjoa, W. (1977) Infect. Immun. 18, 330.

12. Tomoeda, M., Inuzuka, M. Kubo, N. and Nakamura, S. (1968) J. Bact., 95, 1078.

13. Wolberg, G. and DeWitt, C.W. (1969) J. Bact. 100, 730.

14. Stossel, T.P. and Taylor, M. (1976) Manual of Clinical Microbiology. A.S.M. Washington, D.C. pp 148-154.

© 1979 Elsevier/North Holland Biomedical Press
Plasmids of Medical, Environmental and Commercial Importance
K.N. Timmis and A. Pühler, editors

ColV PLASMID-MEDIATED IRON UPTAKE AND THE ENHANCED VIRULENCE OF INVASIVE STRAINS OF *ESCHERICHIA COLI*

PETER H. WILLIAMS AND HELEN K. GEORGE
Department of Genetics, University of Leicester, Leicester LE1 7RH (England)

INTRODUCTION

Iron is an essential trace element for bacterial growth. However, in nature it exists either as insoluble, high molecular weight, polynuclear aggregates[1], or complexed with iron-binding proteins such as transferrin in the tissues and fluids of the bodies of animal hosts[2]. Bacterial growth, therefore, depends upon successful sequestering of iron from these sources, and subsequent efficient transport across bacterial membranes[3]. In *Escherichia coli*, conditions of iron stress induce the synthesis and excretion of the iron-chelating agent enterochelin[4], and the increased production of membrane proteins involved in active transport of the ferric-enterochelin complex[5]. Furthermore, when citrate ions are present in the growth medium, iron may be taken into cells of some strains of *E.coli* as a ferric-citrate chelate[6] by a route which also involves inducible membrane components[5]. In addition, enteric bacteria are able to transport iron complexed with siderophores such as ferrichrome[7,8] excreted into the environment by other micro-organisms.

In this paper we report the existence of a novel iron-sequestering system specified by ColV plasmids. These plasmids determine the production of the specific antibacterial protein colicin V, and are commonly found in strains of *E.coli* isolated from cases of bacteraemia in man and a number of domestic animals[9]. Colicinogenic clinical isolates show a marked reduction in virulence when cured of the resident ColV plasmid, while acquisition of a ColV plasmid, even by laboratory strains of *E.coli*, is accompanied by a significant enhancement of the ability to survive in the tissues and body fluids of an infected host[9,10]. We propose that the operation of an efficient plasmid-promoted iron transport system is an important component of the pathogenicity of colicinogenic *E.coli* strains.

MATERIALS AND METHODS

Bacteria. Strains of *E.coli* K-12 employed in this study are described in

Table 1. Pathogenic *E.coli* strains, which were kindly supplied by H.Williams Smith, were isolated from the blood or internal organs of human (H247), calf (B188), pig (P72) or chicken (F70) subjects suffering from bacteraemia. These bacterial strains carry plasmids designated ColV-H247, ColV-B188, ColV-P72 and ColV-F70 respectively. Cured derivatives of the bacteraemic strains have been isolated using ethidium bromide[11] as the curing agent (strains designated H247V$^-$ etc.)

TABLE 1
E.coli K-12 STRAINS

Designation	Characteristics	Source
KH533	Carries prototype plasmid ColV-K30[a]	K. Hardy
W3110	*thyA rpsL*	I.B.Holland
W3110(ColV-K30)	Derivative of W3110 carrying ColV-K30	Conjugation
W3110(ColV-H247)	Derivative of W3110 carrying ColV-H247	Conjugation
W3110(ColV-P72)	Derivative of W3110 carrying ColV-P72	Conjugation
AN1937	*entA ara lac leu mtl proC rpsL trp tonA thi supE xyl*	G.S.Plastow
AN1937(ColV-K30)	Derivative of AN1937 carrying ColV-K30	Conjugation
AN1937(ColV-H247)	Derivative of AN1937 carrying ColV-H247	Conjugation
AN1937(ColV-P72)	Derivative of AN1937 carrying ColV-P72	Conjugation
1559	*ara lac leu met proC rpsL trp thi his metB arg*	I.B.Holland
GUC6	*tonA tonB thr leu thi supE*	I.B.Holland
LG1316	Derivative of AN1937 having *tonB trp*$^+$ of GUC6	P1 transduction
LG1316(ColV-K30)	Derivative of LG1316 carrying ColV-K30	Conjugation

[a]Plasmid ColV-K30[12] is derepressed for conjugal transfer.

Growth conditions. Defined growth medium was M9 minimal medium[13] containing glucose (0.2% w/v), thiamine (50 µg/ml) and required amino acids (20 µg/ml). Rate of mass increase was determined by absorbance measurements (450 nm) on a Spectronic 20 spectrophotometer (Bausch & Lomb). Immunoglobulin-free blood for use as a bacterial growth medium was obtained from a 3 hr old colostrum-deprived male Jersey calf[10]. Heparin (Sigma, "endotoxin-free") was added (100 units/ml) to prevent coagulation. Bacterial growth was determined by viable counts on MacConkey agar.

Colicin tests. Colicinogenicity was determined by transferring small inocula of individual clones to a lawn of colicin-sensitive bacteria (e.g. strain W3110) on nutrient agar. Zones of inhibition in growing lawns indicated colicin production.

Iron uptake studies. Growth medium was made up as described by Langman *et al.*[14], precautions being taken to minimise the iron content. Strains to be tested were grown overnight at 37C with aeration in iron-depleted medium containing tryptophan and other required nutrients, then inoculated into fresh medium and grown over 5 mass-doublings to a density of approximately 2.10^8 cells/ml. The bacteria were harvested by centrifugation, washed twice with iron-depleted medium lacking tryptophan, and resuspended at a cell density of 5.10^8 cells/ml in the same medium. Carrier-free $^{55}FeCl_3$ (Radiochemical Centre, Amersham, > 50 µCi/mg Fe) was added (1 µCi/ml) to shaken cell suspensions; samples were taken at intervals and washed on membrane filters (Sartorius, pore size 0.45 µm) with medium containing sodium citrate (100 µm) Filters were air-dried, and the level of radioactivity counted in a liquid scintillation counter (Packard model 3255).

Preparation and analysis of outer membranes. Cell envelopes were prepared from sonicated pellets (approximately 10^{10} cells) of bacteria grown in M9 medium as described by Churchward and Holland[15]. Incubation with 0.5% (w/v) Sarkosyl NL97 (Geigy) solubilised inner membrane, and outer membrane material was collected by centrifugation at 40,000 x g for 45 min. Sodium dodecyl sulphate-polyacrylamide gel electrophoresis (SDS - PAGE) was performed using 10% gels with an acrylamide monomer:dimer ratio of 73.3:1. Gels were stained with Coomassie blue, destained with an electrophoretic destainer (Biorad, model 1200), and photographed.

Detection of enterochelin. Bacteria were grown in M9 medium to a density of 2.10^8 cells/ml, and pelleted by centrifugation. Cell-free supernatants were tested for absorbance in the wavelength range 300-400 nm in a Unicam SP800 spectrophotometer.

RESULTS

Effect of transferrin on bacterial growth in M9 medium

Adequate iron for bacterial growth is usually present in defined media, added as an impurity in the component chemicals. However, incorporation of human transferrin (Sigma) at approximately 3 µm in M9 medium was found to limit the growth rate of *E.coli* strains W3110 (fig. 1c), presumably due to the conversion of free iron to a predominantly unavailable complexed form. Addiion of ferric ions in excess of the level required for saturation of the trans -ferrin restored the growth rate to that observed in the absence of the iron-binding protein.

The growth rate of the colicinogenic human bacteraemic strain H247 in M9 medium, however, was unaffected by the same concentration of transferrin that was inhibitory for strain W3110 (fig. 1a). On the other hand, a derivative of strain H247 cured of the ColV plasmid (strain H247V⁻) grew significantly more slowly in the presence of transferrin than when iron was freely available (fig. 1b). Similar results were obtained with colicinogenic bacteraemic strains B188, F70 and P72, and with the cured derivatives B188V⁻, F70V⁻ and P72V⁻ respectively.

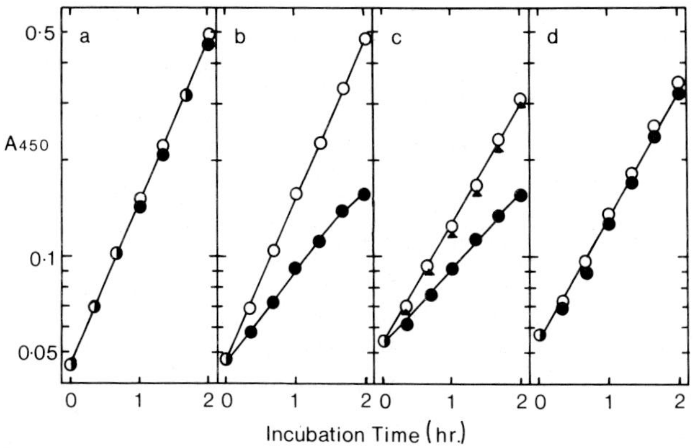

Fig.1 Effect of transferrin on bacterial growth. The rate of mass increase of strains (a) H247, (b) H247V⁻ (c) W3110 and (d) W3110(ColV-H247) was measured in M9 medium (O—O), in medium containing 3 μM transferrin (●—●), or in medium supplemented with both transferrin and 10 μm $FeCl_3$ (▲).

Confirmation of the involvement of ColV plasmids in overcoming the growth limiting effect of transferrin was obtained from a derivative of strain W3110 carrying the ColV plasmid from the bacteraemic strain H247. Acquisition of the plasmid resulted in a sparing of the inhibitory effect of iron-chelation on bacterial growth (fig. 1d). Similar results were observed with strains W3110-(ColV-K30) and W3110(ColV-P72).

Bacterial growth in immunoglobulin-free blood

Samples of blood were inoculated with mixtures of isogenic colicinogenic and non-colicinogenic strains at low cell density, and incubated with vigorous shaking at 37C. Samples taken at intervals were analysed for total bacterial growth and the proportion of colicinogenic bacteria (fig. 2) . In the experiment shown, the colicinogenic strain H247, which was the minority (14%)

component of the inoculum, predominated after 8 hr incubation unless excess iron was added to saturate iron-binding proteins in the blood. These data are consistent with a differential growth-limiting influence of complexed iron on the two strains H247 and H247V$^-$ that comprised the inoculum. The generation time of the ColV$^+$ strain was approximately 35 min, and that of the plasmid-free strain approximately 50 min. When iron was present in

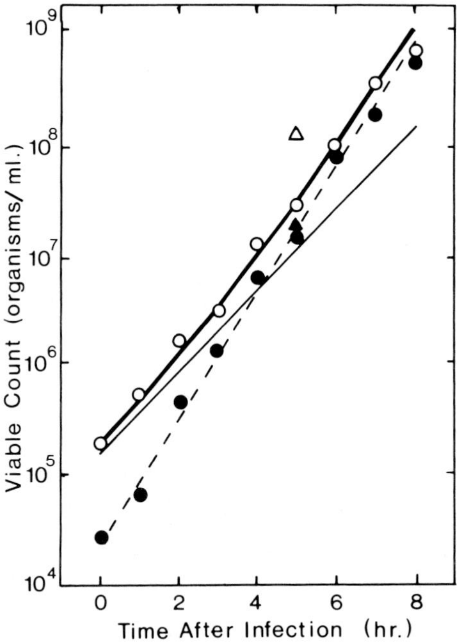

Fig.2. Bacterial growth in blood. Samples of blood were inoculated with a 1:7 mixture of strains H247 and H247V$^-$. Total bacterial growth was estimated by viable counts (○), and the proportion of ColV$^+$ bacteria was determined by testing at least 500 individual colonies from each time point (●) The continuous thin line represents the growth of the non-colicinogenic component of the inoculum. Total (Δ) and ColV$^+$ bacteria (▲) were also determined in blood samples supplemented with $FeCl_3$ (100 μm).

excess the relative proportions of the two strains remained unchanged, indicating that plasmid transfer and colicin killing of the plasmid-free strains were insignificant in the incubation conditions used.

Effect of ColV plasmids on the growth of an enterochelin-deficient mutant

Strain 1559 (*entA*$^+$) was unaffected by the addition of sodium citrate (10 mM) to the minimal growth medium (fig. 3a). On the other hand, strain AN1937, deficient in enterochelin biosynthesis (*entA*) required citrate for growth, as a vehicle for the uptake of iron (fig. 3b). Starting from a nutrient broth culture of strain AN1937, one or two cycles of dilution and growth in minimal medium lacking citrate were required, presumably to exhaust intracellular pools of iron, before the manifestation of growth inhibition as shown in

Figure 3b. A derivative of strain AN1937 carrying the plasmid ColV-K30, however, continued growing normally through several cycles of dilution in the absence of citrate (fig. 3c). Identical results were obtained with derivatives of AN1937 carrying ColV plasmids of bacteraemic strains H247 and P72. By contrast, strain LG1316(ColV-K30), a *tonB* derivative of the *entA* mutant AN1937(ColV-K30), did not grow in the absence of citrate.

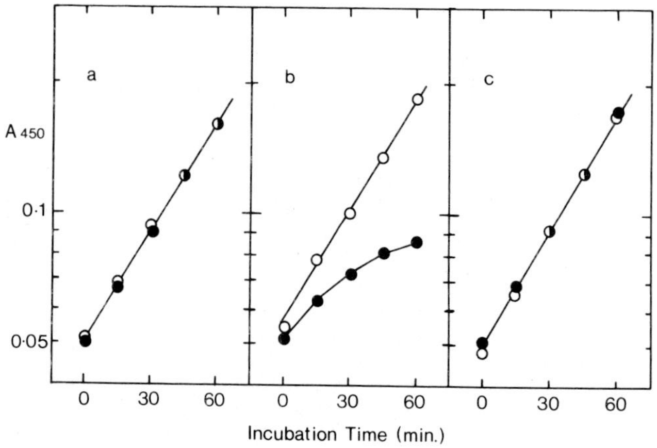

Fig.3. Effect of citrate on bacterial growth. The rate of mass increase of strains (a) 1559, (b) AN1937 and (c) AN1937(ColV-K30) was measured in M9 medium (●—●), or in medium containing 10 mM sodium citrate (○—○).

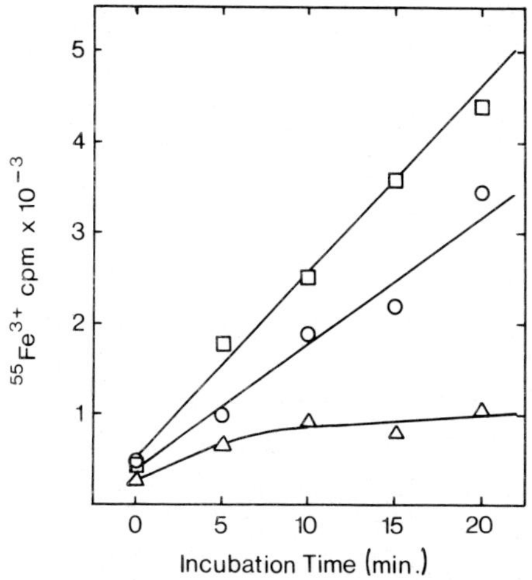

Fig.4. Uptake of ^{55}Fe by bacterial cells. Incorporation of radioactive iron into cells of strains 1559 (○), AN1937 (Δ) and AN1937(ColV-K30) (□), grown in iron-depleted medium.

Uptake of radioactive iron

Measurements of the incorporation of radioactive iron into non-growing bacterial cultures in iron-depleted medium confirmed that the active growth of ColV^{+} strains of the mutant AN1937 in the absence of citrate was due to plasmid-directed restoration of the ability to transport iron into cells (fig.4). In the enterochelin-deficient mutant iron incorporation was limited to a low background level of non-specific uptake. In the ColV^{+} strain, however, iron incorporation was more efficient than in the Ent^{+} control strain 1559.

Induction of outer membrane proteins

In conditions of iron stress, the synthesis of a number of outer membrane proteins is induced in strains of *E.coli*[5]. These can be detected by SDS - PAGE (fig.5). In strain AN1937 growing in conditions where iron was freely available, these proteins were present at low levels (sample a). They were induced, however, when endogenous pools of iron had been exhausted by repeated cycles of dilution and growth in minimal medium lacking citrate (sample b), that is, just prior to growth inhibition as shown in Figure 3. Under exactly the same culture conditions, however, synthesis of the characteristic outer membrane proteins was not induced in the colicinogenic derivative AN1937(ColV-K30) (samples c and d), indicating that the intracellular iron concentration remained sufficiently high to maintain repression. These data show that the presence of the ColV plasmid relieved iron stress in the enterochelin deficient mutant.

Enterochelin biosynthesis

Enterochelin (a cyclic trimer of 2,3-dihydroxy-N-benzoyl-L-serine, DBS) and the products of its hydrolysis, monomer, dimer and linear trimer forms of DBS[3,4] are excreted into bacterial growth media in which iron is unavailable, and can be detected spectrophotometrically[16] by their characteristic absorption maxima in the range 330-342 nm (fig.6). The possibility exists that the complementation of deficiency in iron uptake exhibited by mutant strain AN1937 could be due to plasmid-specified enterochelin production. Table 2 shows the type of absorption spectrum observed with a number of *E.coli* K-12 strains grown in M9 minimal medium with iron freely available, or in medium in which iron was complexed with transferrin (3 µm). Enterochelin and the DBS forms were not observed in the culture medium of a strain wild-type for enterochelin biosynthesis and metabolism unless iron was in the relatively unavailable bound form with transferrin. Obviously the *entA* mutant strain did not produce enterochelin in either circumstance. Furthermore, there was no evidence for the induction of enterochelin synthesis by three ColV^{+} derivatives of strain AN1937.

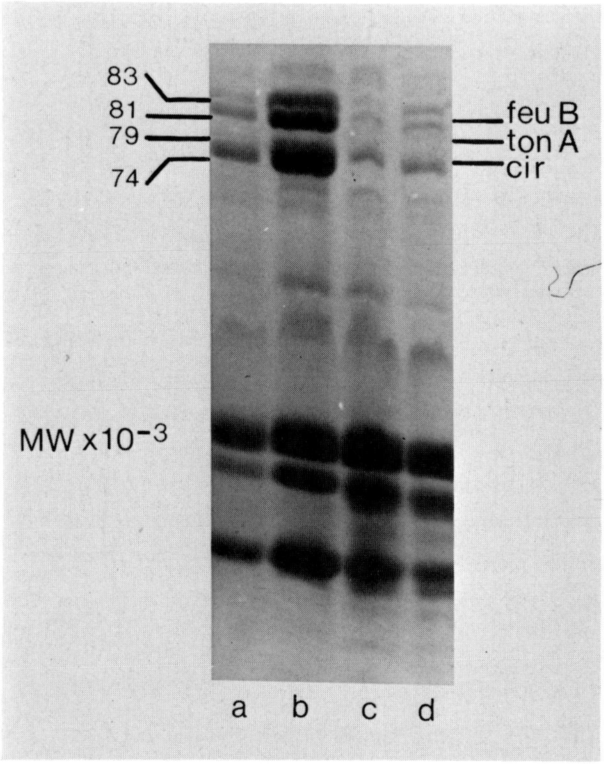

Fig. 5. Induction of specific outer membrane proteins in response to iron stress. Outer membranes of strain AN1937 grown in conditions of iron availability (a) or iron depletion (b), and from strain AN1937(ColV-K30) grown under identical conditions (c and d, respectively) were analysed by SDS-PAGE, and stained with Coomassie blue.

TABLE 2

ABSORPTION SPECTRA (300-400 nm) OF BACTERIAL CULTURE MEDIA[a]

Strain	M9 minimal medium	M9+transferrin[b]	M9+transferrin+$FeCl_3$[b]
W3110 (*entA*$^{+}$)	type a	type b	type a
AN1937 (*entA*)	type a	type a	(not done)
AN1937(ColV-K30)	type a	type a	(not done)
AN1937(ColV-H247)	type a	type a	(not done)
AN1937(ColV-P72)	type a	type a	(not done)

[a]Types a and b absorption spectra shown in Fig.6. [b]3 μM transferrin, 10 μM Fe

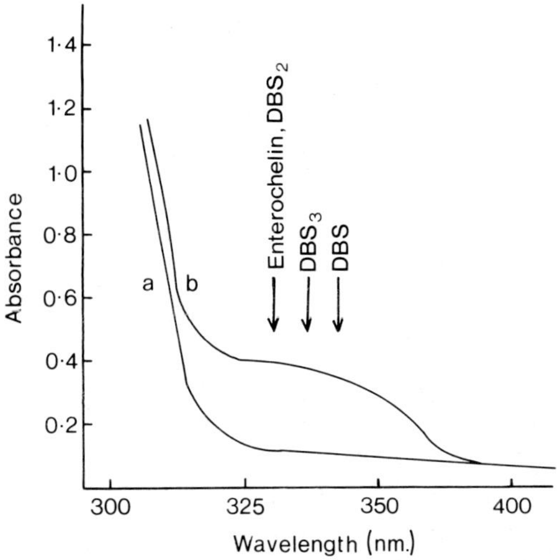

Fig.6 Spectrophotometric analysis of bacterial culture media. Curve a represents the absence, and curve b the presence of enterochelin (330 nm) and DBS monomer (342 nm), dimer (330 nm) and trimer (336 nm).

DISCUSSION

There is considerable evidence in the literature that the concentration of ferric cations in the tissues and fluids of the body is crucial to the outcome of the conflict between establishment of a bacterial infection and its suppression by the host animal[2,17,18]. The body makes significant metabolic changes in response to an infection that more or less effectively deprive invading bacteria of iron that is essential for their growth. For instance, infected subjects may display *inter alia*, decreased intestinal absorption specifically of iron salts[19], or reduced levels of iron in the blood plasma[20]. These natural defence reactions can be overcome in experimental infections by injection of iron compounds directly into the body at a concentration in excess of that required to saturate iron-binding proteins[17]. Furthermore, certain clinical conditions in which iron metabolism is grossly altered are frequently associated with increased susceptibility to bacterial infections[2].

It has been clearly demonstrated that ColV plasmids significantly enhance the virulence of *E.coli* strains which harbour them by increasing their ability to survive and proliferate in the body of an infected animal[10]. In this paper, we present evidence that colicinogenicity represents a selective advantage under these conditions due to an efficient ColV plasmid-mediated iron sequestering mechanism.

The phenomenon was initially demonstrated by subjecting bacterial strains to iron stress. This was achieved, not by depleting medium of iron, but by complexing the available iron with human transferrin, a situation that more closely resembles that *in vivo*. In these conditions the growth rate of a plasmid-free *E.coli* K-12 strain was markedly reduced, and the fact that addition of excess iron relieved growth inhibition confirms that the presence of transferrin in defined medium makes iron relatively unavailable for normal bacterial growth. On the other hand, a number of colicin V-producing bacteraemic strains of *E.coli* from various animals were not inhibited by transferrin. Two pieces of evidence suggest that the ability of these strains to recover adequate iron from the bound form is due to the presence of the ColV plasmid. First, the growth of cured derivatives of the bacteraemic strains was severely limited by the presence of transferrin, and second, the acquisition by a K-12 strain of *E.coli* of plasmids from the bacteraemic strains resulted in insensitivity to transferring-induced iron stress.

In work with pathogenic bacteria it is important to determine if observations made in laboratory experiments, in this case the differential tolerance of $ColV^+$ and plasmid-free strains to iron stress, have relevance *in vivo*. Smith and Huggins found that colicinogenic bacteria predominate in the bloodstream and tissues by the death of an animal infected with a mixture of colicinogenic and non-colicinogenic bacteria, but not because of plasmid transfer or colicin-killing of the plasmid-free cells in the inoculum[10]. In particular, colostrum-deprived neonatal claves have been used to demonstrate the operation of plasmid properties in enhancing survival in the absence of generalised bactericidal and bacteriostatic effects of serum immunoglobulins[10]. By inoculating samples of calf blood with a mixture of $ColV^+$ and non-colicinogenic bacteria and showing that the former have a faster rate of growth, we have shown that relatively small differences in mean generation time, due to differential inhibition by iron stress, could account for the observed effects in experimental infections.

Mutant strain AN1937 of *E.coli* K-12 is deficient in the biosynthesis of enterochelin. It is unable to take up iron, and therefore cannot grow, unless certain siderophores are added to the growth medium. For experimental purposes, we used sodium citrate, the ferric-citrate complex[6] being transported into cells via a citrate-inducible outer membrane protein[5]. However, iron stress causes the induction of a number of other outer membrane proteins, at least two of which are know to be receptors for siderophores[5]. Thus, when enterochelin[4] or ferrichrome[8] are available in the medium, complexed iron can be transported into cells by way of the inducible *feuB* and *tonA* gene products, respectively (the latter is not in fact applicable to strain AN1937, which is mutant at *tonA*). Also induced by iron-depletion[5] are a protein of 83,000 daltons whose function is unclear, and (intriguingly, since we are considering a phenomenon mediated by ColV plasmids) a receptor protein for colicin V (the *cir* gene product, also the receptor for colicin Ib[21]). Furthermore, it has been suggested that the *tonB* gene product is essential for the energy-dependent stage of all the active transport systems that involve these inducible membrane components[22].

Derivatives of strain AN1937 carrying ColV plasmids are able to grow in the absence of exogenously supplied siderophores. Using radioactive iron we have demonstrated that ferric ions efficiently enter the cells of these strains, and analysis of bacterial membranes clearly showed that sufficient iron entered colicinogenic bacteria growing in the absence of citrate to maintain repression of the outer membrane proteins characteristically induced in conditions of iron stress. This suggests that these proteins are not required at high levels for the transport of the ColV plasmid-specified iron-sequestering compound. The *tonB* protein, however, is required, since the presence of ColV plasmis has no effect on the growth of a *tonB entA* double mutant in the absence of citrate. Thus, from the point of view of energy dependence, the plasmid-mediated iron uptake system resembles those previously described in *E.coli*[22].

The integration of a prototype ColV plasmid into the chromosome of *E.coli* K-12 has been reported[23]. The possibility exists, therefore, that the ColV plasmids described in this study are analogous to F'-elements generated by erroneous excision of integrated plasmid DNA, and merely carry host genes specifying enterochelin biosynthesis. Since no enterochelin was excreted into the culture medium by the *entA* mutant carrying a ColV plasmid this suggestion can be excluded. Clearly ColV plasmids specify a novel iron uptake system which, we propose, contributes significantly, though perhaps not entirely, to the pathogenicity of the bacteraemic *E.coli* strains which harbour them. The molecular nature of the iron-sequestering agent, and its route of entry into bacterial cells are currently under investigation.

ACKNOWLEDGEMENTS

This work was funded by grants from the Medical Research Council and the University of Leicester. We thank H. Williams Smith for supplying bacteraemic *E.coli* strains, and we are grateful to R.H. Pritchard, I.B. Holland, V.Darby and G.S. Plastow for bacterial strains and for advice, discussions and encouragement.

REFERENCES

1. Spiro, T.G. and Saltman, P. (1969) Struct. Bonding, 6, 116-156.
2. Weinberg, E.D. (1978) Microbiol. Rev., 42, 45-66.
3. Rosenberg, H. and Young, I.G. (1974) Microbial Iron Metabolism, Academic Press, New York, pp.67-82.
4. O'Brien, I.G. and Gibson, F. (1970) Biochim. Biophys. Acta, 215, 393-402.
5. Hancock, R.E.W., Hantke, K. and Braun, V. (1976) J.Bacteriol. 127,1370-1375.
6. Frost, G.E. and Rosenberg, H. (1973) Biochim. Biophys. Acta, 330,90-101
7. Luckey, M., Pollack, J.R., Wayne, R., Ames, B.N. and Neilands, J.B. (1972) J. Bacteriol., 111, 731-738.
8. Hantke, K. and Braun, V. (1975) FEBS Lett., 49, 301-305.
9. Smith, H.W. (1974) J. Gen. Microbiol., 83, 95-111.
10. Smith, H.W. and Huggins, M.B. (1976) J. Gen. Microbiol., 92, 335-350
11. Bouanchaud, D.H., Scavizzi, M.R. and Chabbert, Y.A. (1968) J. Gen. Microbiol., 54, 417-425.
12. MacFarren, A.C. and Clowes, R.C. (1967) J.Bacteriol., 94, 365-377.
13. Roberts, R.B., Abelson, P.H., Cowie, D.B., Bolton, E.T. and Britten, R.J. (1963) Publication 607, Carnegie Institute of Washington, Washington D.C., p.5.
14. Langman, L., Young, I.G., Frost, G.E., Rosenberg, H. and Gibson, F. (19-72) J. Bacteriol. 112, 1142-1149.
15. Churchward, G.G. and Holland, I.B. (1976) J.Mol.Biol., 105, 245-261.
16. Pugsley, A.P. and Reeves, P. (1976) J. Bacteriol., 127, 218-228.
17. Bullen, J.J., Rogers, H.J. and Griffiths, E. (1974) Microbial Iron Metabolism, Academic Press, New York, pp.518-552.
18. Payne, S.M. and Finkelstein, R.A. (1978) J.Clin. Invest. 61, 1428-1439.
19. Beresford, C.H., Neale, R.J. and Brooks, O.G. (1971) Lancet, i, 568-572.
20. Butler E.J., Curtis, M.J. and Watford, M. (1973) Res. Vet. Sci., 15, 267-269
21. Hantke, K. and Braun, V. (1975) FEBS Lett., 59, 277-281.
22. Frost, G.E. and Rosenberg, H. (1975) J. Bacteriol., 124, 704-712.
23. Kahn, P.L. (1968) J. Bacteriol., 96, 205-214.

II – PLASMIDS OF MEDICAL IMPORTANCE: STRUCTURE AND EPIDEMIOLOGY

© 1979 Elsevier/North-Holland Biomedical Press
Plasmids of Medical, Environmental and Commercial Importance
K.N. Timmis and A. Pühler, editors

GENTAMICIN RESISTANCE PLASMIDS

NAOMI DATTA, VICTORIA HUGHES AND MARILYN NUGENT
Royal Postgraduate Medical School, Du Cane Road, London W12 0HS, England.

INTRODUCTION

Gentamicin has played an important role in the treatment of hospital infections for over 10 years. It is active against a wide range of genera, including *Pseudomonas aeruginosa*; when it was first introduced and for several years thereafter, gentamicin resistance was rare and so it was inevitably much used for treatment of hospital infections with otherwise resistant bacteria and also for urgent treatment of acute cases before an infecting organism had been identified. Increasing usage had led to the emergence of resistance, usually plasmid-determined, and reports of outbreaks of infection in hospitals with gentamicin-resistant bacteria of many genera are now common. The incidence of resistance is very variable in different areas, in different hospitals and at different times within the same hospitals.

We have collected all gentamicin-resistant isolates of bacteria over certain periods in our own hospital, to study the incidence and nature of the resistance, and are also examining gentamicin-resistant strains from other hospitals, to follow the spread of the resistance genes and the plasmids and bacteria that carry them. The findings illustrate on one hand stability of plasmid characteristics over months or years and on the other, very labile interrelationships of plasmids in bacteria causing hospital infections.

HOSPITAL 1

Hammersmith Hospital is a general hospital of about 600 beds that include a minority for patients at special risk of infection in units for the treatment of leukaemia and renal failure. The incidence of gentamicin resistance in bacteria of all genera causing infections in the hospital has never risen above 5% in any year. In *Pseudomonas aeruginosa* the incidence has been low and we have not identified plasmids in that species. Among the enterobacteria, the distribution of gentamicin R plasmids is shown in Table 1. All belonged to one of 3 incompatibility (Inc) groups.

TABLE 1

GENTAMICIN R PLASMIDS IN HOSPITAL 1

Group	R pattern	Bacteria Species & serotype	No.
Inc FII	CmSuHgGmTb	*K. aerogenes* K10	8
	ApCmSuHgGmTb	*K. aerogenes* K21	2
		K. aerogenes NT	1
Inc C	TcCmKmSuGmTb	*E. coli*	2
	ApTcCmKmSuGmTb	*Citrobacter*	1
	ApCmSuKmHgGmTbSmTp	*Klebsiella*	1
	ApSmKmHgGmTb	*Proteus mirabilis*	1
		Providencia	1
Inc W	ApSuTpHgGm	*E. coli*	3
	ApSuHgGm	*Enterobacter*	1
		Citrobacter	2

Abbreviations: Ap, ampicillin; Sm, streptomycin; Tc, tetracycline; Cm, chloramphenicol; Km, kanamycin; Su, sulphonamides; Tp, trimethoprim; Hg, mercuric chloride; Gm, gentamicin; Tb, tobramycin.

Inc FII Gm^r plasmids

Klebsiella aerogenes, of one capsular serotype, K10, with an IncFII plasmid was endemic in the hospital for over a year. The first appearance of this organism was in overwhelming infections in two patients, in adjacent beds and both with very serious diseases, leukaemia and sickle cell anaemia. No other patients were then found to be infected or colonised nor was any environmental source of the strain detected. It reappeared in other patients in different ward blocks but the incidence of infection never reached epidemic proportions. The pattern of resistance determined by the plasmid remained constant except for the addition of ampicillin resistance in some cases; apparently the same plasmid was found in other bacteria (Table 1). These plasmids determined synthesis of the gentamicin-modifying enzyme 2''-0 nucleotidyl transferase (ANT(2''))[1].

Inc C Gm^r plasmids

In the same hospital plasmids of IncC were identified in several bacterial genera rather than one special host and the plasmid-determined patterns of resistance were various (Table 1). These plasmids resemble the pIP55 epidemic plasmid[2] in incompatibility group and gentamicin resistance mechanism (specifying ANT(2'')[1]) but appear to have acquired a variety of resistance genes including in one instance the trimethoprim-streptomycin resistance transposon Tn7[3]. IncC plasmids were identified in the hospital over a 2 year period, but whether they were resident in the hospital or brought in from other hospitals on separate occasions is not clear.

Inc W Gm^r plasmids

In the early 1970s there was resident in the hospital a klebsiella strain carrying a sulphonamide-trimethoprim resistance (SuTp) plasmid of IncW[4,5]. This strain is no longer detected but IncW plasmids determining SuTp and in addition, ampicillin, mercuric chloride and gentamicin resistance are now found, suggesting that the original SuTp plasmid (exemplar R388) has acquired additional resistance genes, probably by transposition (Table 2). Work in progress aims to show whether these genes are carried on transposons. The enzymic nature of the IncW determined gentamicin resistance is not known but apparently differs from that of the IncFII and IncC plasmids, since it confers no tobramycin resistance.

TABLE 2
INC W PLASMIDS IN HOSPITAL 1

Year	R patterns
1974	SuTp
1977	SuTp ApSuTpHgGm ApKmSuTpHgGm ApSuHgGm

HOSPITAL 2

In this hospital in central London there was an outbreak, limited to one ward and ended in a few months, of infection with klebsiella of serotype K16[6]. The klebsiella carried a plasmid of IncM, designated pTH1, with nine resistance genes including that for gentamicin resistance and the trimethoprim-streptomycin transposon[7]. The plasmid was also identified in epidemiologically related strains of *Escherichia coli* and *Citrobacter koseri*. Restriction fragments obtained with *Eco*RI and *Bgl*II were identical in plasmids derived from 14 klebsiella K16 clones from different patients and isolated over a 3 month period. The plasmids derived from the *E. coli* and *C. koseri* strains were similar in molecular weight and incompatibility but showed evidence by restriction enzyme analysis and antibiotic resistance patterns of small differences in their DNA sequences, perhaps of the kinds described by Timmis *et al.*[8] as occurring in plasmid R6-5.

These plasmids determined the synthesis of gentamicin-modifying 3-*N*-acetyltransferase (AAC(3)I). A plasmid of IncM, determining the same enzyme is Rip135 identified in 1972 and carrying resistance to streptomycin, tetracyline, sulphonamides as well as gentamicin[9]. Rip135 has a molecular weight of 45 megadaltons (Mdal) and that of pTH1 is 65 Mdal: *Eco*RI cleavage analysis of two plasmids showed that for 12 of the 14 fragments of Rip135 there was a fragment of pTH1 with equal electrophoretic mobility[2]. The incompatibility properties, resistance patterns, gentamicin modifying enzymes and restriction analysis suggest that pTH1 is the same plasmid as Rip135 having acquired DNA sequences that include extra resistance genes. Plasmids resembling Rip135 and pTH1 in incompatibility and gentamicin modifying enzymes and having all the resistance markers of the latter except the transposon-determined trimethoprim resistance, have been found in bacteria of several genera in hospitals in Melbourne, Australia[10,11]. These observations demonstrate the long-continued association of the gene coding for AAC(3)I with the IncM property and the wide geographical distribution in hospital bacteria of plasmids with these characters.

HOSPITAL 3

This was a hospital in Toronto where there was spread of a gentamicin resistant klebsiella of serotype K22. The epidemiology, the serotype and the multiple resistance pattern all indicated that a single bacterial strain was responsible. But when the gentamicin resistant strain was first identified in the hospital, its resistance were readily transferable to *E. coli* K12 whereas from later isolates, otherwise apparently identical, the gentamicin resistance either did not transfer or did so at barely detectable frequency[12]. We have had the opportunity of studying the plasmids in a selection of the wild strains and some of our findings are shown in Table 3. Gentamicin resistance was carried on an IncFII plasmid. In the same klebsiella strains there was a non-conjugative plasmid determining ampicillin resistance and an IncFI plasmid determining lac^+, and streptomycin, sulphonamide and mercuric chloride resistance. Throughout the 20 month period of observations the non-conjugative ampicillin R plasmid remained constant in its molecular weight and *Eco*RI digest pattern but the FI and FII plasmids varied greatly in both these respects and in their resistance genes and transfer frequencies; the differences were so great that plasmids of the same incompatibility group from sequential isolates would not have been recognised as being closely related but for their epidemiological history. Frequent recombination between the plasmid species in the klebsiella host must have brought about these changes, and it continued in Rec^+ K12 transconjugants when FI and FII plasmids were both present, giving many permutations of linkage groups. We assume that the reduction in transfer

TABLE 3

PLASMIDS FROM KLEBSIELLA SEROTYPE K22, HOSPITAL 3

Isolates in sequential order	Plasmid characters	Tra[x]	Transfer[y] frequency from Klebs	M.wt[z] (Mdal)	Inc group
1	a) Ap	-	10^{-4}	7.4	
	b) TcCmGm	+	10^{-4}	54	FII
	c) TcCmKmGm	+	nt	62	FII
	d) lac^+ SmSuHgTcCmKm	+	10^{-4}	110	FI
2	a) Ap	-	10^{-4}	7.4	
	b) ApTcCmGm	+	10^{-4}	60	FII
3	a) Ap	-	10^{-4}	7.4	
	b) ApTcCmGm	±	10^{-7}	60	FII
4	a) Ap	-	10^{-5}	7.4	
	b) ApTcCmGm	±	10^{-8}	62	FII
	c) lac^+ SmSuHg	±	10^{-8}	40	FI
5	a) Ap	-	10^{-4}	7.4	
	b) ApSmTcCmGm	-	$< 10^{-8}$	46	FII

Abbreviations as in Table 1 except that Gm indicates low level resistance to tobramycin and kanamycin as well as gentamicin: ANT 2''[12]. nt, not tested.

x Plasmids in *E. coli* K12 tested for transfer: +, normal frequency, - undetectable, ± very low frequency.

y Frequency of transfer per donor cell in 1 hr mating.

z M.wt measured in single colony lysates compared with known standards in gel electrophoresis.

The plasmids listed are those transferred to *E. coli* K12. Plasmid b) from strain 5 was mobilised by introduction of IncP plasmid, R751. It is likely that all the wild strains carried both FI and FII plasmids, as all were Hg-resistant and carried a large plasmid, in addition to the FII plasmid, M.wt. approx. 90 Mdal.

FII plasmids throughout the series showed great differences in *Eco*RI digest patterns.

frequency of FI and FII plasmids in the later klebsiella isolates resulted from rearrangement of DNA sequences controlling transfer. The non-conjugative ampicillin R plasmid was still mobilised, showing that the strains still conjugated effectively, but normal transfer of the FI and FII plasmids responsible for conjugation was prevented; the mechanism of this effect is not known.

HOSPITAL 4

Another example of instability of DNA sequences in plasmids carried by one bacterial clone in a hospital environment was seen in hospital 4, in West London. Again a gentamicin resistant klebsiella strain, this time serotype K21, was identified; it could transfer multiple plasmid-determined resistances including gentamicin-resistance to *E. coli* K12. The host strain was recognisable not only by its biotype and serotype but also by its resistance to trimethoprim and nalidixic acid. (It is assumed that the latter resistances were chromosomally determined since they were never transferred and the trimethoprim resistance was of a lower level than that conferred by any known trimethoprim R plasmid.) The strain was introduced into the hospital with one identified patient and its spread to other patients, first in one ward and later in a second ward, was observed[13]. At one point in the sequence of infections the klebsiella strain lost resistance to chloramphenicol and gentamicin. It seemed likely that it had lost a plasmid. A study, still incomplete, of the plasmids transferred to *E. coli* K12 from these klebsiella isolates indicates that the loss of resistance was not due simply to loss of one distinguishable plasmid with retention of others. In the early isolate

TABLE 4

KLEBSIELLA SEROTYPE K21 [$Tp^r Nal^r$], HOSPITAL 4

	Klebsiella resistances	Plasmids transferred to *E. coli* K12	M.wt (Mdal)	Group
1)	ApSmSuTcCmHgGm	a) lac^+ ApTcCmSuHg	105	?
		b) ApCmSuGmHg	85	Inc C
2)	ApSmSuTcHg	lac^+ TcSu	120	Inc C

[$Tp^r Nal^r$] = chromosomal markers

Gm = resistant to GmTb

Cm = low level Cm-resistance

there were two plasmids, one carrying lac$^+$ genes and the other, of IncC, carrying gentamicin resistance (Table 4). The Inc group of the lac$^+$ plasmid is not yet known but it was presumably not IncC since it coexisted in the wild strains with the gentamicin R plasmid. But from the later isolate was transferred a large IncC plasmid carrying lac$^+$ but not gentamicin resistance; a gross molecular rearrangement had presumably transposed lac$^+$ genes to the IncC plasmid and led to loss of gentamicin and chloramphenicol resistance.

CONCLUSIONS

These observations on gentamicin R plasmids in bacteria in several hospitals show stable connexion of the acetylating enzyme, AAC(3)I with the IncM property and of the adenylylating enzyme ANT(2'') with the IncC property; both these correlations were first found in Paris in the early 70s and are still evident today in London hospitals and elsewhere. The enzyme ANT 2'' is also carried by plasmids of FII; an early example was JR66[14,15] from a klebsiella strain isolated in Washington D.C.[16] and this combination of enzyme with Inc group was seen in one of the London hospitals discussed here and in the Toronto hospital. In the bacteria we have studied, all the gentamicin R plasmids belonged to these 3 groups, IncC, M and FII and in addition IncW, the latter probably being a new linkage. This is a narrower range of Inc groups than was seen in parallel studies of trimethoprim R plasmids[3].

In all 4 hospitals studied, a particular klebsiella serotype, different in each case, carrying a gentamicin R plasmid, was transmitted from patient to patient, causing infections of varying severity. The stability or otherwise of plasmid structure during the course of spread in the hospitals showed marked contrasts. In one hospital the plasmid was very stable, altering its restriction digest pattern not at all except when found in other bacterial species. In two other hospitals, the plasmids in the epidemic klebsiellas showed drastic changes in molecular weights and patterns of restriction enzyme cut sites. In each case we have interpreted this as resulting from extensive recombination between a gentamicin R plasmid and a lac$^+$ plasmid both present in the klebsiella, but the effect could equally have been caused by exchanges with unidentified plasmid or phage genomes from other bacteria in the environment. Whatever the cause, such gross rearrangements may provide a means of plasmid evolution different from, and additional to, the acquisition of transposons.

REFERENCES

1. Shannon, K. P. and Casewell, M. W., personal communication.
2. Chabbert, Y-A, Roussel, A., Witchitz, J. L., Sanson-Le Pors, M. J. and Courvalin, P. M. This volume p. 183-193.
3. Richards, H. and Nugent, M. This volume p. 195-198.
4. Datta, N. and Hedges, R. W. (1972) J. Gen. Microbiol., 72, 349-356.
5. Jobanputra, R. S. and Datta, N. (1974) J. Med. Microbiol., 7, 169-177.
6. Casewell, M. W., Dalton, M. T., Webster, M. and Phillips, I. (1977) Lancet, 2, 444-446.
7. Datta, N., Hughes, V. M., Nugent, M. E. and Richards, H. (1979) Plasmid, 2, in press.
8. Timmis, K. N., Cabello, F., Andrés, I., Nordheim, A., Burkhardt, H. J. and Cohen, S. N. (1978) Molec. gen. Genet. 167, 11-19.
9. Witchitz, J. L. and Gerbaud, G. R. (1972) Ann. Inst. Past., 123, 333-339.
10. Davey, R. B. and Pittard, J. (1977) Austral. J. Exp. Biol. Med. Sci., 55, 299-307.
11. Richards, H. and Datta, N. (1979) Plasmid, 2, in press.
12. Rennie, R. P. and Duncan, I. B. R. (1977) Antimicrob. Ag. Chemother. 11, 179-184.
13. McSwiggan, D. A. and Seal, D. V., personal communication.
14. Benveniste, R. and Davies, J. (1971) FEBS Letters, 14, 293-296.
15. Datta, N. and Hedges, R. W. (1973) J. Gen. Microbiol., 77, 11-17.
16. Martin, C. M., Ikari, N. S., Zimmerman, J. and Waitz, A. J. (1971) J. infect. Dis., 124, Supplement 24-29.

© 1979 Elsevier/North-Holland Biomedical Press
Plasmids of Medical, Environmental and Commercial Importance
K.N. Timmis and A. Pühler, editors

RESTRICTION ENDONUCLEASE GENERATED PATTERNS OF PLASMIDS BELONGING TO INCOMPATIBILITY GROUPS I1,C,M and N; APPLICATION TO PLASMID TAXONOMY AND EPIDEMIOLOGY.

YVES A. CHABBERT, AGNES ROUSSEL, JANINE L. WITCHITZ, MARIE-JOSE SANSON-LE PORS and PATRICE COURVALIN .

Unité de Bactériologie Médicale - LA. CNRS 271, Institut Pasteur,75724 Paris Cedex 15, France .

INTRODUCTION

Since their discovery, it has been postulated that *in vitro* selftransferable plasmids are also infectious and epidemic *in vivo* . However the criteria used for discriminating epidemic plasmids from non epidemic plasmids have varied greatly over the past twenty years.

The discovery, in Japan in 1958, of multiresistant strains of *Shigella* and *E. coli* harbouring R-plasmids that were able to transfer *in vitro* between these two bacterial genera seemed sufficient to speculate that these plasmids were also epidemic under the natural conditions. However the detection, in various countries, of transferable plasmids encoding the same resistances showed that a plasmid cannot be identified on the basis of its R-determinants alone. In 1965, the spread of a plasmid called Δ factor by Anderson "et al"[1], could be traced because of the phage type modification which it induced in *S typhimurium* . Since 1970, with the development of plasmid classification into incompatibility groups, it is possible to individualize plasmids according to both, Inc group and newly discovered resistance characters . For example Ingram "et al"[2], in 1969, showed that an Inc P plasmid from *P. aeruginosa* isolated in Birmingham was also detected in *Klebsiella* during the same time . In 1972 Datta and Hedges[3] studied and Inc W plasmid coding for high level resistance to trimethoprim . This plasmid was detected simultaneously in three hospitals in London in isolates of *E. coli* and *Klebsiella* . Finally, between 1969 and 1972 Witchitz and Chabbert[4] traced an IncC plasmid encoding resistance to gentamicin : we will use this plasmid as a model for epidemic plasmids .

To date our knowledge of the physical structure of certain plasmids has greatly improved with the availability of hybridization techniques and restriction endonuclease analysis of DNA. Assuming that plasmids belonging to a given Inc group are composed of a relatively stable "core" into which several transposons can insert at various preferential sites, one may expect important difference among epidemiologically unrelated plasmids but the maintenance of an identical structure in an epidemic plasmid . In order to test these hypotheses its is necessary to compare the structures of various plasmids isolated from wild strains. Therefore we studied the restriction endonuclease patterns of unrelated plasmids belonging to Inc N and I1 groups and of plasmids belonging to Inc groups C and M which were supposed to have been transferred during epidemics .

RESULTS

Unless otherwise stated, the plasmids used are listed in Appendix B in DNA insertion elements, plasmids and episomes[5] .

A) Epidemiologically unrelated plasmids.

1) *EcoR1* restriction patterns of Inc N plasmids . Hybridization studies of plasmids belonging to Inc groups FI,FII,C,M,N,P and W[6,7,8] have clearly demonstrated that plasmids of the same group shared at least 80 % of their DNA sequence and that plasmids of different groups had less than 15 % of their sequence in common. If the number and the location of the restriction endonuclease recognition sequences were consistent with these data it should be possible to classify plasmids merely according to their restriction endonuclease patterns.

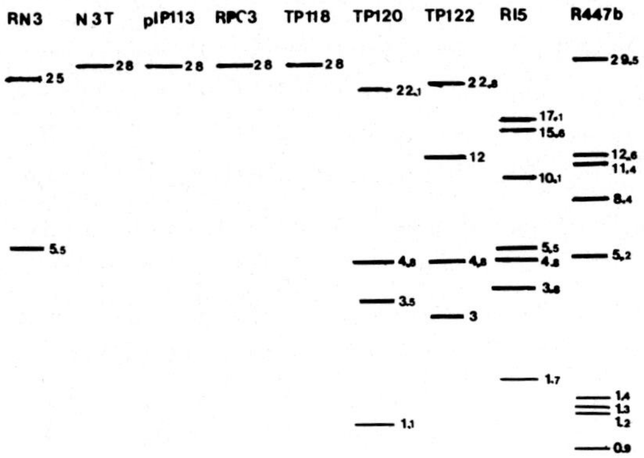

Fig.1. Schematic representation of an agarose gel analysis of *EcoR1* digests of plasmids belonging to the Inc N group. The molecular weights are expressed in Megadaltons.

We have compared, using *EcoR1*, Inc N plasmids RN3 (TcSmSu) and N3T (Tc) from *Shigella*, RCP3 (SmKm) from *E. coli* pIP113 (Tc),from *S. panama*, Tp118 (ApSu) from *S. enteritidis* with R1drd19 (FII) , R55 (C), pIP112 (I1), Sa (W), (Figure 1). Indeed *EcoR1* generates only two to three DNA fragments in N and W plasmids but more than ten in the other plasmids . We have included in our study four additional Inc N plasmids : TP120 (SmSuTc) and Tp122 (ApSuTc) from *S. typhimurium*,

R15 (SmSu) from *P. vulgaris* and R447b (ApKm) from *P. morgani* (Figure 1). These plasmids have from four to eight *EcoR1* sites and the size of the DNA fragments generated vary greatly from one plasmid to another. Therefore Inc N plasmids seem to have very different structures which could be due to transposon insertions in multiple sites .

2) <u>*EcoR1* restriction patterns of Inc I1 plasmids</u> . The restriction patterns observed in seven Inc I1 plasmids have a remarkable number of *EcoR1* generated DNA fragments of similar size (Figure 2). The properties of the plasmids studied are listed in Figure 3. There is no epidemiological relationship between the wild type host strains : they have been isolated in different countries over a period of several years .

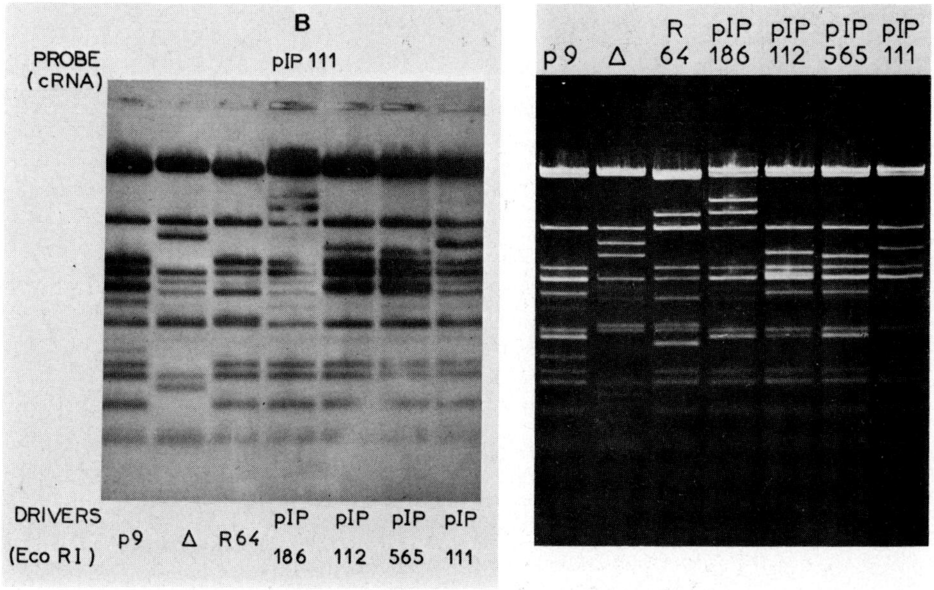

Fig. 2. (A) Agarose gel electrophoresis of *EcoR1* digests of plasmids belonging to the Inc I1 group. Electrophoresis was carried out in a 0.8% agarose horizontal slab gel for 12 h at a potential of 3 V/cm . (B) Analysis of Inc I1 plasmids by hybridization. The DNA fragments were transferred to a nitrocellulose sheet and hybridize to ^{32}P-labeled pIP111 cRNA as indicated at the top (probe). The autoradiogram was exposed for a period of four hours .

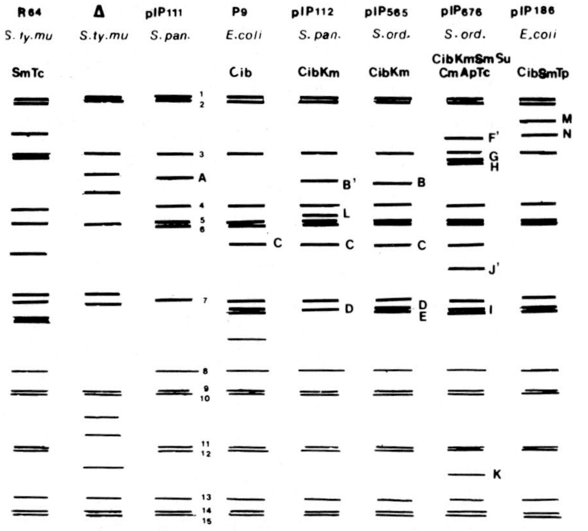

Fig. 3. Schematic representation of an agarose gel analysis of *EcoR1* digests of plasmids belonging to the Inc I1 group. The Inc I1 plasmids share fragments with identical size numbered 1 to 15, except for the Δ factor.Additional fragments,which are not common to the Inc I1 plasmids, are designated with letters.

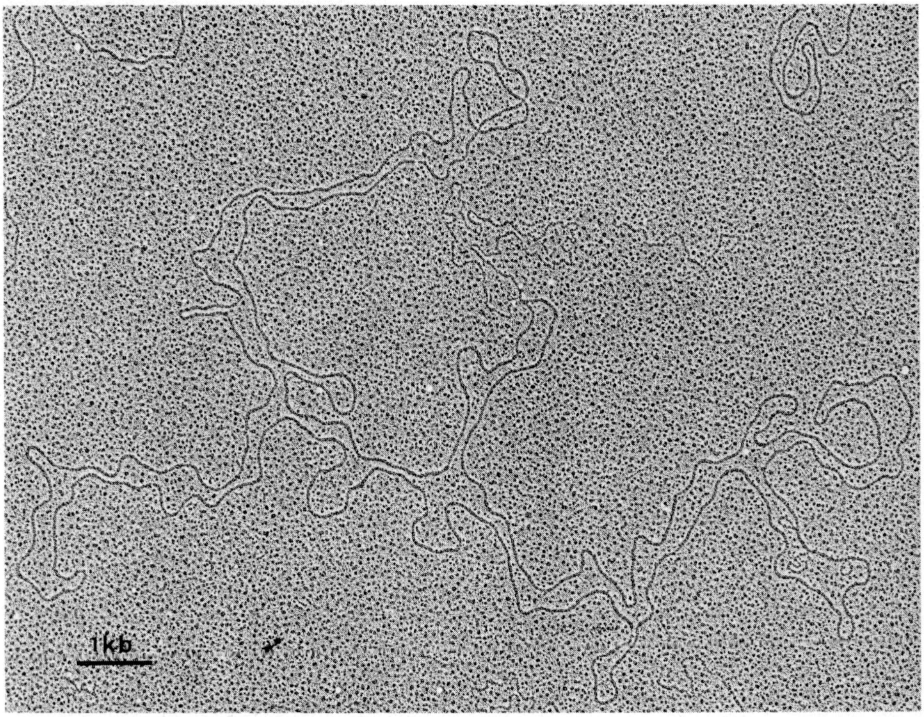

Fig. 4. A heteroduplex between pIP111 and pIP565. The additional DNA on pIP565 is seen as a loop of single-stranded DNA . kb, kilobase.

The restriction endonuclease digests yielded a very characterisitic set of bands. P9, pIP186, pIP112, pIP565[9], and pIP111 have fifteen DNA fragments of similar molecular weight; R64 shares 13 of these fragments . Only the Δ factor differs noticeably : it shares three of the larger fragments but the remaining twelve fragments differ slightly from the corresponding fragments of the above mentioned plasmids . We used the technique described by Southern to detect partial or complete homology between the DNA fragments. *in vitro* ^{32}P-labeled complementary RNA from pIP111, a"transfer factor" with no detectable R-determinant, was used as a probe . The autoradiogram (Figure 2) shows hybridization with the fifteen fragments common to the five first plasmids studied (P9,pIP186,pIP112, pIP565 and pIP111) and also with the thirteen bands common to R64 . Despite their different mobility, the majority of the Δ fragments also hybridizes to the probe . These data suggest that all the Inc I1 plasmids studied have related "cores" and that the analysis of *EcoR1* generated patterns in unable to visualize the differences between these "cores".

In addition to the fifteen described DNA fragments, pIP112, pIP565 and pIP186 possess bands which allow discrimination between these plasmids and which might correspond to the inserted R-determinants (Figures 2,3). The *EcoR1* generated fragment A from the "transfer factor" pIP111 is missing in plasmids pIP112, pIP565 and pIP186 . It is replaced by the four fragments designated B'CDL, BCDE and MNOP respectively . The fact that fragment A disappears suggests that the insertion of the R-determinants took place within this DNA segment . The probe from pIP111 hybridizes with the fragments B' and L of pIP112, with fragments B and C of pIP565 and with fragments M and N of pIP186 (Figure 2). This is consistent with the notion that fragment A of pIP111 is partially homologous with these DNA segments . They therefore constitute the junction fragments between the cores and the R-determinants. pIP112 fragments D and L, pIP565 fragments D and E and pIP186 fragments O and P which do not hybridize correspond to additional central fragments .

Electron microscopy heteroduplex analysis of pIP111 and pIP565 shows a unique single-stranded DNA insertion loop which corresponds to the ColIb and Km characters (Figure 4). The size of this loop (10.1 kb) is in good agreement with the molecular weight of these plasmids, as determined by agarose gel electrophoresis(figure 3). Examination of a large number of such molecules failed to show the presence of an inverted repeat in the additional DNA fragment . This might indicate that the insertion is not due to the presence of inverted repeated sequences or that such structures, if they exist, are too short to be visualized by the technique used . In addition these heteroduplexes showed a small region with unpaired sequences which could exist under different configurations . This region is reminiscent of the $\gamma\delta$ region observed in the fertility factor F and the G segment of the bacteriophages Mu and P1[10]. The location of this region is constant : the distance from the site of insertion of the additional DNA fragment to the proximal end of the unpaired region is 22.3 kb . A similar analysis of plasmids pIP112 and pIP186 showed that the insertion of the R-determinants also occured as a single DNA loop.

Apart from the absence of detectable inverted repeats this structural arrangement is analogous to that reported by Cohen[5] for FII plasmids . Our results implie that Inc I1 plasmids can only be individualized by a detailed study of the structure of their R-determinants .

B) Epidemiologically related plasmid .

1) An epidemic plasmid (ApCmSuGm) of the group Inc C . In november 1969, Witchitz isolated a strain of *Klebsiella* serotype 35 Dd from a patient hospitalized in Paris . This strain harboured an Inc C plasmid (pIP55) with an approximate molecular weight of 130 kb and transferable to *E. coli*. pIP55 codes for a rare type of penicillinase, oxa 3 (bla^+, oxa) and for resistance to chloramphenicol (cat^+), sulfonamide and aminoglycosides (ANT(2")). The latter enzyme was observed for the first time in Europe. One month later, in the same hospital ward, *P. mirabilis, Providencia and P. aeruginosa* strains were isolated from a peritoneal dialysis . These bacteria were also able to transfer to *E. coli* the same characters which are encoded by pIP55 . Within 6 months, the number of strains conferring these resistance characters increased gradually to 8 % of the clinical isolates in the hospital unit. The majority of these strains were isolated in the same hospital ward and belong to the following genera : *Klebsiella*, *P. mirabilis*, *P. aeruginosa*, *Enterobacter*, and *E. coli* . Eight months later strains encoding the same characters were detected in another hospital . The particular traits of this situation reminded us of a classical microbial epidemic : appearance, in a defined place, of a peculiar organism, different from those usually observed, geographical spread and increase of incidence with time . These epidemiological circumstances then strongly suggested a plasmid epidemic . We compared three plasmids from the two first patients isolated in 1969, pIP55 and pIP92 from *Providencia* and pIP64 from *P. aeruginosa* with two plasmids isolated in the same hospital ward in 1972, pIP149 from *Klebsiella* 68 M and in 1975, pIP504 from *E. coli* .

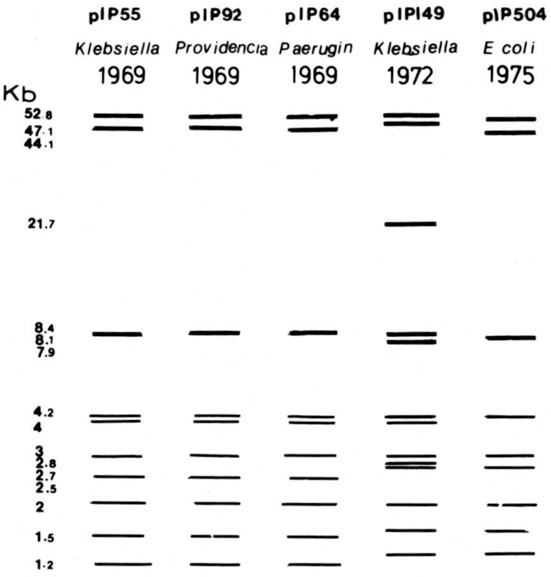

Fig. 5 Schematic representation of agarose gel analysis of *EcoR1* digests of plasmids belonging to the Inc C group. The molecular weights are in kilobases.

The analysis after digestion with *EcoR1* and agarose gel electrophoresis (Figure 5) does not show any detectable difference among the ten fragments generated in pIP55, pIP92 and pIP64 . DNA from pIP149 has three additional fragments and plasmid pIP504 lacks one *EcoR1* fragment . These data indicate a high degree of structural stability among the Inc C plasmids through numerous rounds of replication and transfer *in vivo* .

2) Inc M plasmids. Inc M plasmids are distributed worldwide . The first IncM plasmid identified in our laboratory was isolated from a *S. paratyphi B* which acquired the plasmid in the gut of a *Klebsiella* carrier . In 1971 strains of enterobacteria appeared in France which were able to transfer to *E. coli* an Inc M (SmTcSuGm) plasmid . The mechanism of resistance to Gm (AAC(3)type I) is different from that discussed above . In this case we were unable to detect the focus of the outbreak and to trace its spread. We studied plasmids encoding the same characters isolated from bacteria in various hospitals . The fact that this mechanism of resistance to aminoglycosides was only rencently discovered is in favor of a plasmid epidemic . We have compared plasmids pIP135 isolated in 1971 from an *Enterobacter* and pIP151 isolated in 1972 from a *Klebsiella*

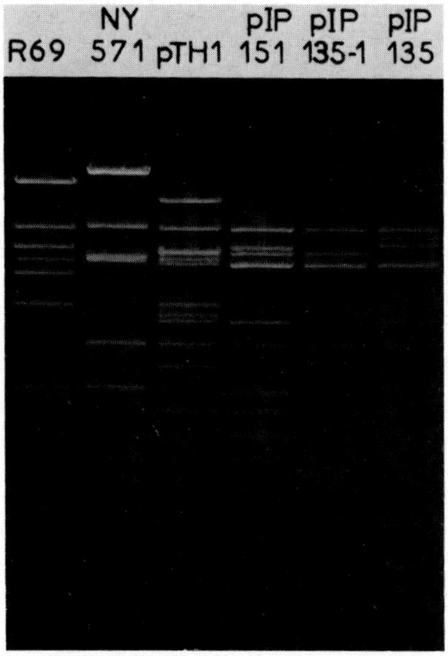

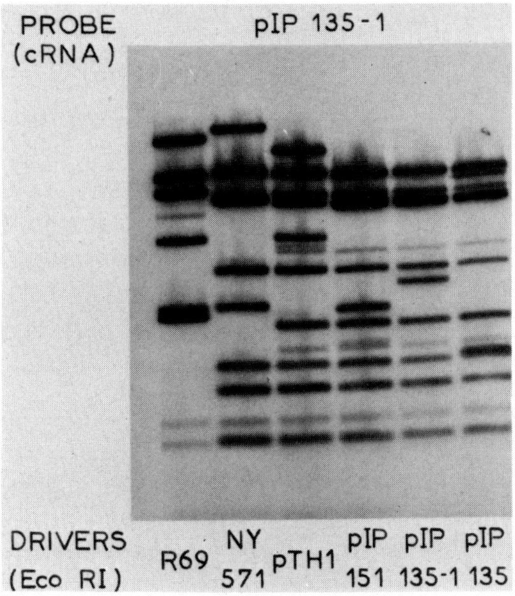

Fig. 6. (Left) Agarose gel electrophoresis of *EcoR1* digests of plasmids belonging to the Inc M group . (Right) Analysis of Inc M plasmids by hybridization. Same conditions as in Fig. 2 .

with two Inc M plasmids (NY571 and NY2147) isolated in New York in 1973 and 1974 . The two latter plasmids encode resistance to ampicillin and gentamicin (ANT(2")). The molecular weight of these plasmids is approximatively 70 kb. Analysis by agarose gel electrophoresis following digestion with *EcoR1* (Figure 6) shows that pIP135 and pIP151 are very similar(thirteen out of fourteen fragments are common to both plasmids); plasmids NY571 and NY2147 have nine out of ten fragments in common . However pIP135 and pIP571 *EcoR1* patterns are very different from those of NY571 and NY2147 . Recently we have included in our study plasmid pTH1 inc M (ApTcCmGmSmTp) isolated by Datta in 1979 in London (Personal communication) . The analysis of *EcoR1* fragments and hybridization experiments according to Southern indicate that pTH1 is related to pIP135 and pIP151 but very different from NY571 and NY2147 (Figure 6) .

DISCUSSION

The low percentages of DNA-DNA homology observed between transferable plasmids from Gram-negative bacteria belonging to the major incompatibility groups FII, I1,C,M and N suggest the existence of DNA "cores" which differ largely between incompatibility groups. Subsequently, the restriction endonuclease patterns of plasmids from different Inc groups vary greatly. Under certain circumstances plasmids can be differentiated by their restriction endonuclease patterns but it appears difficult to establish a classification based on this technique alone . On the other hand, within each Inc group, the high percentage of total homology and the results obtained with the Southern technique show that plasmidic "cores " are closely related. However, the results obtained with eight plasmids of group Inc N and six plasmids of group Inc I1 indicate that the structural arrangements are different from one Inc group to another.

In vitro studies with various transposons (e.g. Tn3, Tn5 and Tn10) indicate that insertion can occur into many sites[5] . As a consequence the respective positions of the various transposons should be very different from one plasmid to another and the restriction endonuclease generated DNA fragments should vary greatly in plasmids belonging to the same Inc group . This seems to be the case for the four Inc N plasmids studied . On the other hand, the similarity of the patterns obtained with Inc I1 plasmids indicate the existence of a common DNA "core" . This notion was confirmed by electron microscopy heteroduplex analysis . The various R-determinants and the gene(s) for colicin Ib production are inserted in a hot spot in the same *EcoR1* fragment. This structure is similar to that reported by Cohen[5] for Inc FII plasmids but the organisation and the mode(s) of insertion of the R-determinants are probably different . Therefore Inc I1 plasmids can only be characterized by a detailed study of the structure of the R-determinants .

Two situations can be found in epidemiologically related plasmids.

1) The three Inc C (ApSuSmGm) plasmids harboured by three different bacterial genera (*Klebsiella, Providencia and Pseudomonas*) isolated at one month interval from two infected patients have indistinguishible *EcoR1* restriction patterns. It seems therefore that certain plasmid structures are relatively stable under the natural conditions. This structural identity is a very strong argument in favor of a parental relationship between these plasmids.

2) Numerous Inc C plasmids encoding the same resistance characters have been isolated later in the same hospital ward. Two of them were isolated three and six years after the outbreak of the epidemy and differ only little from the

initial plasmid . Similarly there are only minor differences between Inc M (SmTcGm) plasmids isolated from *Klebsiella and Enterobacter* strains in Paris at a one year interval and other Inc M (ApGm) plasmids isolated in New York. However, there are also small differences between Inc C (CibKm) plasmids isolated during two independant epidemics, *S. panama* (France) and *S. ordonez* (Africa). When minor differences between plasmids are to be interpreted the singularity of certain of their characters and the epidemiological circumstances of their isolation must be taken into account .

More comprehensive observations are required in order to accumulate a desirable degree of information on plasmid variability under the natural conditions. Nethertheless it seems that this variability is less important than expected from the results obtained on the multiplicity of transposon insertion sites . The relatively low variability could reflect the fact that certain structures have more selective advantages than others due to their stability during replication and transfer *in vivo* . If, for practical reasons, it is necessary to construct *in vitro* plasmids with selective advantages, it would be important to know the exact arrangement of the more stable "natural" structures .

ACKNOWLEDGEMENTS

We are extremely grateful Cécile Carlier for heteroduplex analysis, Guy Gerbaud for genetic analysis, Ekkehard Collatz for critical reading of the manuscript, and Odette Rouelland for secretarial assistance. This work was supported by a grant from the Institut National de la Santé et de la Recherche Médicale (CRL 78.4.133.1) .

REFERENCES

1. Anderson, E.S., Humphreys, G.O. and Willshaw, A. (1975) J. Gen. Microbiol. 91, 376-382.
2. Ingram, L.C., Richmond, M.H. and Sykes, R.B. (1973) Antimicrob. Agents and Chemother. 3, 279-288.
3. Datta, N. and Hedges, R.W. (1972) J. Gen. Microbiol. 72, 349-355.
4. Witchitz, J.L. and Chabbert, Y.A. (1972) Ann. de l'Institut Pasteur(Paris) 123-3, 67-78
5. Bukhari, A.I., Shapiro, J.A., and Adhya, S.L. (1977) DNA insertion elements, plasmids and episomes Cold Spring Harbor Laboratory, Cold Spring Harbor, New York, pp. 601-677.
6. Falkow, S. (1975) Infectious Multiple Drug Resistance . London: Pion.
7. Grindley, N.D.F., Humphreys, G.O. and Anderson, E.S. (1973) J. of Bacteriol. 115, 387-398 .

8. Roussel, A.F. and Chabbert, Y.A. (1978) J. Gen. Microbiol. 104, 269-276)

9. Roussel, A., Carlier, C., Gerbaud, G., Chabbert, Y.A., Croissant, O. and Blangy, D. (1979) Molec. gen. Genet. 169, 13-25 .

10. Broker, T.R., Chow, L.T., Soll, L. (1977) DNA insertion elements, plasmids and episomes (Bukhari, Shapiro, Adhya, eds) pp. 575-580. New York : Cold Spring Harbor Laboratory .

Plasmids of Medical, Environmental and Commercial Importance
K.N. Timmis and A. Pühler, editors

THE INCIDENCE AND SPREAD OF TRANSPOSON 7

Hilary Richards and Marilyn Nugent
Royal Postgraduate Medical School, Du Cane Road, London W.12.

INTRODUCTION

Trimethoprim is an antibacterial drug much used both in and out of hospitals because of its effectiveness, broad spectrum of activity and lack of toxicity. It has, up to now, always been used in combination with a sulphonamide, for example as co-trimoxazole, in clinical medicine. However, sulphonamide resistance is common[1] and the spread of a transposon carrying trimethoprim resistance genes could have serious implications on the future use of trimethoprim in clinical and veterinary medicine.

TRANSPOSON 7

A transposon encoding resistance to trimethoprim and streptomycin was identified in 1976[2] and called transposon 7. Tn7 was originally recognised on R483 and has a molecular weight of 9 Md. A map of Tn7 for the restriction enzymes EcoR1, HindIII and BamHI is shown in figure 1.

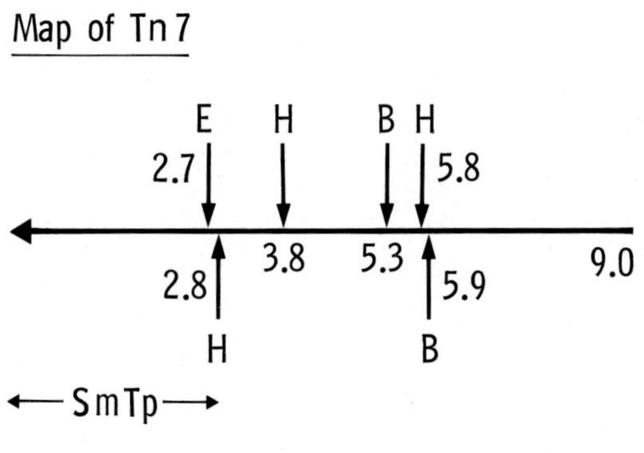

Fig. 1. EcoR1 (E), HindIII (H) and BamHI (B) cut sites on Tn7.

The enzymes HindIII and BamHI are particularly useful in identifying the presence of Tn7 on a plasmid as they give internal fragments characteristic of the transposon. Figure 2 shows the characteristic fragments of 2 and 1 Md upon HindIII digestion of plasmids carrying Tn7.

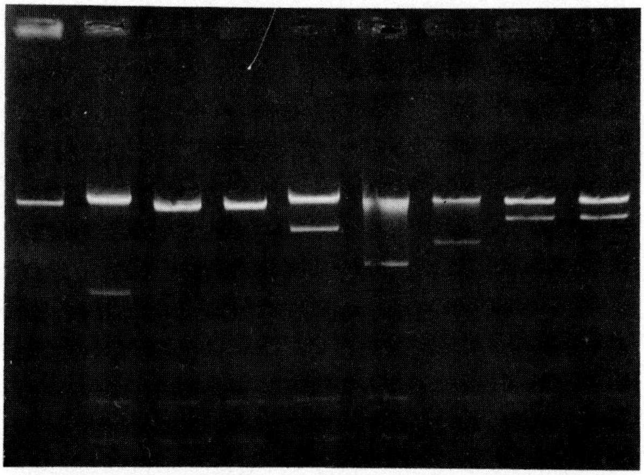

Fig. 2. Track 1 is RP4 which has one cut site for HindIII. Tracks 2-9 are HindIII digests of RP4 carrying transposons indistinguishable from Tn7, isolated from different replicons.

THE SPREAD OF TRANSPOSON 7

In 1977, Barth and Datta[3] examined the Hammersmith Hospital collection of over 1,000 R plasmids and found only two others carrying a transposon indistinguishable from Tn7. The two plasmids, R721 and pBW1 came from different serotypes of *Escherichia coli* isolated in different counties in England, one from a human infection and the other from an animal. The first three plasmids on which Tn7 was identified all determined I pili and were therefore classified as belonging to the I incompatibility complex although they were assigned to different subgroups.

Further examples of I plasmids carrying Tn7 have been identified in strains of *Salmonella typhimurium*, isolated from different animal infections in different regions of England[4]. Plasmid-determined trimethoprim resistance is still uncommon among salmonella strains but is increasing in frequency and the evidence suggests that the spread of transposon 7 is contributing to the increase.

The first report of the spread of Tn7 to an incompatibility group in nature, other than the I complex, was its identification on a multiple resistance plasmid of incompatibility group M[5]. This plasmid came from *Klebsiella aerogenes* which had caused an outbreak of infection in a urological ward of a London hospital[6].

The IncM plasmid carrying the transposon had also spread to environmentally related *E. coli* and to *Citrobacter koseri*. As well as the IncM plasmid all the wild vector bacteria contained cryptic plasmids none of which were ever shown to transfer to *E. coli* K12. Curing of the IncM plasmid from the wild strains by acridine orange treatment never resulted in loss of the transposon and it was concluded that the transposon was present in the chromosome and/or the other plasmids, as well as in the IncM plasmid[5].

During recent work a ^{32}PColE1:Tn7 probe was made *in vitro* and tested for hybridization with unlabelled plasmid DNA which had been transferred to cellulose nitrate filters by the Southern blot technique[7]. By this means Tn7 homology was identified in the IncM plasmid but not in any of the cryptic plasmids. Therefore when the IncM plasmid was lost, the transposon must have been retained in the chromosome. Thus, the *K. aerogenes*, *E. coli* and *C. koseri* each carried two copies of the Tn7 transposon, one in the IncM plasmid and one in the chromosome and they each carried at least one other cryptic plasmid which did not contain the transposon.

In Hammersmith Hospital we have been collecting plasmids conferring resistance to the newer antibacterial drugs, including trimethoprim. Table 1 shows the range of incompatibility groups in which plasmid-borne trimethoprim resistance has so far been identified.

TABLE 1

TRIMETHOPRIM RESISTANT PLASMIDS IN HAMMERSMITH HOSPITAL

Range of Incompatibility groups:	
1974	W
1977	W C I FII N P X

The trimethoprim resistance W plasmids isolated in 1974, of which R338 is an example do not carry Tn7 and it is unlikely that those isolated in 1977 carry the transposon. However, Tn7 has been identified on an IncC plasmid (pHH1190) and on an Iδ plasmid (pHH1189) see Table 2. Experiments are in progress to determine if plasmids of the other incompatibility groups also carry Tn7.

TABLE 2

NATURALLY OCCURRING PLASMIDS CARRYING Tn7

Plasmid	Incompatibility Group	Host	Origin
R483	Iα	*E. coli*	Kent
R721	Iδ	*E. coli*	London
pBW1	Iδ	*E. coli*	Hertfordshire
pHH1268	Iα	*S. typhimurium* 204	Surrey
pHH1269	Iδ	*S. typhimurium* 49	Surrey
pTH1	M	*K. aerogenes*	London
pHH1189	Iδ	*E. coli*	London
pHH1190	C	*E. coli*	London

The evidence is that Tn7 is spreading amongst plasmids of different incompatibility groups and between different bacteria and this is comparable to the spread of the ampicillin transposon. As yet the spread of Tn7 has not been described outside England, however laboratory evidence shows that Tn7 is readily transposable amongst replicons, so the continued therapeutic use of trimethoprim is likely to result in the world-wide spread of Tn7.

REFERENCES

1. Grüneberg, R. N. (1976) J. clin. Path, 29, 292-295.
2. Barth, P. T., Datta, N., Hedges, R. W. and Grinter, N. J. (1976) J. Bacteriol, 125, 800-810.
3. Barth, P. T. and Datta, N. (1977) J. gen. Microbiol, 102, 129-134.
4. Richards, H., Datta, N., Sojka, W. J. and Wray, C. (1978) Lancet, 2, 1194-1195.
5. Datta, N., Hughes, V. M., Nugent, M. E. and Richards, H. (1979) Plasmid, in press.
6. Casewell, M. W., Dalton, M. T., Webster, M. and Phillips, I. (1977) Lancet, 2, 444-446.
7. Southern, E. M. (1975) J. Mol. Biol, 98, 503-517.

© 1979 Elsevier/North-Holland Biomedical Press
Plasmids of Medical, Environmental and Commercial Importance
K.N. Timmis and A. Pühler, editors

RAF PLASMIDS IN STRAINS OF ESCHERICHIA COLI AND THEIR POSSIBLE ROLE IN ENTEROPATHOGENY

RÜDIGER SCHMITT, RALF MATTES, KURT SCHMID AND JOSEF ALTENBUCHNER

Lehrstuhl für Genetik, Universität Regensburg, Universitätsstr.31
D-8400 Regensburg (Germany)

INTRODUCTION

Enteropathogenic strains of Escherichia coli are frequently found in connection with diarrhoeal diseases in animals and man[1]. Diarrhoea is caused by two types of enterotoxins, a heat labile toxin (LT) and a heat stable toxin (ST), both of which are plasmid-encoded[2]. Manifestation of the disease also requires certain bacterial surface antigens like K88 antigen in porcine neonatal diarrhoea[3], which facilitates colonisation of the anterior small intestine[4]. The production of K88 and the ability to utilise raffinose (Raf) as sole carbon source are plasmid-mediated and are frequently borne by one plasmid. Instead of K88 antigen, the production of H_2S (Hys) may be linked to Raf[5]. K88 and Hys were in no case found together on one plasmid. This report describes biochemical and immunochemical studies together with genetic and physical mapping of Raf plasmids and their relation to K88 and Hys, respectively. The following points will be covered.

1. Metabolism and regulation of plasmid-encoded raffinose utilisation: the raf operon.
2. Immunochemical relationships and classification of Raf plasmids.
3. Deletion mapping and restriction enzyme analysis of the raf region.
4. IS1-mediated transposition of raf-hys and raf-K88 elements and insertion mutagenesis.
5. Cloning of raf and hys determinants.
6. The possible role of Raf plasmids in enteropathogeny.

RESULTS AND DISCUSSION

1. The 83 Md-conjugative Raf plasmid pRSD2[6] was used as prototype in these studies. Three inducible, plasmid-borne functions mediate the peripheral metabolism of raffinose[7], namely, a specific permease (gene rafB), an α-galactosidase (gene rafA) and an invertase (gene rafD) as illustrated in Figure 1.

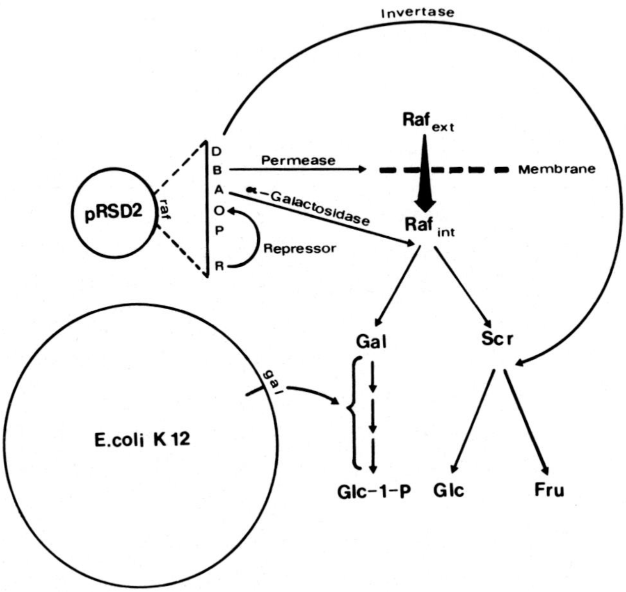

Fig.1. Peripheral catabolism of raffinose by plasmid-borne (pRSD2) and chromosomal (E.coli K12) functions.

These are co-ordinately controlled by the repressor gene rafR[8]. α-Galactosidase (mol.wt. 4 x 82,000) and invertase (mol.wt. 2 x 54,000) were purified to homogeneity and characterised. Two membrane-bound proteins of mol.wts. 65,000 and 42,000, respectively, were identified in a minicell system[9] and have been assigned to Raf permease. It can, however, not be excluded that only one of these proteins bears the transport activity. The biochemical properties and specificities of these plasmid-encoded functions differ markedly from homologous chromosomal functions.

The genetic loci rafA, rafB and rafD, respectively, were identified by point mutations and insertions of phage Mu. Mu-induced polarity indicated that they are part of a "raf operon". Mutations in rafR lead to constitutive expression of all raf functions suggesting their negative control by the regulatory gene. The existence of a diffusable raf repressor has been verified by complementation studies.

2. Antiserum against purified α-galactosidase of pRSD2-1[7,10] served to establish immunochemical relationships among the α-galactosidases of 39 Raf plasmids from various sources and countries. The double-diffusion technic of Ouchterlony[11] was applied to extracts of preinduced cells (to ensure high levels of α-galactosidase). A typical experiment is shown schematically in Figure 2.

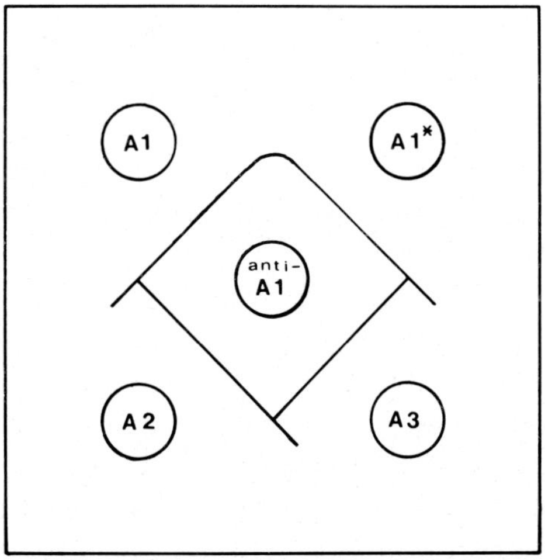

Fig.2. Agar plate double-diffusion of α-galactosidase-containing cell extracts against pRSD2-1-α-galactosidase antiserum (anti-A1). Extracts of the reference strain (A1*) harbouring pRSD2-1 and of three other plasmid-containing strains (A1, A2 and A3, respectively) were applied in the peripheral wells. Precipitation lines were formed after 16 h diffusion at 37°C. All 39 Raf^+ strains were tested accordingly.[10]

Fusing precipitation lines indicate serological identity, spurs reveal partial differences in the antigenic determinants[11]. This and similar experiments with all available Raf plasmids lead to the following conclusions:

- the α-galactosidases of all 39 Raf plasmids (from chicken, piglets and man) are cross-reacting and, hence, closely related;
- spur-formation revealed three serotypes, related in the order A1 - A2 - A3. Class A1 (31 strains) is most abundant, whereas A2 (3 strains) is restricted to samples from Sweden and A3 (5 strains) to samples from England and USA;
- the following combination of markers was always found in correlation to the serotypes and has been used for classification of these Raf plasmids:

A1	Hys <u>or</u> K88		(A1-plasmids)
A2	Hys	-	(A2-plasmids)
A3	-	-	(A3-plasmids).

These results suggest a common origin of all plasmid-borne *raf* determinants. The three serological classes of α-galactosidase may have evolved after recombination with certain pathogenic factors (K88, Hys), which lead to biological and geographical isolation.

3. Mapping of the *raf* region of pRSD2 was accomplished by deletion analysis with a λd*raf* transducing phage. These results were extended lateron by insertion mutagenesis with *Tn*5, by transposition and by cloning of the *raf* region (see 4. and 5.).

A λd*raf* hybrid phage containing the entire *raf* operon of pRSD2 (but missing the regulatory gene *raf*R) was isolated by employing a host strain (harbouring pRSD2) with a deletion of the λ*att* locus[12]. The original isolate λd*raf* has most of its head and tail genes substituted by DNA of pRSD2 (approx. 15 Md), whereas thermoinducibility owing to c_{I857}[13] remained intact. Transduction of *E.coli* with λd*raf* lead to thermosensitive Raf^+ derivatives which contained the defective prophage integrated between chromosomal gal and bio markers (Fig. 3). Heat induction and appropriate selection lead to deletions extending into the *raf* operon from the operator-distal end as indicated in Figure 3. Successive inactivation of invertase (rafD), permease (rafB) and α-galactosidase (rafA) by deletions have demonstrated, that the order of genes in the operon is raf (O,P)-A-B-D.

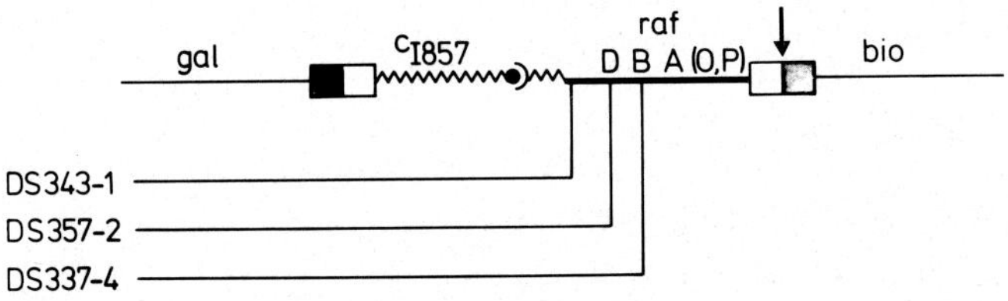

Fig.3. Portion of E.coli chromosome (——) with prophage λdraf integrated between gal and bio. Hybrid attachment sites formed by integration are boxed. Deletions extending from λ (/\/\/\) into raf (▬▬) are marked by lines and strain numbers.

These deletion mutants were again lysogenised with λnin (Fig. 3, arrow), a thermoinducible lambda derivative with a 15%-deletion of the genome[14]. Transducing lines each carrying one of the raf deletions were selected and their DNAs subjected to restriction enzyme analysis. Figure 4 shows the resulting restriction maps obtained with endoR. BamHI, BglII, EcoRI, HindIII, SalI, SmaI and XhoI. The raf operon is located next to the phage attachment site and comprises a maximum of 4.8 Md as defined by deletion DS343-1.

4. Translocation of raf genes into the genome of phage P1 was possible and lead to specifically transducing P1raf hybrids[15]. Transducing P1raf particles were produced with a frequency of 10^{-7} to 10^{-8}. Raf-Hys plasmids of class A1 (see 2.) lead to P1raf-hys hybrids indicating that these two markers are translocated together. Similarly, P1raf-K88^{+} hybrids were selected from the corresponding Raf-K88 plasmids. In contrast, Raf-Hys plasmids of class A2 yielded P1raf hybrids deficient in hys indicating that translocation of raf does not include hys. Raf plasmids of serotype A3 lead to P1 raf hybrids (Table 1).

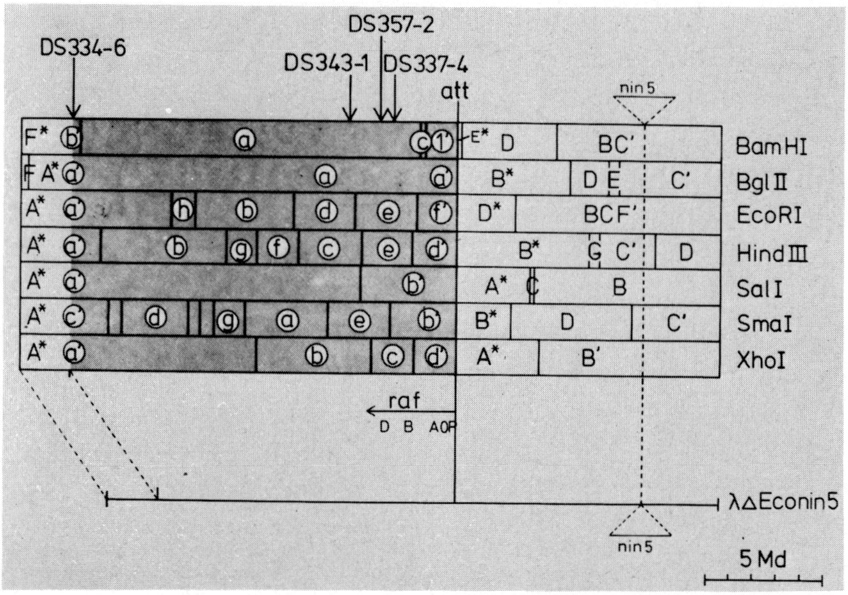

Fig.4. Restriction map of λdraf. Lettering of restriction fragments according to size. Upper case letters = λ fragments, lower case letters = plasmid DNA-fragments" asterices and primes mark original fragments of λ or plasmid DNA, which were altered by deletion or integration of foreign DNA, respectively. The extension of the deletions, which define the location and sequence of genes in the raf operon, is marked by arrows. Plasmid-DNA is indicated by shading.

TABLE 1

CHARACTERISATION OF P1 HYBRIDS DERIVED FROM VARIOUS RAF PLASMIDS

Plasmid	α-Gal serotype	Plasmid-borne markers	Markers translocated to P1	Integration site in P1	Size of insertion in P1
D1021	A1	Raf Hys	Raf Hys	IS1	26 Md
D1022	A1	Raf Hys	Raf Hys	IS1	nd*
D1148	A1	Raf Hys	Raf Hys	IS1	nd
D1073	A2	Raf Hys	Raf	IS1	32 Md
SF711(152)	A1	Raf K88	Raf K88	IS1	41 Md
V200	A1	Raf K88	Raf K88	IS1	nd
X26	A3	Raf	Raf	IS1	nd
F170	A3	Raf	Raf	IS1	nd

*nd = not determined

The unusual formation of these P1 hybrid phages has been rationalised by restriction enzyme and heteroduplex analyses (Fig. 5) as follows.

- All Raf plasmids tested (Table 1) contain the raf region on an element which is flanked by direct repeats of IS1. Translocation of this element into the single IS1 sequence of P1 DNA (at map unit 20[16]) ensues by recA-dependent recombination. IS1 sequences in P1raf hybrids were identified by heteroduplices formed by the appropriate restriction fragments (containing IS1) and Tn9 known to contain a Cm^R determinant bound by IS1 elements[18] (Fig. 5).
- Depending on the donor plasmid, the size of the raf element translocated to P1 ranges between 26 Md and 41 Md. A1-plasmids also carry the Hys or K88 markers on this element, whereas A2-plasmids do not.
- Restriction maps of various raf-hys, raf-$K88^+$ and raf elements bound by direct repeats of IS1 show a "region of homology" around the raf operon (approx. 10 Md), but differ in the other portions (Fig. 6).

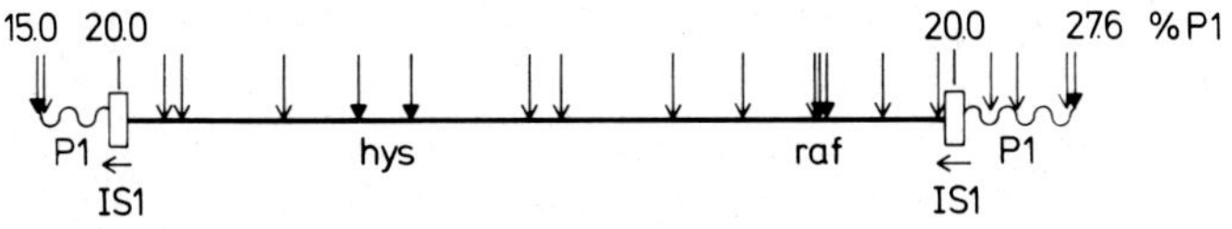

Fig.5. Restriction map of the hybrid portion of P1raf-hys. A 26 Md-element flanked by direct repeats of IS1 (boxes) was translocated from pRSD2 to the IS1 site of P1 at map unit 20[16]. Numbers indicate P1 map units[17]. BamHI (↓) and EcoRI sites (↓) are marked by arrows. The locations of raf and hys regions were determined by Tn5 insertions (not shown).

Transposon Tn5[19] was used to analyse the fine structure of the raf and hys genes of pRSD2. For mapping of raf genes a specifically transducing P1raf phage[20], for hys genes a suitable PstI fragment cloned into the conjugational vector pRSD1[6,21] were employed.

Insertions in *raf* characterised by inactivation of the respective enzyme(s) were subjected to restriction analysis. The results verified the relative gene order and polarity of transcription shown above (see 2.). Insertions in *raf*R lead to constitutive expression of the structural genes; *raf*R is located proximal to the *raf* promoter.

Insertion mutagenesis of *hys* yielded at least two closely linked genes for H_2S-production encoding a thiosulfate reductase (*hys*A) and a tetrathionate reductase (*hys*B). A polar effect of insertions in *hys*A on the expression of *hys*B suggests a common transcription of the two genes. The *hys* genes are separated from the *raf* region of pRSD2 by at least 10 Md of DNA; this latter portion has little or no homology with the corresponding region of other translocatable *raf* elements[22]. The relative locations of *raf* and *hys* are shown in Figure 6. The K88 location on the physical map of a *raf*-K88 element shown in Figure 6 was deduced from cloning data of Mooi *et al.*[23] (see 5.).

5. *In vitro* cloning of *raf* and *hys* determinants into pBR322[24] has been applied to reveal the relative position of these genes within the IS1-bound DNA segment. The *raf* structural genes are contained within a 4 Md-*Sal*I fragment. After cloning of this fragment the *raf* genes were expressed constitutively suggesting either loss of the *raf*R gene or promoter fusion. This *Sal*I fragment belongs to the "region of homology" (see 4. and Fig. 6) and is found on various Raf plasmids.

The *hys* genes are comprised in a 8 Md-*Pst*I fragment, which is located approximately 10 Md apart from the *Sal*I fragment encompassing *raf* genes. This *Pst*I fragment is absent from Hys^- deletions indicating that *raf* and *hys* genes are not closely linked.

Cloning studies revealed the same *Sal*I fragment in the Raf-K88 plasmid pRI8801[23]. $K88^+$ is located on a 7.7 Md-*Hind*III fragment as shown by these cloning studies. A fragment of identical size and fine structure was found in the Raf-K88 plasmid SF711[25] analysed in our laboratory. This *Hind*III fragment is part of the IS1-flanked DNA segment, which contains the *raf* and $K88^+$ genes. The *Hind*III fragment ($K88^+$) is located 16 Md apart from the raf region (Fig. 6).

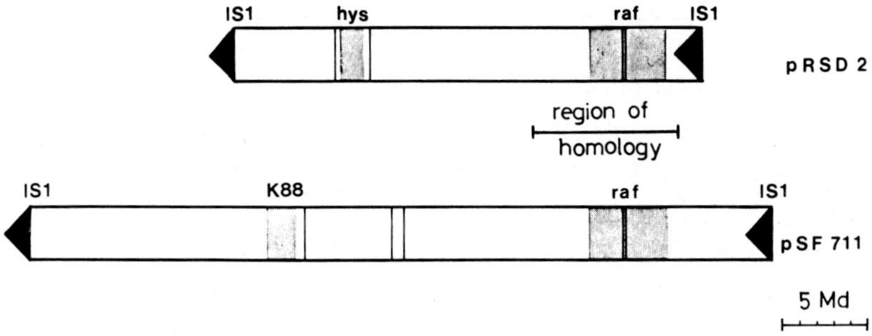

Fig.6. BamHI restriction maps of raf-hys (pRSD2) and rafK88^{+} elements (pSF711). Maps were established after translocation to P1 (Fig. 5). Vertical lines mark BamHI sites, shading indicates locations and extend of raf, hys and K88^{+} determinants. The "region of homology" around raf is indicated.

Since K88^{+} genes are located centrally on the fragment[23], the distance raf - K88^{+} ranges about 20 Md in plasmid SF711 (Fig. 6).

It is possible that K88^{+} and hys determinants represent independent transposable elements which were integrated into the IS1-bound element carrying raf. This model is supported by the observation that A2-plasmids contain hys determinants separate from IS1-bound raf elements.

6. The widespread distribution of Raf plasmids and their fragment occurrance in enteropathogens suggests a selective advantage of the raf determinants. When feeding a mixed population of Raf^{+} and Raf^{-} strains to healthy chicken, no selective enrichment of Raf^{+} strains in the animals intestines was found[26]. This may well be different under pathological conditions. Diarrhoeal diseases in man (traverler's disease) and in newborn domestic animals (neonatal diarrhoea) caused by toxigenic strains of E.coli are best understood. Manifestation of the disease requires host-specific surface antigens (K88 in piglets, K99 in calves), which facilitate colonisation of the upper small intestine[27], and the production of either or both of two types of enterotoxin, one heat stable (ST) and the other heat labile (LT). These produce a short-

lived massive salt and water secretion from mucosal cells of the small intestine.

The frequent combination of K88$^+$ and raf determinants in A1-plasmids and their location on an element flanked by direct repeats of IS1 provides a "molecular explanation" for the occurrence of raf genes in pathogenic strains of E.coli. Raf plasmids contained in such strains have acquired K88$^+$ and raf determinants simultaneously by IS1-mediated translocation.

We postulate that Raf$^+$ strains have also a direct selective advantage under pathological conditions. This advantage resides in their potential to utilise raffinose (present in vegetary foods) as source for the synthesis of capsular material. Acid polysaccharide K antigens of E.coli containing galactose, glucose, N-acetylglucosamin and colitose are known to interfere with the host defense by preventing phagocytosis[28] thus providing a considerable advantage for the pathogenic strains. Similar reasons may hold for the presence of sucrose and lactose plasmids in enteropathogens[29]. It has been reported that under growth-limiting conditions prevailing in diarrhoea changes in the composition of the bacterial cell wall occur, which lead to increased resistance to host immune defenses[28]. This observation can be reconciled with our finding that copy numbers of Raf plasmid pRSD2 and, hence, gene dosage effects increase fourfold when growth is limited[6].

We conclude that the raf genes of all Raf plasmids are closely related and have evolved from a common ancestor. The structural genes form an operon located on IS1-bound elements, which facilitate translocation into the IS1 sites of other replicons. The selective advantage of raf determinants in enteropathogenic strains resides in their combination with colonisation factors (and possibly other pathogenic traits) and in their potential to establish bacterial resistance against phagocytosis by cell wall production from raffinose under growth-limiting conditions.

ACKNOWLEDGEMENT

This investigation was supported by the Deutsche Forschungsgemeinschaft.

REFERENCES

1. Magalhães, M. and Véras, A. (1977) J.gen.Microbiol. 99, 445-447.
2. Gyles, C.L., So, M. & Falkow, S. (1974) J.Infect.Dis. 130, 40-49.
3. Ørskov, I., Ørskov, F., Sojka, W.J., Leach, J.M. (1961) Acta Pathol.Microbiol.Scand. 53, 404-422.
4. Smith, H.W. and Parsell, Z. (1975) J.gen.Microbiol. 87, 129-140.
5. Ørskov, I. and Ørskov, F. (1973) J.gen.Microbiol. 77, 487-499.
6. Burkardt, H.-J., Mattes, R., Schmid, K. & Schmitt, R. (1978) Mol.Gen.Genet. 166, 75-84.
7. Schmid, K. and Schmitt, R. (1976) Eur.J.Biochem. 67, 95-104.
8. Schmitt, R., Schachinger, K. & Schmid, K. (1977) Hoppe-Seylers Z.Physiol.Chem. 358, 302.
9. Reeve, J. (1977) Mol.Gen.Genet. 158, 73-79.
10. Schmid, K., Ritschewald, S. and Schmitt, R. (1979) J.gen. Microbiol. (submitted).
11. Ouchterlony, O. (1967) In D.M. Weir (ed.), Handbook of experimental immunology. Blackwell Scientific Publications, Oxford, England.
12. Schrenk, W.J. and Weisberg, R.A. (1975) Mol.Gen.Genet. 137, 101-107.
13. Sussman, R. and Jacob, F. (1962) Compt.Rend.Acad.Sci. 254, 1517-1519.
14. Murray, E. and Murray, K. (1975) J.Mol.Biol. 98, 551-564.
15. Mattes, R. and Schmitt, R. (1977) Hoppe-Seylers Z.Physiol. Chem. 358, 275-276.
16. Iida, S., Meyer, J. and Arber, W. (1978) Plasmid 1, 357-365.
17. Bächi, B. and Arber, W. (1977) Molec.Gen.Genet. 153, 311-324.
18. MacHattie, L.A. and Jackowski, J.B. (1977) pp. 219-228. In: DNA Insertion Elements, Plasmids and Episomes (A.I. Bukhari, J.A. Shapiro and S.L. Adhya eds.). Cold Spring Harbor Laboratory, New York.
19. Berg, D.E. (1977) pp. 205-212. In: DNA Insertion Elements, Plasmids and Episomes (A.I. Bukhari, J.A. Shapiro and S.L. Adhya eds.). Cold Spring Harbor Laboratory, New York.
20. Mattes, R., Altenbuchner, J. and Schmitt, R. (1978) Hoppe-Seylers Z.Physiol.Chem. 359, 1118-1119.
21. Mattes, R. (in preparation).
22. Mattes, R., Seiderer, E. and Schmitt, R. (in preparation)
23. Mooi, F.R., deGraaf, F.K. and van Embden, J.D.A. (1979) Nucl. Acid.Res. (in press).

24. Bolivar, F., Rodriguez, R.L., Greene, P.J., Betlach, M.C., Heyneker, H.L., Boyer, H.W., Crosa, J.H. and Falkow, S.(1977) Gene 2, 95-113.

25. Shipley, P.L., Gyles, C.L. and Falkow, S. (1978) Infect.Immun. 20, 559-566.

26. Smith, H.W. and Huggins, M.B. (1978) J.Med.Microbiol. 11, 471-492.

27. Smith, H.W. and Linggood, M.A. (1971) J.Med.Microbiol. 4, 467-485.

28. Smith, H. (1977) Bacteriol.Rev. 41, 475-500.

29. Johnson, E.M., Wohlhieter, J.A., Placek, B.P., Sleet, R.B. and Baron, L.S. (1976) J.Bacteriol. 125, 385-386.

© 1979 Elsevier/North-Holland Biomedical Press
Plasmids of Medical, Environmental and Commercial Importance
K.N. Timmis and A. Pühler, editors

INTRAMOLECULAR AMPLIFICATION OF THE TETRACYCLINE RESISTANCE DETERMINANT OF TRANSPOSON Tn 1771 IN ESCHERICHIA COLI

F. SCHÖFFL and H.J. BURKARDT
Institute of Microbiology, University Erlangen, Egerlandstr. 7, D-852o Erlangen, F.R.G.

INTRODUCTION

Transposons are defined as genetic determinants residing on discrete DNA-segments capable of transposition to other replicating units[1]. Previous investigations have shown that in general the drug resistance transposons are flanked at their termini by short inverted repeats[2] with the exception of the chloramphenicol transposon Tn9 which is bordered by direct repetitive IS1[3]. Resistance determinants flanked by direct repeats are known to produce multiple copies, as it was shown for certain R-plasmids in Proteus mirabilis[4] and Streptococcus faecalis[5].

Recently, a new tetracycline transposon (Tn1771), originated from the clinically isolated E.coli strain UR12644[6] was identified[7]. In addition to the property of transposition we found that in the presence of high concentrations of tetracycline Tn1771 is also capable of gene amplification in E.coli[8]. In order to characterize the molecular basis of the gene amplification process the amplified and a reduced structure of Tn1771 were analysed. A model for the molecular structure of the transposon is presented, suggesting that there are three repetitive DNA segments on Tn1771. To prove this model the existence and orientation of the small repeated DNA sequences was examined.

PHYSICAL CHARACTERIZATION OF Tn1771

Tn1771 was originally dedected as residing on a spontaneously formed non-conjugative plasmid (pFS402) conferring tetracycline resistance[7]. Transposition of Tn1771 was selected from pFS402 to the plasmid pFS2, a resistance transfer plasmid carrying no further drug resistance determinants[9]. The newly formed conjugative plasmid with the Tn1771 insertion was called pFS202. The molecular characteristics of the plasmids involved in the construction of pFS202 are depicted in Table 1. From these data a molecular weight of 7.1 Mdals can be deduced for the DNA of Tn1771, having a density of 1.72 g/ml.

TABLE 1

MOLECULAR AND GENETIC PROPERTIES OF PLASMIDS

Plasmid	Phenotype[a]	Molecular weight (Mdals)	Density (g/ml)
pFS402[b]	Tra^-, Tc^R	12.8	1.712
pFS2[c]	Tra^+	18.9	1.700
pF202[d]	Tra^+, Tc^R	26	1.706
pFS203[e]	Tra^+, Tc^S	22.4	1.703

[a] Abbreviations: Tra: transfer properties; Tc^R: tetracycline resistant; Tc^S: tetracycline sensitive.

[b] carries Tn1771[7]

[c] naturally occuring[8]

[d] constructed by transposition of Tn1771 from pFS402 to pFS2[7]

[e] spontaneously formed Tc^S derivative of pFS202[8]

Restriction enzyme analysis of the plasmids pFS402 and pFS202 led to the physical map of Tn1771 shown in Fig. 1B. This map is characterized by the nearly symetrical location of three EcoRI sites, two of them located close to the ends and the third one in the middle of the transposon. This shows that Tn1771 is cut out nearly completely from the insertion site of the host plasmids by EcoRI treatment. In this way two Tn1771 specific DNA fragments of 3.5 Mdals and 3.6 Mdals are generated as indicated by the fragment patterns of pFS202 and pFS402 obtained after gel electrophoresis of EcoRI digests (Fig. 1A, lane B and lane C). The 3.5 Mdals fragment carries two additional SmaI sites, whereas the 3.6 Mdals fragment is cut once each by HindIII and PstI, respectively (Fig. 2B).

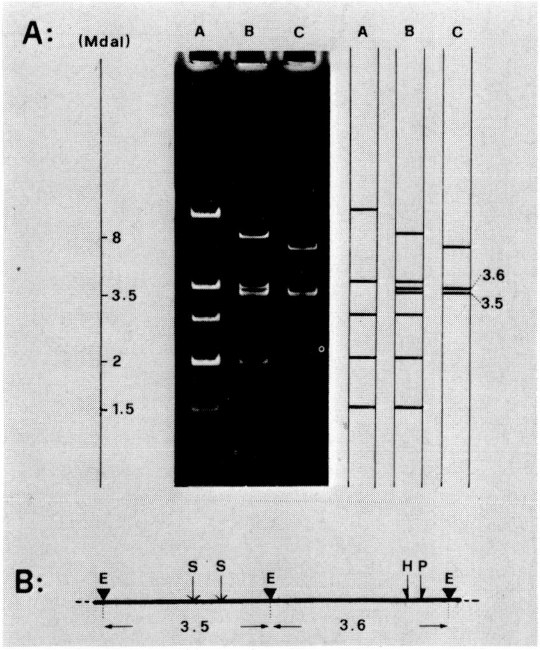

Fig. 1A) Agarose gel electrophoresis of EcoRI digests of the plasmids pFS202 (B), pFS402 (C) and the reference plasmid pFS201 (A) 1B) Restriction map of Tn1771. Cleavage sites are E: EcoRI, S: SmaI, H: HindIII, P: PstI. Tn1771 fragments are indicated in the diagrams of the patterns by their moleculare weight designations given in Mdals.

AMPLIFICATION AND LOSS OF THE RESISTANCE DETERMINANT OF Tn1771

Examination of the tetracycline resistance levels of E.coli strains harboring plasmid pFS202 indicated that the minimal inhibitory concentration can be varied by pregrowth of cells in tetracycline containing medium. Ordinaryly, pFS202 exhibits an inducible tetracycline resistance level of 50-80 ug/ml tetracycline. However, after extended growth of the bacteria in media containing higher tetracycline concentrations (50-200 ug/ml), resistance levels up to 500 μg/ml are expressed. On the other hand loss of tetracycline resistance was observed at a frequency of 10^{-2} - 10^{-3}, occuring in cultures grown in drug-free medium.

The molecular basis of these drug resistance phenomena was investigated by the analysis of plasmids, isolated from E.coli

CSH51(pFS202) grown either in the presence or absence of 100 μg/ml tetracycline. Plasmid samples from tetracycline untreated cells, pFS202(-) and from tetracycline treated cells pFS202(+) were examined by agarose gel electrophoresis. Simultaneously the variant plasmid pFS203, isolated from tetracycline sensitive segregants of E.coli CSH51(pFS202) was investigated in the same way. The results of agarose gel electrophoresis are shown in Fig. 2. pFS203 and pFS202(-) plasmids migrate as a prominent band (lane B, C), representing covalently closed circular (ccc) DNA, whereas pFS202(+) bands at several positions (more than 10) in the gel (lane D).

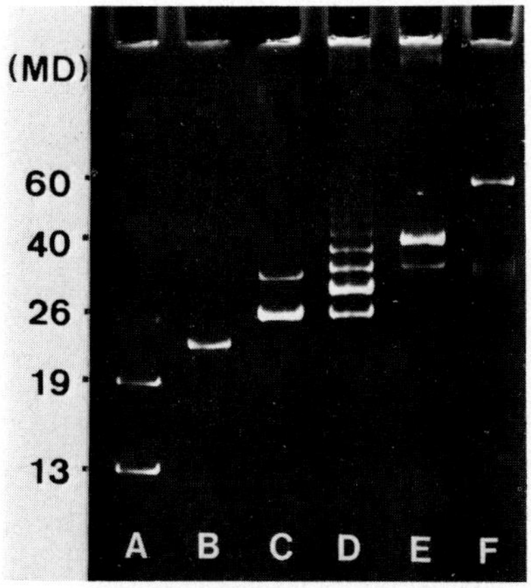

Fig. 2:
Agarose gel electrophoresis of plasmids.
A, E, F: marker plasmids,
B: pFS203,
C: pFS202(-),
D: pFS202(+).
ccc-plasmids were isolated by dye buoyant density centrifugation[11] of sarcosyl lysates of E.coli cells[12].

The molecular weights of the plasmid bands were determined by the use of additional molecular weight markers (lane A, E, F) assuming a logarithmic correlation between molecular weight and the relative electrophoretic mobility of ccc-plasmids[10]. When compared with its parental plasmid pFS202(-) the tetracycline sensitive plasmid pFS203 can be shown to be deleted by about 3.5 Mdals. The neighbouring size classes of pFS202(+) differ from each other also by an average molecular weight of about 3.5 Mdals.

The restriction analysis of these plasmid fractions demonstrats that a 3.5 Mdals deletion of Tn1771 in pFS203 comprises the same DNA fragment which on the other hand is amplified on pFS202(+).

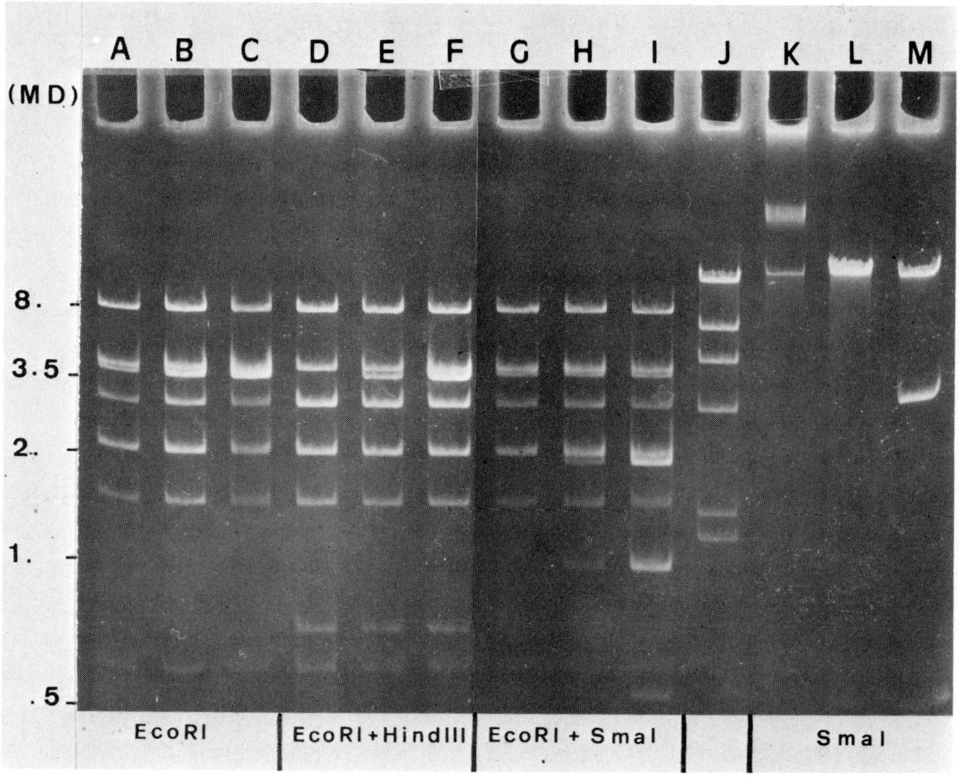

Fig. 3: Agarose gel electrophoresis of restriction endonuclease digests of the plasmids pFS203 (A, D, G, K), pFS202(-) (B, E, H, L), pFS202(+) (C, F, I, M). λ-HindIII digest (J) was used as molecular weight marker.

Fig. 3 shows the fragment patterns of plasmids digested by EcoRI (lane A-C), EcoRI/HindIII (lane D-F), EcoRI/SmaI (G-I) and SmaI (lane K-M). The EcoRI digests of all three plasmids samples show nearly identical fragment patterns. Only the Tn1771 specific fragments, banding at the 3.5 Mdals position in the gel differ in DNA concentration. Plasmid pFS203 lacks this fragment as is clearly demonstrated by the absence of SmaI cleavage sites (lane G and K). In contrast pFS202(+) contains a higher nonstoichiometric amount of this fragment. The double digestion experiments using EcoRI/HindIII and EcoRI/SmaI revealed that the amplified EcoRI fragment of pFS202(+) is cut by SmaI (lane I), whereas the 3.6 Mdals EcoRI fragment, which is cut one time by HindIII, remains non-amplified (lane H). This 3.6 Mdals EcoRI-fragment of Tn1771 is also retained in pFS203. It should be noted that the restriction sites of the endonucleases SmaI and HindIII are located only on the transposon segment of the plasmid molecule. These results indicate that only about one half of the transposon DNA is involved in the deletion formation causing tetracycline sensivity as well as in gene amplification correlated with the enhanced tetracycline resistance.

The direct repetition of the amplified 3.5 Mdals EcoRI fragment on pFS202 was deduced from the appearance of a characteristic 3.1 Mdals fragment present in the SmaI digest of pFS202(+) but not in the SmaI digest of pFS202(-) (Fig. 3, lane L and lane M).

The results of the restriction analysis are summarized in Fig.7, showing the physical maps obtained for the amplified (B) and deleted (C) structure of Tn1771.

The tandem arrangement of the 3.5 Mdals EcoRI fragment (Fig.7B) represents only the first step in the amplification process. Further amplification leads to an increase in the number of directly repeated units. The loss of exactly this part of the transposon (Fig. 7C) corresponds with tetracycline sensitivity of pFS203 indicating that this segment carries the tetracycline resistance determinant. By cloning experiments it was directly shown that only the 3.5 Mdals EcoRI fragment is associated with tetracycline resistance (Schöffl, unpublished). The enhanced drug resistance levels mediated by pFS202(+) can be therefore explained by gene dosage effect, due to an intramolecular amplification of the resistance determinant.

DETERMINATION OF COPY NUMBERS OF MULTIPLE R-DETERMINANTS

The numbers of intramolecularly amplified r-determinant copies of pFS202(+) can be deduced from the pattern of plasmid bands of the agarose gel electrophoresis, shown in Fig. 2, lane C. Multiple r-determinant copies ranging from 1 to more than 10 are indicated by the various size classes of plasmids. Since the DNA density of Tn1771 (1.72 g/ml) differs from that one of the host plasmid pFS2 (1.700 g/ml), the gene amplification could be studied also by measuring the density shift of pFS202 during resistance amplification.

Fig. 4 shows the density profiles of pFS202(+) in panel A and pFS202(-) in panel D obtained by CsCl gradient centrifugation in

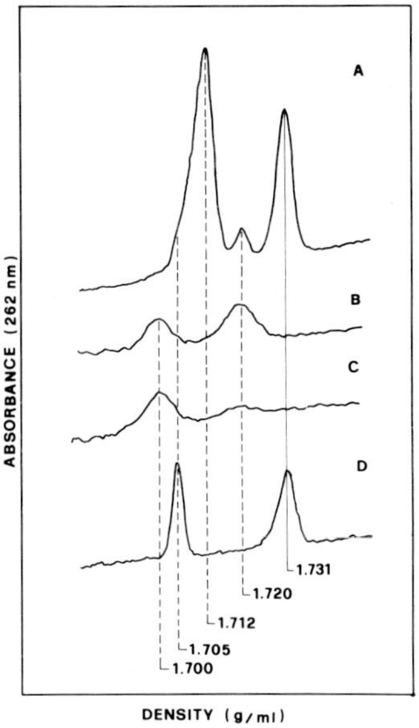

Fig. 4: DNA density profiles of undigested and EcoRI treated plasmid fractions obtained after CsCl gradient centrifugation in the analytical ultracentrifuge.
A: pFS202(+),
B: pFS202(+)/EcoRI,
C: pFS202(-)/EcoRI,
D: pFS202(-).
The peaks at 1.732 g/ml density represent marker DNA.

an analytical ultracentrifuge. The profiles of the EcoRI digests of the same plasmid fractions are depicted in panel B and C. It was calculated that the density difference between the undigested plasmids (pFS202(-): 1.706 g/ml and pFS202(+): 1.712 g/ml) reflects a maximum increase in the amplification of the 3.5 Mdals fragment of about 8-9 copies. The density profile obtained for the EcoRI digest of pFS202(+) (Fig. 4, panel B) confirms selective amplification of Tn1771 specific DNA fragments by an increased amount of DNA banding at 1.72 g/ml in comparison to the profile of the EcoRI digest of pFS202(-) (panel C). DNA fragments banding at 1.700 g/ml represent the host plasmid DNA. From the DNA-ratios host-plasmid/Tn1771, an average copy number of 5 r-determinants per plasmid molecule was calculated for the amplified plasmid fraction.

The fact, that obviously no fixed copy number is selected by treatment with defined concentrations of tetracycline may account for the spontaneous occurence of recombinational processes leading to a variety of r-determinant copy numbers in amplified pFS202. In this context, the small DNA-peak present at 1.72 g/ml in Fig. 4 (panel A) could represent recombinative intermediates of the amplification process since it was only observed in pFS202(+) fractions. Such intermediates consisting of circular r-determinant DNA of Tn1771 are probably not capable of autonomeous replication.

MOLECULAR STRUCTURE OF Tn1771

In analog to the models developed for the amplification of certain re-determinants of NR1 in Proteus mirabillis[4] and for the production of multiple copies of the tetracycline r-determinant of pAMα1 in Streptococcus faecalis[5], we suggest that the intramolecular multiplication in pFS202 is due to small directly repeated DNA sequences flanking the amplifiable region of Tn1771. Recombinational events occuring at these sites of homology would result ir gene amplification. The proposed direct repeats on Tn1771 most probably carry an EcoRI site as is indicated by the accurate deletion of the 3.5 Mdals EcoRI-fragment on pFS203.

The position and relative orientation of direct repeats on Tn1771 is indicated in the transposon-model shown by arrows no. 1 and 2 underlining the EcoRI sites (Fig. 7A). The existence of a further EcoRI site at the right-hand terminus of Tn1771 suggests a third repetitive sequence mapping at this position (arrow 3).

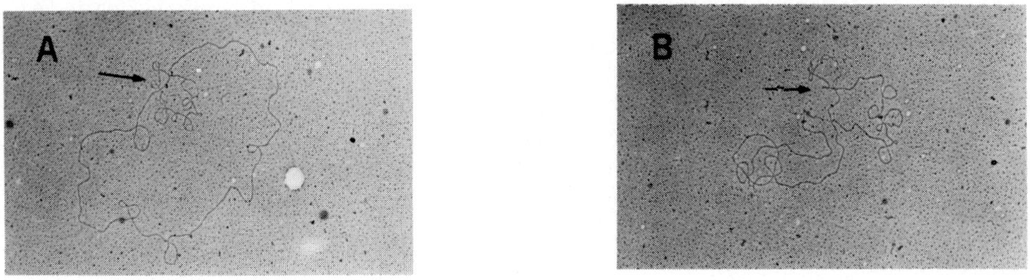

Fig. 5: Electron micrographs of pFS202 homoduplex molecules
A) pFS202 with a 7 Mdals underwound loop
B) pFS202 with a 3.5 Mdals underwoud loop

The inverted repeats, depicted in Fig. 7 are in accordance with electron-microscopic observations, showing two different types of "snap back" structures of pFS202 homoduplex molecules. Both structures are featured by very small (about 100 base pairs) inverted repeats, "looping out" in either a small (about 3.5 Mdals) or a large (about 7 Mdals) underwound DNA region (Fig. 5 A, B).

These structures can be explained by the intramolecular "snap back" due to self-annealing of the inverted repeats no. 2 and 3, respectively 1 and 3 of Tn1771 (see Fig. 7A) before complete renaturation. This interpretation was confirmed by the results of the electron-microscopic heteroduplex studies using pFS202 and λ::Tn1771 molecules. Fig. 6 shows an electron-micrograph taken from a DNA heteroduplex molecule representing a single stranded fragment of pFS202 hybridized with single stranded linear λ::Tn1771. The double stranded DNA region comprises 3.5 Mdals accompanied at one side by an underwound loop of nearly the same size. This structure can be interpreted by a complete annealing of the r-determinant of Tn1771, followed by a "snap back loop" between the inverted repeats no. 2 and 3 (see Fig. 7) flanking the second half of the transposon.

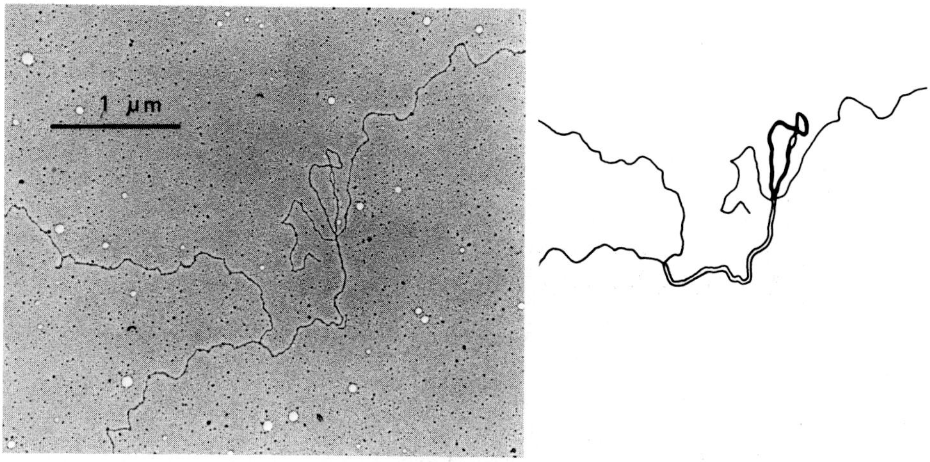

Fig.6: Electron micrograph of a DNA-heteroduplex formed between a pFS202-fragment and λ::Tn1771. The methods applied are essentially as described by Burkardt et al[13]. (λ::Tn1771 was constructed by transposotion of Tn1771 from pFS202 to a λ-phage carrying the deletion b515, b519, nin5[14]). Single line: heterologous DNA, double line: homologous DNA (r-determinant of Tn1771), thick line: underwound DNA ("snap back loop" of half the Tn1771-DNA).

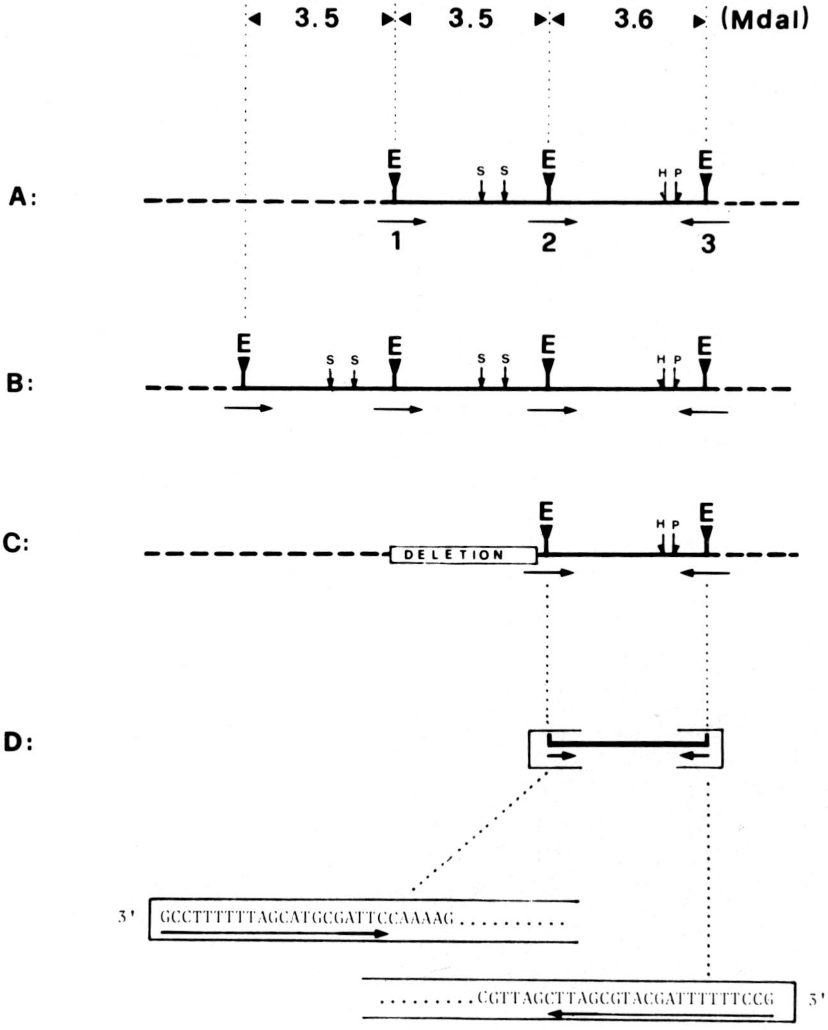

Fig. 7: Physical maps of Tn1771 structures.
A: normal structure, present on pFS202(-),
B: amplified structure present on pFS202(+),
C: reduced structure found on pFS203.
D: The DNA sequences read from both ends of the 3.6 Mdals EcoRI fragment of Tn1771. DNA sequencing was carried out basicly according to the method described by Maxam and Gilbert[15] using 3' labeled fragments.

In order to confirm the proposed arrangement of the interted repeats no. 2 and 3, we performed DNA sequencing analysis with focus on the regions surrounding the EcoRI sites. Preliminary data obtained so far for the DNA sequences at the ends of the 3.6 Mdals EcoRI fragment of Tn1771 indicate the existence of inverted repeats present at these positions. The sequences starting from the 3'-ends of the EcoRI cleveages sites are identical for the first 21 bases, as is shown in Fig. 7D.

These short inverse repetitive sequences probably represent only one part of the inverted repeats which may extend to the adjacent EcoRI fragments. Further DNA sequencing at the ends of the 3.5 Mdals EcoRI fragment of Tn1771 and the connecting host plasmid fragments should enable us to determine the complete extension of the repeated sequences.

CONCLUDING REMARKS

The 7.1 Mdals transposon Tn1771 is capable of intramolecular gene amplification in E.coli comprising the r-determinant region residing on a 3.5 Mdals fragment of pFS202. Tandem-repeated multiple r-determinant copies were found to be generated if cells were grown in the presence of high concentrations of tetracycline. To our knowledge this is the first case where a transposable genetic element for tetracycline resistance shows intramolecular gene amplification in E.coli. However, very similar results were obtained for the Tc-determinants genes of pRSD1, where an equally sized fragment of 3.5 Mdals is amplified during tetracycline stress of E.coli cells[18]. Furthermore it is indicated that this region of pRSD1 is encompassed by a transposable element showing similar physical characteristics as Tn1771 (R.Schmitt, personal communication). It is intended to clarify the molecular relationship between these transposons by DNA hybridization studies.

The identity of Tn1771 with the previous described tetracycline transposon Tn10 can be excluded since they differ in significant physical characteristics. Both the restriction map of Tn10[17] and its molecular structure, characterized by large inverted repeats at the ends[18], differ from those ones of Tn1771.

The gene amplification of Tn1771 in E.coli has its parallel in Streptococcus faecalis where a 2.65 Mdals unit of repetition was identified on the Tc^R plasmid pAMα1[5]. It has been shown that the

Tc-genes are flanked by identical sequences in direct repeats, which are probably involved in recombinational amplification of the r-determinant region. For the amplifiable segment of Tn1771 a similar arrangement of short direct repeats is proposed. Their location at the position of the EcoRI sites of the 3.5 Mdals fragment was deduced from the observation of spontaneous deletion and amplification processes obviously using these joints. The involvement of the cellular recombination system in these processes is indicated by their $recA^+$ dependence in E.coli (Schöffl, F. in preparation).

The existence of a third repetitive DNA segment on Tn1771, inversly orientated to the direct repeats, is indicated by the formation of two different "snap back structures" shown by electron microscopical homoduplex and heteroduplex studies. Taking into consideration that inverted repeats may account for transposition properties of a transposon, it can be predicted that also the reduced transposon (Fig. 7C) should be capable of transposition. Tn1771 is an example of a genetic element where possibly the same target sites (repetitive DNA sequences) are involved in the $recA^+$-independent transposition and a $recA^+$-dependent gene amplification.

ACKNOWLEDGEMENTS

We thank V.Pirrotta for providing a working situation in his laboratory and C.Tschudi for his helpful instructions in DNA sequencing. Furthermore we thank A.Pühler and W.Heumann for helpful discussions, Brigitte Scheel-Schneider for excellent technical assistance, W.Lotz for critical reading and Anneliese Hofmann for typing the manuscript.

This work was supported by grants of the Deutsche Forschungsgemeinschaft (Pu 28/8).

REFERENCES

1. Campell, A. et al. (1977) DNA Insertion Elements, Plasmids and Episomes (Bukhari, A.J. et al. ed.) 15-22, Cold Spring Harbor Laboratory.

2. Kleckner, N. et al. (1977) J. Molec. Biol. 125-159.

3. Mac Hattie, L.A. and Jackowski, J.B. (1977)DNA-Insertion Ele-

ments, Plasmids and Episomes (Bukhari, A.J. et al. ed.) 219-228, Cold Spring Harbor Laboratory.

4. Rownd, R.H. et al.(1975) Microbiology-1974(Schlesinger, D.ed.) 76-94, American Society for Microbiology Washington, D.C.
5. Yagi, Y. and Clewell, D.B. (1977) J. Bact. 129, 400-406.
6. Traub, W.H. and Kleber, I. (1975) Path. Microbiol. 43, 10-16.
7. Schöffl, F. (1979) submitted for publication.
8. Schöffl, F. and Pühler, A. (1979) Genet. Res., in press.
9. Schöffl, F. et al. (1977) In: Plasmids, Medical and Theoretical Aspects (Mitsuhashi et al. ed.) 161-170, Springer Verlag, Berlin.
10. Hansen, J.B. and Olson, R.H. (1978) J. Bact. 135, 227-228.
11. Radloff, et al. (1967) Proc. Natl. Acad. Sci. USA 57, 1514-1520.
12. Bazaral, M. and Helinski, D.R. (1968) J. Molec. Biol. 36, 185-194.
14. Davidson, N. and Sybalski, W. (1971) In: The Bacteriophage Lambda (Hershey, A.D. ed.) 42-82, Cold Spring Harbor Laboratory.
15. Maxam, A.M. and Gilbert, W. (1977) Proc. Natl. Acad. Sci. USA, 74, 560-564.
16. Mattes, R. et al. (1979) Molec. Gen. Genet. 168, 173-184.
17. Kleckner, N. et al. (1978) Genetics 90, 427-461.
18. Ptashne, K. and Cohen, S.N. (1975) J. Bact. 122, 776-781.

© 1979 Elsevier/North-Holland Biomedical Press
Plasmids of Medical, Environmental and Commercial Importance
K.N. Timmis and A. Pühler, editors

MULTIPLE INTEGRATION AND AMPLIFICATION OF TRANSPOSABLE DNA SEQUENCES IN HAEMOPHILUS INFLUENZAE R PLASMIDS.

R. Laufs, G. Jahn, H. Kolenda and P.-M. Kaulfers.
Institute of Medical Microbiology and Immunology, University of Hamburg, Martinistrasse 52, D 2000 Hamburg 20, Federal Republic of Germany

INTRODUCTION

The emergence of R factors in H.influenzae is of great clinical concern, since H.influenzae causes serious infections including meningitis, epiglottitis, pneumonia and otitis media. The conjugative Haemophilus influenzae R factors which have been described are closely related to each other and have more than 60% of their base sequences in common independent of their geographical origins and their antibiotic resistance markers (1,2). This paper describes the molecular nature of H.influenzae R factors specifying for two resistances and the localization of the resistance genes within the plasmids. The effect of antibiotics on these R factors was investigated.

MATERIALS AND METHODS

Bacterial strains. The bacterial strains and plasmids used are listed in Table 1.

Table 1. Bacterial strains and their plasmids.

Strain	Plasmid present	Mol wt of plasmid ($\times 10^6$)	Resistance pattern[a]	Source/reference
H.influenzae				
HK539	pHK539	36	Tc,Ap	W.Kilian, Aarhus
REV124	pREV124	32	Tc,Ap	J.Acar, Paris
HC234	pR1234	34	Tc,Cm	van Klingeren et al. (7)
KRE5367	pKRE5367	30	Ap	Laufs et al. (2)
FR16017	pFR16017	33	Tc	Kaulfers et al. (3)

[a]Ap, ampicillin; Tc, tetracycline; Cm, chloramphenicol.

Methods for the molecular characterization of the R factors. Media, isolations of unlabeled and labeled plasmid DNA, DNA-DNA duplex studies, agarose gel electrophoresis of DNA, conjugation, transformation and electron microscope DNA heteroduplex and homoduplex analysis have been recently described in detail (2,3).

RESULTS

We have recently shown that the 30 Mdal H.influenzae R factor pKRE5367 (Ap^r) and the 33 Mdal H.influenzae R factor pFR16017 (Tc^r) have an almost identical plasmid core and are different only due to the fact that pKRE5367 contains the ampicillin resistance transposon (TnAp) and that pFR16017 contains the tetracyline resistance transposon (TnTc). The transposons are integrated at different sites in the plasmid as was shown in heteroduplex molecules between pKRE5367 and pFR16017 (3). Now we have found two H.influenzae R factors in clinical isolates which encode Ap resistance as well as Tc resistance (pHK539 and pREV124). The presence of TnAp in these R plasmids was demonstrated using the 5.5 Mdal plasmid RSF1030 which contains the whole TnAp (4) as a molecular probe for DNA-DNA duplex studies. It had 53% of its base sequences in common with pHK539 and 51% with pREV124, indicating that both H.influenzae R plasmids contained the transposable ampicillin element. As a molecular probe for the detection of tetracycline resistance genes the 9.9 Mdal plasmid pKT007 was used which carries a 2.8 Mdal DNA sequence of Tn10 (5). It showed 28% homology with pHK539 and 25% homology with pREV124 indicating the presence of TnTc in these plasmids. As negative and positive controls the R plasmids pKRE5367 (Ap^r) and pFR16017 (Tc^r) were used, since all these H.influenzae R factors have essentially the same plasmid core. The base homology between the 30 Mdal R factor pKRE5367 and the 36 Mdal R factor pHK539 was found to be 80% and between pKRE5367 and the 32 Mdal pREV124 to be 67%. This polynucleotide sequence homology detected using the single strand specific endonuclease, S1, of Aspergillus oryzae for the analysis of the DNA-DNA duplexes (6) was confirmed by electron microscope experiments. They revealed that TnAp and TnTc in pHK539 as well as in pREV124 were not integrated at different sites in the plasmid core but that TnA in both plasmids was integrated in the inverted repeats of TnTc. Figure 1 shows a self-annealed pREV124 molecule.

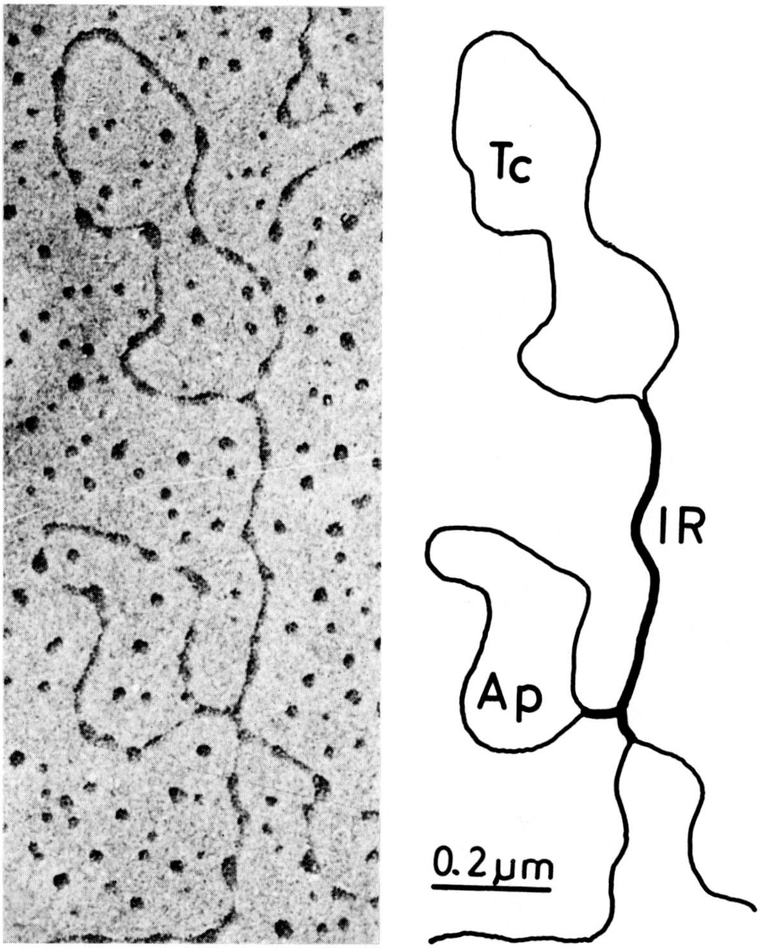

Fig. 1. Electron micrograph of formamide-spread single stranded, self-annealed pREV124 (Tc^r, Ap^r) molecule showing intramolecular sequence homology representing the Tc-Ap transposon . IR, inverted repeat sequence; Ap, ampicillin transposon; Tc, tetracycline transposon.

The 34 Mdal H.influenzae R plasmid pR1234 (7)which specifies Tc resistance combined with chloramphenicol (Cm) resistance had 63% of its base sequences in common with pKRE5367. Electron microscope heteroduplex and self-annealing experiments showed that the Tc and Cm resistance genes were located in a single composite DNA sequence which showed similar characteristics as those found in pHK539 and in pREV124. The Cm resistance specifying genes were found to be integrated in the insertion elements of TnTc and flanked at both sides by long inverted repeats (Fig.2.).

After several passages of pR1234 (Tc^r, Cm^r) on solid media containing 10 ug tetracycline ml^{-1}, the MICs determined in liquid media had raised not only from 10 ug to 30 ug tetracycline ml^{-1}, but at the same time from 10 ug to 40 ug chloramphenicol ml^{-1}. Strain HC234 (pR1234) now showed two plasmid DNA bands in the gel. The smaller plasmid was identical with that shown in Figure 2 and had a molecular size of 34.6 ± 1.4 Mdal. The second plasmid was 7.5 ± 0.4 Mdal larger in size. This size difference corresponded with the size of the Tc-Cm specifying structure and self-annealing experiments showed indeed that the larger plasmid carried the Tc-Cm structure integrated twice.

After subcultering single colonies of the tetracycline resistant H.influenzae strain FR16017 (pFR16017) on solid medium containing 10 ug/ml^{-1} tetracycline, the subsequent gel electrophoresis of the cleared lysate revealed now the presence of three plasmid DNA bands. The difference of the molecular size between each of these classes of plasmid was about 6 Mdal which corresponded well the 6.3 Mdal molecular size of the tetracycline transposon TnTc (3). This finding was due as shown by self-annealing experiments to the multiple integration of TnTc in pFR16017. TnTc was found to be integrated up to four times in pFR16017.

The further passage of the H.influenzae isolate FR16017 on Tc containing medium induced a second type of change of the plasmid pFR16017. The loop of TnTc was found to be amplified in steps of 4-5 Mdal. This size corresponded with that of the TnTc loop. The multiple integration of TnTc as well as the amplification of the loop in TnTc was correlated with higher MICs against Tc. The plasmid heterogeneity was found to be monoclonal.

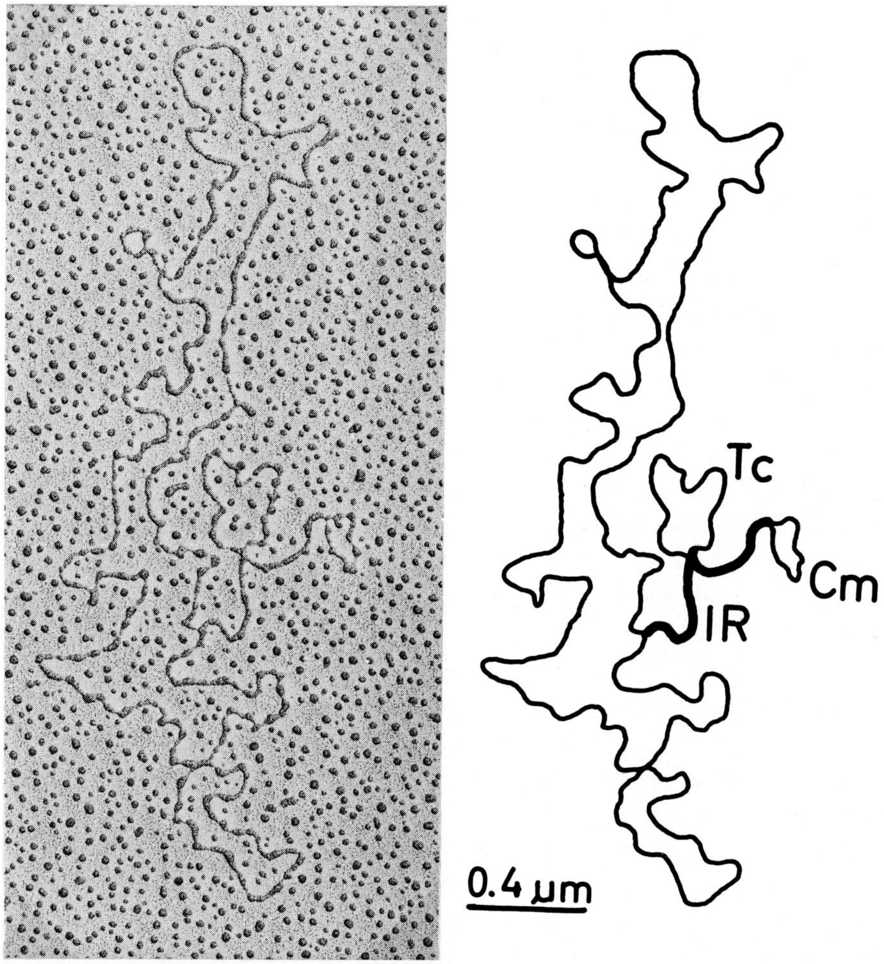

Fig. 2. Electron micrograph of a self-annealed pHK539 (Tc^r, Cm^r) molecule showing the transposon specifying combined tetracycline and chloramphenicol resistance. IR, inverted repeat; Tc, tetracycline transposon; Cm, cholramphenicol transposon.

DISCUSSION

The molecular characterization of the three H.influenzae R plasmids each of them specifying two resistances revealed that in each plasmid the genes for the two resistances were integrated in a single composite DNA sequence. Each of the three R plasmids contained the tetracycline transposon TnTc and in each plasmid a second resistance was integrated close to the termini of the inverted repeats of TnTc. The ampicillin transposon TnA was found in electron microscope analysis of self-annealed pHK539 molecules to be integrated in the double-stranded stem of TnTc close to its end towards the single-stranded Tc loop while in pREV124 the TnA was found to be integrated in the double-stranded stem of TnTc close to its end towards the plasmid core. The self-annealed plasmid DNA of pR1234 (Tc^r, Cm^r) showed the Cm genes integrated over long inverted repeats in the stem of TnTc at a site similar to the integration site of TnA in pHK539. The inverted repeats of TnTc and those flanking the Cm loop had the same molecular size and it seems possible that they are identical. In all of the R plasmids from H.influenzae examined the inverted repeats of TnTc were used for the integration of the second resistance specifying transposon and we name this observation "pick-a-back phenomenon". The termini of TnTc and TnA served as hotspots for the integration events.

The growth of the H.influenzae R plasmid pHK539 (Tc^r, Cm^r) and pFR16017 (Tc^r) on the Tc containing medium resulted in the multiple integration of the resistance transposons. In the case of pHK539 the transposon composed out of TnTc (Tn10) and TnCm (Tn9) (8) linked with long inverted repeats was found to be integrated at two different sites of the same plasmid molecule. The Tc selection pressure resulted in the case of pFR16017 in the multiple integration of TnTc which was found to be integrated up to four times. The multiple integration of the transposons did not result in a change of the size of the plasmid cores.

The prolonged cultivation of pFR16017 on Tc containing medium resulted in the amplification of the loop in TnTc, which was found to be amplified in steps up to 6 times of its normal size. Multiple integration as well as amplification of the transposon were monoclonal and correlated with higher MICs.

ACKNOWLEDGEMENTS

This work was supported by the Deutsche Forschungsgemeinschaft, Bonn-Bad Godesberg, W. Germany. We thank Mrs. I. Richter for excellent technical assistance.

REFERENCES

1. Elwell, L.P., Saunders, J.R., Richmond, M.H. and Falkow, S. (1977) J. Bacteriol. 131, 356-362.
2. Laufs, R. and Kaulfers, P.-M. (1977) J. Gen. Microbiol. 131, 277-286.
3. Kaulfers, P.-M., Laufs, R. and Jahn, G. (1978) J. Gen. Microbiol. 105, 243-252.
4. Heffron, F., Sublett, R., Hedges, R., Jacob, A. and Falkow, S. (1975) J. Bacteriol. 122, 250-256.
5. Timmis, K.N., Cabello, F. and Cohen, S.N. (1978) Mol. Gen. Genet. 162, 121-137.
6. Crosa, J.H., Brenner, J. and Falkow, S. (1973) J. Bacteriol. 115, 904-911.
7. van Klingeren, B.J., van Embden, D.A. and Dessens-Kroon, M. (1977) Antimicrob. Agents Chemother. 11, 383-387.
8. Rosner, J.L. and Gottesman, M.M. (1977) DNA insertion elements, plasmids, and episomes, Cold Spring Harbour Laboratory, pp. 213-218.

© 1979 Elsevier/North-Holland Biomedical Press
Plasmids of Medical, Environmental and Commercial Importance
K.N. Timmis and A. Pühler, editors

STRUCTURAL ELEMENTS OF THE R1 PLASMID AND THEIR REARRANGEMENT IN DERIVATIVES OF IT AND THEIR MINIPLASMIDS

DIETMAR BLOHM
BASF Aktiengesellschaft, Hauptlaboratorium
D6700 Ludwigshafen (West-Germany)

R1 is a conjugative R plasmid, mediating antibiotic resistance to ampicillin (Ap), chloramphenicol (Cm), kanamycin (Km), streptomycin (Sm) and sulfonamides (Su). It was discovered by Datta and Meynell in 1966 and was the parent plasmid for R1drd-19, a derepressed mutant[1], which is largely used in studies of the genetics and physiology of this kind of medically important infectious heredity material.

Just as the also extensively investigated R-plasmids R100 (NR1 or R222) and R6-5, it belongs to the incompatibility group F_{II} and shares with them extensive sequence homology, as shown by heteroduplex analysis[2].

R1 consists of two parts, connected by IS-1 sequences. The RFT part codes for the transfer functions and the r-det contains different transposons as e. g. Tn 3 and Tn 4[3].

It is the parent plasmid for deletion mutants such as the isolated RTF or R1drd-16. The latter has lost all resistance markers besides Km. R1drd-19 also gives rise to different multicopy mutants[4]. These spontaneously generate miniplasmids as e. g. plasmids of the Rsc group, which all contain Tn 3 and the rep A function[5].

For accurate localization of structural elements on the plasmid molecules restriction mapping is very helpful . The first restriction map obtained for one of these large R plasmids was determined in 1976 by Tanaka et al. for R100 using the enzyme EcoRI[6]. By mapping of R6-5 with EcoRI and HindIII it was shown, that R100 and R6-5 have a series of restriction fragments in common[7]. The EcoRI restriction pattern of R1drd-19 differs from both to an extent which is shown in figure 1.

R1drd-19 (this work)	R100 (Tanaka et al.)	R6-5 (Timmis et al.)
(A) 11.7	(A) 13.5	(E-1) 16.5
(B) 11.5	(B) 8.5	(E-2) 8.8
(C) 8.5	(C) 7.55	(E-3) 8.7
(D) 6.3	(D) 7.2	(E-4) 7.2
(E) 5.1	(E) 5.0	(E-5) 5.2
		(E-6) 4.8
(F) 4.3	(F) 4.0	(E-7) 4.0
(G) 3.5	(G) 3.5	(E-8) 3.5
	(H) 3.2	(E-9) 3.1
(H) 2.8	(I) 2.7	(E-10) 2.7
(I) 1.85	(J) 1.8	
(J) 1.75		
(K) 1.35		
(L) 1.30	(K) 1.1	(E-11) 1.2
(M) 0.95	(L) 1.0	(E-12) 1.1
(N) 0.85	(M) 0.75	(E-13) 0.85
(O) 0.55		
(P) 0.15		
(Q) 0.10		
62.5	59.3	67.5

Fig. 1.
EcoRI fragments of R1drd-19, R100 and R6-5, arranged according to their size independent from its degree of homology. Molecular weights are given in Megadalton.

Using the restriction-endonucleases BamHI, EcoRI, HindIII and SaII a physical map was drawn up from fragments of R1 or of its derivatives, obtained by partial digestion with each of the enzymes, by double and triple digestion of the DNA with two or three enzymes and in part by isolation of individual bands from preparative agarose gels and renewed cutting of the fragments obtained, with a further enzyme.

The first molecule which could be mapped was the RTF. Using a computer program an EcoRI and SaII fragment order could be established which is contained in figure 2.

Based on this structure the Km region, present in R1drd-16, and the Ap-Su-Sm-Cm region of the Km sensitive multicopy plasmid pKN 102 (previously named R1drd-19B2) could be mapped. Appropriate maps are also shown in figure 2.

The evidence, that R1drd-16, pKN102 and RTF are homologous to the parent plasmid R1, differing from it by different deletions, is given by comparing the restiction pattern of R1 and those of its derivatives, obtained with different enzymes.

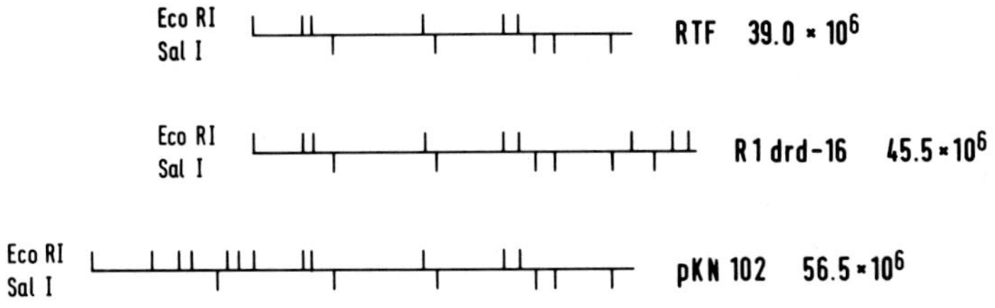

Fig. 2.
Physical maps and molecular weights in Dalton of RTF, R1drd-16 and pKN102 obtained from fragments generated by digestion with EcoRI and SalI.

In figure 3 such a comparison is shown, using EcoRI.

The same results are obtained from digestion pattern generated with other enzymes, and even double fragmentation with EcoRI and SalI shows that most fragments are identical. The only exception are fragments from ring closing elements and those corresponding

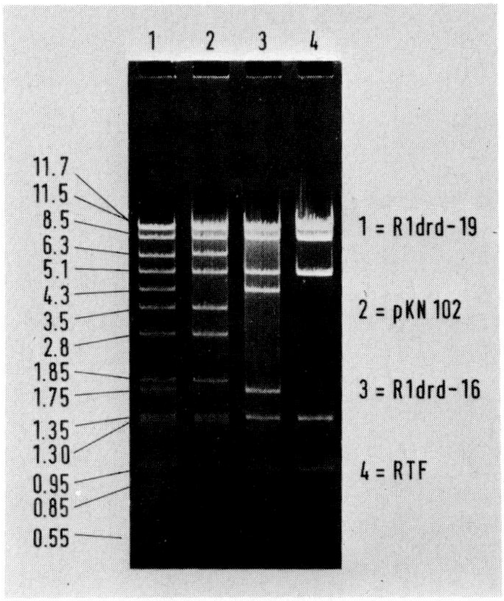

Fig. 3.
Comparison of the EcoRI restriction pattern of R1drd-19, pKN102, R1drd-16 and RTF, obtained after electrophoresis in 0.9 % agarose gels.

to the deletions.

Therefore it seems reasonable to combine the maps of RTF, R1-drd-16 and pKN102 in order to form the R1drd-19 map, which is shown in figure 4.

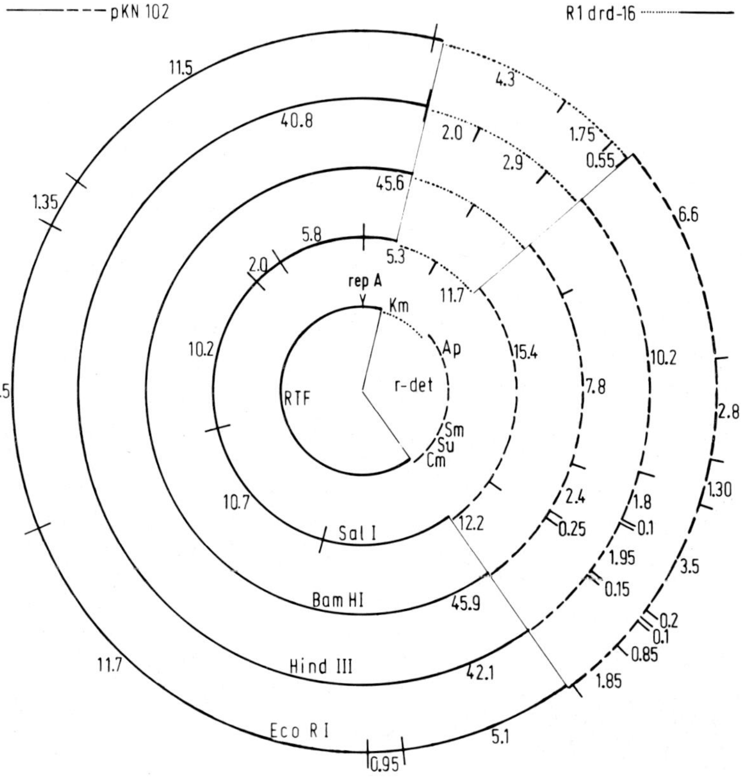

Fig. 4.
Physical map of R1drd-19, combined from maps of RTF, R1drd-16 and pKN102. Molecular weights are given in Megadalton. Fragments differing in size are labelled outside the circles for pKN102 and inside them for R1drd-16.

The best known functional structure of R plasmid molecules are IS-sequences, transposons and the antibiotic resistance genes located on them, the rep A and inc regions, some of the tra functions and sequences coding for small RNAs.

Although in R100 and R6-5 a rep B region has been characterized it is still uncertain, whether R1 carries a second replication origin or not.

If one assumes that this hypothetical R1 rep B origin needs functions from different parts of the molecule, in vivo recombination could be a chance to obtain it on a miniplasmid rather than by fragment cloning. In vivo recombination also allows to test, whether miniplasmids with other structures than Tn 3 can be obtained from the pKN102 plasmid[8].

Therefore pKN102 DNA was cleaved with HindIII and ligated without adding any other DNA and transformed in E. coli C. By selection on Ap, Cm or Sm alone or on combinations of these markers in which one of them is lost, transformants were obtained, which contain a large variety of different plasmids.

Many of these pKN102 descendants were unstable and were lost or they split into molecules of different size classes during propagation.

Only those were finally analyzed, which were stably maintained in the course of investigation in respect to their molecular size and to all their resistance markers on which they were originally isolated. A few of these plasmids turned out to be very similar with each other or to plasmids isolated completely independent from this experiment.

pDB202 for instance is identical with Rsc 11 in as much as it contains only an additional sequence of about 0.2 Megadalton, connected to that part of Rsc 11 which is originally a part of the RTF[9].

26 plasmids were analyzed in more detail. They have molecular weights between 4.7 and about 43 Megadalton. Two of them mediate only Sm resistance, five only Cm resistance, twelve only Ap resistance and seven plasmids mediate resistance to Ap and Sm.

By detailed restriction mapping of a few of them, rearrangement of the known structures could be estimated. pDB20 for example is a plasmid mediating only Cm resistance and consists of a complete RTF connected to the cat gene region, which is characterized by EcoRI fragments of 1.85 and 0.85 Megadalton. Other Cm plasmids contain more than half of the RTF as shown for pDB3 in figure 5.

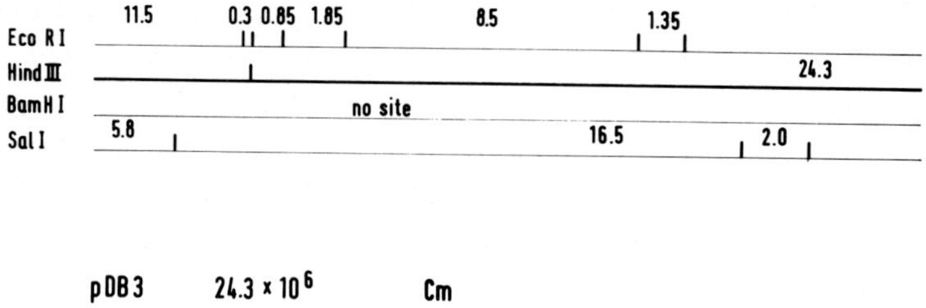

Fig. 5.
Physical map of pDB3. For drawing it in a linear scale the cut size was chosen arbitrarily.

Not only the length of the region which is connected to the rep A function on the RTF varies, but also the sequences adjacent to the cat-gene region are cut out at different sites. That is shown in figure 6 using pDB1 and pDB2 as examples.

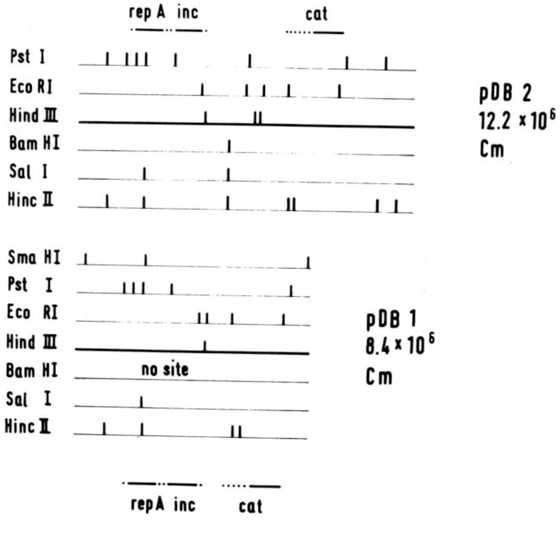

Fig. 6.
Physical map of pDB1 and pDB2 drawn up for 7 restriction-endonucleases. The structure of the left part of both plasmids is identical with structures described for the rep A and the inc regions of the RTF.

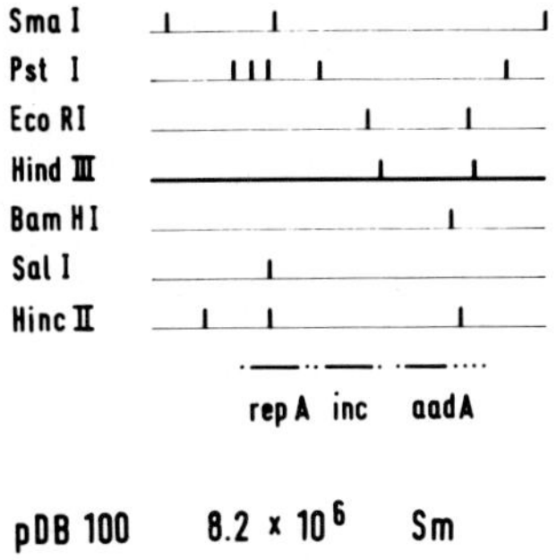

Fig. 7.
Physical map of pDB100. The extention of the aadA-gne is limited on the left side because the HindIII cut site can be shown to belong already to sequence of the RTF.

Like most of the plasmids isolated in the course of these experiments, the resistance marker regions are joined in these plasmids to the rep A region near to IS-1b[8]. Nevertheless the recombination event occurs not always on one of the endpoints of the IS-1 position. This can also be shown in pDB100, which contains the aadA-gene of R1, known to be a part of Tn 4, and is in this case as a separate unit recombinated with the rep A sequence.

On the restriction map of pDB100, presented in figure 7, the

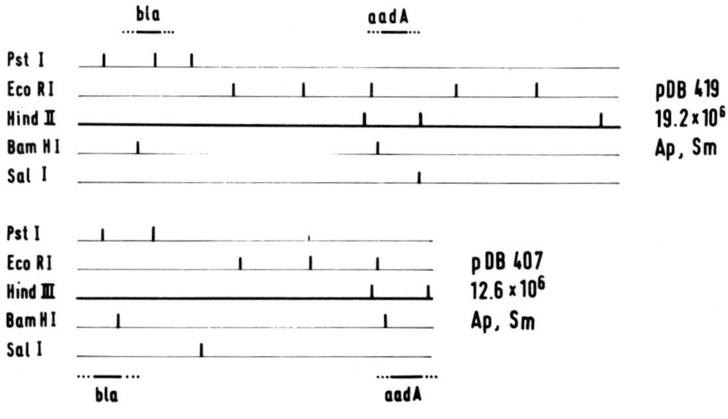

Fig. 8.
Physical map of pDB407 and pDB419. The bla-gene region is characterized by the BamHI cut site in the 1.85 Megadalton Pst-fragment and the aadA region by relative positions of the EcoRI, HindII and BamHI cut sites, which are also found in pDB100.

aadA-gene region is seen to be joined almost immediatly to the HindIII cut site, which is adjacent to the rep A sequence.

In all these plasmids the rep A structure is easily detectable because a series of well characterized restriction fragments are typical for it.

In pDB407 and pDB419, mediating both Ap and Sm resistance, these fragment patterns could not be detected.

Their restriction maps, shown in figure 8, prove that they actually do not contain any region similar to the rep A or inc region of R1.

According to these restriction maps, the bla and aadA gene regions are probably differently orientated in these plasmids. The localization of the replication origin of these plasmids is still unknown.

Preliminary results show, that the copy number of pDB407 and

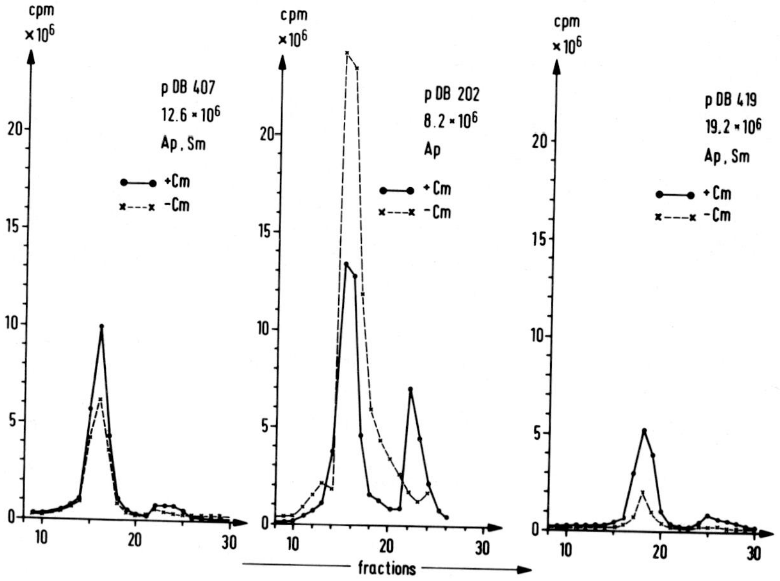

Fig. 9.
^{3}H-thymidine incorporation in pDB407 and pDB419 with and without Cm treatment, compared to pDB202 which carries the R1 rep A function

pDB419 is much smaller than that of similar plasmids containing the rep A function of pKN102. The replication of pDB407 and pDB 419 seems to be relatively insensitive against Cm, if it is measured as ^{3}H-thymidine incorporation by counting CsCl-Ethidiumbromide gradients prepared from cleared lysates.

In figure 9 these gradients are compared with those prepared from E. coli C harbouring the pDB202, which is almost identical with Rsc11, as already described. Although gelelectrophoresis of enzyme digested pDB407 or pDB419 DNA gives homogenous fragment patterns, electron microscopic measurements show, that a significant number of molecules is smaller than the main part, suggesting a partial instability of these structures.

REFERENCES

1. Meynell, E. and Datta, N. (1967) Nature (Lond.), 214, 885-887
2. Sharp, P.A., Cohen, S.N. and Davidson, N. (1973) J. Molec. Biol. 75, 235-255
3. Kopecko, D.J., Brevet, J. and Cohen, S.N. (1976) J. Molec. Biol. 108, 333-360
4. Uhlin, B.E., Gustafsson, R., Molin, S. and Nordström, K. (1978) DNA synthesis: present and future (Kohiyama, M. and Molineux I., eds.) Plenum Publishing Corporation, New York
5. Luibrand, G., Blohm, D., Mayer, H. and Goebel, W. (1977) Molec. gen. Genet. 152, 43-51
6. Tanaka, N., Cramer, J.H. and Rownd, R.H. (1976) J. Bact. 127, 619-636
7. Timmis, K.N., Cabello, F. and Cohen, S.N. (1978) Molec. gen. Genet. 162, 121-137
8. Goebel, W., Lindenmaier, W., Schrempf, H., Kollek, R. and Blohm, D. (1977) R-factors: Their properties and possible control (Drews, J. and Högenauer, G., eds.) Springer, Berlin-Heidelberg-New York
9. Ohtsubo, E., Rosenbloom, M., Schrempf, H., Goebel, W. and Rosen, J. (1978) Molec. gen. Genet. 159, 131-141

The main part of this work was carried out at the Gesellschaft für Biotechnologische Forschung mbH, 3300 Braunschweig-Stöckheim

III – PLASMIDS AND ANTIBIOTIC SYNTHESIS

© 1979 Elsevier/North-Holland Biomedical Press
Plasmids of Medical, Environmental and Commercial Importance
K.N. Timmis and A. Pühler, editors

PLASMIDS IN STREPTOMYCES COELICOLOR AND RELATED SPECIES

D. A. HOPWOOD, M. J. BIBB, J. M. WARD, AND J. WESTPHELING
John Innes Institute, Norwich, NR4 7UH (England)

INTRODUCTION

Interest in Streptomyces plasmids at this symposium scarcely needs justification; the actinomycetes produce several thousand antibiotics. The number of those having significant commercial application, in human and veterinary medicine and in agriculture, was put at about 70 in 1977 and is steadily increasing[1]. Glucose isomerase, one of the most important microbial enzymes in commerce because of its key role in the production of corn syrup, is also usually produced from Streptomyces spp.

Plasmids are involved in the determination of several characteristics of streptomycetes[2], including antibiotic production in many species[3]. Their role in the latter can be direct, as in the case of methylenomycin A in S. coelicolor A3(2) (see below) but it is probably more often indirect[3,4].

This paper summarises information on plasmids in S. coelicolor A3(2), still genetically by far the best characterized streptomycete[5,6], and in its close relative S. lividans strain 66[7]. Although still at an early stage compared with studies on certain plasmids of Gram-negative bacteria, our results, together with those on S. reticuli described by H. Schrempf and W. Goebel at this meeting, give a glimpse of the variety of interesting and relevant phenomena awaiting further investigation in Streptomyces, several of which, at least, are qualitatively different from comparable phenomena already known in E. coli.

CHROMOSOME MOBILIZATION BY THE TWO SEX FACTORS OF STREPTOMYCES COELICOLOR A3(2)

Development of knowledge about the two sex plasmids of S. coelicolor A3(2), particularly in relation to chromosomal recombination, is documented elsewhere[5,8,9]. The original isolate contained both autonomous plasmids, SCP1 and SCP2, giving a chromosomal recombination frequency ("fertility") of about 10^{-5} to 10^{-6} amongst pairs of genetically marked derivatives of the wild-type. Either plasmid can be lost, leaving the other in residence, reducing the fertility by not more than an order of magnitude if SCP1 is lost, or less if SCP2 is eliminated. Only when both plasmids are absent from both parents in a cross is the fertility markedly reduced, to 10^{-7} to 10^{-8}, although not totally abolished. The fertility of $SCP1^+$ X $SCP1^-$ crosses is not greater than that of $SCP1^+$ X $SCP1^+$ crosses, whereas $SCP2^+$ X $SCP2^-$ combinations are some two orders of magnitude more fertile than $SCP2^+$ X $SCP2^+$ crosses. This result might suggest a phenomenon akin to that of entry exclusion in *E. coli* for SCP2 but not SCP1. However, the effects of such a phenomenon, together with possible plasmid incompatibility and factors specific to the mycelial coenocytic growth habit of streptomycetes, are at present hard to evaluate.

The apparent frequency of SCP1 or SCP2 transfer from a $plasmid^+$ into a $plasmid^-$ strain during the course of a cross ("apparent" because of an unquantified contribution to the $plasmid^+$ progeny by secondary transfers from the originally infected individuals) is nearly 100%, irrespective of the presence or absence of the other plasmid. Both plasmids are therefore, by definition, self-transmissible.

SCP1 certainly interacts with the chromosome, giving integrated donors of several kinds which can give extremely fertile crosses with $SCP1^-$ cultures: up to 100% of the progeny of such matings are recombinants in respect of chromosomal markers near to the site of SCP1 integration. SCP1-prime plasmids, analogous to F-prime plasmids in *E. coli*, also occur. On the other hand, SCP2 has yet to be shown to interact with the chromosome. A class of variants of SCP2, called $SCP2^*$, is known which confers considerably enhanced fertility on its host compared with that found in strains with the parent SCP2 plasmid, but since no preferential donation of any chromosomal regions has been detected, evidence of direct interaction with the chromosome is lacking.

The fertility of $SCP2^*$ X $SCP2^*$ crosses is similar to that of $SCP2^*$ X $SCP2^-$ crosses (about 10^{-3}). In this respect the $SCP2^*$ variant resembles the "wild-type" form of SCP1. $SCP2^*$ strains also resemble $SCP1^+$ strains in showing a marked "lethal zygosis" reaction (see below) against a strain lacking the corresponding plasmid, whereas the parent $SCP2^+$ strains show a very much weaker reaction against $SCP2^-$ strains.

SCP1 AND METHYLENOMYCIN

SCP1 is particularly interesting because of its association with the antibiotic methylenomycin A. SCP1-containing cultures of *S. coelicolor* produce methylenomycin A and are resistant to it, while $SCP1^-$ *S. coelicolor*

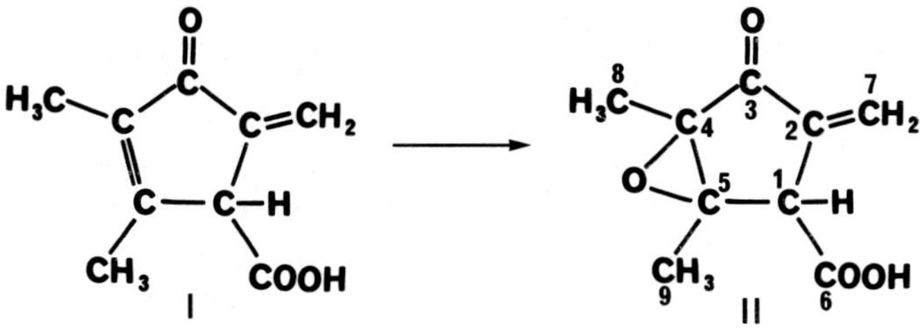

Fig. 1. Desepoxy-4,5-didehydromethylenomycin A (I) and methylenomycin A (II).

strains, like nearly all other Streptomyces spp., fail to produce the antibiotic and are sensitive to it[10,11]. The evidence that SCP1 carries structural genes for all the biosynthetic enzymes uniquely involved in methylenomycin A synthesis has been reviewed[3]. It consists of the findings that point mutations (mmy) blocking methylenomycin production invariably map on SCP1 rather than the chromosome and that SCP1, when transferred to S. lividans or S. parvulus (see below), confers on them the ability to make methylenomycin A.

We do not yet know the precise roles of the SCP1-linked methylenomycin genes, or their exact number. The final step in the synthesis is epoxidation of desepoxy-4,5-didehydromethylenomycin A (I) to methylenomycin A (II) (Fig. 1), and the mmy mutants point to the existence of at least four or five earlier but as yet unidentified steps[11]. The origin of the nine carbon atoms of I and II is only partially known: on the basis of ^{13}C NMR labelling after feeding ^{13}C enriched acetates, four (C-6, C-1, C-5, C-9) probably are derived from acetate via an acetoacetate unit, with two others (C-4, C-8) also coming from acetate[13] (U. Hornemann, personal communication), but the origin of the other three carbon atoms and the steps involved in assembling the cyclopentane ring are unknown. Efforts to determine the complete gene-enzyme relationships in this biosynthesis are certainly warranted since this remains the only proven example of an extensive series of plasmid-linked anabolic structural genes. In several other cases of plasmid involvement in antibiotic production in streptomycetes, the plasmids apparently act in some way on the expression of chromosomally located structural genes[3,4], since mutations having the properties expected of simple interruptions in the biosynthetic pathway are found to map on the chromosome while loss of plasmids results in a failure to produce the antibiotic, or to produce it at normal levels, under particular conditions.

PHYSICAL CHARACTERIZATION OF SCP2

SCP2 is readily isolated on CsCl-ethidium bromide density gradients after conventional cleared lysate procedures[9,14]. Characterized by electron microscopy and by the analysis of restriction endonuclease digests on agarose gels, the plasmid has a molecular mass of 18-20 Md. A considerable proportion of dimer molecules and occasional trimers are found. SCP2* has not been distinguished physically from SCP2.

Our results[9] indicate the following numbers of recognition sites for commonly used restriction enzymes: EcoRI, 1: HindIII, 1; BamHI, 5; PstI (SalPI), 4; and SalGI, more than 20. Schrempf and Goebel[14] reported similar results (their pSH1 = SCP2) except that only 4 BamHI and 3 PstI sites were recognised, together with many sites for SmaI and HincII.

PHYSICAL STUDIES ON SCP1

SCP1 has proved very difficult to isolate; no trace of it is seen by the standard procedures that readily reveal SCP2[9,14]. Gentle methods for isolation of large plasmids, greater than 100 Md[15-18], have also failed to isolate SCP1. Many variations of these procedures were tried, including use of a wide range of ionic and non-ionic detergents, several pronases, protease K, RNase and heat treatments, in various sequences and combinations, but DNA corresponding to SCP1 was not isolated (J. Westpheling unpublished).

These procedures depend on a supercoiled ccc configuration for plasmid isolation and differences among them are in methods of cell lysis and enrichment for plasmid DNA. Methods which do not depend on SCP1 existing as a supercoiled ccc DNA molecule were also tried. They included use of neutral sucrose gradients of various concentrations and non-ionic density matrices such as metrizamide, isolation of nucleoids and their gentle dissociation[19,20], lysis directly on to agarose gels[21], and techniques for relaxation complex isolation[22,23]. SCP2 and a Ti plasmid in *Agrobacterium tumefaciens*, when present as an additional control, were successfully isolated by all these procedures, but SCP1 was not.

Recently, a new lysis procedure has been developed which is qualitatively different from those used previously in *Streptomyces* (J. Westpheling, unpublished). The cellular material is completely solubilized with high concentrations of lysozyme and EDTA, followed by SDS in the presence of diethylpyrocarbonate, a nuclease inhibitor. After heating to 60°C the lysates are absolutely clear and relatively non-viscous.

Concentrated Tris-HCl buffer (pH 12) is then added which effects a small basic pH change, and chromosomal DNA is removed by high salt-SDS precipitation. By this method SCP1$^+$ SCP2$^-$ strains yield a band on CsCl-ethidium bromide density gradients which has the same position and density with respect to chromosomal DNA as the SCP2* band in SCP1$^-$ SCP2* cultures. This band is absent from SCP1$^-$ SCP2$^-$ strains. The material has a u.v. absorbance spectrum characteristic of nucleic acids, and incorporates ^{3}H-thymidine in vivo, but unambiguous characterization of the DNA as a unique plasmid species representing SCP1 has yet to be achieved. Since there are other genetically defined plasmids in various Streptomyces spp., refractory to isolation by commonly used methods, the perfection of a method for the reproducible isolation of SCP1 might have applications beyond S. coelicolor.

DNA-DNA hybridization has recently been applied to detect DNA sequences specific to SCP1-containing strains (J. Westpheling and M. J. Byers, unpublished). The first series of experiments involved a negative selection for SCP1 sequences. Total DNA from S. coelicolor SCP1$^+$ SCP2$^-$ was labelled in vitro with ^{125}I and repeatedly reacted with unlabelled total DNA from an S. coelicolor SCP1$^-$ SCP2$^-$ strain until no further reaction was detected. Remaining sequences from the SCP1$^+$ strain which reacted only with DNA from this strain and not with DNA from the SCP1$^-$ strain represented approximately 1.6% of the S. coelicolor genome. In a second series of experiments a positive selection for SCP1 sequences was made when total S. coelicolor SCP1$^+$ SCP2$^-$ ^{125}I-DNA was reacted with unlabelled total DNA from SCP1$^+$ SCP2$^-$ S. parvulus ATCC 12434. The latter DNA was shown to have very low sequence homology with S. coelicolor DNA and of those sequences which did form hybrids, only a small proportion had high melting stability. These represented 1.6% of the total S. coelicolor SCP1$^+$ DNA. These data may be interpreted as a firm indication that SCP1 exists as DNA and represents approximately 1.6% of the total S. coelicolor DNA. 1.6% is equivalent to 100 Md of DNA, assuming a genome size of 7.09 c 10^9 daltons[24].

TRANSFER OF SCP1 AND SCP2 INTO S. LIVIDANS AND S. PARVULUS

No ccc DNA has been detected by standard procedures in normal cultures of S. lividans strain 66 (John Innes Stock No. 1326) or S. parvulus ATCC 12434 by standard procedures[9,14]. Genetic recombination occurs in both these strains[25] (D. A. Hopwood and H. M. Wright, unpublished), but it is not clear whether or not this depends on the activity of sex plasmids.

Thus there is as yet no firm evidence, physical or genetical, for the normal existence of plasmids in the strains.

Hopwood and Wright[25] transferred the SCP1 plasmid to *S. lividans* in matings with SCP1$^+$ *S. coelicolor* A3(2). Detection was based on the expectation that SCP1$^+$ spores of *S. lividans* arising from the mating and developing in a dense lawn of the original (SCP1$^-$) *S. lividans* parent (with selection against the *S. coelicolor* donor) would give rise to small zones of inhibition of the lawn through the production of the SCP1-specific antibiotic (now known to be methylenomycin A). Thus each SCP1$^+$ microcolony produced a "pock" from the centre of which a clone of SCP1$^+$ *S. lividans* could be recovered. It was later observed that transfer of certain *mmy* mutants of SCP1 from *S. coelicolor* to *S. lividans* could also be detected by "pock" formation; thus the "pocks" were not related to methylenomycin production. Their development was found to depend on efficient plasmid transfer during growth of the confluent plate culture - transfer-defective mutants of SCP1[26] did not elicit the reaction, nor did SCP1$^+$ spores of *S. coelicolor* in a lawn of SCP1$^-$ *S. lividans* or SCP1$^+$ *S. lividans* in a lawn of SCP1$^-$ *S. coelicolor* (D. A. Hopwood and H. M. Wright, unpublished) - and it was concluded that partial inhibition of the SCP1$^-$ mycelium was caused by entry of the SCP1 plasmid, a phenomenon possibly analogous to lethal zygosis in *E. coli*[27].

SCP1 can readily be transferred also to *S. parvulus*, the original basis of detection being selection for methylenomycin resistant *S. parvulus* colonies arising from a mating with SCP1$^+$ *S. coelicolor* A3(2)[8]. Following the isolation of SCP1-prime plasmids carrying prototrophic alleles of chromosomal genes, selection for transfer of the plasmid between the three species can readily be made[8].

SCP2*, like SCP1, can be transferred by mating from *S. coelicolor* into *S. parvulus* or *S. lividans*, and detected by the presence of "pocks" due to the lethal zygosis elicited by this variant of SCP2 (M. J. Bibb and J. M. Ward, unpublished).

The apparent frequency of transfer of SCP1 in matings from *S. coelicolor* into *S. lividans* or *S. parvulus* is about 1-10%[25] (D. A. Hopwood and H. M. Wright, unpublished). Transfer back from either species to *S. coelicolor* occurs at a frequency several orders of magnitude lower. In spite of a failure to detect restriction-modification systems differentiating the three strains in tests with several bacteriophages[25,28], or type II restriction endonuclease activity in extracts of *S. coelicolor* or *S. lividans*[28], this

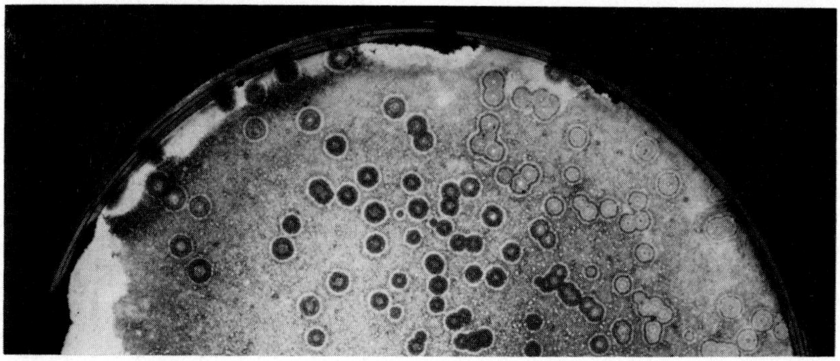

Fig. 2. "Pocks" of lethal zygosis caused by spores carrying SLP1.1 in a background of S. lividans strain 1326.

result suggested the presence of a restriction system in S. coelicolor. It has recently been confirmed by interspecific crosses between S. coelicolor and S. griseus[29] and by the much reduced frequency of transformation of S. coelicolor protoplasts (see below) by SCP2* DNA isolated from S. lividans or S. parvulus (M. J. Bibb and J. M. Ward, unpublished).

A NOVEL FAMILY OF PLASMIDS FROM S. LIVIDANS

Recently, a new class of Streptomyces plasmids has been revealed by studies of S. lividans 1326 (M. J. Bibb, J. M. Ward and D. A. Hopwood, manuscript in preparation). Although this strain normally shows no evidence of plasmid DNA, occasional "pocks" were seen when large populations of spores were grown into confluent cultures on agar plates. Cultures derived from the centres of such "pocks" elicited lethal zygosis (ltz^+) when replica plated to lawns of the parent 1326 culture. Moreover the ltz^+ phenotype was transferred efficiently in matings of such cultures with genetically marked derivatives of S. lividans 1326 (Fig. 2). Thus, by analogy with SCP1 and SCP2*, there was genetic evidence for a transmissible plasmid, designated SLP1, in cultures derived from strain 1326 itself.

Many separate "pocks" arising from different clones of strain 1326 were picked and purified. The resulting cultures differed in stability - the proportion of ltz^- colonies in the culture which varied from 0 to over

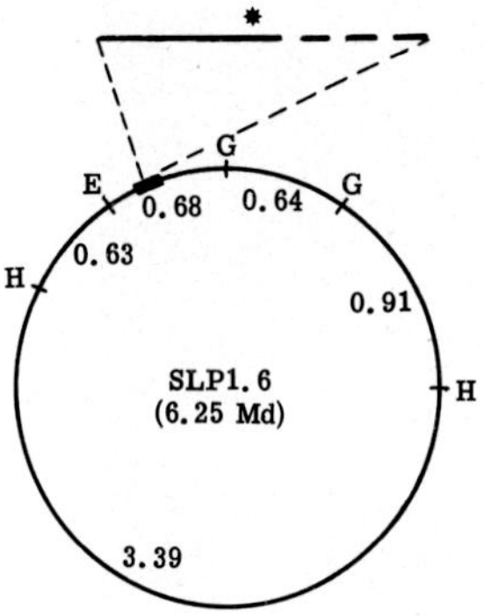

E = EcoRI
H = HindIII
G = SalGI
B = BamHI
P = SalPI (PstI)

Plasmid	Size (Md)	Restriction Sites E	H	G	B	P	Extra segment*
SLP1.6	6.25	1	2	2	0	0	None
SLP1.4	6.4	1	3	2	0	0	H
SLP1.3	6.88	1	3	3	0	0	H G
SLP1.1	7.13	1	3	3	0	1	H GP
SLP1.5	7.28	1	3	3	0	1	H GP
SLP1.2	8.23	1	3	3	1	2	H GP PB

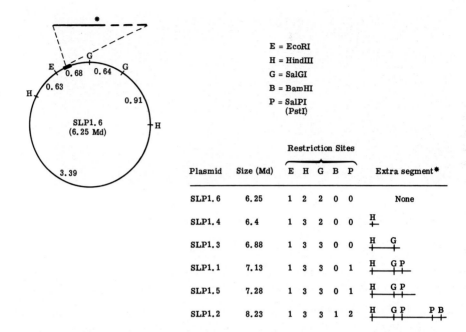

Fig. 3. Molecular sizes and restriction maps of the SLP1 family of plasmids. All the plasmids have the basic restriction map shown for SLP1.6; other members of the family have an extra segment of DNA in the region indicated on the SLP1.6 map.

90% - and in the precise appearance of the "pocks" which they produced on strain 1326 under standard conditions. Thus there was evidence of variation between different isolates of SLP1, but all were related by the criterion that a culture carrying any one variant was resistant to lethal zygosis exerted by a strain carrying any other variant.

Evidence for physical variation came from experiments to isolate ccc DNA from the putative SLP1$^+$ variants. Many of the *ltz*$^+$ isolates yielded no detectable plasmid DNA by the standard procedure[9], but others did so, and there was some correlation between "pock" phenotype and the presence or absence of isolable ccc DNA. Restriction endonuclease analysis of the plasmid DNA isolated from various independent SLP1$^+$ cultures showed that each contained a uniform class of molecules, but that different "pock" types typically yielded plasmids differing in size and restriction pattern. The first *ltz*$^+$ culture to be studied, strain M170, yielded a plasmid which had been called SLP1. This was re-named SLP1.1 and later variants were designated SLP1.2, SLP1.3, etc.

Fig. 3 summarizes physical data on six variants, SLP1.1 - SLP1.6. Of these, the first three were isolated from strain 1326 in the manner described. The other three were from cultures obtained by mating S. lividans SLP1.1$^+$ or SLP1.2$^+$ with S. coelicolor, isolating ltz$^+$ variants of S. coelicolor from the crosses, and mating these back with S. lividans 1326 to yield ltz$^+$ derivatives of S. lividans. SLP1.4 came from such crosses initiated with SLP1.1, and SLP1.5 and SLP1.6 from crosses initiated with SLP1.2.

SLP1.6 is the smallest member of the series so far isolated, with a molecular mass of 6.25 Md estimated from agarose gel electrophoresis of restriction fragments. It has a single EcoRI site, two HindIII sites, two SalGI sites and no sites for SalPI (PstI) or BamHI. The other variants differ from SLP1.6 in having an extra segment of DNA of varying size, inserted always at the same position. As we ascend the size series, the inserted DNA grows by the acquisition of extra segments, in one direction (clockwise in Fig. 3); thus certain variants acquire additional restriction sites.

The simplest explanation for these novel findings is that S. lividans harbours an integrated plasmid which can occasionally "loop-out" from the chromosome, bringing with it a variable segment of DNA from one side of the integration site. The minimum size for a functional autonomous SLP1 plasmid has yet to be determined. Physical evidence for integrated sequences corresponding to SLP1 in strain 1326 is currently being sought. In the original 1326 strain, lethal zygosis functions are unexpressed, since the strain is sensitive to lethal zygosis exerted by SLP1.1 - SLP1.6 and is itself ltz$^-$. The state of the SLP1 DNA in the large class of ltz$^+$ derivatives of strain 1326 which fail to reveal ccc DNA by the standard procedure is unknown. The significance of the fact that SLP1.4, SLP1.5, and SLP1.6 arose after crosses between SLP1$^+$ S. lividans and S. coelicolor is also uncertain since we cannot be sure that these plasmids do not represent new isolates of the plasmid coming from strain 1326 itself.

THE ROLE OF SEX PLASMIDS IN STREPTOMYCES

Of the earlier characterized self-transmissible plasmids in Streptomyces, SCP1 (see above) and the plasmid of S. reticuli (H. Schrempf and W. Goebel, this symposium) appear to be large (50 Md or more), while SCP2 is about 19 Md in mass. These plasmids are thus at least as large as the smallest conjugative plasmids of E. coli, and other Gram-negative species, suggesting that conjugation in Streptomyces might be a process requiring as many gene products as that determined by enteric and pseudomonad sex plasmids[9]. However, the recent discovery of the SLP1 family of apparently self-transmissible plasmids, the smallest only 6.25 Md, contradicts this view and makes the functional analysis of conjugation in these complex Gram-positive organisms of considerable interest.

Hyphal fusion leading to the formation of heterokaryons between auxotrophic derivatives of S. coelicolor A3(2) occurs at an appreciable frequency which appears not to depend on the presence of either SCP1 or SCP2 (D. A. Hopwood and H. M. Wright, unpublished). Heterokaryons are operationally defined as prototrophic mycelia giving rise to the two auxotrophic parental genotypes on subculture but not (or only very rarely) to true recombinants, but it is not certain that chromosomes of different genotype actually reside in the same compartment of the septate heterokaryotic hyphae; if they did, it is hard to imagine why, in the absence of nuclear membranes, they would not frequently recombine as they do when brought together by protoplast fusion[30]. Thus the interesting possibility exists that recombination in Streptomyces might depend on two events: rather generalized fusion events between mycelia to produce heterokaryons, followed by processes promoted by sex plasmids and resulting in the interaction of unlike genomes within the same mycelium. However, these ideas are for the present only speculative.

Until recently, Streptomyces was the only Gram-positive genus known to harbour conjugative plasmids, but these have now been found in streptococci[31,32]. Known streptococcal sex plasmids are relatively large (at least 17 Md), but nothing appears to be known about their role in conjugation. It will be interesting to see whether plasmid-mediated conjugation in the Gram-positive streptomycetes and streptococci is functionally similar, or whether the mycelial growth habit of the streptomycetes sets them apart from the non-mycelial bacteria, both Gram-positive and Gram-negative.

TRANSFORMATION OF PROTOPLASTS BY PLASMID DNA

Protoplasts of Streptomyces spp. can readily be prepared by treatment with lysozyme of mycelium grown in a medium containing a critical (sub-lethal) concentration of glycine[33]. Protoplasts of most species can be regenerated without difficulty; this is true of S. coelicolor, S. lividans and S. parvulus. Such protoplasts, when exposed briefly to SCP2* or SLP1 DNA in the presence of about 20% (w/v) polyethylene glycol, take up the DNA and the transformants can be recognized by testing spores harvested from a confluent regenerated culture for their ability to elicit lethal zygosis when grown in a lawn of the appropriate plasmid$^-$ strain[34]. Alternatively, in order to assess the frequency of transformation per protoplast, individual colonies arising from a diluted transformation mixture can be replica-plated to a lawn of the plasmid$^-$ culture to detect the ltz$^+$ phenotype. Such tests revealed that, remarkably, at least 20% of regenerated protoplasts of S. coelicolor or S. parvulus were transformed with SCP2* DNA[34] under optimal conditions, and similar proportions of S. lividans protoplasts took up SLP1 DNA, the highest frequency of transformation recorded in an individual experiment being 85% with SLP1.2 (M. J. Bibb, unpublished). In these experiments, the ratio of plasmids to protoplasts is about 5000 to 1, or a yield of transformants of about 10^6 per microgram of DNA.

These results were obtained with samples of natural plasmid DNA, largely in the supercoiled ccc configuration. In preliminary experiments with SCP2* DNA which had been cleaved with EcoRI and re-ligated to yield ccc but non-supercoiled molecules, the frequency of transformation of S. coelicolor protoplasts was reduced only by about an order of magnitude (M. J. Bibb and C. Thompson, unpublished). This result agrees with those obtained recently in Bacillus subtilis[35], where a 20-50 fold reduction was seen.

THE IMPORTANCE OF LETHAL ZYGOSIS

The lethal zygosis phenomenon, leading to "pock" formation when plasmid$^+$ spores are grown in a predominantly plasmid$^-$ lawn, may or may not turn out to be intrinsically interesting. However, there is no doubt that it is of great technical importance in the recognition and handling of Streptomyces sex plasmids. It is shown by both S. coelicolor plasmids under suitable conditions. It was instrumental in the recognition of the SLP1 family of plasmids, and the occurrence of spontaneous "pocks" has also led to the discovery of plasmid DNA in an unidentified streptomycete resembling S. olivaceus (J. Westpheling, unpublished). Lethal zygosis will doubtless reveal other plasmids in a variety of other species. The "pock" phenotype is very sensitive to growth conditions and different media are required to maximize its expression in different strains. For example, the R2 protoplast regeneration medium is particularly good for SCP2 and SLP1 "pocks"; so much so that even SCP2 itself, and not merely the SCP2* variant, can be seen to give faint "pocks" on this medium.

The resolution of the lethal zygosis test to recognize transformants[34] should not be underestimated. A single plasmid$^+$ individual can be recognized amongst the largest population of spores that can be spread on an agar plate (about 10^9) so, although visual, the detection power for the plasmid is as great as in the case of any antibiotic resistance marker.

PROSPECTS FOR IN VITRO GENETIC MANIPULATION

Gene cloning in Streptomyces (and in related genera of antibiotic-producing actinomycetes such as Nocardia, Micromonospora, Streptosporangium and Actinoplanes) will prove invaluable in the analysis, understanding and manipulation of antibiotic production[3].

With the availability of a highly efficient transformation procedure for plasmid DNA in Streptomyces and several potential vector plasmids, the development of recombinant DNA procedures for use within and between the actinomycetes seems assured. For S. lividans, at least, the SLP1 family of plasmids seems particularly attractive in view of their rather small size and the fact that any restriction sites within the variable segment of DNA must be available for insertion of foreign DNA without inactivation of essential plasmid replication and maintenance functions. Non-essential sites on SCP2* must also exist.

The copy number of SCP2* (or SCP2) per chromosome of S. coelicolor appears to be low, estimates of 1-2[9] and 3-4[14] having been obtained. Good copy number estimates for the SLP1 plasmids in S. lividans have not yet been made, but they probably exist in at least a few copies per chromosome (M. J. Bibb, unpublished). Low copy number will be desirable for some DNA cloning operations in actinomycetes; for example in the analysis of the organization of antibiotic biosynthetic genes. Another application of recombinant DNA techniques which would probably require a low copy number vector concerns the generation of new antibiotic structures by the introduction of genes coding for various potential antibiotic side-chains or for the addition or removal of functional groups on the molecules. Other applications of in vitro genetic manipulation include the insertion of genes coding for appropriate hydrolytic enzymes to allow the culture to be grown on cheaper carbohydrate sources or the amplification of gene products involved in rate-limiting biosynthetic steps.

ACKNOWLEDGEMENTS

M. J. Bibb (whose present address is Department of Genetics, Stanford University School of Medicine, Stanford, California 94305, U.S.A.) gratefully acknowledges financial support from the National Research Development Corporation.

REFERENCES

1. Hopwood, D. A. (1978) Current Adv. Plant Sci., 31, 467-481.
2. Chater, K. F. (1979) Genetics of Industrial Microorganisms (Ed. O. K. Sebek) American Society for Microbiology, Washington, D.C., pp. 123-133.
3. Hopwood, D. A. (1978) Ann. Rev. Microbiol., 32, 373-392.
4. Okanishi, M. (1979) Genetics of Industrial Microorganisms (Ed. O. K. Sebek) American Society for Microbiology, Washington, D.C., pp. 134-140.
5. Hopwood, D. A., Chater, K. F., Dowding, J. E. and Vivian, A. (1973) Bacteriol. Rev., 37, 371-405.
6. Hopwood, D. A. and Merrick, M. J. (1977) Bacteriol. Rev., 41, 595-635.
7. Lomovskaya, N. D., Mkrtumian, N. M., Gostimskaya, N. L. and Danilenko, V. N. (1972) J. Virol., 9, 258-262.
8. Hopwood, D. A. and Wright, H. M. (1976) Second International Symposium on the Genetics of Industrial Microorganisms (Ed. K. D. Macdonald) Academic Press, London, pp. 607-619.
9. Bibb, M. J., Freeman, R. F. and Hopwood, D. A. (1977) Mol. Gen. Genet., 154, 155-166.
10. Wright, L. F. and Hopwood, D. A. (1976) J. Gen. Microbiol., 95, 96-106.
11. Kirby, R. and Hopwood, D. A. (1977) J. Gen. Microbiol., 98, 239-252.
12. Hornemann, U. and Hopwood, D. A. (1978) Tetrahedron Lett., 33, 2977-2978.
13. Hornemann, U. and Hopwood, D. A. (1979) Antibiotics, Biosynthesis Vol. IV (Ed. J. W. Corcoran) Springer-Verlag, Berlin, Heidelberg, New York (In press).
14. Schrempf, H. and Goebel, W. (1977) J. Bacteriol., 131, 251-258.
15. Palchaudhuri, S. and Chakrabarty, A. L. (1976) J. Bacteriol., 126, 410-416.
16. Currier, T. C. and Nester, E. W. (1976) Analytical Biochem., 76, 431-441.
17. Van Larabeke, N., Genetello, C., Hernalsteens, J. P., De Picker, A., Zaenen, I., Messens, E., Van Montagu, M., Schell, J. (1977) Molec. Gen. Genet., 152, 119-124.
18. Hansen, J. B. and Olsen, R. H. (1978) J. Bacteriol., 135, 227-238.
19. Worcel, A., and Burgi, E. (1972) J. Mol. Biol., 71, 127-147.
20. Guillen, N., Le Hegarat, F., Fleury, A., and Hirshbein, L. (1978) Nucleic Acid Res., 5, 475-489.
21. Meyers, J. A., Sanchez, D., Elwell, L. P. and Falkow, S. (1976) J. Bacteriol., 127, 1529-1537.
22. Clewell, D. B. and Helinski, D. R. (1970) Biochemistry, 9, 4428-4440.
23. Womble, D. D. and Rownd, R. M. (1977) J. Bacteriol., 131, 145-152.
24. Benigni, R., Petrov, P. A. and Carere, A. (1975) Appl. Microbiol., 30, 324-326.
25. Hopwood, D. A. and Wright, H. M. (1973) J. Gen. Microbiol., 77, 187-195.

26. Kirby, R. (1976) Ph. D. Thesis, University of East Anglia, Norwich.

27. Skurray, R. A. and Reeves, P. (1973) J. Bacteriol., 113, 58-70.

28. Chater, K. F. (1978) Nocardia and Streptomyces (Ed. M. Modarski, W. Kurylowicz and J. Jeljaszewicz) Gustav Fischer Verlag, Stuttgart, pp. 303-310.

29. Lomovskaya, N. D., Voeykova, T. A. and Mkrtumian, N. M. (1977) J. Gen. Microbiol., 98, 187-198.

30. Hopwood, D. A. and Wright, H. M. (1978) Mol. Gen. Genet., 162, 307-317.

31. Jacob, A. E. and Hobbs, S. J. (1974) J. Bacteriol., 117, 360-372.

32. Hershfield, V. (1979) Plasmid, 2, 137-149.

33. Okaniski, M., Suzuki, K. and Umezawa, H. (1974) J. Gen. Microbiol., 80, 389-400.

34. Bibb, M. J., Ward, J. M. and Hopwood, D. A. (1978) Nature, 274, 398-400.

35. Chang, S. and Cohen, S. N. (1979) Mol. Gen. Genet., 168, 111-115.

Plasmids of Medical, Environmental and Commercial Importance
K.N. Timmis and A. Pühler, editors

FUNCTIONS OF PLASMID GENES IN STREPTOMYCES RETICULI

H. SCHREMPF and W. GOEBEL
Institut für Genetik und Mikrobiologie der Universität Würzburg,
Röntgenring 11, 8700 Würzburg, West-Germany

ABSTRACT

Streptomyces reticuli is a well sporulating strain which produces a macrolide antibiotic and melanin. This strain contains one large plasmid with a molecular weight of $48\text{-}49 \times 10^6$ daltons. After treatment with eliminating agents different variants have been obtained which have either lost the extrachromosomal DNA or contain deleted plasmids. Genetical and biochemical studies of these variants reveal that at least some of the structural and/or regulatory genes required for the expression of sporulation, melanin synthesis and antibiotic production are plasmid borne.

INTRODUCTION

Strains of Streptomyces have been used in the past 40 years to produce antibiotics. During this time most effort was concentrated on the isolation and chemical characterization of new antibiotics[1,2]. Till now very little is known on the genetics of these secondary metabolites and the regulation of their production on the molecular level.

Genetical data have revealed that plasmid genes may code for structural and/or regulatory genes of some antibiotics. The structural genes for the synthesis of methylenomycin (S. coelicolor A3(2)) are located on the genetically charaterized SCP1-plasmid[3]. Plasmid genes seem also be required for the expression of antibiotics such as chloramphenicol (S. venezuelae)[4], kasugamycin (S. kasugaensis)[5] or holomycin (S. clavuligerus)[6]. However the synthesis of oxytetracyclin (S. rimosus) is determined by structural genes of the chromosome[7].

Recently methods have been developed to isolate plasmid DNA from Streptomyces strains successfully[8]. Extrachromosomal DNAs ranging from $8\text{-}100 \times 10^6$ daltons have been characterized[9,10]. However in many cases the function of the plasmid genes is yet unknown.

This report describes the identification of covalently closed circular DNA from S. reticuli. Analysis of different variants which have either lost the plasmid DNA or contain deleted plasmids indicates that plasmid genes are involved in the expression of secondary metabolites.

MATERIAL AND METHODS

The details of the materials used and the employed methods will be described elsewhere (manuscript in preparation).

RESULTS

Isolation of variants being affected in the synthesis of secondary metabolites

Streptomyces reticuli is a well sporulating strain, which produces melanin in the presence of tyrosine. In addition this strain synthesizes the macrolide antibiotic leukomycin (Zähner, personal communication).

Different variants can be obtained either spontaneously or after treatment with plasmid-eliminating agents such as acridine orange or ethidium bromide.

Variants of type I have lost the ability to produce melanin, yet are still synthesizing the antibiotic and spores. Variants of type II do not sporulate. However they produce melanin and the antibiotic to about the same amount as the wild type.

Variants of type IV do not produce spores and melanin.

Variants of type III and V synthesize neither spores nor melanin nor the antibiotic.

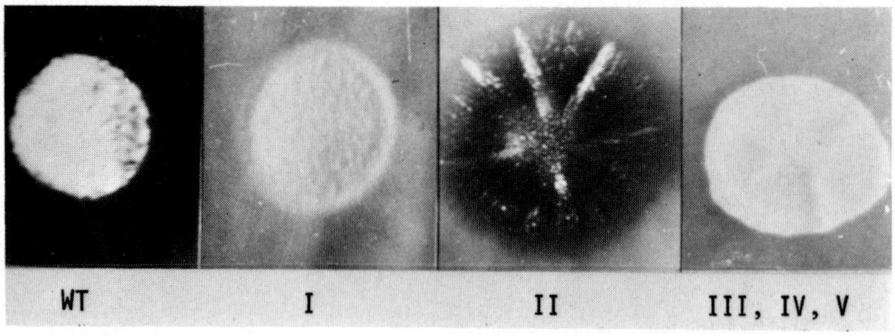

Fig. 1 Colonies of the wild type and the variants

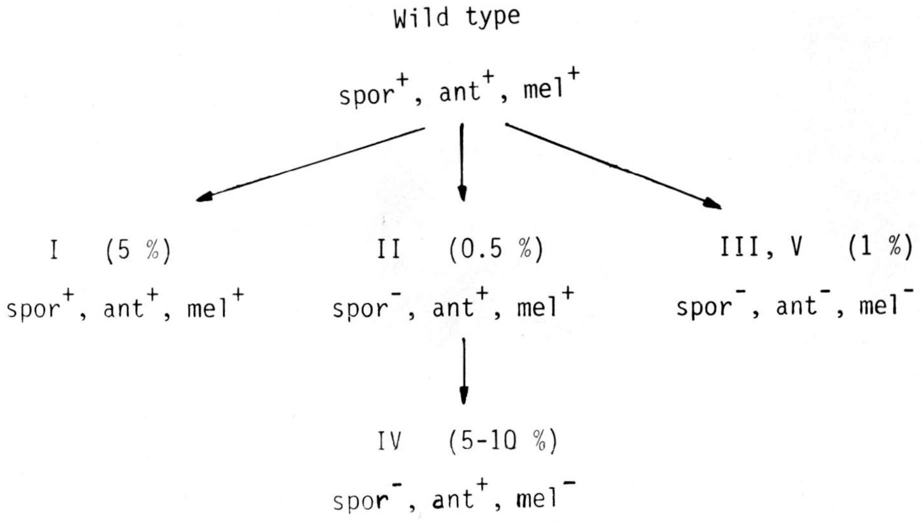

Fig. 2 Frequencies with which the different variants were obtained after treatment with acridine orange

Variants of the types I, III are stable; no revertants to melanin or antibiotic producing strains were obtained out of 10^5 colonies tested. Variants of type II, however, exhibit rather unstable phenotypes. Depending on the variant 5 - 10 % of the cells in a growing culture loose the capability of excreting melanin. These segregants are genetically stable. In order to exclude the possibility that segregants have been arisen from a heterokaryotic mycelium, the mycelium of each strain was converted to protoplasts. Regenerated strains obtained from these protoplasts showed again unstable properties.

These data indicate the possible involvement of plasmid genes in the expression of sporulation, melanin and antibiotic production and (or) the regulation of these properties.

Isolation and Characterization of extrachromosomal DNA

In order to isolate the presumptive plasmid DNA conditions had to be found to lyse the strain. Mycelium grown in normal minimal or complete media was difficult to lyse by lysozyme. Growth in the presence of 25 % sucrose and 0.25 % glycine resulted in fine, well dispersed mycelium. After the addition of 2 mg/ml lysozyme protoplasts could be obtained within 10 - 15 minutes. Effective lysis was performed after the addition of the detergent dodecylsulfate. After addition of sodium chloride, a clearing spin was performed[8]. The supernatant was centrifuged in a CsCl-gradient in the presence of ethidium bromide. Closed circular DNA could be separated from the chromosomal DNA. The CCC DNA represents about 2 % of the total DNA.

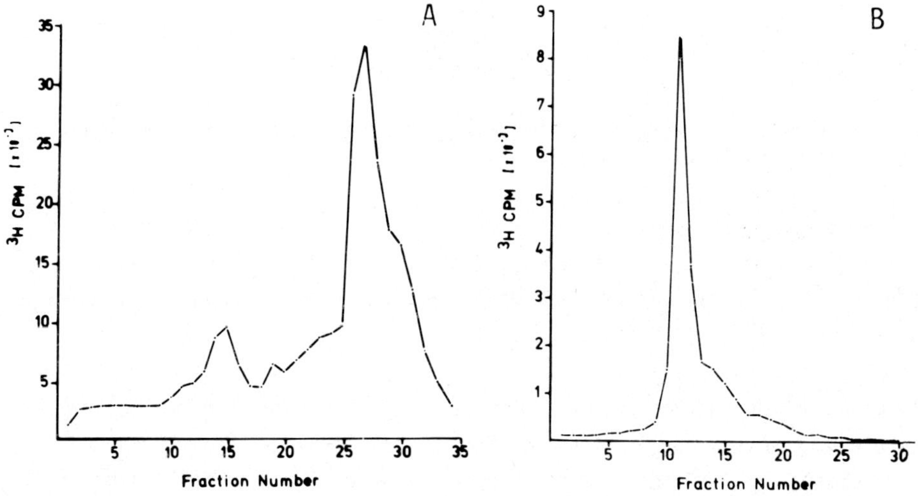

Fig. 3 Centrifugation to equilibrium of the CsCl-Etbr gradient containing a cleared lysate of the wild type (A) or of a variant of type V (B). Strains were grown in the presence of ^{3}H-thymidine.

Sedimentation analysis and electron microscopy revealed that the plasmid DNA consists of one species of covalently closed circular DNA which has a molecular weight of about $48\text{-}49 \times 10^6$ daltons. Assuming a molecular weight for the chromosomal DNA of 5×10^9 daltons the copy number per chromosome[11] for this plasmid can be calculated to 2.

Analysis of the DNA of the different variants revealed some interesting features. Strains of type V which are not sporulating and do not synthesize melanin and the antibiotic contain no covalently closed circular DNA (Fig. 3B).

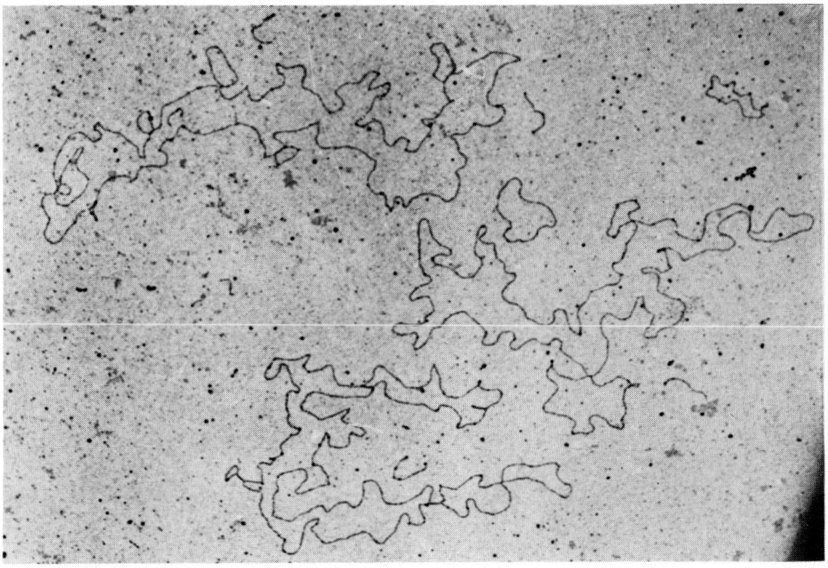

Fig. 4 Electron micrograph of open circular DNA molecules of the wild type strain. ColE1-DNA was used as internal length standard.

However from all the other variants (types I, II, III and IV) extrachromosomal DNA could be isolated and was characterized by electron microscopy.

Colonies of type I, which are white sporulating and antibiotic producing, but have lost the ability to excrete melanin, contain one species of CCC DNA which has lost 1-2x10^6 daltons of the original plasmid (Fig. 5).

Variants of the type III are phenotypically identical with those of type V. However, these variants still contain plasmid DNA, which is rather heterogeneous. CCC DNA slightly larger (50x10^6 daltons) than the original plasmid DNA of the wild type and in addition smaller CCC DNAs with molecular weights of 47, 44-45x10^6, 8-10 and 13x10^6 daltons were isolated. Other variants contain in addition plasmids of about 15-16x10^6 daltons. The ratio between the different plasmid DNA species is varying from strain to strain.

Variants of type II contain at least two classes of plasmids. Their molecular weights range between 45x10^6 and 47-48x10^6 daltons. These plasmids seem to have lost 2 to 5x10^6 daltons of DNA compared to the wild type plasmid.

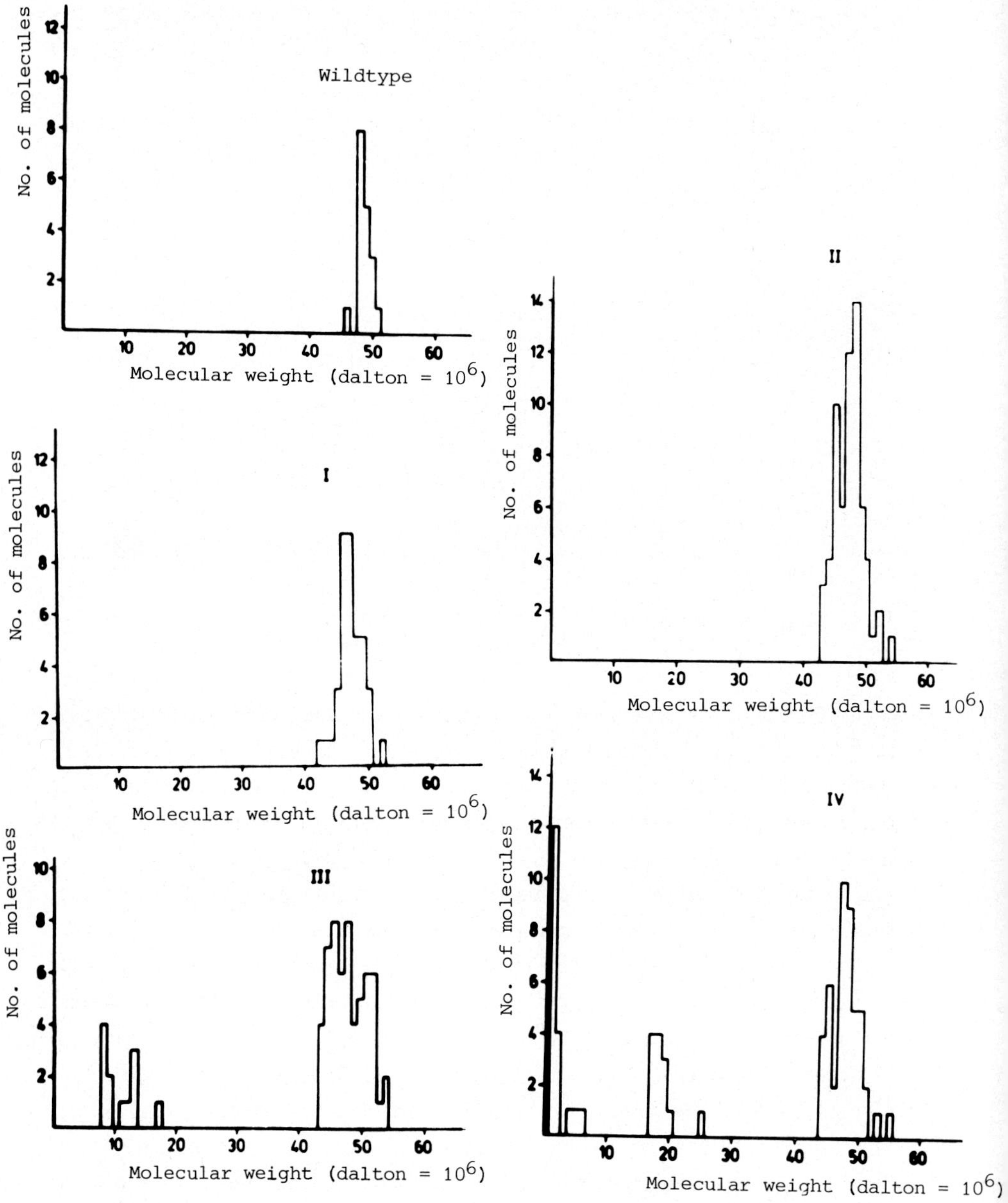

Fig. 5 Size distributions of plasmid molecules isolated from different variants

Variants of type IV contain in addition to the large plasmid smaller plasmids (MG 18-18.5x10^6, 4x10^6 and 1-2x10^6 daltons).

Analysis of the plasmids by the restriction enzyme SalI

The plasmid DNA of the wildtype has one cleavage site for the restriction enzyme BamHI and PstI respectively, no cleavage site for HindIII and 4 recognition sites for EcoRI. As SalI cleaves the plasmid of the wild type strain into more than 20 fragments, it was chosen for a more detailed characterization of the CCC DNA of the different variants.

Comparing the SalI digestion patterns of plasmid DNA isolated from the wild type with that of a mel$^-$ strain (type I), it is obvious that the CCC DNA of the variant has lost the second SalI fragment (MG 3x10^6 daltons) and a fragment of a molecular weight of 2.2x10^6 daltons appears. This result indicates that a deletion (MG 0.8x10^6 dalton) within the second SalI fragment leads to the failure to produce melanin.

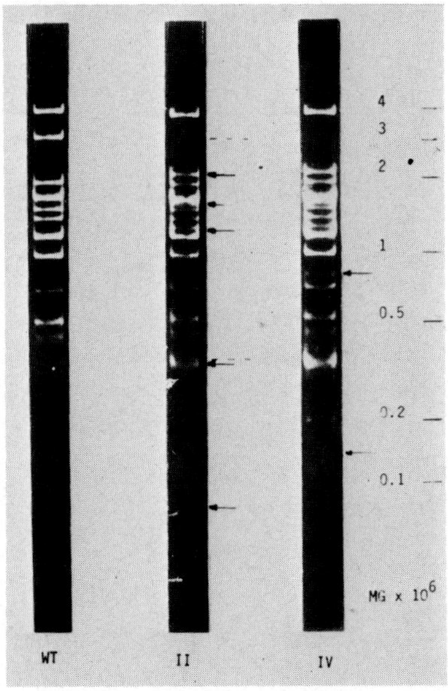

Fig. 6 SalI digestion patterns of plasmid DNA isolated from the wild type and some variants (lost --- or additional ← fragment).

Variants of type II are nonsporulating but synthesize melanin and the antibiotic. The largest CCC DNA species of this type has lost 300 base pairs from the largest SalI fragment. The second SalI and a smaller fragment of about 450 base pairs are missing. There are 5 additional fragments.

Variants of type IV which arise spontaneously from type II variants have lost the ability to produce melanin. The largest plasmid DNA is cleaved by SalI to nearly the same fragments as the plasmid DNA from type II strains. The digestion patterns differ only in two additional fragments of 1.140 and 180 base pairs respectively. Presumably due to these insertions melanin is not produced. The SalI digests of the plasmids of the variants often contain small amounts of additional fragments. These data suggest that plasmid DNA of a certain size consists of a heterogenous population of DNA molecules.

These may arise by rearangements within a small number of plasmid molecules.

All strains have also been tested for their resistance against the antibiotic isolated from the wild type strain. All strains which no longer produce antibiotic substances inhibitory to Micrococcus flavus have also lost the resistance against this antibiotic.

Transfer of the plasmid

The plasmid is a transfer factor. After a cross between an auxotrophic wild type strain and a prototrophic plasmid-less strain exconjugants appear

ACCEPTOR STRAIN	DONOR STRAIN	EXCONJUGANTS	FREQUENCY
TYPE V PROTOTROPH	WILDTYPE AUXOTROPH	$SPOR^+$, MEL^+ ANT^+, PROTOTROPH	50 - 100 %
TYPE II $SPOR^-$, ANT^+ MEL^+	TYPE I $SPOR^+$, ANT^+ MEL^-	$SPOR^+$, ANT^+ MEL^+	50 - 100 %
WILDTYPE HIS^-	WILDTYPE PRO^-	PRO^-, HIS^-	0,2 %

Fig. 7

which are prototrophic, melanin- and antibiotic-producing and sporulating. The frequency of the transfer is about 50 - 100 %. Similar crosses between colonies of type II and type I give rise to exconjugants which have the same properties as the wild type strain. In comparison recombination with chromosomal markers - possibly due to intergration of the plasmid-occurs at a frequency of about 0.2 %.

DISCUSSION

Our data suggest that either some of the structural genes and (or) genes for the regulation of the antibiotic production, antibiotic resistance, melanin production and sporulation are plasmid borne.

From our results one may also speculate that in highly differentiated microorganisms such as Streptomycetes deletions, insertions and possibly transpositions leading to phenotypic alterations occur more frequently than the loss of the whole plasmid which is the more frequent observed event in E. coli.

Up to now the existence of plasmids in many different Streptomyces strains has been predicted from the loss of various properties, which occurred either spontaneously or after treatment with plasmid eliminating agents. We have analysed several of these strains. In some of these strains CCC DNA has not been found. These data may indicate that the observed phenotypic alterations are due to variations in the DNA sequences of the chromosome similar to those described above. Such events may also create strains with many unstable properties.

ACKNOWLEDGEMENTS

Ch. Bonnet is thanked for skilful technical assistance, F. Woebcken for measuring the contour length of the DNA molecules and E. Appel for typewriting the manuscript. This work was supported by a grant from the Deutsche Forschungsgemeinschaft (SFB 105 - A13).

REFERENCES

1. Cocoran, J.W. and Hahn, F. (1967) Antibiotics, Springer-Verlag
2. Omura, S. and Nakagawa, A. (1975) J. of Antibiotics 28, 401-433
3. Kirby, R. and Hopwood, D.A. (1977) J. gen. Microbiol. 98, 239-252
4. Akagawa, H. et al. (1975) J. Gen. Microbiol. 90, 336-346
5. Okanishi, M. et al. (1970) J. Antibiot. 23, 45-47
6. Kirby, R. (1978) Fed. Eur. Microbiol. Soc. Lett. In press

7. Boronin, A.M. and Midlin, S.Z. (1971) Genetika, 7, 125-131

8. Schrempf, H. and Goebel, W. (1977) Bacteriol. 131, 251-258

9. Schrempf, H. and Goebel, W. (1978) Genetics of the Actinomycetales, Gustav Springer Verlag

10. Schrempf, H. and Goebel, W. manuscript in preparation

11. Sermonti, G. and Puglia, A.M. (1974) In Proceedings, 2nd International Symposium on Genetics of Industrial Microorganism. Sheffield. Academic Press Inc., London

IV – DEGRADATIVE PLASMIDS

© 1979 Elsevier/North-Holland Biomedical Press
Plasmids of Medical, Environmental and Commercial Importance
K.N. Timmis and A. Pühler, editors

MOLECULAR STUDIES ON PSEUDOMONAS PLASMIDS.

Paul Broda, Susan A. Bayley, Robert G. Downing, Clive J. Duggleby and David W. Morris, Department of Molecular Biology, Edinburgh University, King's Buildings Edinburgh EH9 3JR, Scotland.

An interesting feature of the pseudomonads is their extreme nutritional diversity, which is a characteristic of individual strains as well as of the group as a whole[1]. A number of strains (most of them isolated by enrichment culture) are able to degrade compounds that are uncommon in nature, because they carry degradative plasmids (see Chakrabarty, this volume). Among such plasmids is TOL, which encodes enzymes that degrade toluene and the xylenes[2,3,4]. The host strain, *P.putida*(arvilla) mt-2, is the archetype of a series of *Pseudomonas* strains, all isolated independently by enrichment culture on *m*-toluate (an intermediate of *m*-xylene breakdown). In all cases this catabolic function was plasmid-specified and could be lost after growth on benzoate, an intermediate of the plasmid-encoded *meta* pathway, but which may also be broken down by the chromosomally encoded *ortho* pathway. The reason for this lies in the modes of regulation of the two pathways[2]. Only the *meta* pathway is expressed, when the plasmid is present, whereas loss of the plasmid-borne function (by loss of either the whole plasmid or a specific segment[5]) leads to faster growth *via* the *ortho* pathway. Thus "cured" strains overgrow their Tol^+ siblings. This technique has been of great experimental value.

Among the plasmids from these independently-isolated strains are some that are very closely related to TOL itself, on the basis of size and the distribution of endonuclease cleavage sites (see Broda, this volume). The molecular relationships of the group as a whole will be discussed later. One particularly interesting TOL plasmid, from an isolate designated MT14, undergoes an increase in size after serial subculture in *m*-toluate[6]. Although the mechanism has not yet been fully characterised, it seems analogous to the "transition" phenomenon described by Rownd and his colleagues[7].

Studies on the TOL plasmid pWWO. We have chosen to study the original TOL plasmid (designated pWWO) in some detail. We are also examining the relationships between degradative plasmids and other plasmids. In the former study, the first step was to construct an endonuclease cleavage map of pWWO. Its size (78 megadaltons or 117 kilobases) and the large number of cleavage products given by all of the endonucleases tested (usually more than 30 fragments) relationship with a number of other TOL plasmids[13]. More detailed analyses reveal that those plasmids fall into clusters; members of such groups demonstrate close sequence homology with each other. One cluster comprises pWWO and its close relatives referred to earlier. Other clusters have also been observed and characterised. Even the large (273 kilobase) IncP-2 plasmid OCT, which encodes the dissimilation of alkyl rather than aromatic compounds, contains sequences which hybridise with all the other degradative plasmids so far examined. Most of these homologous sequences are confined to a particular small region (or two regions) of OCT.

We have also examined the extent of homology which exists between *Pseudomonas* plasmids of different and the same incompatibility groups. In some cases, homology between plasmids can be ascribed to the presence of known transposons. Thus, the relationship occurring between the IncP-1 plasmid RP1 and the IncP-10 plasmid R91, both isolated from the same host strain, is due to both plasmids possessing the transposable element Tn*1*. However, the most interesting plasmids are those of the IncP-9 group, which includes both degradative plasmids (e.g. NAH and TOL) and resistance plasmids (e.g. R2 and pMG18).

TABLE 1.
Extent of homology (as percentages) of four IncP-9 plasmids, determined by hybridisation of whole plasmid DNA. The apparent inconsistency between the data of the NAH-TOL and TOL-NAH tests is probably due to a repeated sequence on NAH that is homologous with TOL (Morris, unpublished data). Data are the means of eight separate determinations.

	Labelled DNA samples			
	TOL	NAH	R2	pMG18
plasmid size (kb)	117	72	73	100
TOL	100±6	12±2	14±3	20±2
NAH	40±3	100±2	5±1	2±1
R2	11±3	4±1	100±3	67±2
pMG18	17±4	6±2	98±3	100±2

Although the degradative plasmids and the R plasmids have no known common function there was significant homology between them. Hybridisation suggested that these plasmids shared a small "core" sequence with additional segments that distinguish the different plasmids. Thus even where plasmids code for different functions and are found in bacteria from quite different habitats meant that this was a substantial undertaking. Using the two enzymes available that gave the fewest cuts (*Xho*I: 10; *Hind*III: 19), and starting with the deleted plasmid pWWO-8 (one of the TOL derivatives referred to above that has undergone a specific excision event) this was achieved using a number of methods in conjunction (ref.8; Downing and Broda, unpublished results). We had found that there were many more small fragments than would obtain from a random distribution of cleavage sites; it now emerged that the sites giving rise to these small fragments were clustered.

With this map it was possible to show that the DNA lost by pWWO-8 was composed of a single segment. There exists the interesting possibility that this segment in the parent plasmid is bounded by a directly repeated sequence (Downing, unpublished results). Such sites would allow excision by the classical Campbell mechanism.

Three different groups have reported that the Tol function can be translocated onto the R plasmid RP4[9,10,11]. We have confirmed that such RP4-Tol hybrid plasmids have a larger Tol segment than that lost in the formation of pWWO-8 (57 kilobases compared with less than 45), although in each case a similar segment is involved. It overlaps the 45 kb segment at both ends and in two of the three hybrid molecules the segment is inserted into the tetracycline-resistance gene of RP4.

Transposons have been used extensively to generate insertion mutations. From such mutant plasmids families of deletion mutants can then often be generated in turn. We have examined a series of pWWO derivatives carrying Tn401 (which specifies carbenicillin resistance) provided by P. Williams and C. Franklin. Interestingly (but disappointingly from the point of view of our analysis of pWWO) 11 of 12 independently isolated $Tol^{+}Cb^{r}$ mutants had the transposon inserted into the 10 kb *Hind*III fragment, which contains one end of the 45 kb segment that is excised. Further, of 11 Tol^{-} derivatives all had lost this segment, but in only three cases had the deletion extended further. Since the information obtained from such data is very limited, we are now studying the derivatives of other transposons. This should aid in the construction of a genetic map to which the physical map may be correlated.

Molecular relationships of Pseudomonas plasmids. Molecular hybridisation, after Southern transfer[12], showed that many degradative plasmids share common DNA sequences. Plasmids NAH and SAL, which encode the degradation of napthlene and salicylate respectively (with naphthalene being broken down *via* salicylate), are very closely related and, furthermore, show significant (soil and hospitals) incompatibility is a valid criterion of relationship. The evolutionary history of the plasmids must be considered in the context of these results.

REFERENCES.

1. Stanier, R.Y., Palleroni, N.J. and Doudoroff, M. (1966) J.Gen.Micro. 154-273.
2. Williams, P.A. and Murray, K. (1974). J.Bact. 120, 416-423.
3. Wong, C.L. and Dunn, N.W. (1974). Genet.Res. 23, 227-232.
4. Duggleby, C.J., Bayley, S.A., Worsey, M.J., Williams, P.A., and Broda,P. (1977). J.Bact. 130, 1274-1280.
5. Bayley, S.A., Duggleby, C.J., Worsey, M.J., Williams, P.A., Hardy, K.G. and Broda, P. (1977). Molec.Gen.Genet. 154, 203-204.
6. Broda, P., Bayley, S.A., Duggleby, C.J., Heinaru, A., Worsey, M.J. and Williams, P.A. (1978) in: Microbiology 1978. D. Schlessinger, Ed. American Society for Microbiology.
7. Perlman, D. and Rownd, R.H. (1975). J. Bact. 123, 1013-1034.
8. Downing, R.G., Duggleby, C.J., Villems, R. and Broda, P. (1979) Molec.Gen. Genet. 168, 97-99.
9. Nakazawa, T., Hayashi, E., Yokota, T., Ebina, Y. and Nakagawa, A. (1978) J.Bact. 270-277.
10. Jacoby, G.A., Rogers, J.E., Jacob, A.E. and Hedges, R.W. (1978). Nature, 274, 179-180.
11. Chakrabarty, A.M., Friello, D.A. and Bopp, L.H. (1978) Proc.Natl.Acad. Sci.,Wash. 75, 3109-3112.
12. Southern, E. (1975). J.Molec.Biol. 98, 503-517.
13. Heinaru, A.L., Duggleby, C.J. and Broda P. (1978). Molec.Gen.Genet. 160, 347-351.

Plasmids of Medical, Environmental and Commercial Importance
K.N. Timmis and A. Pühler, editors

PLASMIDS SPECIFYING p-CHLOROBIPHENYL DEGRADATION IN ENTERIC BACTERIA

P.F. KAMP AND A.M. CHAKRABARTY*
General Electric Research & Development Center, Schenectady, New York 12301 (U.S.A.)

INTRODUCTION

Polychlorinated biphenyls (PCBs) are recognized as ubiquitous contaminants in aquatic and to a lesser extent terrestrial ecosystems[1]. Public concern over the widespread occurrence of PCBs stems from their persistence, bioaccumulation in food chains, and toxicity to certain living organisms[2]. The presence of chlorine substituents is known to retard the microbial degradability of various chlorinated organic compounds[3], and while biphenyl molecules with one or two chlorine atoms are known to be slowly biodegradable, molecules with large number of chlorine atoms appear to be quite recalcitrant to microbial attack[4]. We have isolated facultatively anaerobic enteric bacterial strains from the PCB-contaminated areas of Hudson river sediment that can slowly utilize Aroclor 1242 and p-chlorobiphenyl (pCB) as the sole source of carbon. In this report, we present evidence that suggests that the genes encoding biodegradation of p-chlorobiphenyl in a strain of *Klebsiella pneumoniae* are borne on a plasmid. In addition, such strains are shown to express plasmid-associated hydrocarbon degradative genes transferred from *Pseudomonas* species.

MATERIALS AND METHODS

Organisms. PCB-degrading organisms were isolated from the PCB-contaminated areas of the Hudson river after enrichment on Aroclor 1242. Samples were inoculated in a mineral salts medium, initially with biphenyl and Aroclor 1242 and later with Aroclor 1242 alone, and incubated on a shaker at 30°C. After growth for two to three weeks, the liquid culture was streaked on minimal plates with 0.1% p-chlorobiphenyl (pCB). Growth of single colonies were observed after six or seven days at 30°C. These single colonies were respotted on minimal pCB plates, and colonies exhibiting growth in five to seven days were characterized and maintained in the stock culture collection. Some of the relevant properties of the strains are shown in Table 1.

---

* Present address: Department of Microbiology, University of Illinois Medical Center. 835 South Wolcott St., Chicago, Illinois 60612, U.S.A.

Curing and transformation. The procedure for curing plasmids by treatment with mitomycin C is essentially as described by Rheinwald et al[5]. Transformation procedure with plasmid DNA has also been described[6].

Plasmid DNA isolation, agarose gel electrophoresis and electron microscopy. Plasmid DNA was isolated from pCB^+ and pCB^- *K. pneumoniae* strains essentially as described by Currier and Nester[7]. Agarose gel electrophoresis was carried out using 0.7% (w/v) agarose gel in 0.04 M Tris, pH 7.4, 0.005 M sodium acetate, pH 8.30, and 0.001 M EDTA, pH 8.20 (E buffer)[8]. The samples were adjusted to 10% sucrose-0.025% Bromophenol Blue before loading on the gels. The electrophoresis was run at 60 volts for 4 hours at room temperature, the gel was stained with ethidium bromide (0.5 μg/ml in E buffer) and photographed using a Vivitar No. 25 (A) red filter.

For electron microscopy, plasmid DNA samples were spread according to Davis et al[9], stained with uranyl acetate, and rotary shadowed with platinum. Contour lengths were measured using a Numonics Electronic Graphics Calculator (model 240-117).

Characterization of acidic metabolites. pCB^+ and pCB^- *K. pneumoniae* strains were grown in 500 ml minimal media with 0.005% yeast extract and 0.1% pCB at 30°C for 10 days. The cultures were acidified with concentrated HCl to pH 1.0, and extracted twice with hexane followed by two extractions with a 1:1 mixture of hexane and ether. The pooled organic phase was treated twice with 1.0 N NaOH. The resulting aqueous phase was acidified with Conc. HCl to pH 1.0 and extracted twice with ether. The ether phase was dried over anhydrous Na_2SO_4 and rotary evaporated to a final volume of 1.0 ml. The acidic metabolites were methylated with 3.0 ml BF_3-methanol at 60°C for 10 hr. After cooling 30 ml of hexane was added and washed twice with saturated NaCl. The hexane layer was dried over anhydrous Na_2SO_4 and evaporated under N_2 to a fixed volume of 0.3 ml. The acidic metabolites were then analyzed by gas chromatography-mass spectrometry (GC-MS).

GC-MS was performed on a Varian MAT III system equipped with a 10 ft x 2 mm glass column, packed with 3% OV-17 on Gas Chromosorb Q. The temperature was programmed from 50-300°C at a rate of 20°/min. GC-MS grating temperatures were: injector 280°C, jet separator 280°C, and ion source 250°C. The electron energy was 70 ev and the mass spectra were exponentially scanned from 20-600 A.M.U. with a speed of 3.7 sec/decade.

Preparation of cell-free extracts and enzyme assays. Cells were grown in minimal salts media with benzoate (1 mg/ml) or m-toluate (1 mg/ml) as sole sources of carbon. In some experiments, for induction of TOL enzymes by m-xylene, cells were grown on minimal agar plates with m-xylene in vapor phase. The cells were harvested at the end of the exponential phase, washed twice with 50 mM phosphate buffer, pH 7.5 and disrupted for three times at 0°C with a Bronson sonicator for a total of 30 sec with 1 min intervals in between. Cell debris was pelleted at 14000 rpm for 45 min, and the supernatant was used for assaying the enzymes.

All enzyme assays were carried out spectrophotometrically in 1 ml cuvettes of 1 cm path length. Catechol 1,2-oxygenase was assayed by measuring the rate of formation of *cis, cis*-muconate as described by Hegeman[10]. Catechol, 2,3-oxygenase was assayed by determining the rate of accumulation of 2-hydroxymuconic semialdehyde essentially as described by Feist and Hegeman[11].

Table 1. LIST OF BACTERIAL STRAINS

Strain	Collection No.	Genotype	Plasmid	Derivation/Ref.
P. putida PpG1	AC2	*trpB615*		Reference 16
"	AC1002	*met-1,ade-101*	pAC8[a]	Reference 16
P. putida mt-2	AC131	*wt*	TOL	Reference 16
K. pneumoniae	AC901	*wt*	pAC21,cryptic	
"	AC902	*pro-1*	pAC21,cryptic	
"	AC904	*wt*	cryptic	Curing by MC[b]
"	AC911	*wt*	pAC21,pAC8	
", KP1	AC915	*wt*		F. Ausubel
S. marcescens	AC906	*wt*	unknown	
"	AC907	*wt*	unknown	Cured by MC
"	AC913	*wt*	pAC8,unknown	

a. pAC8, previously referred as RP4-TOL (ref. 16).
b MC, mitomycin C

RESULTS

The problems of environmental pollution with PCBs are widespread and have been discussed from various viewpoints[12]. In U.S., there is an estimated 450 million pounds of PCBs in landfills, air, water, soil and bottom sediments and 750 million pounds of PCBs presently in use[13]. Since anaerobes and facultativ anaerobes comprise the vast majority of river sediment microbiota, there must have been strong selective pressure on them to detoxify such compounds. Yet, biodegradation of hydrocarbons is known to need participation of molecular oxygen, so that facultative anaerobes will have a better chance of biodegradation of PCBs under conditions of appropriate oxygen tension. We have therefore been interested in isolating facultative anaerobes from the PCB-contaminated areas of Hudson river sediments to examine the evolution of PCB and other hydrocarbon degradative competence. By repeated enrichment on Aroclor 1242, it was possible to isolate from the Hudson river sediments near Fort Edward two cultures that could grow slowly with Aroclor 1242 and utilize pCB as a sole source of carbon. The taxonomic characterization led to identification of the two strains as Klebsiella pneumoniae (AC901) and Serratia marcescens (AC906) (Table 2). In addition to pCB, both cultures could utilize benzoate as a sole carbon source. K. pneumoniae AC901 could also utilize p-hydroxybenzoate as a sole source of carbon; however, S. marcescens AC906 was incapable of utilizing p-hydroxybenzoate. While both cultures grew well with glucose under anaerobic conditions, neither could utilize benzoate (or p-hydroxybenzoate for AC901) under similar conditions. Neither could utilize biphenyl as a sole carbon source.

Table 2. CHARACTERIZATION OF pCB^+ STRAINS AS K. PNEUMONIAE AND S. MARESCENS

Test[a]	Bacterial strains		
	AC901	AC915[a]	AC906
Indole	+	+	-
Citrate	+	+	+
Motility	-	-	+
Lysine decarboxylase	+	+	+
Nitrate Reductase	+	+	-
Ornithine decarboxylase	-	-	+
Triple sugar iron agar	A/A gas	A/A gas	K/A
Voges proskauer	+	+	+
Oxidase	-	-	-
MacConkey agar	+	+	+
Pigment	-	-	Red

[a] AC915, a known strain of K. pneumoniae Kpl, has been included for comparison.

To determine if the pCB degradative pathway is specified by plasmid-borne genes, we attempted to isolate cured derivatives of AC901 and AC906. Treatment of growing cells with mitomycin C resulted in the appearance of segregants (at a frequency of 10^{-3} per cell per generation) that could no longer grow with pCB, even when incubated for periods up to three weeks. In order to see if such segregants can accumulate intermediates of pCB degradation, both the pCB^+ (AC901) and the pCB^- (AC904) K. pneumoniae strains were incubated for ten days in mineral salts medium with 0.005% yeast extracts and 0.1% pCB as the sole source of carbon. The acidic metabolites were then extracted and characterized by gas chromatography-mass spectrometry (Fig. 1). While several acidic metabolites such as p-chlorobenzoic acid and non-chlorinated aliphatic acids with molecular formula $C_9H_{11}O_2$ could be detected in the supernatant of the pCB^+ strain, none could be detected in the pCB^- segregant. The segregant therefore appears to be incapable of attacking pCB.

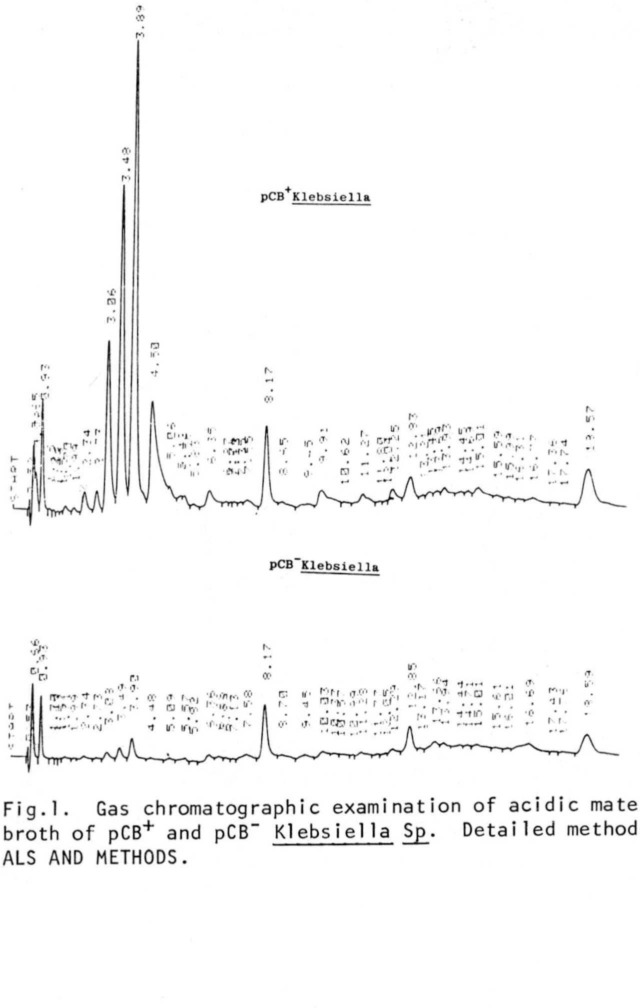

Fig.1. Gas chromatographic examination of acidic materials extracted from the broth of pCB^+ and pCB^- Klebsiella Sp. Detailed methods are given under MATERIALS AND METHODS.

An examination of the plasmid profiles of these two strains showed interesting differences. The pCB^+ AC901 strain showed the presence of predominantly three types of plasmids in the agarose gel, while the pCB^- segregant AC904 showed the presence of the smaller two types (Fig. 2).

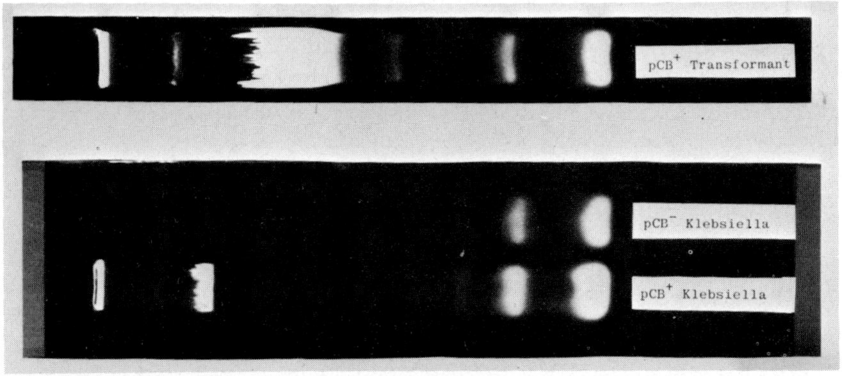

Fig.2. Electrophoretic separation on agarose gel of plasmid DNA isolated from pCB^+ and pCB^- Klebsiella pneumoniae AC901 and AC904 and a pCB^+ transformant of AC904.

Electron microscopic examination (Fig. 3) demonstrated the occurrence of covalently-closed circular forms corresponding to 65, 5.30 and 3.50 million dalton (Mdal) molecular masses in K. pneumoniae AC901 while the larger plasmid was conspicuously absent in AC904.

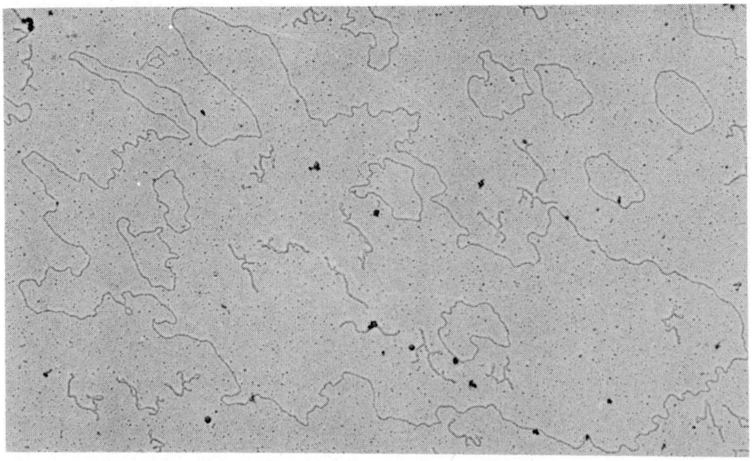

Fig.3. The occurrence of different forms of covalently closed circular molecules extracted from pCB^+ and pCB^- Klebsiella pneumoniae AC901 and AC904. The large molecule (31.5 μm) was absent in the plasmid DNA isolated from the pCB^- segregant (AC904).

The 65 Mdal plasmid was demonstrated to be present in the transformant (Fig. 2), when AC904 was transformed to pCB$^+$ with the plasmid DNA isolated from AC901. The pCB plasmid could also be transferred from AC902, a proline-requiring mutant of AC901, to AC904 with a frequency of about 10^{-7} per donor cell during mating in liquid broth and subsequent plating on minimal pCB plates. Thus genetic and physical evidence clearly suggest that the pCB degradative pathway is specified by the 65 Mdal plasmid, designated as pAC21.

Similar to the *Klebsiella* strain, the *Serratia* strain AC906 could be cured of its ability to grow with pCB by mitomycin C. The resulting pCB$^-$ strain AC907 still retained its ability to utilize benzoate. Unlike the *Klebsiella* strain, it was not possible to isolate any plasmid DNA from either AC906 or AC907. This may be due to technical difficulties in isolating plasmids from this *Serratia* strain and not indicate any real absence of pAC21 type of plasmids, since we have been unable to isolate known plasmids such as pAC8 that have been introduced into AC906.

Expression of Hydrocarbon Degradative Genes in AC901 and AC906. The occurrence of enteric bacteria such as *Klebsiella* or *Serratia* species capable of utilizing hydrocarbons such as pCB or aromatic compounds such as benzoate in the PCB-contaminated areas of the Hudson river sediments is interesting, since the laboratory strains of these bacteria are normally not known to be capable of hydrocarbon degradation. Reports from a number of laboratories[14-16] also indicate that most enteric bacteria are incapable of expressing hydrocarbon degradative genes when transferred from *Pseudomonas* in the form of plasmids. It was therefore of interest to determine if the enteric bacteria such as AC901 or AC906 may have evolved specific capabilities that allow them to utilize hydrocarbons such as pCB, or aromatic compounds such as benzoate. Since benzoate is oxidized through either the *ortho* pathway or the *meta* pathway in different microorganisms[17], it was also of interest to determine the nature of the specific pathway for the metabolism of benzoate in these two species. The results in Table 3 clearly indicate that both *K. pneumoniae* AC901 and *S. marcescens* AC906 utilize benzoate *via* the *ortho* pathway when grown with benzoate as a sole source of carbon. None of the enzymes of the *meta* pathway were detectable while the presence of all the enzymes of the *ortho* pathway was demonstrated. (P.F. Kamp, D. Babbit, L.N. Ornston, D.W. Ribbons and A.M. Chakrabarty, manuscript in preparation.)

Table 3. SPECIFIC ACTIVITIES OF AN ORTHO PATHWAY ENZYME (CATECHOL 1,2-OXYGENASE) AND A META PATHWAY ENZYME (CATECHOL 2,3-OXYGENASE) IN P. PUTIDA, K. PNEUMONIAE AND S. MARCESCENS.

Bacteria	Growth Substrate	Specific Activities[a] of 1,2-Oxygenase	Specific Activities[a] of 2,3-Oxygenase
P. putida AC2	benzoate	0.154	<0.1
" AC131[b]	benzoate	0.15	53.0
" AC1002	m-toluate	<0.01	41.0
K. pneumoniae AC901	benzoate	0.476	0.16
" AC904	benzoate	0.35	<0.1
" AC911	m-toluate	<0.01	18.0
S. marcescens AC906	benzoate	0.16	<0.1
" AC907	benzoate	0.175	<0.1
" AC913	m-toluate	-	25.0

[a] Specific activities have been expressed as µmoles substrate consumed or product formed per min per mg protein.

[b] AC131 is *P. putida* arvilla (mt-2) which is known to oxidize benzoate primarily by the *meta* pathway, while AC2 (*P. putida* strain PpG1) is known to metabolize benzoate by the *ortho* pathway.

It is possible to transfer RP4-TOL (pAC8) to both the pCB^+ *Klebsiella* and *Serratia* strains, selecting for kanamycin or carbenicillin resistance. When such cells were plated on minimal toluate plates, toluate-positive colonies came up with a frequency of about 10^{-7} to 10^{-8} per cell plated. Such TOL^+ derivatives (AC911 and AC913) could grow slowly with xylenes, toluene or *m*-toluate as sole sources of carbon and demonstrate *meta* pathway enzyme activities at 40 to 60% of the *P. putida* levels (Table 3). In contrast, $pAC8^+$ derivatives of laboratory strains of *E. coli* or *S. marcescens* demonstrate *meta* pathway enzyme activities varying from 5 to 25% of *P. putida* levels, and are incapable of growing with xylenes, toluene or *m*-toluate as sole sources of carbon[18].

DISCUSSION

The spontaneous and mitomycin C-induced curing of the pCB^+ character, the loss of a 65 Mdal plasmid band in pCB^- segregants and the reappearance of this plasmid in the segregants when transformed to pCB^+ by the plasmid DNA isolated from the pCB^+ wild type - all point out to the pCB degradative genes as part of the 65 Mdal plasmid in K. pneumoniae AC901. Similar to K. pneumoniae AC901, the S. marcescens strain AC906 can be cured of its pCB$^+$ character by treatment with mitomycin C, although in this instance it has not been possible to isolate any plasmid DNA from either the pCB$^+$ parent or the pCB$^-$ variant. It is known that many Serratia strains produce extracellular nucleases, which degrade the DNA during isolation, so that nuclease-negative mutants must be used for detection of plasmid DNA[19]. It would be interesting to determine if, as with the pAC21 plasmid, the pCB degradative pathway in the Serratia species is also coded by plasmid-borne genes, and whether the plasmid may have a complete or partial homology with the pCB-degradative plasmid in K. pneumoniae AC901.

The loss of expression of TOL, when transferred as part of the recombinant plasmid pAC8 from Pseudomonas species into E. coli, Salmonella typhimurium, or laboratory strains of S. marcescens or Agrobacterium tumefaciens, is intriguing. The full expression of such genes in Pseudomonas, when the pAC8 plasmid is transferred from E. coli or laboratory strains of S. marcescens, suggests that pseudomonads presumably have chromosomally-coded functions necessary for their expression. It is therefore interesting that it is possible to isolate pAC8$^+$ pCB$^+$ K. pneumoniae and S. marcescens mutants that can express TOL genes, albeit inefficiently, and grow with toluene as a sole source of carbon. It was not possible to isolate Tol$^+$ colonies, even when 10^{10} cells of pAC8$^+$ E. coli, Salmonella or laboratory strains of Serratia, with or without mutagen treatment, were plated on minimal toluate plates. Since the pCB$^+$ K. pneumoniae or S. marcescens strains sport only mutant colonies capable of slow growth with toluene, it appears that full expression of hydrocarbon degradative genes in the enteric bacteria requires multienzyme functions. It would be interesting to determine the nature of genes and the biochemical mechanisms allowing hydrocarbon degradation in the enteric bacteria.

SUMMARY

Facultatively anaerobic strains belonging to Enterobacteriaceae such as Klebsiella pneumoniae or Serratia marcescens, capable of utilizing p-chlorobiphenyl (pCB) as a sole source of carbon and producing acidic metabolites such as p-chlorobenzoate, have been isolated from the PCB-contaminated sediments of the Hudson river. Both the Klebsiella and the Serratia strains can lose their ability to utilize pCB on treatment with mitomycin C. Isolation of plasmid DNA from the pCB$^+$ and pCB$^-$ K. pneumoniae cells and their agarose gel and electron microscopic characterization, as well as transformation and conjugation experiments, suggest that a conjugative 65 million dalton plasmid, pAC21, codes for the pCB degradative pathway. Unlike known members of Enterobacteriaceae, the evolvant pCB^+ strains of Klebsiella and Serratia produce mutant colonies that can functionally express TOL plasmid genes transferred from Pseudomonas, and utilize xylenes and toluene as sole sources of carbon.

ACKNOWLEDGEMENTS

The work has been supported in part by a grant from the National Science Foundation (PCM 77-25450). We thank T.M. Su and A. Fritz for cultures, Denise Friello for help in electron microscopy and R. Davis of Bender Laboratories for taxonomic characterization.

REFERENCES

1. Risebrough, R.W., Rieche, P., Herman, S.G., Peakall, D.E., and Kirven, M.N. (1968) Nature, 220, 1098-1102.

2. Kimbrough, R.D. (1974) CRD Crit. Rev. Toxicol. 2, 445-460.

3. Kaufman, D.D. and Kearney, P.C. (1976) Herbicides: Physiology, Biochemistry, Ecology (L.J. Audus, ed.) Vol. 2. Academic Press, New York, pp. 29-64.

4. Alexander, M. (1969) Soil Biology, Reviews of Research, UNESCO, 9, 209-240.

5. Rheinwald, J.G., Chakrabarty, A.M., and Gunsalus, I.C. (1973). Proc. Natl. Acad. Sci. USA, 70, 885-889.

6. Chakrabarty, A.M., Mylroie, J.R., Friello, D.A. and Vacca, J.G. (1975) Proc. Natl. Acad. Sci. USA, 72, 3647-3651.

7. Currier, T.C. and Nester, E.W. (1976) Anal. Biochem. 76, 431-441.

8. Hayward, G.S. and Smith, M.G. (1972) J. Mol. Biol. 63, 383-395.

9. Davis, R.W., Simon, M.N. and Davidson, N. (1971). Methods in Enzymology (L. Grossman and K. Moldave, eds.) Vol. 21, Academic Press, New York, pp. 413-428.

10. Hegeman, G.D. (1966) J. Bacteriol. 91, 1140-1154.

11. Feist, C.F. and Hegeman, G.D. (1969) J. Bacteriol. 100, 869-877.

12. Conference Proceedings (1976) National Conference on Polychlorinated Biphenyls. No. 19-21, Chicago, Ill., EPA 560/6-75-004.

13. Environmental Protection Agency (1976) Federal Register 41, No. 143, pp. 30468-30477.

14. Nakazawa, T., Hayashi, E., Yokota, T., Ebina, Y. and Nakazawa, A. (1978) J. Bacteriol. 134, 270-277.

15. Jacoby, G.A., Rogers, J.E., Jacob, A.E. and Hedges, R.W. (1978) Nature 274, 179-180.

16. Chakrabarty, A.M., Friello, D.A. and Bopp, L.H. (1978) Proc. Natl. Acad. Sci. USA, 75, 3109-3112.

17. Ornston, L.N. and Parke, D. (1977) Current Topics in Cellular Regulation (Horecker, B.L. and Stadtman, E.R., eds.) Vol. 12, Academic Press, New York, pp. 209-262.

18. Ribbons, D.W., Wigmore, G.J. and Chakrabarty, A.M. (1979) Soc. Gen. Microbiol. Quart. 6, 24-25.

19. Timmis, K.N. and Winkler, U. (1973) J. Bacteriol. 113, 508-509.

Plasmids of Medical, Environmental and Commercial Importance
K.N. Timmis and A. Pühler, editors

EVOLUTION AND SPREAD OF PESTICIDE DEGRADING ABILITY AMONG SOIL MICRO-ORGANISMS

J.M. PEMBERTON, B. CORNEY AND R.H. DON
Department of Microbiology, University of Queensland, St. Lucia, 4067. Australia.

INTRODUCTION

Chlorinated phenoxyherbicides such as 2,4-dichlorophenoxyacetic acid (TFD) and 2,4,5-trichlorophenoxyacetic acid (TFT) have been in continual, widespread agricultural use for almost forty years[1]. Only recently has there been concern expressed over possible genetic, carcinogenic and teratogenic effects of these synthetic pesticides[2,3]. Debate over chemical pollution, with its attendant biological hazards, caused by these and closely related herbicides has resulted in a flurry of scientific and governmental activity in the areas of pesticide evaluation and regulation. Conflict between economic benefit and potential biological hazard has been highlighted in the debate over the continued use of both TFD and TFT. How this conflict is eventually resolved will depend on the collection and evaluation of a wide range of data on pesticide parameters such as environmental persistence and mammalian toxicity.

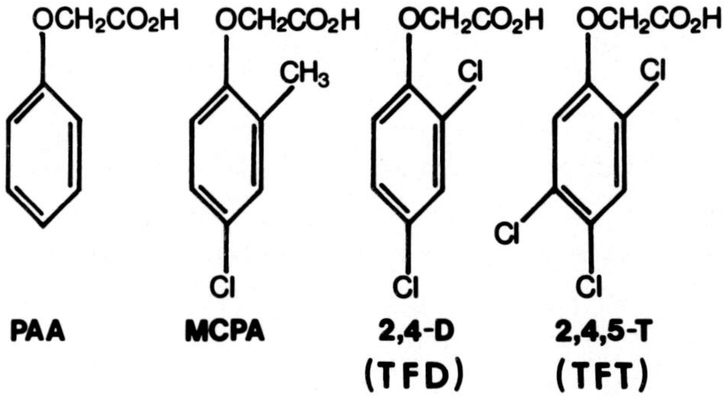

Fig. 1. Structure of phenoxyacetic acid (PAA), 2,4-dichlorophenoxyacetic acid (TFD), 2,4,5-trichlorophenoxyacetic acid (TFT) and 2-methyl,4-chloro,phenoxyacetic acid (MCPA).

Environmental persistence of TFD is not a problem because of the widespread occurrance of a variety of soil micro-organisms having the ability to degrade this synthetic molecule[4,5]. By contrast TFT persists for long periods of time and its environmental fate is unknown[6,7]. Recently we reported the isolation of a strain of *Alcaligenes paradoxus* in which the *tfd* genes are borne as part of a conjugative plasmid[8,9], providing a mechanism for the distribution of *tfd* genes through microbial populations. Subsequent biophysical and genetic analyses of a variety of TFD degrading soil micro-organisms carried out in this laboratory have revealed that each isolate carried a single, conjugative TFD plasmid. Such plasmids encoded the degradation of TFD but not the parental compound phenoxyacetic acid (PAA). However we have isolated mutant TFD plasmids which encoded degradation of both TFD and PAA. In this paper we present data demonstrating that this broadened substrate specificity resulted from separate *tfd* and *paa* genes on the plasmid molecule. Two distinct events appeared to have occurred in the evolutionary process. First, the plasmid *tfd* gene was **duplicated**. Second, one of the *tfd* genes was altered by mutation to give rise to the *paa* gene.

MATERIALS AND METHODS

Bacterial and plasmid strains. Bacterial strains used in this study are listed in Table 1.

Strain JMP116 is a strain of *Alcaligenes paradoxus*[9], while JMP134 has been identified as *Alcaligenes eutrophus*. Strains JMP130, 133, 135 and 141 have not been taxonomically identified. Strains JMP502 and JMP516 are strains of *Escherichia coli*, a kind gift of Dr. R.W. Hedges, Royal Postgraduate Medical School, London.

All strains were preserved by lyophilisation. This procedure was particularly necessary in view of the instability of the Tfd$^+$ phenotype. Those strains in routine use were kept under sterile distilled water at 5 C. Plasmid bearing strains were routinely checked for retention of plasmid borne markers.

TABLE 1

BACTERIAL AND PLASMID STRAINS

Strain Number	Chromosomal[a] Genotype	Plasmid	Plasmid[b] phenotype	Reference/source
JMP116		pJP1	Tfd^+, Inc P-1	9
JMP130		pJP2	Tfd^+,Paa^+	This paper
JMP133		pJP3	Tfd^+, Inc P-1	This paper
JMP134		pJP4	Tfd^+, Inc P-1	This paper
JMP135		pJP5	Tfd^+, Inc P-1	This paper
JMP141		pJP6	Tfd^+	This paper
JMP181		pJP4-1	Tfd^+,Paa^+	Spontaneous Paa^+ derivative of JMP134
JMP221				Cured derivative of JMP181[c]
JMP222	*str-201*			Spontaneous Str^r mutant of JMP221
JMP225		pJP4-11	Tfd^+,Paa^+,Inc P-1, Tra^-	This paper
JMP246		pJP4-11 S-a	Tfd^+,Paa^+, Su, Km, Cm	This paper
JMP502		RP1	Km, Tc, Cb, Inc P-1	9
JMP516		S-a	Su, Km, Cm, Inc W	9
JMP700		pJP4-2	Paa^+,Inc P-1	This paper

[a]Genotypic symbols follow those proposed by Demerec *et al*[10]. *str* denotes streptomycin resistance.

[b]Phenotypic symbols follow those proposed by Novick *et al*[11]. Tfd^+,2,4-dichlorophenoxyacetic acid degradation; Paa^+, phenoxyacetic acid degradation; Tra^-, transfer defective. Resistance to the following antibiotics: Su,sulphonamide; Km, kanamycin; Cb,carbenicillin; Tc,tetracycline; Cm,chloramphenicol.

[c]JMP221 is a spontaneous male-specific bacteriophage (PR11) resistant mutant of JMP181 cured of pJP4-1.

Bacteriophages. Male-specific bacteriophages PR11[8,9] which plaques on P and W Inc plasmid bearing strains and PRR1[12] which plaques on Inc P-1 plasmid bearing strains were used. Bacteriophage sensitivity was determined by spotting, cross-streaking, plaque formation or increase in phage titre.

Media and cultural conditions. Peptone yeast extract (PYE)agar and broth; 2,4-dichlorophenoxyacetic acid bromocresol purple liquid medium (TFD-BCPM); 2,4-dichlorophenoxyacetic acid eosin methylene blue agar (TFD-EMB); and Gilardi mineral salts medium were prepared as previously described[8,9]. Concentrated solutions (10 mg/ml) of analytical grade carbon sources were prepared in distilled water, adjusted to pH 7.0 with either concentrated HCl or NaOH and chloroform sterilised. Sterile carbon sources were always added after media was autoclave sterilised to the following final concentrations: 2,2-dichloroproprionic acid (DAL), 500 μg/ml; PAA, 500 μg/ml; 2-methyl, 4-chloro, phenoxyacetic acid (MCPA), 500 μg/ml; TFT, 300 μg/ml; phenol, 250 μg/ml; 2,4-dichlorophenol, 250 μg/ml.

Unless otherwise stated, all cultures were incubated at 32°C. Cultures were aerated by shaking at 350 rpm in a New Brunswick G24 Environmental Incubator shaker.

Curing experiments. Curing with mitomycin C was performed as previously described[9].

Transfer on solid medium. Transfer on solid medium was performed using the method of Haas and Holloway[13].

Surface exclusion and incompatibility. Surface exclusion and incompatibility were determined by the method of Datta[14].

Isolation of plasmid DNA. Plasmid DNA was isolated by the method of Guerry *et al*[15] and Hansen and Olsen[16].

Electronmicroscopy. Plasmid DNA was prepared by the method of Kleinschmidt and Zahn[17, 18, 19].

RESULTS

Isolation of Tfd$^+$ strains. Fifty independent soil samples were screened for the presence of TFD degrading micro-organisms by growth in TFD-BCPM medium. The size of the soil sample was critical, a minimum sample weight of 1g (dry weight) was required to obtain consistent isolation of Tfd$^+$ strains from a particular soil sample. The method employed favoured isolation of micro-organisms which completely degraded TFD rather than those strains which simply co-metabolised the substrate.

A positive reduction in the level of TFD was observed in thirty eight (76%)

of the fifty samples. Although most positive samples gave a positive reaction after a single subculture in TFD-BCPM medium, three successive subcultures were required before a sample was deemed positive or negative. Twenty-two of the thirty-eight positive cultures yielded single Tfd^+ colonies after subculture on TFD-EMB agar.

Seventeen of the twenty-two strains produced Tfd^- segregants at high levels (> 1%) when subcultured in PYE broth for 15-20 generations in the absence of TFD. The five remaining Tfd^+ strains were retained for further study. These strains were JMP130, JMP133, JMP134, JMP135 and JMP141. Included in this study was JMP116, a strain of *Alcaligenes paradoxus* which harbours the pesticide plasmid pJP1 whose properties were the subject of previous communications[8,9].

Phenotypic properties of Tfd^+ isolates. Data given in Table 2 show that JMP134, like JMP116(pJP1), was sensitive to the male-specific bacteriophage PR11, indicating the presence of a plasmid related to pJP1. Although pJP1 has been placed in the Inc P-1 group, JMP116 was not sensitive to PRR1, a male-specific bacteriophage specific for Inc P-1 plasmids. Failure to express PRR1 sensitivity has been observed in strains of *Rhodopseudomonas sphaeroides* carrying Inc P-1 plasmids RP4, R702 and R751 (Pemberton and Tucker, unpublished data). Hence sensitivity to PRR1 appears to depend on the genetic background in which the P-1 plasmid occurs.

TABLE 2

PHENOTYPIC PROPERTIES OF Tfd^+ ISOLATES

Strain Number	Sensitivity to Male-Specific Bacteriophage[a]		Ability to Degrade Various Pesticides and Related Molecules[b]				
	PRR1	PR11	PAA	MCPA	TFD	TFT	DAL
JMP116	-	+	-	+	+	-	+
JMP130	-	-	+	+	+	-	+
JMP133	-	-	-	+	+	-	+
JMP134	-	+	-	+	+	-	+
JMP135	-	-	-	+	+	-	+
JMP141	-	-	-	+	+	-	+

[a]Sensitivity to male-specific bacteriophage was determined as either plaque formation or as an increase in phage titre

[b]Abbreviations: PAA, phenoxyacetic acid; MCPA, 2-methyl,4-chloro,phenoxyacetic acid; TFD, 2,4-dichlorophenoxyacetic acid; TFT, 2,4,5-trichlorophenoxyacetic acid; DAL, dalapon,2,2-dichloropropionic acid.

Although all six strains degraded the non-persistent pesticides MCPA, TFD and DAL, none showed any activity against the persistent pesticide TFT. An interesting feature of one strain, JMP130, was its ability to degrade the structurally-related, parental molecule phenoxyacetic acid (PAA) in addition to the chlorinated derivatives MCPA and TFD.

Curing of the Tfd$^+$ phenotype. Successive subcultures of JMP116, JMP130, JMP133 and JMP134 in the absence of TFD produced few (< 0.1%) Tfd$^-$ segregants; JMP135 and JMP141 were less stable giving 0.1-0.2% Tfd$^-$ clones after 15-20 generations in PYE broth. Spontaneously cured derivatives of JMP135 and JMP141 were retained for further study.

Since strain JMP134, like JMP116(pJP1), could be cured of its Tfd$^+$ phenotype by selection of PR11 resistant clones, it was concluded that the *tfd* gene(s) of JMP134 was plasmid borne.

Both JMP130 and JMP133 were cured of their Tfd$^+$ phenotype by passage in the presence of sub-lethal concentrations of the curing agent mitomycin C. After a single passage, 0.1-0.2% of clones re-isolated on PYE agar had lost the Tfd$^+$ phenotype.

It was concluded from the spontaneous, irreversible loss of the Tfd$^+$ phenotype, the loss of the Tfd$^+$ phenotype in PR11 resistant clones and the curing of the Tfd$^+$ by mitomycin C that each of the six strains possessed a TFD plasmid(s).

Intrastrain transfer of TFD plasmids. Using streptomycin or naladixic acid resistant, cured (Tfd$^-$) derivatives of each of the six strains, overnight matings were performed on solid media.

TABLE 3

INTRASTRAIN TRANSFER OF THE Tfd$^+$ PHENOTYPE

Donor	Donor Phenotype	Frequency of Tfd$^+$ Transfer Per Donor Cell
JMP116	Tfd$^+$	1×10^{-2}
JMP130	Tfd$^+$Paa$^+$	1×10^{-2} [a]
JMP133	Tfd$^+$	2×10^{-2}
JMP134	Tfd$^+$	5×10^{-2}
JMP135	Tfd$^+$	1×10^{-3}
JMP141	Tfd$^+$	1×10^{-3}

[a]Note that there was 100% co-transfer of the Tfd$^+$Paa$^+$ phenotype irregardless of the marker selected

Intrastrain transfer of the Tfd^+ phenotype occurred at a level of between 0.1-5.0% per donor cell (Table 3) indicating that conjugative plasmids were involved in the transfer of this phenotype. An interesting feature of the transfer and curing experiments was that JMP130 showed co-transfer and co-curing of both the Tfd^+ and Paa^+ phenotypes. It was concluded from these results that JMP130 carried a plasmid which conferred the ability to degrade both the synthetic molecule TFD and its naturally-occurring, structurally related counterpart PAA. In each Tfd^+ isolate we were unable to separate the conjugative and Tfd^+ phenotypes indicating that each strain carried a single conjugative, TFD plasmid.

Isolation of plasmid DNA. Using the method of Guerry *et al*[15] a single size class of plasmid was isolated from JMP116, pJP1, $58 \pm 2 \times 10^6$; JMP133,pJP3, $60 \pm 2 \times 10^6$; and JMP134,pJP4, $50 \pm 1.5 \times 10^6$. Many degradative plasmids have high molecular weights in excess of 150×10^6 which cannot be isolated by the method of Guerry *et al*[15]; strains JMP130, JMP135 and JMP141 may carry such plasmids.

A method has been developed by Hansen and Olsen[16] which allows the isolation of extremely large plasmids. When this method was used, only JMP130 yielded plasmid DNA. The plasmid, pJP2, had a molecular weight of $150 \pm 4 \times 10^6$. Both JMP135 and JMP141 have plasmids unextractable in detectable quantities using either method, this may be due to high molecular weight.

Surface exclusion and incompativility. A variety of plasmids representing many incompatibility groups were tested for transfer into the Tfd^+ strains. Only P and W Inc group plasmids were transferred at detectable frequencies (Table 4). The Inc W group plasmid S-a was readily transferred from *E. coli* to the Tfd^+ strains and back again into *E. coli*. Each of the TFD plasmids co-existed stably in the same cell as S-a indicating that none of these plasmids belonged to the Inc W group.

Incompatibility testing was more difficult when the Inc P-1 plasmid RP1 was used. Powerful surface exclusion greatly reduced the level of transcipient formation such that transfer appeared to occur into spontaneously cured (Tfd^-) derivatives of the Tfd^+ strains when selection was made for the incoming (RP1) plasmid. This difficulty was overcome by simultaneously selecting for a marker carried by the resident plasmid (Tfd^+) and the incoming plasmid (Kan^r).

Transcipients carrying both plasmids were then tested for retention of the two markers in non-selective conditions.

TABLE 4

SURFACE EXCLUSION AND INCOMPATIBILITY PROPERTIES OF THE TFD PLASMIDS

Donor[a]	Recipient	Surface Exclusion Ratio[b]	Incompatibility Ratio[c]
JMP516(S-a)	JMP116(pJP1)	1	20/20
	JMP130(pJP2)	1	20/20
	JMP133(pJP3)	1	20/20
	JMP134(pJP4)	1	20/20
	JMP135(pJP5)	1	20/20
	JMP141(pJP6)	1	20/20
JMP502	JMP116(pJP1)	10^6	0/20
	JMP130(pJP2)	1	19/20
	JMP133(pJP3)	10^4	0/20
	JMP134(pJP4)	10^6	0/20
	JMP135(pJP5)	10^4	2/20
	JMP141(pJP6)	10^1	20/20

[a]Transfer of S-a into both plasmid bearing and cured strains varied between 10-100% per donor cell. Transfer of RP1 into the cured derivatives varied between 1-10% per donor cell.

[b]Surface Exclusion Ratio was derived by dividing the frequency of Tfd^+Kan^r transcipients, when a cured strain was used as the recipient, by the frequency of Tfd^+Kan^r transcipients where a TFD plasmid bearing strain was used as a recipient.

[c]Incompatibility Ratio was derived as the fraction of clones retaining the Tfd^+Kan^r phenotype after a single passage (15-20) generations in PYE broth. Segregant clones always lost the Tfd^+ marker, but retained the Kan^r marker.

Transfer of the Inc P plasmid RP1 was greatly reduced into four of the six TFD plasmid bearing strains, while transfer was normal into cured derivatives of these strains. Hence plasmids pJP1, pJP3, pJP4 and pJP5 exhibit surface exclusion towards RP1. Incompatibility testing (Table 4) places pJP1, pJP3, pJP4 and pJP5 into the Inc P-1 group. Further testing of pJP2 and pJP6 was not possible due to lack of transfer of suitable Inc group plasmids.

Isolation of Paa^+ mutants of JMP134. Although JMP134(Tfd^+Paa^-) identified as a strain of *Alcaligenes eutrophus* (Dr. L.I. Sly, personal communication), did not degrade PAA, mutants capable of growth on PAA occurred at a frequency of 10^{-6}, indicating spontaneous mutation of a single gene. The majority

(>99%) of the Paa$^+$ mutants retained their Tfd$^+$ phenotype, suggesting that this broadened substrate specificity, like that of the amidase gene of *Pseudomonas aeruginosa*, was encoded in a single mutant gene. One such mutant Tfd$^+$Paa$^+$ strain, JMP181, was examined in more detail.

Clones of JMP181(Tfd$^+$Paa$^+$) isolated as resistant to PR11 ($\emptyset^r$: frequency 10^{-5}) which had lost the plasmid became simultaneously Paa$^-$Tfd$^-$ indicating that the newly acquired mutant phenotype (Paa$^+$) was encoded by a mutant derivative of pJP4. Strains cured of the mutant plasmid, designated pJP4-1, did not show detectable levels ($<10^{-9}$) of backmutation to either Tfd$^+$ or Paa$^+$. Similarly, cured derivatives of both JMP134 and JMP181 re-infected with pJP4-1 became phenotypically Tfd$^+$Paa$^+$ confirming the linkage of the *paa* gene to other genes carried by pJP4-1.

Construction of a deletion map of pJP4-1. Unlike other male-specific bacteriophages, PR11 appears to attach to the cell surface and not to sex pili[16]. This view that pili were not involved in the attachment of PR11 to cells harbouring P and W Inc plasmids was strengthened by the finding that PR11 resistant plasmid mutants retained their transfer proficiency (Tra$^+$).

Deletion mutants of pJP4-1 were obtained (frequency 10^{-8}) by screening $\emptyset^r$Paa$^+$ or $\emptyset^r$Tfd$^+$ strains, which retained surface exclusion (Sex$^+$), for loss of transfer (Tra$^-$) and/or degradative (Tfd$^-$ or Paa$^-$) functions. Clearly these strains were deletion mutants since they had simultaneously lost two independent plasmid functions while retaining at least two of the remaining plasmid conferred phenotypes. The groups of deletion mutants (Table 5) were used to construct a deletion map of pJP4-1 (Figure 2).

While the deletion map was a reflection of the order of genes on pJP4-1, the data was not considered sufficient to establish actual map distances. More extensive deletion mapping of pJP4-1 has been undertaken in an attempt to identify whether this plasmid possesses a similar arrangement of transfer genes to those possessed by the F factor of *E. coli*. In addition attempts will be made to determine precisely the number of degradative genes carried by pJP4-1.

TABLE 5

PHENOTYPES OF STRAINS HARBOURING DELETION MUTANTS OF THE TFD PLASMID pJP4-1

Selected Marker	Possible Phenotypes of Strains[a]				Number in Each Class
	Tfd	Sex	Paa	Tra	
Tfd^+	+	+	+	+	17/50
	+	+	+	-	7/50
	+	+	-	+	19/50
	+	+	-	-	7/50
Paa^+	+	+	+	+	14/23
	+	+	+	-	9/23
	-	+	+	+	0/23
	-	+	+	-	0/23

[a]Phenotypic symbols: $Tfd^{+/-}$, ability/inability to degrade TFD; Sex^+, exhibits surface exclusion towards RP1; $Paa^{+/-}$, ability/inability to degrade PAA; $Tra^{+/-}$, ability/inability to transfer pJP4-1.

An interesting feature of the results was that the Tfd^+ and Paa^+ phenotypes were encoded in separate genes carried by pJP4-1.

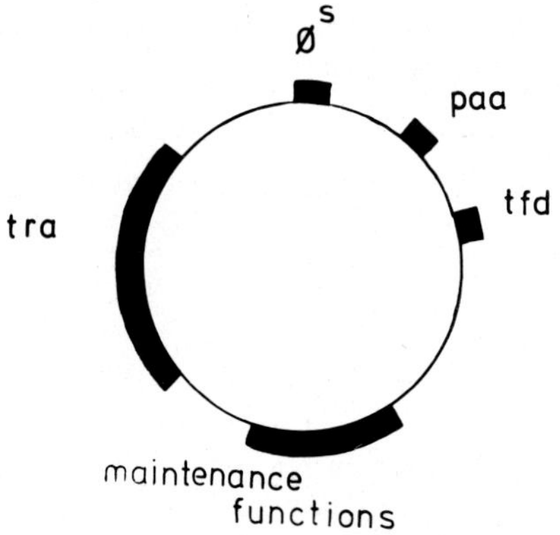

Fig. 2. Deletion map of TFD plasmid pJP4-1, constructed using the data in Table 5.

Duplication and gene mutation. Although a few (0.1%) Tfd^-Paa^+ plasmids were isolated in the initial selection for Paa^+ mutants of JMP134, the majority (>99%) of the mutants were stably Tfd^+Paa^+ during many subcultures in the absence of both TFD and PAA. Since the selection for these mutants was made only on PAA minimal medium, these data suggest that there was an overwhelming selection for cells carrying duplications of the *tfd* gene. Similarly, selection of Tfd^+ mutants from a strain JMP700 (pJP4-2, Tfd^-Paa^+) carrying only the *paa* gene resulted in a majority (>99%) of cells which possessed the dual phenotype (Tfd^+Paa^+; frequency 10^{-6}). The size of the duplicated region in broth cultured Tfd^+Paa^+ cells was 1.6 μm or less since length measurements of plasmids carrying the *tfd* gene (pJP4) or the *paa* gene (pJP4-2) alone did not differ significantly in size from the plasmid carrying both the *paa* and *tfd* genes (pJP4-1; 24 ± 0.8 μm).

It was concluded from these data that a single gene was duplicated and then one copy altered by mutation.

DISCUSSION

Extensive, unrestricted use of synthetic molecules such as antibiotics and pesticides has resulted in the widespread development of microbial populations capable of degrading many of these molecules. Unlike the development of antibiotic resistance, which is an undesirable occurrance, the evolution and spread of pesticide degrading genes has a beneficial effect by removing potentially dangerous pollutants from the environment.

From data presented in this paper we have concluded that evolution of the ability to degrade the chlorinated phenoxyherbicide 2,4-dichlorophenoxyacetic acid occurs by a process of plasmid borne gene duplication and mutation (Figure 3).

If a single *(paa)* gene was duplicated then one copy altered by mutation, the high frequency (10^{-6}) of this event would suggest that the gene alteration was due to a single step mutation. Under these circumstances the rate of gene duplication must have been close to 100%. This conclusion is supported by the finding that mutation to either Tfd^+ or Paa^+ gave mutants with dual substrate specificity (Tfd^+Paa^+) rather than single Tfd^+Paa^- or Tfd^-Paa^+ substrate specificity which would be expected if there was only a single copy of the plasmid borne gene. Such levels of gene duplication have been reported by Yaji and Clewell[20] for a transposon like genetic element encoding resistance to tetracycline.

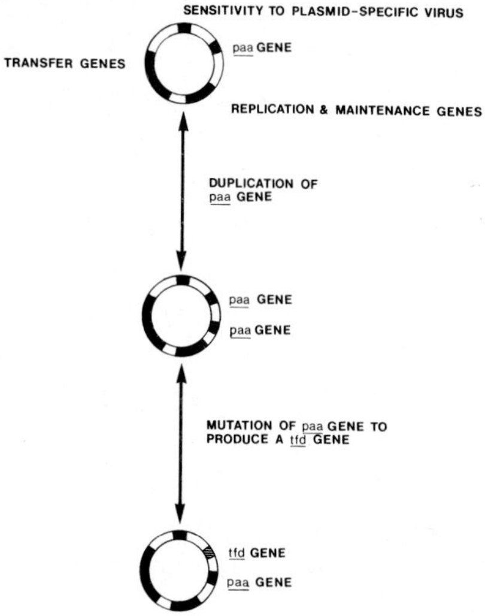

Fig. 3. Evolution of the *tfd* gene from a plasmid borne *paa* gene.

Plasmids carrying duplications of the antibiotic genes were selected by growth in sub-lethal concentrations of tetracycline. All such duplications, whether they occur on a plasmid or on the main chromosome, are characteristically unstable[21,22] and removal of the selective conditions results in loss of the duplicated regions. In contrast Tfd^+Paa^+ strains which carry duplications of the *tfd* and *paa* genes remain stable in the absence of TFD and PAA.

Although a number of structural and regulatory genes are involved in the degradation of chlorinated phenoxyacetates, segregants of JMP134 lacking pJP4 still retain the ability to degrade 2,4-dichlorophenol, the product of the first step in the degradative pathway[9]. This result suggests that the *tfd* gene carried by pJP4 encodes the enzyme controlling the first step in the degradative pathway.

Neither the wild type nor mutant strains of the TFD plasmids showed any activity against the persistent pesticide TFT. A survey of fifty soil samples, thirty-eight of which were shown to harbour organisms capable of degrading TFD, failed to show detectable degradation or co-metabolism of TFT over a period of six months (Pemberton, unpublished data). Hence a major difference between non-persistent (e.g. TFD) and recalcitrant (e.g. TFT) pesticide molecules appears to be the ability to elicit the rapid evolution and spread of pesticide plasmids encoding their degradation.

ACKNOWLEDGEMENTS

This investigation was supported by grant D2 75/15203 from the Australian Research Grants Committee. R.H.D. holds a postgraduate award from the Australian Government.

REFERENCES

1. Fletcher, W.W. (1978). The Pest War. Basil Backwell-Oxford.
2. Firestone, D. (1977). In Chlorinated Phenoxy Acids and their Dioxins. Ecol. Bull. (Stockholm), 27, 39-52.
3. Courtney, K.D., Gaylor, D.W., Hogan, M.D., Falk, H.L., Bates, R.R. and Mitchell, I. (1970). Science, 168, 864-866.
4. Alexander, M. (1969). Soil. Biol. Rev. Res. Nat. Resour. Res.,(UNESCO), 9, 209-240.
5. Kearney, P.C., and D.D. Kaufman (1972). In Degradation of Synthetic Organic Molecules in the Biosphere. Natl. Acad. Sci., Washington, D.C. 166-189.
6. Jensen, H.L. (1957). Can. J. Microbiol., 3, 151-164.
7. Newman, A.S., Thomas, J.R. and Walker, R.L. (1952). Proc. Soil. Sci. Amer., 16, 21-24.
8. Pemberton, J.M. and Fisher, P.R. (1977). Nature, 268, 732-733.
9. Fisher, P.R., Appleton, J., and Pemberton, J.M. (1978). J. Bacteriol. 135, 798-805.
10. Demerec, M., Adelberg, E.A., Clark, A.J. and P.E. Hartman (1966). Genetics, 54, 61-76.
11. Novick, R.P., Clowes, R.C., Cohen, S.N., Curtiss, R., and N. Datta. (1976). Bacteriol. Rev. 40, 168-189.
12. Stanisich, V.A. (1974). J. Gen. Microbiol., 84, 332-342.
13. Haas, D. and Holloway, B.W. (1976). Molec. Gen. Genet. 144, 243-251.
14. Datta, N. (1977). In R. Factor. Ed. I. Mitsuhashi. University Park Press, pp 255-272.
15. Guerry, P., LeBlanc, J., and Falkow, S. (1973). J. Bacteriol., 116, 1064-1066.
16. Hansen, J.B. and Olsen, R.H. (1978). J. Bacteriol., 135, 227-238.
17. Kleinschmidt, A., and Zahn, R.K.(1959). Z. Naturforsch., 14b, 770-779.
18. Pemberton, J.M. and Clark, A.J. (1973). J. Bacteriol., 114, 424-433.
19. Pemberton, J.M. (1974). J. Bacteriol., 119, 748-752.
20. Yaji, Y. and Clewell, D.B. (1976). J. Mol. Biol., 102, 583-600.
21. Anderson, R.P., Miller, C.G. and Roth, J.R. (1976). J. Mol. Biol. 105, 201-218.
22. Kleckmer, N., Roth, J.R. and Botstein, D. (1977). J. Mol. Biol. 116, 125-159.

Plasmids of Medical, Environmental and Commercial Importance
K.N. Timmis and A. Pühler, editors

INVOLVEMENT OF PLASMIDS IN THE BACTERIAL DEGRADATION OF LIGNIN-DERIVED COMPOUNDS

M.S. SALKINOJA-SALONEN, E. VÄISÄNEN and A. PATERSON
Department of General Microbiology, University of Helsinki, SF-00100 Helsinki 10 Finland

INTRODUCTION

Lignin is a renewable resource of aromatic carbon skeletons. When lignin is degraded by microorganisms, a variety of simple and complex aromatic compounds arises. Fungi are known to be active in degrading lignin[1]. Bacteria are usually thought to degrade the smaller aromatic molecules arising from fungal attack or industrial processes such as manufacture of cellulose. Lignin-degrading microorganisms are studied intensively in the hope of developing lignin-based biotechnical processes such as production of chemicals or microbial proteins.

It has been shown recently that bacteria can also be active in degrading lignin into CO_2[2,3]. Bacteria are more amenable than the multinucleate fungi for genetic research. With a knowledge of the genes involved in the degradation of lignin by bacteria, one could place mutations in the relevant genes in such a way that desired end products will emerge. This might open a possibility for biotechnically preparing defined chemical substances from lignin. This requires more knowledge than is available at present on the physiology and genetics of the bacteria involved.

We isolated 363 strains of bacteria from decaying wood and from pulp mill effluent, environments rich in lignin. Except for one gram-positive strain, these were gram-negative or -variable, mostly aerobic or facultative rod-shaped bacteria. Most strains utilized a broad spectrum of aromatic compounds as carbon source. Many strains were genetically unstable: they pleiotropically lost their ability to grow on several methoxylated aromatic acids[4]. We studied the strain K17 in detail and describe the nature of its genetic instability in the present paper. The results support the view that unstable pathways are coded for by plasmid-borne genes. If the same turns out to be true for the other cases of genetical instability we observed, plasmids would play a central role in bacterial metabolism of lignin and its degradation products.

MATERIALS AND METHODS

Media. K salt medium contains per litre 2 g of NH_4Cl, 3,8 g of $K_2HPO_4 \cdot 3H_2O$, 2,1 g of KH_2PO_4, 300 mg of $MgSO_4 \cdot 7H_2O$, 100 mg of NaCl, 10 mg of $CaCl_2$ and 0,01 mg of $FeSO_4$. To prepare nitrogen-free media, NH_4Cl was omitted. pH was set at 6,7. Carbon sources were added from a filter-sterilized stock solution concentrated 50 times to a final concentration of 5 mM, except for the alcohols of which 10 ml/litre was added. α-conidendrin and naphthalene are insoluble and were added as solid powder (1 g/l). Antibiotics were dissolved in sterile water and filter-sterilized. The concentrations were 10 μg/ml of tetracycline, 50 μg/ml of kanamycine, 100 μg/ml of carbenicillin, erytromycin, nalidixic acid and streptomycin each. For solid media, 10 g/l of agar was added.

Electron microscopy and Warburg respirometry were performed as described elsewhere[5,6].

Kinetic measument of O_2 uptake was performed at room temperature with an oxygen probe (Orion Research) fitted to a 10-ml cell. Readings were displayed by a digital pH meter (Orion 701A) with automatic correction for temperature and salinity. In each measurement, 50 to 100 μl of washed cell suspension or 100 to 250 μl of 10'000 x g supernatant of sonicated cells were used. The buffer was aerated K salt (pH 6,7) or 50 mM MES (pH 5,5). For washed cell suspension, 5 mM substrate was used and 0,1 mM for cell-free extracts. NADH or NADPH were added equimolarly to the oxidizable aromatic substrate (0,1 mM).

Protein was measured by the biuret method using lysozyme as reference protein.

Matings were performed in liquid by mixing 12,5 ml of stationary culture of the recipient with 2,5 ml of exponential phase donor culture and incubating overnight stationary in a 500-ml infusion bottle to insure aerobiosis. Filter mating was performed by sucking through a 0,45-μm membrane filter 4-5 ml of donor and recipient cultures. Turbidity of the donor culture was 1/5 of that of the recipient culture. The filter was incubated on a plate made of K salts solution with 1 g/l of yeast extract. The cells were then washed off the filter, diluted and plated on selective media.

Strains. The strain K17 was described by Sundman[6] and, as the strain 33, by de Smedt and de Ley[8]. Strains C58C1 (RP4) and LBA57 (met^-, ile^-) of *Agrobacterium tumefaciens*, used for infecting K17 with RP4, were obtained from Dr. P. Hooykaas[9].

RESULTS

Vanillic, veratric, ferulic, syringic and protocatechuic acids were used as the sole carbon source in a minimal medium by 58 bacterial strains isolated from spent bleach liquor of a kraft pulp mill (Table 1). Structures of the compounds are given in Figure 1. These metabolic pathways seem to offer some selective advantages for bacteria that thrive in spent bleach liquor.

After growth in spent bleach liquor containing 1 % of sodium dodecyl sulphate, most of the strains pleiotropically lost their ability to grow on most of the methoxylated aromatic acids and on parahydroxybenzoic acid. Growth on protocatechuic acid was affected less often. All strains were purified repeatedly as single colonies. A passage of growth on a mineral medium supplemented with 1 g/l of yeast extract did not reverse the curing.

We did not make further efforts to check if the colonies were a syntrophic associate rather than single species. Another possible explanation could be the curing of the plasmids that code for the relevant metabolic pathways.

The strain K17 was found to be unstable in its ability to use the following compounds: α-conidendrin, parahydroxybenzoic acid, salicylic acid, isovanillic acid, vanillic acid, veratric acid, benzoic acid, methanol and ethanol. K17 is a gram-negative, nonfluorescent, peritrichously flagellated rod-shaped bacterium (Fig. 2). It uses a great number of different compounds as source of carbon and energy, including methanol, a C-1 compound (Fig. 1, Table 2), but does not need added nitrogen compounds in growth medium (Table 6).

The parent strain is thus quite multipotent, but mutants with a more restricted spectrum of utilizable carbon sources were frequently obtained. Such mutants fell into three categories (Table 2). Frequency of the mutants varied depending on the medium used for cultivation (Table 3).

Tables 2 and 3 show that mutants that had lost their ability to grow on methanol and ethanol were most frequently encountered. Such mutants grew less than the parent strain on α-conidendrin. Using cells that had α-conidendrin as sole carbon source we measured the 18-h O_2 uptake in a Warburg respirometer and found that the wild-type strain 11 consumes 380 to 440 μl of O_2 in oxidizing one μmole of α-conidendrin and the strain 101 uses only 200-220 μl. The wild-type culture with α-conidendrin remains colourless whereas that of mutant 101 turns brown within 1 to 2 days. Degradation of α-conidendrin is probably incomplete in the strain 101, a class I mutant.

TABLE 1

GROWTH PROPERTIES OF 58 BACTERIAL STRAINS ISOLATED FROM SPENT BLEACH LIQUOR OF A KRAFT PULP MILL

The strains, all gram-negative and rod-shaped, were grown in spent bleach liquor solidified with 10 g/l of agar. Growth tests were performed on solid media made of a K salt solution and 5 mM of carbon source.

Test	Number of strains with positive results	
	Fresh strains	Strains treated with 1 % sodium dodecyl sulphate
Sole source of carbon utilized		
Vanillic acid	58	5
Veratric acid	58	4
Ferulic acid	58	12
Syringylic acid	58	3
Parahydroxybenzoic acid	50	1
Metahydroxybenzoic acid	9	9
Salicylic acid	0	0
Benzoic acid	0	
Phenol	0	
Protocatechuic acid	58	24
Fluorescence on King B agar	5[a]	5
Total number of strains	58	38

[a]254 and 360 nm

Mutants of classes II and III differ from those of class I in that they have also lost their ability to utilize salicylic acid (II) or benzoic and parahydroxybenzoic acid and α-conidendrin (III). In a Warburg respirometer the mutants of class III showed no uptake of oxygen with α-conidendrin within 48 h.

In all cases curing was irreversible and pleiotropic. No intermediate mutant types were encountered. Mutants of class III were least frequent. In order to establish whether they represent 'hot spot' mutation (plasmid deletion) or whether they are conventional mutants, the parent was treated with mitomycin C, which has been shown to cure *Ps. putida* for plasmid-mediated degradation of salicylic acid and naphthalene[10]. Added to the medium, 10 μg/ml of mitomycin C, after 2-day incubation, led to the appearance of colonies which no longer grew on parahydroxybenzoic acid as sole carbon source. None of the 22 mutants selected as parahydroxybenzoic acid negatives, grew on α-conidendrin.

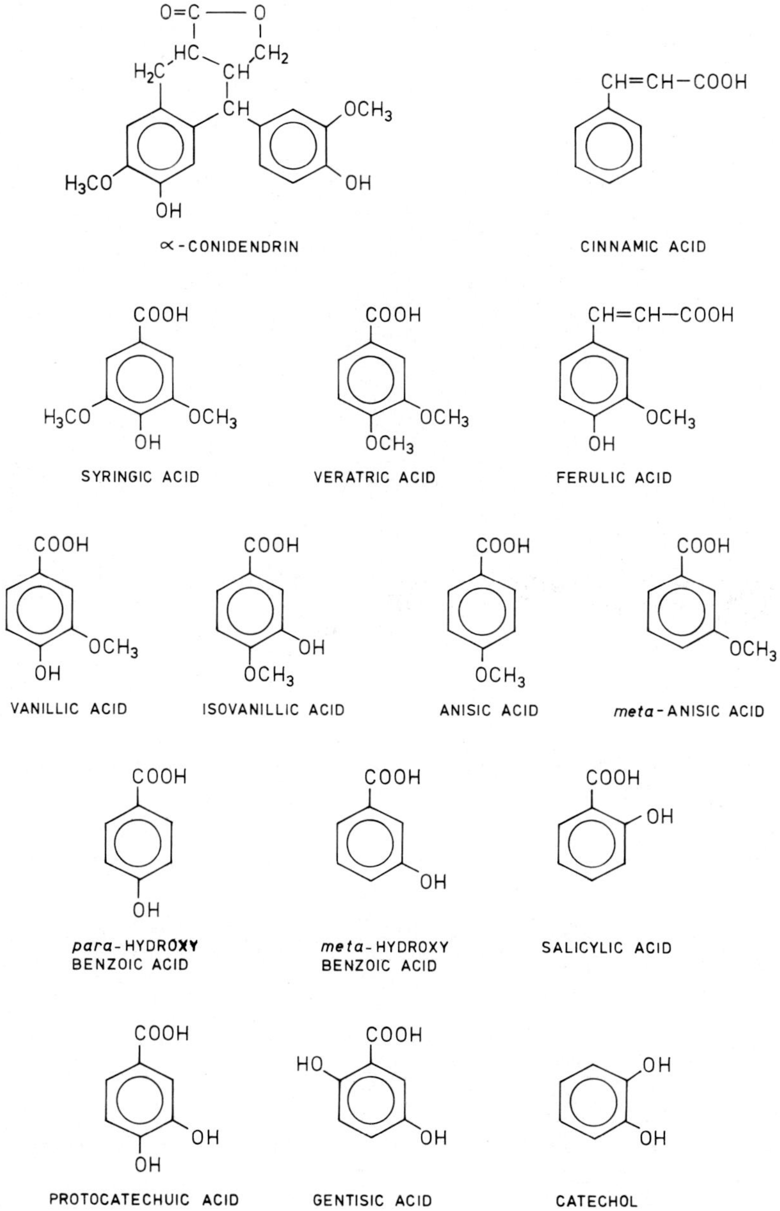

Fig. 1. Aromatic substrates used in this work.

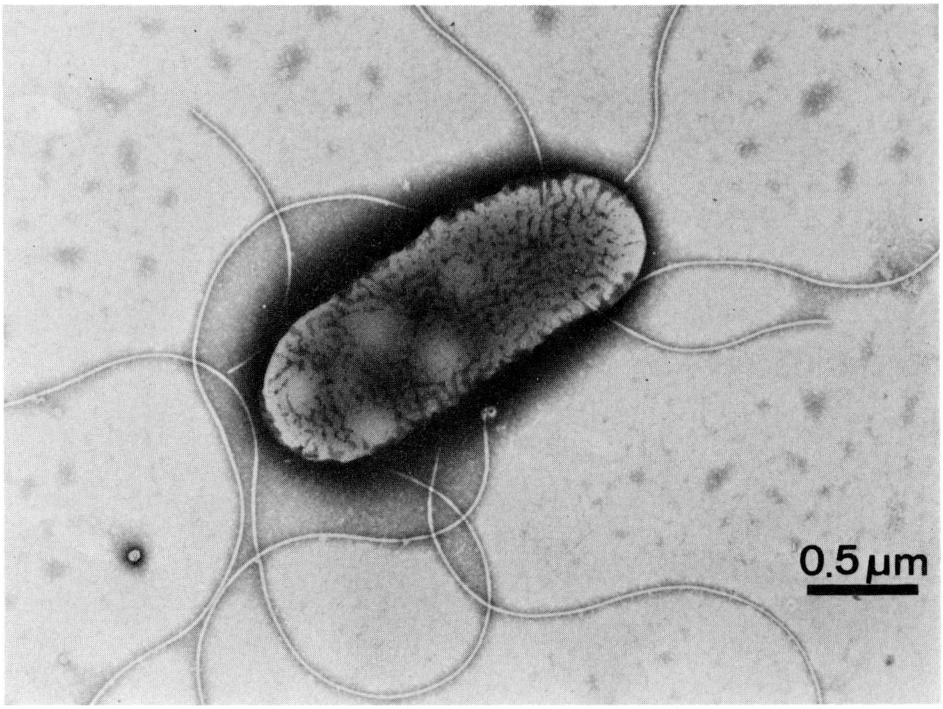

Fig. 2. A cell of the strain K17 showing the peritrichously located flagellae. The preparation was stained with phosphotungstate.

All grew on many other carbon sources, such as phenylacetic acid, salicylic and metahydroxybenzoic acids, and thus were not just auxotrophic mutants. The reverse was also true: when 30 α-conidendrin-negative mutants were selected in another mitomycin C experiment, none of them grew on parahydroxybenzoic acid either. The results have been described in detail by Salkinoja-Salonen and Sundman[4].

We grew the wild type K17 and mutants of class I on α-conidendrin, parahydroxybenzoic acid, glucose, ethanol and salicylic acid and measured the ability of washed cell suspensions to oxidize a number of aromatic carboxylic acids. Results for whole cells are summarized in Table 4 and for cell-free extracts in Table 5.

TABLE 2

GROWTH PROPERTIES OF THE STRAIN K17 AND ITS CURED DERIVATIVES

Strain No.	Phenotype	Good growth on
11-20	Wild	Methanol, ethanol, n-propanol, b-butanol, isovanillic, para- and metahydroxybenzoic, salicylic, veratric, anisic and phenylacetic acids, protocatechuic and gentisic, acetic, citric and succinic acids, all amino acids, α-conidendrin and naphthalene
101-110	Mutant, class I	As for Strain 11, except no growth on methanol, ethanol, vanillic, anisic and veratric acids and poor growth on α-conidenrin, n-propanol and n-butanol
170-180	Mutant, class II	As for Strains 101-110, except no growth on salicylic acid
201-210	Mutant, class III	As for Strains 101-110, except no growth on benzoic, parahydroxybenzoic and isovanillic acids and α-conidendrin
1771-1775	Mutant, class IV	As for Strains 201-210, except no grow on salicylic acid

It is seen that α-conidendrin-grown cells of the wild type K17 (strain 11) oxidize vanillic, isovanillic, anisic, meta-anisic and veratric acids. Results with cell-free extracts show that the enzymes involved are mixed-function oxygenases that use either NADH or NADPH as cosubstrate. Contents of the oxygen electrode measurement vessel were collected, extracted with diethylether (HCl added to pH 2), evaporated and taken up in 100 µl of ether. Then they were run on thin-layer plates in benzene/methanol/propionic acid (88:8:4 by vol.) and isopropanol/methanol/chloroform (60:25:15 by vol.). The product formed from meta-anisic acid was identified by Rf as metahydroxybenzoic acid, from anisic acid as parahydroxybenzoic acid and from isovanillic acid as protocatechuic acid. Therefore the enzymes involved in the attack on these compounds are O-demethylases.

Results shown in Table 4 suggest that the mutant 101 lacks activity on methoxy groups in the meta position. Similar results were obtained with other class-I-type mutants.

TABLE 3

SPONTANEOUS CURING OF THE BACTERIUM K17

K17 was grown in liquid medium on a shaker with 1 g/l of glucose, 10 ml/l of ethanol, 5 mM of benzoic acid or 1 g/l of yeast extract as the sole carbon source. Every other day the cultures were diluted 5 times with fresh medium. After 10 days the cultures were plated on K salts medium supplemented with 1 g/l of yeast extract. After 3 to 5 days of incubation single colonies were tooth-picked from the plates, suspended in 0,2 ml of sterile water and streaked on a series of plates with different carbon sources. Numbers in the table indicate the number of mutants found. The total number of colonies tested is given in the parentheses.

Carbon source in liquid medium	Number of mutant colonies Class I	Class II	Class III
Yeast extract		3 (250)	2 (250)
Ethanol	2 (32)	15 (32)	0 (32)
Glucose	7 (18)	0 (18)	0 (18)
Benzoic acid	61 (71)	24 (71)	0 (71)

Mutants of class II fail to grow on salicylic acid (Table 2). Salicylic acid induces in the wild type enzymes for the oxidation of salicylic, meta-anisic and gentisic acids, but not of catechol. Mutant 177 grows on gentisic acid, but does not oxidize the other two substrates even if salicylic acid is added to its growth medium. These results suggest that mutant 177 lacks the genes for a hydroxylase converting salicylic acid into gentisic acid.

Mutants of class III do not grow on benzoic or parahydroxybenzoic acids or on α-conidendrin. We showed that these mutants lack an enzyme that hydroxylates parahydroxybenzoic acid into protocatechuic acid[4]. The enzyme is a mono-oxygenase that accepts as cosubstrate both NADH and NADPH. The same enzyme, or some other enzyme induced coordinately with it, O-demethylates anisic acid into parahydroxybenzoic acid.

As was mentioned above, the bacterium K17 is independent of added nitrogen source in growth medium. We measured the acetylene reducing capacity of the wild type strain and cured derivatives thereof. Table 6 shows the results. It is seen that all strains actually do reduce acetylene to ethylene and are thus true nitrogen fixers. Curing did not significantly alter their capacity to reduce acetylene. No ethylene was formed by the gram-positive *Nocardia sp.* which was run as control.

If the curing phenomena in the bacterium K17 are based on plasmids that code for the metabolic activities described above, one should be able to physically isolate the plasmids concerned and to restore in the cured mutants metabolic

TABLE 4

COORDINATE INDUCTION IN THE BACTERIUM K17 AND ITS CURED MUTANTS FOR OXIDATION OF AROMATIC COMPOUNDS

Bacteria were grown in liquid medium with parahydroxybenzoic or salicylic acid, α-conidendrin, glucose or ethanol as sole carbon source. Stationary-phase cells were harvested by centrifugation, washed once with K salt solution and resuspended in K salt solution (1/30 of the culture volume). Oxygen uptake was measured at pH 6,7 by oxygen electrode as described in the text.

Strain	Mutant	Carbon source in	Oxygen uptake µg O_2 h^{-1} mg $protein^{-1}$								
No.	class	growth medium	*i*van[x]	van	anis	*m*anis	ver	phben	prot	gen[a]	sal
11	wild	α-conidendrin	58	21	46	48	66	0	14		
101	I	α-conidendrin	0	0	450	0	58	0	93		
11	wild	phben	45	109	136	53	210	630	410	0	0
101	I	phben	13	0	71	0		632	218	0	0
11	wild	glucose			84			114	29		0
101	I	glucose			0			0			
11	wild	ethanol	80	60	27	64	0	53	0	0	130
11	wild	salicylic acid				152				158	39
101	I	salicylic acid				0				145	13

[a]Measured at pH 5,5 (0,05-M MES buffer).
[x]*i*van = isovanillic acid, van = vanillic acid, anis = anisic acid, *m*anis = meta-anisic acid, ver = veratric acid, phben = parahydroxybenzoic acid, prot = protocatechuic acid, gen = gentisic acid, sal = salicylic acid.

TABLE 5

OXIDATION OF AROMATIC METHOXYLATED ACIDS BY CELL-FREE EXTRACTS OF K17 STRAIN 11

50 ml of culture grown on parahydroxybenzoic acid was spun off and taken up in 5 ml of 50 mM MES buffer at pH 6,5 and sonicated 4 times for 30 sec. by a MSE disintegrator. Suspension was allowed to cool in spiritus-ice mixture between sonications for 60 sec. Sonicated mixture was spun for 30 min. at 10'000 rpm at 4 ^{o}C. Supernate was decanted and used for measurement by oxygen electrode.

Substrate added	O_2 uptake μg O_2 h^{-1} mg $protein^{-1}$ Cosubstrate added		
	None	NADH	NADPH
Parahydroxybenzoic acid	25	616	535
Protocatechuic acid	2068	2100	
Meta-anisic acid	20	429	
Anisic acid	10	63	
Salicylic acid	0	0	
Veratric acid	14	566	440
Isovanillic acid	15	42	
None	2	10	8

activities by reintroducing the plasmid. Using N-methyl-N-nitro-N'-nitrosoguanidine we prepared amino acid auxotrophs from the bacterium K17. We obtained nonleaky auxotrophs for leucine only and found that the bacterium was cured by the mutagen treatment into a class-I-type mutant. As recipients we used mutants of classes II, III and IV resistant to nalidixic acid, streptomycin or erytromycin.

Preliminary experiments showed conjugation at very low (10^{-7}) frequency, if at all. Therefore, a self-fertile resistance transfer factor RP4 was introduced into the strain 101 leu^- by conjugation with *Agrobacterium tumefaciens*. Results from typical matings are shown in Table 7.

A reasonable fraction of the transconjugants that had received RP4 appeared to have also received genes that specify for growth on parahydroxybenzoic acid (phben) or on salicylic acid (sal) or on both. Since phben and sal were transferred independently, they probably are separate plasmids. However, no transconjugants were found of the phenotype sal^+ $phben^-$.

TABLE 6

ACETYLENE REDUCTION BY THE BACTERIA K17

Wild-type and cured mutants were grown for 7 days in 20-ml serum bottles on N-free slants made of K salts agar supplemented with 0,5 mg/l of $CuSO_4 \cdot 5H_2O$, 0,1 mg/l of H_3BO_3, 2,3 mg/l of $MnCl_2 \cdot 4H_2O$ and 2,5 mg/l of Na_2MoO_4. Glucose (1 g/l) was added as carbon source and 2 ml of acetylene (0,1 atm) was injected through rubber seal. The amount of ethylene was measured after 40 h by gas chromatography using a Porapak Q column operated at 50 ^{o}C.

Strain	Phenotype	Ethylene formed nmoles mg^{-1} protein
11	Wild	518
101	Cured mutant, class I	433
177	Cured mutant, class II	123
202	Cured mutant, class III	417
Nocardia sp.	See J. Trojanowski *et al.*[2]	1,8

DISCUSSION

Pulp mill effluents and decaying saw dust are abundant sources of bacteria that metabolize vanillic acid and other aromatic acids (Fig. 1). These acids are generally considered to be model compounds in the study of the microbiological degradation of lignin. In the laboratory, however, many such bacteria tend to lose their ability to utilize vanillic, isovanillic, anisic, meta-anisic, veratric, ferulic, syringic and parahydroxybenzoic acids. Such a loss (curing) can be potentiated by treating the bacteria with sublethal concentrations of sodiumdodecylsulphate (Table 1) or mitomycin C[4].

Curing is known to be caused by plasmid-located genes in the case of degradation by *Pseudomonas* of salicylic acid, naphthalene, camphor, metatoluic acid, xylene, octane (for a review see Charkabarty[10]), alkylbenzenesulphonate[11] and nicotine-nicotinate metabolism[12] and by *Alcaligenes paradoxus* of the herbicide 2,4-D[13]. In this paper we show results that support the existance in the bacterium K17 of three sets of plasmid-coded pathways: (1) ability to utilize methanol or ethanol as sole source of carbon and to metabolize aromatic acids with a methoxyl group in the meta position; (2) ability to hydroxylate parahydroxybenzoic acid into protocatechuic acid and to O-demethylate aromatic ring in the para position; (3) ability to metabolize salicylic acid.

The bacterium K17 is rather unusual: it fixes molecular nitrogen and uses a variety of carbon sources, which range from a C1-compound methanol to α-conidendrin, a complicated lignan. It was originally[6] designated as an atypical

TABLE 7

CONJUGAL TRANSFER IN THE BACTERIUM K17 OF GENES CODING FOR THE UTILIZATION OF SALICYLIC AND PARAHYDROXYBENZOIC ACIDS

Resistance transfer factor RP4 was introduced into the strains 101 leu⁻ and 177 of K17 by mating with the strain LBA 57 (RP4) of *Agrobacterium tumefaciens*. Transconjugants were selected on K salts medium with salicylic acid or parahydroxybenzoic acid as carbon source, supplemented with 10 μg/ml of leucine and 50 μg/ml of kanamycin. Transconjugants were also resistant to carbenicillin and to tetracycline and were used as donors in the matings listed below.

Donor	Recipient	Selection	Transconjugant phenotypes[a]
101 leu^- (RP4)	1775 nal^r	$carb^r$ leu^+	$phben^+$ nal^r (9/30) $phben^+$ sal^+ nal^r (7/30)
177 (RP4)	201 str^r	$carb^r$ sal^+	$phben^+$ str^r (85/90)
177 (RP4)	201 str^r	$carb^r$ $phben^+$	phben str^r (0/90)

[a]For abbreviations see Table 4.

Agrobacterium sp. However, taxonomic studies of de Smedt and de Ley[8] showed that it is not an Agrobacterium. We analyzed its lipopolysaccharide[14] and found that it deffers from that of Agrobacteria[15] in that it has heptose but no 2--keto-3-deoxyoctulsonic acid (KDO). The name of Lignobacter will be proposed to K17[16].

We have quite suggestive evidence for plasmid-coded genes involved in the bacterium K17 in the hydroxylation of parahydroxybenzoic acid, in 4-O-demethylation and in the metabolism of salicylic acid. Since the phben and sal determinants are independently transferred in conjugation, they probably are coded for by separate plasmids. Activities in parahydroxybenzoic acid hydroxylase and 4-O-demethylase probably depend on a single plasmid. They are induced coordinately (Table 4) and possibly are due to only one enzyme. Such dual activity has been shown for 4-O-demethylase of *Pseudomonas putida* by Bernhardt *et al.*[17].

Salicylic acid in K17 seems to be metabolized through gentisate and not through catechol (Table 4) and, consequently, the plasmid carrying the sal determinant in K17 must be different from the sal plasmid in *P. putida* described by Chakrabarty[18].

We do not yet have results to show on conjugal transfer in K17 of the third group of curable genetic determinants, which specify for growth on methanol and ethanol and on substrates that require O-demethylation at the meta position of the aromatic ring. These genes are particularly easy to cure by e.g. the mutagens we used for making auxotophic mutants. Since the strain K17-11 is not self-fertile, we infected it with RP4 by mating it with the strain C58C1 (RP4) of *Agrobacterium tumefaciens*. In mating with the strain 201 ery^r (mutant, class III, $phben^-$, ben^-) K17-11 (RP4) was used as donor. We then selected erytromycin -resistant transconjugants with parahydroxybenzoic acid as the carbon source and obtained them at a frequency per donor of about 10^{-5}. One part of them were of the phenotype $methanol^+$ $ethanol^+$ ben^+, another part ben^+ only. However, at the moment we have no means of establishing whether these were true transconjugants of the strain 201 ery^r or erytromycin-resistant mutants from the donor.

Curing for the ability to grow on methanol and ethanol was always concomitant with the loss of the meta-O-demethylating activity, i.e. the ability to oxidize vanillic and meta-anisic acid. Such coupling to C1 metabolism of aromatic catabolism may seem odd at first sight. However, it is known[17,18], and was also observed by us, that bacterial O-demethylases are mono-oxygenases and that the methoxyl group is cleaved off as formaldehyde. Formaldehyde is also the first product of methanol assimilation in methylotrophic bacteria[19,20]. Methanol dehydrogenase is known to be active also on ethanol, n-propanol and higher alcohols[20], so that its absence would explain the lack of growth on both methanol and ethanol and the poor growth on n-propanol of the cured class I mutants. The location of the methanol dehydrogenase gene on plasmid in the bacterium K17 remains, however, a hypothesis until it is shown that methanol dehydrogenase activity is present in cell-free extracts of the wild type and absent in the cured mutants of class I.

ACKNOWLEDGEMENTS

Authors acknowledge with thanks the active participation of Barbara Bushwell, Mirja Laukkanen, Raili Paasivuo and Matti Hämeranta. The work was funded by Suomalainen Tiedeakatemia and the Foundation of Maj and Tor Nessling.

REFERENCES

1. Ander, P. and Eriksson, K.-E. (1978) Progr. Industr. Microbiol., 14, 1-58.
2. Trojanowski, J., Haider, K. and Sundman, V. (1977) Arch. Microbiol., 114, 149-153.
3. Haider, K., Trojanowski, J. and Sundman, V. (1978) Arch. Microbiol., 119, 103-106.
4. Salkinoja-Salonen, M.S. and Sundman, V. (1979) *In* Lignin Biodegradation, Microbiology, Chemistry and Applications, ed. by T.K. Kirk, CRC Press, Inc., Cleveland, Ohio, in press.

5. Meadow, P., Wells, P.L., Salkinoja-Salonen, M. and Nurmiaho, E.-L. (1978) J. Gen. Microbiol., 105, 23-28.

6. Sundman, V. (1964) J. Gen. Microbiol., 36, 185-201.

7. Herriot, R.M. (1941) Proc. Soc. Exp. Biol. Med., 46, 642.

8. de Smedt, J. and de Ley, J. (1977) Int. J. Syst. Bacteriol., 27, 222-240.

9. Hooykaas, P.J.J., Klapwijk, P.M., Nuti, M.P., Schilperoort, R.A. and Rörsch, A. (1977) J. Gen. Microbiol., 98, 477-484.

10. Chakrabarty, A.M. (1976) Ann. Rev. Genet., 10, 7-30.

11. Sagoo, G.S. and Cain, R.B. (1979) Soc. Gen. Microbiol. Quart., 6, 17.

12. Thacker, R., Rørvig, O., Karlson, P. and Gunsalus, I.G. (1978) J. Bacteriol. 135, 289-290.

13. Fisher, P.R., Appleton, J. and Pemberton, J.M. (1978) J. Bacteriol., 135, 798-804.

14. Salkinoja-Salonen, M.S. and Boeck, R. (1976) Abstracts of the 8th Meeting of the North West European Microbiological Group, p. 150.

15. Salkinoja-Salonen, M.S. and Boeck, R. (1978) J. Gen. Microbiol., 105, 119-125.

16. Salkinoja-Salonen, M.S. Paterson, A. and Sundman, V. (1979) Manuscript.

17. Bernhardt, F.-H., Erdin, N., Staudinger, H.J. and Ullrich, V. (1973) Eur. J. Biochem., 35, 126-134.

18. Ribbons, D.W. and Harrison, J.E. (1972) *In* Degradation of synthetic organic molecules in the biosphere, Natl. Acad. Sci. USA, Washington, pp. 98-115.

19. van Verseveld, H.W. and Stouthamer, A.H. (1978) Arch. Microbiol., 118, 21-26.

20. Patel, R.N., Hou, C.T. and Felix A. (1978) J. Bacteriol., 133, 641-649.

V – PLASMIDS INVOLVED IN BACTERIA: PLANT INTERACTIONS

Plasmids of Medical, Environmental and Commercial Importance
K.N. Timmis and A. Pühler, editors

PLASMIDS AND THE RHIZOBIUM-LEGUME SYMBIOSIS

A.W.B. JOHNSTON, J.E. BERINGER, J.L. BEYNON, N. BREWIN, A.V. BUCHANAN-WOLLASTON and P.R. HIRSCH*
John Innes Institute, Colney Lane, Norwich. NR4 7UH (UK)

INTRODUCTION

Biological nitrogen fixation is responsible for the conversion of c.10^8 tonnes of atmospheric nitrogen to ammonia per annum and the contribution by *Rhizobium* in the root nodules of legumes accounts for c. 20% of the total[1]. Thus the credentials of *Rhizobium* as an organism of economic importance are not in question.

Plasmid DNA can be isolated from various *Rhizobium* species. Some of the published reports have demonstrated relatively small plasmids of molecular weight (M.W.) in the range $3\text{-}6 \times 10^7$ (refs 2-7), but more recently it has become clear that *Rhizobium* strains can contain large plasmids of M.W. in excess of 10^8 (refs 8 and 9). The recognition of such plasmids is of especial interest because of the findings that large plasmids have an important role in the oncogenicity of *Agrobacterium tumefaciens* [10,11], another member of the Rhizobiaciae. In this paper we will consider whether and to what extent plasmids are responsible for the symbiotic ability of *Rhizobium*.

To demonstrate the involvement of plasmids in this process it is necessary to have genetic evidence to support any physical studies on the number, the size and the structure of resident plasmids. Such genetic evidence can take a number of forms.

1. If in a symbiotically defective mutant there is a concomitant loss of, or a physically detectable change in a plasmid this points to that plasmid's involvement. Prakash *et al*[9] isolated a non-nodulating (non-infective) mutant of *R. leguminosarum* and found that one of three large plasmids present in the wild type parent was missing in the defective mutant. We have found (P.R.H. - unpublished observations) that a non-infective mutant of *R. leguminosarum* had the same number of plasmids (three) as its parent but that one of them had suffered a large deletion.

* Max Plank Institute für Zuchtungsforschung, 5, Koln-Vogelsang 30, W. Germany.

2. The instability of nodulation ability, either spontaneous or induced provides support for plasmid involvement. For example, Higashi[12] found that the ability of R. trifolii to initiate infection of clover was lost at high frequency following treatment by acridine orange, an agent known to induce curing of plasmids in other bacteria.

3. Transfer of a symbiotic phenotype at frequencies greater than those found for chromosomal markers implicates the role of a transmissible plasmid as a determinant of symbiotic characters. Higashi[12] obtained evidence for the transfer of host specificity from R. trifolii to R. phaseoli. These two species normally specifically nodulate Phaseolus and Trifolium respectively. More recently we have shown[13] that the ability to nodulate peas can be transferred at high frequency (c. 10^{-2}) from a strain of R. leguminosarum (the normal occupant of pea root nodules) to strains and species of Rhizobium that are incapable of nodulating this host.

It seems clear that plasmids can determine at least some of the steps in the symbiosis and in this paper we shall describe in more detail the involvement of transmissible plasmids in the ability to nodulate peas.

TRANSFER OF THE ABILITY TO NODULATE PEAS

Indications of a transmissible plasmid in R. leguminosarum stemmed from studies on the bacteriocinogenic properties of a number of field isolates of this species[14]. Almost all the strains studied produced at least one bacteriocin, these falling into three classes based on the estimates of their molecular size. Three of the isolates, strains 248, 306 and 309 that made medium molecular weight bacteriocins transferred their production to non-producing strains of R. leguminosarum at high frequencies (c. 10^{-2}) that were about 10^{6}-fold greater than the transfer of known chromosomal alleles. The transmissible plasmids in strains 248, 306 and 309 were termed pRL1JI, pRL3JI and pRL4JI respectively.

We transferred pRL1JI into the genetically well-characterised[15] R. leguminosarum strain 300. We then inserted the transposon Tn5 (which confers kanamycin resistance) into pRL1JI using the method already described[13, 16]. Briefly, the method relies upon the fact that P-1 incompatibility group plasmids into which the bacteriophage Mu has been inserted can transfer but fail to become established in Agrobacterium[17] or Rhizobium[16, 18]. A Mu-containing derivative of the gentamicin-resistant P group plasmid pPH1JI was constructed in Escherichia coli and into this was subsequently inserted Tn5. An E. coli strain carrying this hybrid plasmid, termed pJB4JI, was crosses to a derivative of R. leguminosarum strain 300 that carried pRL1JI and selection was made for the transfer of kan to the recipient. Such selection picks out those bacteria in which Tn5 has left the 'suicide' plasmid and inserted into the genome of R. leguminosarum.

About 0.3% of such progeny were auxotrophic and it was shown that the insertion of Tn5 was responsible for these mutations[16]. In addition one clone, strain T3, was found to transfer kan to other Rhizobium strains at frequencies of c. 10^2. Furthermore strain T3 failed to produce the medium bacteriocin suggesting that Tn5 had inserted into the bacteriocinogenic gene itself. This was confirmed by the finding that the transducing phage RL38 (ref 19) co-transduced at high frequency kan and the inability to produce medium bacteriocin from strain T3 to another strain of R. leguminosarum that contained the native pRL1JI plasmid. The presence of the transposon in pRL1JI facilitated the detection of its transfer. We have transferred, by selection of kan, the Tn5-marked pRL1JI derivative (termed pJB5JI) to a non-infective mutant (strain 6015) of R. leguminosarum and also to strains of Rhizobium species that nodulate legumes other than the pea; most of our work on interspecific transfer has been with strains of R. phaseoli and R. trifolii. In these cases all kan trans-conjugants tested could nodulate peas showing that some 'nodulation' genes are located on pJB5JI or are co-transferred at high frequency with this plasmid[13].

In most cases the interspecific transconjugants retained their ability to nodulate their normal hosts but at a reduced level compared to the parental recipient strains. These transconjugants were also poorer in the nodulation of peas than were strains of R. leguminosarum. However, about 10% of interspecific transconjugants had lost the ability to nodulate their normal hosts but were able to nodulate peas as effectively as did R. leguminosarum. It appears therefore that the ability of a strain to nodulate hosts of different cross-inoculation groups results in the impairment of symbiotic performance on both.

We have found that when bacteriocin production was transferred to strain 6015 from the field isolate 248 all the transconjugants tested nodulated peas. This showed that transfer of the native plasmid pRL1JI was also accompanied by the transfer of nodulation ability.

Properties of non-infective strain 6015

In much of the work to be described below we assayed the transfer of nodulation ability to the mutant strain 6015. Before giving the results we will briefly describe the history and some properties of this strain.

It was isolated following UV treatment (to c. 1% survival) of the R. leguminosarum strain 897 (phe-1 trp-12 str-37) which is itself a derivative of strain

300. Only one out of 600 survivors of this treatment that were tested on peas was stably non-infective. A rifampicin-resistant derivative of this strain was subsequently isolated and this strain was termed 6015 (ref 13). This strain has never been found to induce nodules on peas. Its appearance and growth on plates is normal, and we have found no change in phenotype, apart from its symbiotic defect.

Strain 300 and derivatives of it contain three large plasmids (ref 9 and P.R.H. unpublished observations) with molecular weights of about 204, 165 and 100×10^6 (ref 9). Strain 6015 has three plasmids, two of which correspond to those in the parental strain. However the largest of the plasmids in the infective strain was not found in strain 6015 but there was instead a smaller plasmid. This suggests that the symbiotic defect in strain 6015 is due to a deletion in the largest of the native plasmids of strain 300.

Transfer of nodulation mediated by other plasmids

We have mentioned already that three field isolates of *R. leguminosarum* transferred medium bacteriocin production at high frequencies. The three plasmids implicated in such transfer, pRL1JI, pRL3JI and pRL4JI seemed to be very similar as judged by a number of criteria[14].

(a) They each transfer at similar frequencies (c. 10^{-2}) and their presence is associated with the mobilization of chromosomal alleles at low (c. 10^{-8}) frequencies.

(b) Co-transduction of Tn5 and the inability to produce medium bacteriocin occurred at high frequency when phage RL38 was propagated on strain T3 and was used to infect derivatives of strain 300 carrying pRL1JI, pRL3JI and pRL4JI. In all three cases the transductants transferred *kan* at high frequencies. However no transduction of *kan* was detected when a strain 300 derivative carrying none of these plasmids was used as a recipient. These results indicate that there is DNA homology between these plasmids at least in the vicinity of the insertion of the transposon in the bacteriocin gene but that there is no corresponding homologous DNA in the genome of the native strain 300.

(c) When pJB5JI was transferred into derivatives of strain 300 carrying pRL1JI, pRL3JI or pRL4JI the transfer frequency was c. 10-fold lower than when the recipient did not carry any of these plasmids and the transconjugants lost the ability to make medium bacteriocin. This indicates that the three plasmids are of the same incompatibility group.

(d) Derivatives of strain 300 carrying any one of the three plasmids are resistant to the medium bacteriocin produced by all three plasmids.

Given these similarities we were surprised to find that when bacteriocin production was transferred to strain 6015 from strains carrying pRL3JI and pRL4JI none of the transconjugants (10 from each cross) could nodulate peas.

However the presence of pRL3JI and pRL4JI in the donor strain did promote the transfer of nodulation ability to strain 6015 but at much lower frequencies than did pRL1JI. To demonstrate this we had to select directly for the transfer of nodulation ability using the technique of end-point dilution analysis[20] that we now describe.

We know from reconstruction experiments that the presence in an inoculum of less than 10 infective R. leguminosarum cells even in the presence of 10^7 non-infective cells of strain 6015 is sufficient for nodulation of peas. We took advantage of this ability of the plant to select rare nodulating individuals by doing crosses using a prototrophic derivative of strain 6015 as the recipient and in which the donor strain was eliminated following the cross by exposure to selective antibiotics. Aliquots of the cross mixtures were then used to inoculate peas at various dilutions and by determining the greatest dilution at which nodules were induced it was possible to obtain an estimate of the frequency of transfer of nodulation ability.

When the wild-type strain 300 was used as the donor in such crosses we detected no transfer of nodulation ability (frequency 10^{-6}). However when derivatives carrying pRL1JI, pRL3JI or pRL4JI were used as donors transfer was found to occur at frequencies of c. 10^{-2}, 10^{-6} and 10^{-6} respectively. Thus the presence of the last two plasmids is associated with the mobilization of nodulation ability but at frequencies c. 10^4-fold less than is found with pRL1JI. We do not know the basis for the difference in behaviour between the plasmids. Two models that are as yet untested are as follows.

(a) pRL1JI is a single plasmid that contains bacteriocin and nodulation genes and hence co-transfers the two phenotypes at high frequency. In contrast, pRL3JI and pRL4JI do not contain nodulation genes, but can pick them up and transfer them; by selecting for the transfer of infectivity in the crosses described above we would (by this model) have selected for the formation of such derivatives of pRL3JI and pRL4JI.

(b) This second model says that none of these three plasmids carries all the genes required for nodulation but that they can mobilize or are associated with the mobilization of nodulation genes that are present on a different plasmid. Perhaps pRL1JI can mobilize such a plasmid at high frequency but the other two can only do so with relatively low efficiencies.

Bacteria isolated from nodules induced by inoculating peas with the cross mixtures described above were tested for their ability to produce medium bacteriocin. For all three crosses (i.e. with pRL1JI, pRL3JI and pRL4JI in the donor) the great majority of the bacteria isolated produced medium bacteriocin. Thus when selection was made for nodulation ability there was a great enrichment for the transfer of bacteriocin production. However the fact that a minority of the nodule occupants did not produce medium bacteriocin but were still infective shows that the two phenotypes of infectivity and bacteriocin production can be separated.

The remainder of this paper will be largely concerned with recent findings on the behaviour of pRL1JI and derivatives of this plasmid.

Transfer to other species:

Following the earlier observation that nodulation ability could be transferred from strain T3 by selection of the Tn5-marked derivative of pRL1JI (pJB5JI) we have used strain T3 as a donor strain in crosses using various strains and species as recipients.

Following nitrosoguanidine mutagenesis of a derivative of the R. leguminosarum strain 300 a number of ineffective mutants were isolated[21] (i.e. that nodulated but failed to fix nitrogen). We transferred pJB5JI into five such mutants and found that transconjugants of two of them were able to induce effective nodules indicating that genes for effectiveness as well as those for nodulation ability could be transferred at high frequency.

So far we have described cases where pJB5JI was transferred to R. leguminosarum or to species of Rhizobium that are very closely related to it[22]. We have since attempted to transfer pJB5JI to other members of the Rhizobiaciae that are less closely related. Strain T3 was used as a donor in crosses with strains of R. meliloti, (which nodulates alfalfa), R. japonicum (which nodulates soybean) and A. tumefaciens. In all three cases kan transfer was found to occur at frequencies in the range 10^{-4} - 10^{-6}. These values were much greater than the spontaneous mutation rates to kan but were less than the frequency of transfer of pJB5JI to strains of R. leguminosarum. From each of the three crosses 10 kan progeny were tested on peas; in no case were nodules induced. In the light of the comparatively low frequency of transfer of kan it is not clear whether the failure to transfer nodulation ability was due to the lack of expression of R. leguminosarum nodulation genes in these foreign backgrounds or to the ability of the nodulation genes to be transferred and become established intact in these recipients. In any event it seems that the transfer of the ability to nodulate peas may only be observed if the recipient strain is closely related to *R. leguminosarum*.

Isolation of symbiotically defective mutants

If genes for nodulation and effectiveness can be transferred at high frequency it follows that plasmid-linked symbiotically defective mutants may be isolated. We have succeeded in isolating such mutants using a protocol that enriches for the insertion of Tn5 into transmissible plasmids.

We used a derivative of *R. leguminosarum* strain 300 that harboured pRL1JI and which also contained a Tn5-induced auxotrophic mutation (*ade-92*) as a donor to another marked derivative of strain 300 that did not contain pRL1JI or Tn5. Selection was made for the transfer of *kan*. The rationale was that such a transfer would be a two-step process involving the transfer in the donor strain of Tn5 from the chromosome into a transmissible plasmid followed by high frequency transfer of the Tn5-marked plasmid to the recipient. Following such crosses, *kan* derivatives of the recipient arose at frequencies of c.10^{-7} and the majority of these then transferred *kan* at high frequencies.

We have examined 100 independent clones in which Tn5 was transferred at high frequency. When transferred to a symbiotically proficient strain of *R. leguminosarum*, none of these marked plasmids caused a detectable defect in the nodulation of peas indicating either that no symbiotic mutants had been induced or perhaps that more than one copy of the symbiotic gene was present.

We attempted to distinguish these possibilities by crossing each of the clones with strain 6015 and testing transconjugants on plants. We hoped that by such means we would at least detect mutants in the transmissible plasmids in which Tn5 was inserted into one of the genes corresponding to those deleted in strain 6015. Most of the transconjugants could nodulate peas effectively but in 15 cases nodulation ability was not co-transferred with *kan*. These latter could be due either to the insertion of Tn5 into a transmissible nodulation gene that corresponds to one of those that are deleted in strain 6015 or to the insertion of Tn5 into another cryptic plasmid that does not carry nor co-transfers nodulation genes.

In addition another 8 clones did transfer *kan-r* and nodulation ability but the nodules induced by the transconjugants were ineffective. This indicates that strain 6015 is not only defective in nodulation but also lacks functions required for effectiveness.

The expression of symbiotic defects in strain 6015 but not in the symbiotically proficient strain suggests that in the latter case there are present wild type alleles corresponding to the Tn5-induced mutations present on the transmissible plasmid.

CONCLUSIONS

Recent work has clearly established not only that many strains of Rhizobium contain plasmids but also that at least in some strains plasmids carry some of the genes required for effective nodulation. We do not know the relative number of 'symbiosis' genes that are plasmid-linked compared to those on the chromosome nor the extent to which Rhizobium plasmid-borne genes have an involvement with symbiotic properties.

Given the large size of some of the plasmids that have been isolated and the fact that several strains contain more than one species of plasmid the analysis of their role in symbiosis may not be simple. The use of transposons to mark such plasmids should aid the analysis, by facilitating selection for plasmid transfer by conjugation and perhaps by transformation. In addition regions close to the insertion may be transferred by transduction. Isolation of plasmid-cured strains should also be simplified by looking for antibiotic sensitive derivatives. Radioactively labelled transposons could be used as hybridization probes and thus facilitate the allocation of the sites of Tn insertions to particular plasmids and perhaps to specific regions within them.

The study of Rhizobium plasmids is clearly at an early stage. Nevertheless the essential tools for genetic and physical analysis are now available. Given the importance of the process of nodulation, it is to be hoped that these studies will provide new insights into the contribution that plasmids make and so improve our understanding of the complex relationship between Rhizobium species and their legume hosts.

REFERENCES

1. Burriss, R.H. (1977) Genetic Engineering for Nitrogen Fixation, Plenum, New York and London, pp. 9-18.
2. Sutton, W.D. (1974) Biochim. Biophys. Acta. 336, 1-10.
3. Klein, G.E., Jemison, P., Haak, R.A. and Mathijsee, A.G. (1975) Biochim. Biophys. Acta. 44, 357-361.
4. Tshitenge, G., Luyundula, N., Lurquin, P.F. and Ledoux, L. (1975) Biochim. Biophys. Acta. 414, 357-361.
5. Dunican, L.K., O.Gara, F.R. and Tierney, A.B. (1976) Symbiotic Nitrogen Fixation in Plants, Cambridge University Press, pp 77-90.
6. Zurkowski, W. and Lerkiewicz, Z. (1976) J. Bacteriol. 128, 481-484.
7. Olivares, J., Montoya, E. and Palamares, A. (1977) Recent Developments in Nitrogen Fixation, Academic Press, London, New York and San Francisco, pp 375-385.
8. Nuti, M.P., Ledeboer, A.M., Lepidi, A.A. and Schilperoort, R.A. (1977) J. gen. Microbiol, 100, 241-248.
9. Prakash, R.K., Hooykaas, P.J.J., Ledeboer, A.M., Kijne, J., Schilperoort,

R.A., Nuti, M.P., Lepidi, A.A., Casse, F., Boucher, C., Julliot, J.S. and Denarie, J. (1979) Proc. 3rd Int. Symp. Nitrogen Fixation, Madison, in the press.

10. Van Larebeke, N., Engler, G., Holsters, M., van den Elsacker, S., Zaenen, I., Schilperoort, R.A. and Schell, J. (1974) Nature (Lond.), 252, 169-170.

11. Watson, B., Currier, T.C., Gordon, M.P., Chilton, M.-D. and Nester, E.W. (1975) J. Bacteriol., 123, 255-264.

12. Higashi, S. (1967) J. gen. appl. Microbiol., 13, 391-403.

13. Johnston, A.W.B., Beynon, J.L., Buchanan-Wollaston, A.V., Setchell, S.M., Hirsch, P.R. and Beringer, J.E. (1978) Nature (Lond.), 276, 634-636.

14. Hirsch, P.R. (1979) J. gen. Microbiol., in the press.

15. Beringer, J.E., Hoggan, S.A. and Johnston, A.W.B. (1978) J. gen. Microbiol., 104, 201-207.

16. Beringer, J.E., Beynon, J.L., Buchanan-Wollaston, A.V. and Johnston, A.W.B. (1978) Nature (Lond.), 276, 633-634.

17. Van Vliet, F., Silva, B., van Montagu, M. and Schell, J. (1978) Plasmid, 1, 446-455.

18. Boucher, C., Bergeron, B., Barate de Bertalimo, M. and Denarie, J. (1977) J. gen. Microbiol., 98, 253-263.

19. Buchanan-Wollaston, A.V. (1979) J. gen. Microbiol., in the press.

20. Vincent, J.M. (1970) A Manual for the Practical Study of Root Nodule Bacteria, IBP Handbook No 15., Blackwell, Oxford pp 1-164.

21. Beringer, J.E., Johnston, A.W.B. and Wells, B. (1977) J. gen. Microbiol., 98, 339-343.

22. Graham, P.H. (1964) J. gen. Microbiol. 35, 511-517.

Plasmids of Medical, Environmental and Commercial Importance
K.N. Timmis and A. Pühler, editors

PRESENCE OF LARGE PLASMIDS AND USE OF Inc P-1 FACTORS IN RHIZOBIUM

Francine CASSE, M. DAVID, P. BOISTARD, J.S. JULLIOT, C. BOUCHER, Lise JOUANIN*, T. HUGUET* and J. DENARIE

Laboratoire de Génétique des Microorganismes, I.N.R.A., Route de Saint-Cyr, 78000, Versailles, France

Laboratoire de Biologie Moléculaire Végétale, Faculté des Sciences, Université Paris-Sud, 91405, Orsay, France

INTRODUCTION

Rhizobia are soil bacteria which can establish a nitrogen-fixing symbiosis with Legumes (see reviews 1, 2). Schematically the symbiotic properties of *Rhizobium* can be classified as follows : (i) recognition of a specific leguminous root (host specificity), (ii) induction of root-nodule organogenesis (infectivity) and (iii) conversion of bacteria into nitrogen-fixing bacteroids within a differentiated nodule (effectiveness). This complex process is likely to be controlled by numerous genes. It is of both theoretical and practical importance to know whether these genes are dispersed or clustered in the bacterial genome, and whether they are chromosomal or extrachromosomal.

Rhizobium and *Agrobacterium*, the only two genera of the *Rhizobiaceae* family, both represent bacterial strains known to influence cell division control in their host plant. In *A. tumefaciens* oncogenicity is controlled by large Ti plasmids (see Schell *et al*., Schilperoort *et al*., this volume). In *A. rhizogenes* the ability to promote secondary root proliferation seems to be also controlled by plasmids[3]. Nuti *et al*.[4] showed the presence of plasmids of high molecular weights in six strains of *Rhizobium*. In the first part of this communication we shall show that the presence of large plasmids (more than 85×10^6 daltons) is a general feature in *Rhizobium* ; we shall then describe the use of Inc P-1 group plasmids for devising genetical tools required for the analysis of the respective roles of chromosomal and plasmid genes in the control of the symbiotic properties of *R. meliloti*.

PRESENCE OF LARGE PLASMIDS IS A GENERAL FEATURE IN RHIZOBIUM

A simple procedure for isolation and characterization of large plasmids

Currier and Nester[5] have described a procedure for isolation of large plasmids of up to 160×10^6 daltons, based on neutral SDS lysis and alkaline denaturation of DNA followed by neutralization ; removal of most proteins and denatured DNA was obtained by phenol treatment in presence of 3% NaCl. We applied this procedure to four R. meliloti strains : L5-30, V7, 41 and RCR 2011. Using cesium chloride-ethidium bromide (CsCl-EtBr) gradient ultracentrifugation followed by electron microscopy, plasmids were found in strains L5-30 (91×10^6 daltons) and 41 (140×10^6) but not in strains V7 and RCR 2011. We then modified this method in an attempt to isolate CCC DNA of molecular weight higher than 160×10^6 and handle a large number of small volume cultures in screening experiments[6]. The detailed procedure has been published elsewhere[6]. The main modification consisted in lysing bacteria directly in an alkaline SDS buffer (pH 12.3), thus avoiding the DNA shearing step prior to alkaline denaturation and providing an immediate protein inactivation effect, which may reduce enzymic degradation of plasmids. DNA shearing was also reduced in subsequent steps by decreasing the phenol extraction to 2 min, eliminating chloroform extraction, and replacing magnesium phosphate by sodium acetate in the ethanol precipitation step to facilitate resuspension of the DNA pellet. Crude DNA extracts were then submitted to 0.7% agarose gel electrophoresis according to Meyers et al.[7].

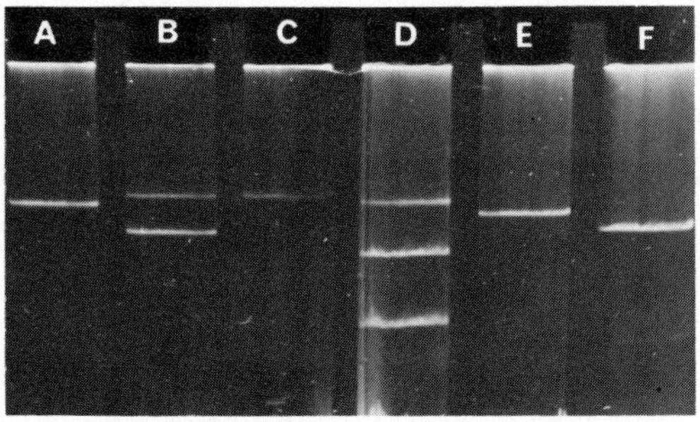

Fig. 1. Agarose gel electrophoresis of ethanol-precipitated plasmid DNA from crude lysates. (A) pMG5 (280 Md) from P. putida PpS 1240, (B) pTi-C58 (120 Md) and pAt-C58 from A. tumefaciens C58, (C) pAt-C58 from A. tumefaciens C58-C9, (D) pMG1 (312 Md) from P. putida PpS 1239, pRme L5-30 (91 Md) and RP4 (36 Md) from R. meliloti L5-30 (RP4), (E) pRme V7 (≃ 200 Md) from R. meliloti V7, (F) pRme-41 (140 Md) from R. meliloti 41.

The above extraction procedure was developped using strains carrying large plasmids of known MW. Electrophoresis of crude DNA extracts of high MW plasmids does not show two limitations encountered with small plasmids of MW lower than 20×10^6 : the open circular form of high MW plasmids does not penetrate the gel and cannot be confused with CCC DNA, and the CCC DNA bands are clearly separatedfrom the smear of residual linear DNA[7]. The plasmids pMG1 and pMG5 from Pseudomonas putida, reported to have MWs of 312 and 280 x 10^6 respectively[8], were easily isolated (Fig. 1, lane A and upper band in lane D). This extraction procedure revealed the presence in R. meliloti strain V7 of a plasmid of around 200 x 10^6 daltons (Fig. 1, lane E) which could not be isolated by the Currier and Nester procedure. Surprisingly, in A. tumefaciens C58 known to carry a Ti plasmid of 120 x 10^6 daltons, two bands were observed (Fig. 1, lane B). The fact that strain C58-C9 cured of the Ti plasmid showed only the upper band (Fig. 1, lane C) suggested that strain C58 contains, in addition to the Ti, a very large cryptic plasmid. This plasmid has been since isolated and characterized. Its contour length measurement indicates a MW higher than 250 x 10^6 (J. Ellis and J. Schell, pers. comm.) and its Sma I cleavage pattern is different from that of Ti[9]. This plasmid was not detected by earlier plasmid isolation techniques used for Agrobacterium. One advantage of agarose gel electrophoresis is that in a single strain different plasmids of relatively similar MWs can be separated ; this was the case for Ti and one additional cryptic plasmid of A. tumefaciens B6-806, both of which are around 125 x 10^6 daltons and cannot be distinguished by electron microscopy[6].

Five strains of R. meliloti (L5-30, 102F51, 12, 1322 and 41), showing only one CCC DNA band of varying relative electrophoretic mobility were chosen for plasmid electron microscopy studies . Contour length measurements provided the following MW values : pRme L5-30 (91 ± 2 x 10^6), pRme 102F51 (93 ± 4 x 10^6), pRme 12 (107 ± 5 x 10^6), pRme 1322 (121 ± 4 x 10^6) and pRme 41 (140 ± 6 x 10^6). For these different plasmids, the logarithm of the relative mobility (RM) of the CCC DNA band (mean of three gels), and the logarithm of MW deduced from contour length measurement, show a very significant linear correlation (r = -0.98). The equation of the regression line is

$$\log MW = - 2.54 \log RM + 4.49$$

and can be used to estimate the MW of plasmids within the range 90 - 140 x 10^6.

This chemical extraction procedure is simple and can be performed by one person on 20 to 40 strains in a single experiment. It was shown to be efficient with Agrobacterium, Rhizobium, Escherichia coli, Klebsiella pneumoniae, Azotobacter vinelandii, Pseudomonas putida and P. mors-prunorum.

Screening for large plasmids in Rhizobium

Alkaline sucrose gradients followed by reassociation kinetics [4, 10], ultra-centrifugation in CsCl-EtBr gradients followed by electron mycroscopy[6], and agarose gel electrophoresis[6] have been used to look for large plasmids in 49 Rhizobium strains, including most strains of fast growing rhizobia on which genetical and physiological studies have been performed : 27 R. meliloti, 10 R. leguminosarum, eight R. trifolii, two R. phaseoli and two slow-growing.

rhizobia (Table 1). Plasmids of high MW were found in all these strains except two R. meliloti : S26 and A145. These exceptions are likely due to a procedural limitation because both these strains were only analyzed by agarose gel electrophoresis of crude extracts, which also gave a negative result for strain RCR 2011 in which a plasmid of 260 x 10^6 has however been found using different procedures (M. Nuti and A. Pühler, pers. comm.). It is worthy of note that among the 50 Rhizobium strains studied, only two were shown to carry, in addition to their large plasmid(s), a plasmid of MW lower than 85 x 10^6 : R. meliloti 311 (pRme 311 a ≃ 30 x 10^6) and 102F28 (pRme 102F28 a : 73 x 10^6). No very small plasmids that might be suitable as cloning vehicles for genetic engineering were found.

TABLE 1

LIST OF RHIZOBIUM STRAINS SCREENED FOR PRESENCE OF LARGE PLASMIDS

Rhizobium species	Electron microscopy	Reassociation kinetics	Agarose electrophoresis of crude extracts
R. meliloti	[a]L5-30, 41 1322, 12, 102F51, V7	[c]RCR 2001 [e]RCR 2011	[b]L5-30, 102F51, V7, 12, 1322, 41 U45, B251, U54, 311, Ls2a, 102F28, 3DoA20a, S33, Balsac, S14, 54032, I1, RF22, Ve8, Sa10, Lb1, RCR 2011, S26, A145, [d]Rm11
R. leguminosarum	[c]RCR 1001	[c]A171, JB897 [a]RCR 1016 RCR 1001	[b]JB897, A171, [d]JB248, JB336, LPR 1248 RCR 1017, RCR 1044, 128c53
R. trifolii		[a]RCR 5 RCR 0402	[d]LPR 5002, RTHY-1, RCR 5, RCR 33, RCR 212, RCR 221, RCR 0401, RCR 0402
R. phaseoli		[c]RCR 3605	[d]9001
R. japonicum		[a]RCR 3407	
Cowpea group		[a]RD1	

[a]Nuti et al., 1977 ; [b]Casse et al., in press ; [c]Prakash et al., in press ; [d]Nuti, Casse and Dénarié, unpublished results ; [e]M. Nuti, unpublished results.

In most strains of R. meliloti investigated, only one plasmid of more than 85 x 10^6 daltons could be detected[6] ; nevertheless, eight strains showed more than one CCC DNA band : Ve8, RF22, Balsac, Ls2a, S14, 311, I1 and 102F28 (see Fig. 2).

For most plasmids the range of MW, estimated by contour length measurements or electrophoretic mobility in agarose gels, was 90 - 200 x 10^6.

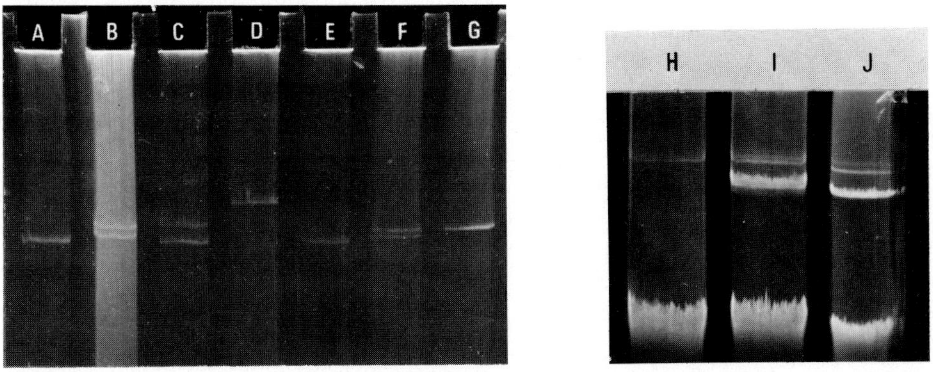

Fig. 2. Agarose gel electrophoresis of ethanol-precipitated plasmid DNA from crude lysates. Left : *R. meliloti* strains (A) pRme-54032 (104 Md), (B) pRme-Bal a (95 Md) and pRme-Bal b (107 Md), (C) pRme-RF22 a (93 Md) and pRme RF22 b (118 Md), (D) pRme-V7 (≃ 200 Md), (E) pRme L5-30 (91 Md) (F) pRme-Ve8 a (93 Md) and pRme-Ve8 b (103 Md), (G) pRme 12 (107 Md). Right : *R. leguminosarum* strains (H) LPR 180 a rough non nodulating derivative of A171 (I) strain A171, (J) JB897 an effective *phe*, *trp*, *str* derivative of JB300. For *Rhizobium* plasmid nomenclature see 6, 10.

Electrophoresis of crude DNA extracts of *R. leguminosarum* gave a different picture (Fig. 2, lane I, J) : at least three CCC DNA bands were observed in strains JB248, JB336, JB897, A171, 1044 and 1248 (Nuti, Casse and Dénarié, unpublished results). The comparison of their relative mobility with that of CCC DNA of known molecular weights (pRme 41 : 140×10^6 ; pRme V7 : 197×10^6 ; pMG1 : 312×10^6 and pMG5 : 280×10^6) indicates that in all these strains at least one of the DNA bands has a low electrophoretic mobility corresponding to a MW higher than 200×10^6. In strain LPR 180, a rough derivative of *R. leguminosarum* A171 which was shown to be cured of its smallest plasmid, the upper bands were still present (Fig. 2, lanes H and I), indicating that in *R. leguminosarum*, as in *A. tumefaciens*, different bands probably correspond to different plasmids[6, 10]. This suggests the existence of different incompatibility (Inc) groups among *R. leguminosarum* plasmids. Compatibility between different large plasmids has been shown in *A. tumefaciens*, where in addition to the Ti, very large plasmids are present. Their role in the interaction with the host plant is unknown[6, 9].

In the case of *R. leguminosarum* large plasmids of different "Inc" groups might be involved in the symbiotic relationship. Curing strain A171 of a plasmid of 100×10^6 daltons (pRle-A171 a) by heat treatment is associated with a loss of infectivity on Pisum sativum[10], whereas at least one of the largest plasmids (MW > 150 x 10^6) hybridizes with structural nif genes of K. pneumoniae, suggesting that it carries the Rhizobium genes controlling nitrogen fixation (F.C. Cannon and M. Nuti, pers. comm.). However in R. leguminosarum A171 if more than one plasmid seems to be involved in the control of symbiotic characters, it is not known whether each plasmid carries a whole set of genes controlling different symbiotic characters on a given host (infectivity, nitrogen fixation), or if genes controlling the relationship on a given host are dispersed on different plasmids. In fact, the presence of nif genes has not been looked for on the pRle-A171a plasmid, and if genes carried by this plasmid are required for nodulation on Pisum sativum, it is not known whether the other larger plasmids present in the strain control ability to nodulate other hosts. Johnston et al. (this volume) showed that a self-transmissible plasmid of R. leguminosarum confers the ability to nodulate P. sativum. The strains used in their studies, R. leguminosarum JB 248 and JB 300 (Table 1, Fig. 2, lane J) contain large plasmids [6, 10], one of which is likely to be involved in the control of infectivity on this host. This ability to nodulate peas can be transferred to R. phaseoli and R. trifolii ; the transconjugants are still able to induce nodule formation on their respective hosts (Johnston et al., this volume). If we suppose that nodule induction on bean and clover is also controlled by plasmids, extension of the host range could mean that plasmids controlling infectivity on these different hosts belong to different "Inc" groups. In R. meliloti, which is generally found to carry only one plasmid, it is not known whether this single plasmid controls nodule formation and nitrogen fixation.

Restriction patterns of R. meliloti plasmids

To investigate whether these plasmids have a highly conserved structure restriction patterns were performed on plasmids of eight strains of R. meliloti chosen because of their various geographical origin, L5-30 and 41 (Europe), V7, 12 and 54032 (Canada), 102F51 (USA), U54 (Uruguay) and 1322 (New Zealand). Electrophoresis of their crude DNA extracts showed only one CCC DNA band suggesting that they carried only one plasmid with the following MWs : L5-30 (91 x 10^6), 102F51 (93 x 10^6), 54032 (104 x 10^6), 12 (107 x 10^6), 1322 (121 x 10^6), 41 (140 x 10^6), U54 (153 x 10^6) and V7 (197 x 10^6). Preparative CCC DNA isolation was performed using the Currier and Nester[5] procedure as modified by Dr M.D. Chilton (pers. comm.).

DNA was treated with restriction endonucleases EcoR1, Hind III which cause only one cut in RP4 and are therefore used for in vitro cloning into this promiscuous plasmid and Sma I which was shown to produce a relatively simple cleavage pattern of A. tumefaciens Ti plasmids[11]. Comparison of the different restriction patterns showed that although a few possible common bands can be seen in some fingerprints, the patterns are markedly dissimilar (see for example Fig. 3). These results contrast with those observed in A. tumefaciens strains of the octopine group, for which Ti plasmids have been shown to share a large number of restriction DNA fragments[9], and with those observed for R plasmids of the same incompatibility group (Chabbert et al., this volume). DNA hybridization measurements will be required for precise estimation of homologies.

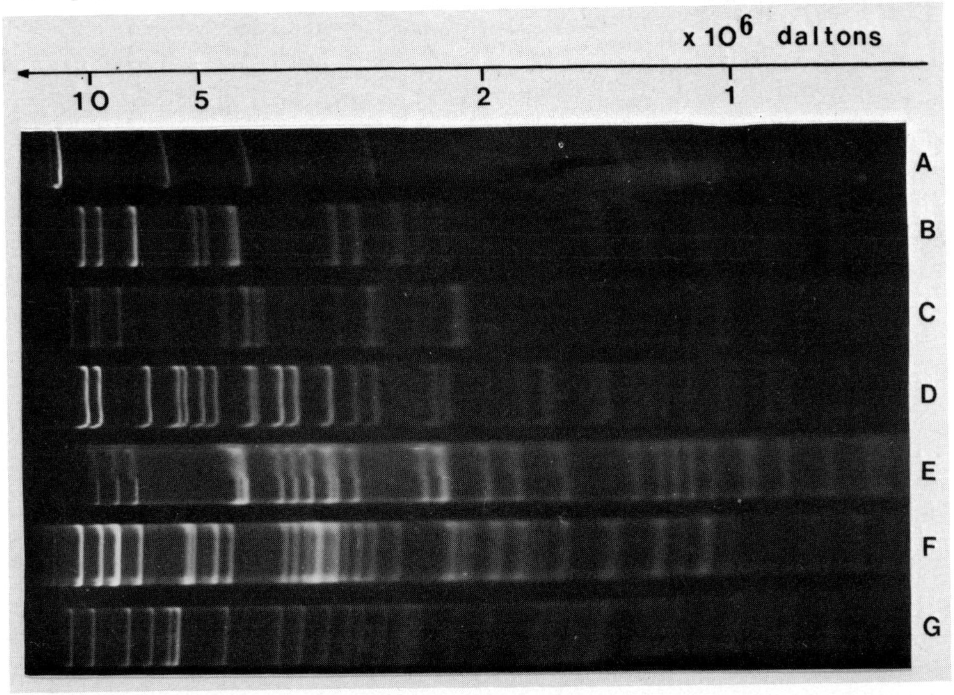

Fig. 3 Fingerprints of plasmids of R. meliloti strains. Plasmids (1 μg) were digested with Hind III (2 units, Biolabs) in 12 mM Tris-HCl pH 7.4, 4,2 mM $MgCl_2$, 36 mM NaCl for 6 hours at 37°C. Samples were electrophoresed on 1% agarose vertical gel in Tris-EDTA buffer ; channel A : λ DNA digested with Hind III, B : 54032, C : 102F51, D : 12, E : L5-30, F : 41, G : 1322.

USE OF Inc P-1 PLASMIDS FOR DEVISING GENETIC TOOLS

Large plasmids of Rhizobium are likely to be involved in the control of some symbiotic traits, however mutations responsible for symbiotic defects in R. leguminosarum have been mapped on the chromosome (J.E. Beringer, pers.

comm.). It is therefore necessary to devise tools to analyze the respective roles of chromosome and plasmids in the control of symbiotic properties. We have used promiscuous Inc P-1 group R plasmids in R. meliloti to introduce Tn5 transposon, to develop a chromosome mapping system and as a vector for in vitro recombination. R. meliloti was chosen because its symbiotic properties can be easily tested in test tubes, genetic transfer systems such as conjugation and transduction are available, and some strains seem to carry only one large plasmid.

Transfer of the transposon Tn5 to R. meliloti

Beringer et al.[12] proposed to use drug resistance transposons as an insertion mutagen in Rhizobium in order to induce symbiotic mutations tagged by antibiotic resistance markers and therefore easy to map. To ensure that transposons are inserted into the Rhizobium genome it was necessary to devise a technique for eliminating the vector plasmid. When the temperate bacteriophage Mu was inserted into RP4, the transfer frequency of the plasmid from E. coli to R. meliloti was reduced about 10^5 fold and most of the plasmids that were transferred were altered[13]. Van Vliet et al.[14] have shown that the same applies to transfer of RP4::Mu from E. coli to A. tumefaciens and used this inability of RP4::Mu to become established to introduce transposons into these species. Using a P-1 plasmid controlling resistance to gentamycin and spectinomycin Beringer et al.[12] constructed a new transposon vector carrying Mu, and used it to introduce Tn5, a transposon relatively non-specific in its insertion and controlling resistance to kanamycin, into R. leguminosarum, R. trifolii and R. phaseoli. We used this plasmid, pJB4JI, to transfer Tn5 to two strains of R. meliloti which seem to carry only one large plasmid. The E. coli donor J53 met, pro, nal (pJB4JI) was crossed with R. meliloti L5-30 str-1[15] and with a non-slimy variant of R. meliloti 41 str[16] (kindly provided by Dr A. Kondorosi) on filter membranes on TY agar medium[12]. To avoid siblings, matings were limited to two hours, which is less than the generation time of the recipient strains. The mixture was then spread on a selective medium containing streptomycin (500 μg/ml) to counterselect the E. coli donor, kanamycin (300 μg/ml) and neomycin (200 μg/ml) to select Tn5 transfer. On this medium no colonies of either the donor or the recipient were detected. With strain L5-30, 99 per cent of the clones having inherited Tn5 were sensitive to gentamycin (20 μg/ml). Electrophoresis of crude DNA extracts confirmed that gentamycin sensitive $Tn5^+$ recipient clones did not carry the vector plasmid pJB4JI. With strain R. meliloti 41 among 1200 $Tn5^+$ transconjugants only three clones were still gentamycin resistant (120 µg/ml) and nine were auxotrophic mutants. Many thousands of gentamycin sensitive Tn5$^+$ clones were purified and are being tested on lucerne seedlings grown aseptically in test tubes[17] to detect Tn5-induced symbiotically-defective mutations. The purpose of this large scale experiment, undertaken in cooperation with Dr A. Kondorosi in Szeged, is to isolate a large number of mutants altered in various steps of symbiotic process. The next step will be to locate the mutations, i. e. Tn5, on the chromosome or on the plasmid.

Oriented transfer of the R. meliloti chromosome

Chromosome mapping has been performed in R. meliloti using Inc P-1 sex factors R68 - 45[16] and RP4[18]. Mobilization of the chromosome by those sex factors occurs from multiple sites and recombination frequencies are relatively low (10^{-4} - 10^{-6}) whereas Tn5 transposition frequency is high. Lastly, wild-type RP4 or R68-45 cannot be used for mapping Tn5 since they already specify kanamycin resistance. Thus an oriented chromosome transfer system mediated by a kanamycin sensitive sex factor is needed for mapping Tn5-induced mutations.

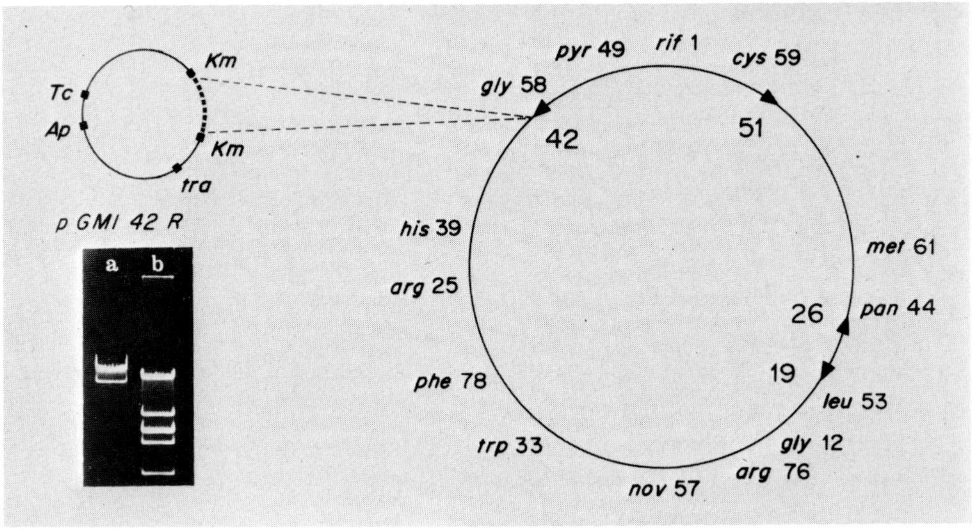

Fig. 4. Right : origin and orientation of chromosome transfer promoted by four in vitro constructed hybrid plasmids in R. meliloti RCR 2011 (for assignment of marker locations see ref. 18 and 21). Left : schematic map and agarose gel electrophoresis of Hind III cleaved hybrid plasmid pMGI 42 R, showing in addition to the RP4 linear molecule a DNA fragment of 7.4 10^6 (lane a). Lane b : Eco R1 cleaved lambda DNA.

In E. coli we have shown that insertion of two Mu prophages, one in the chromosome and one in RP4, providing a DNA homology between the two replicons, promotes an oriented transfer of the chromosome[19] ; however in Rhizobium the poor expression of Mu makes the detection of Mu insertion in the chromosome difficult[13, 19]. Therefore, to create a DNA homology between the Rhizobium chromosome and RP4, we turned to in vitro genetic recombination. RP4 has only one site sensitive to Hind III, which is located in the gene controlling kanamycin resistance[20]. This property is of particular interest both for further Tn5 mapping and for the detection of hybrid plasmids. Four out of 12 in vitro constructed RP4-primes, in which R. meliloti RCR 2011 chromosome fragments were inserted, promoted a polarized transfer of the R. meliloti chromosome (Fig. 4), as shown by the gradient of transfer of various markers and by the genetic constitution of recombinants[21]. Inheritance of markers was due to their integration into the chromosome of the recipient, as shown by the mobilization of recessive markers. For R. meliloti 41 a system of oriented transfer of the chromosome is being developped using R-primes constructed in vivo (A. Kondorosi and Kiss, pers. comm.) as well as in vitro. The same in vitro recombination procedure[21] is currently used to clone the large plasmids of strains L5-30 and 41 into RP4.

Towards a system for cloning of Rhizobium nif genes

In contrast to free-living nitrogen-fixing bacteria such as K. pneumoniae and A. vinelandii, the Rhizobium species are unable to grow on nitrogen-free media. Therefore detection of Rhizobium nif mutations involves massive screening procedures[22]. By contrast, cloning of Rhizobium nif genes on a suitable vector will allow their introduction into bacteria amenable to genetic analysis of nitrogen fixation and their subsequent transfer back to Rhizobium, thereby making possible a study of their functional organization. Recently, using crude DNA preparations of various Rhizobium species, Page[23] reported the transformation of Nif^- mutants of A. vinelandii to Nif^+ ; this result suggests that Rhizobium nif genes can be expressed in Azotobacter, and that Azotobacter Nif^- mutants could be used to clone and study Rhizobium nif genes. However Page[23] was unable to transform A. vinelandii with purified homologous or heterologous DNA. We therefore tried to devise a procedure for transforming Azotobacter with purified plasmid DNA.

Cohen et al.[24] described a method for transformation of E. coli with circular DNA by treatment of recipient cells with 30 mM $CaCl_2$. P. putida cells treated with 100 mM $CaCl_2$, could be transformed with RP1 plasmid DNA.[25] Table 2 shows that treatment by $CaCl_2$ of A. vinelandii UW, and of a Nif^- derivative UW3[26], resulted in an efficient transformation with purified RP4. The optimal concentration of $CaCl_2$ was 200 mM. Transformation frequencies of A. vinelandii by RP4 isolated from Azotobacter were high and similar to those observed with a restriction deficient mutant of E. coli. A. vinelandii could thus be a suitable host for direct cloning of nif genes. However when RP4 was isolated from E. coli, transformation frequencies of A. vinelandii UW were much reduced (about 10^3 fold) suggesting the presence of a restriction-modification system in this strain. Cloning of Rhizobium nif genes should thus be facilitated by the isolation of a restriction-deficient mutant of A. vinelandii.

TABLE 2

TRANSFORMATION OF E. COLI AND A. VINELANDII CELLS FOR DRUG-RESISTANCE CHARACTERS SPECIFIED BY RP4 DNA

Bacterial strains	RP4 DNA isolated from	RP4 DNA (μg/ml)	Treatment $CaCl_2$	Transformation frequency*
E. coli C600 rK^- mK^-	E. coli C600 (RP4)	5	30 mM	5×10^{-5}
	A. vinelandii UW	5	30 mM	1.7×10^{-5}
A. vinelandii UW Nif^+		0	200 mM	$< 10^{-10}$
	A. vinelandii UW	5	30 mM	1.3×10^{-8}
	A. vinelandii UW	5	200 mM	10^{-5}
	idem + DNAse	5	200 mM	$< 10^{-10}$
	E. coli C600	5	200 mM	10^{-8}
A. vinelandii UW3 Nif^-	A. vinelandii UW	2	200 mM	7×10^{-5}

*Azotobacter transformants were selected on C medium[13] containing ampicillin (200 μg/ml) and tetracycline (10 μg/ml) ; E. coli transformants were selected on L medium[21] with ampicillin (50 μg/ml) and tetracycline (10 μg/ml).

ACKNOLEDGEMENTS

We thank Drs J.E. Beringer, L.M. Bordeleau, W.J. Brill, A. Kondorosi, H.M. Meade, M. Nuti, J. Shapiro and J. Schell who provided bacterial strains. We also acknowledge Dr D. Tepfer and Margaret E. Buckingham for their critical reading of the manuscript. This work was supported by grant 78 7 448 of D.G.R.S.T. and Grant 30 63 of C.N.R.S.

REFERENCES

1. Hardy, R.W.F. and Silver, W.S. (1977) A Treatise on Dinitrogen Fixation, Sect. III, John Wileys, New York, pp. 1 - 675.
2. Dénarié, J. and Truchet, G. (1979) Physiol. vég. in press.

3. Moore, L.W., Warren G., Strobel, G. (1978) in Proc. IVth Intern. Conf. on Plant Pathogenic Bacteria, INRA ed., Angers, pp. 127-131.

4. Nuti, M.P., Ledeboer, A.M., Lepidi, A.A. and Schilperoort, R.A. (1977) J. gen. Microbiol., 100, 241-248.

5. Currier, T.C. and Nester, E.W. (1976) Analyt. Biochem. 76, 431-441.

6. Casse, F., Boucher, C., Julliot, J.S., Michel, M. and Dénarié, J. (1979) J. gen. Microbiol., in press.

7. Meyers, J.A., Sanchez, D., Elwell, L.P. and Falkow, S. (1976) J. Bacteriol., 127, 1529-1537.

8. Hansen, J.B. and Olsen, R.H. (1978) J. Bacteriol., 135, 227-238.

9. Chilton, M.D., Gordon, M.P., Mc Pherson, J., Saiki, R.K., Thomashow, M., Nutter, R.C., Gelvin, S.B., Montoya, A.L., Merlo, D.J., Yang, F.M., Garfinkel, D., and Nester, E.W. (1979) in Emergent Techniques for the Genetic Improvment of Crops, Rubenstein, I. et al. ed., University of Minnesota Press, Minneapolis, in press.

10. Prakash, R.K., Hooykaas, P.J.J., Ledeboer, A.M., Kijne, J., Schilperoort, R.A., Nuti, M.P., Lepidi, A.A., Casse, F., Boucher, C., Julliot, J.S. and Dénarié, J. (1979) in Proc. 3rd Intern. Symp. on Nitrogen Fixation, Newton, W.E. and Orme-Johnson, W.H., eds, University Park Press, Madison, in press.

11. Sciaky, D., Montoya, A.L. and Chilton, M.D. (1978) Plasmid, 1, 238-253.

12. Beringen, J.E., Beynon, J.L., Buchanan-Wollaston, A.V. and Johnston, A.W.B. (1978) Nature, 276, 633-634.

13. Boucher, C., Bergeron, B., Barate de Bertalmio, M. and Dénarié, J. (1977) J. gen. Microbiol., 98, 253-263.

14. Van Vliet, F., Silva, B., Van Montagu, M. and Schell, J. (1978) Plasmid, 1, 446-445.

15. Dénarié, J., Truchet, G. and Bergeron, B. (1976) in Symbiotic Nitrogen Fixation in Plants, Nutman, P.S. ed, Cambridge Univ. Press, Cambridge, pp. 47-61.

16. Kondorosi, A., Kiss, G.B., Forrai, T., Wincze, E. and Banfalvi, Z. (1977) Nature, 268, 525-527.

17. Vincent, J.M. (1970) A Manual for the Practical Study of Root-nodule Bacteria. Blackwell Scientific Publications, Oxford, pp. 1-164.

18. Meade, H.M. and Signer, E.R. (1977) Proc. Nat. Acad. Sci. USA, 74, 2076-2078.

19. Dénarié, J., Rosenberg, C., Bergeron, B., Boucher, C., Michel, M., Barate de Bertalmio, M. (1977) in DNA insertion elements, plasmids and episomes, Bukhari A.I., Shapiro J.A. and Adhya S.L., eds, Cold Spring Harbor Laboratory, pp. 507-520.

20. Barth, P. (1979) Plasmid, 2, 130-136.

21. Julliot, J.S. and Boistard, P. (1979) Molec. gen. Genet., in press.

22. Maier, R.J. and Brill, W.J. (1976) J. Bacteriol., 127, 763-769.

23. Page, W.J. (1978) Can. J. Microbiol., 24, 209-214.

24. Cohen, S.N., Chang, A.C.Y. and Hsu, L. (1972) Proc. Nat. Acad. Sci. USA, 69, 2110-2114.

25. Chakrabarty, A.M., Mylroie, J.R., Friello, D.A. and Vacca, J.G. (1975) Proc. Nat. Acad. Sci USA, 72, 3647-3651.

26. Bishop , P.E. and Brill, W.J. (1977) J. Bacteriol., 130, 954-956.

© 1979 Elsevier/North-Holland Biomedical Press
Plasmids of Medical, Environmental and Commercial Importance
K.N. Timmis and A. Pühler, editors

CHARACTERS ON LARGE PLASMIDS IN RHIZOBIACEAE INVOLVED IN THE INTERACTION WITH PLANT CELLS

R.A.SCHILPEROORT, P.J.J.HOOYKAAS, P.M.KLAPWIJK, B.P.KOEKMAN, M.P.NUTI*, G.OOMS & R.K.PRAKASH
Department of Biochemistry, State University of Leiden, Wassenaarseweg 64, Leiden (The Netherlands)
* Istituto di Chimica Agraria, Università di Padova, Via Gradenigo 6, Padova (Italy)

INTRODUCTION

The bacterial family of Rhizobiaceae comprises the root nodule inducing Rhizobia as well as the tumour inducing Agrobacteria. Agrobacteria and the fast-growing Rhizobia (*R.meliloti, R.trifolii, R.phaseoli* and *R.leguminosarum*) are very closely related, whereas the slow-growing Rhizobia (*R.lupini* and *R.japonicum*) are distant from them [1].

Both Rhizobia and Agrobacteria induce plant cells to proliferate. Their mode of infection, however, is very different. Rhizobia invade plant cells via the root system and subsequently induce the formation of root nodules. Agrobacteria are not invasive; they induce unlimited cell proliferation at woundsites leading to the formation of tumours, which are called 'crown galls' since in nature they have been found most frequently at the root crown.

It is well established now, that genes located on a large plasmid present in all virulent Agrobacteria code for the tumour inducing capacity of these bacteria. Recent results suggest, that also genes which code for functions involved in the synbiosis between Rhizobia and their plant hosts are plasmid-borne.

THE FORMATION OF ROOT NODULES

Agrobacteria are able to induce tumours on many plant species, whereas Rhizobia only induce root nodules on their specific hosts among the leguminous plants. Their hostrange has been the basis for their classification in species (*e.g. R. trifolii* induces nodules on clovers and *R.japonicum* on soybean).

Rhizobia attach to plant root hairs possibly after recognition of a lectin[2] produced by the plant, whereafter in many cases an infection thread is formed, which passes through the cortex cells. The bacteria present in the infection thread, trigger the cortex cells to divide, and this results in the formation of a root-nodule. Via the infection thread the bacteria eventually penetrate into the plant cells where they change shape and therefore are called bacteroids. The area of infected cells in the nodule is called bacteroid-tissue and here the process of symbiotic nitrogen-fixation takes place after leghemoglobin has been synthesized. The globin molecules originate from the plant [3], but the nitrogenase is of bacterial origin. Leghemoglobin most likely protects nitrogenase from being inactivated by oxygen.

CROWN GALL

Agrobacteria induce tumours on most dicotyledonous plants; only a few strains are known to have a more limited host-range [4,5]. Wound-sites are required for tumour-induction. The bacteria penetrate into intracellular spaces and into injured cells, multiply here, and attach to specific sites at the cell walls of the adjacent healthy cells [6,7]. They do not penetrate into the latter, but transform these into tumour cells most probably by transferring a piece of DNA (called T-DNA) into these cells [8]. This eventually gives rise to the formation of crown gall tumours.

Crown gall tumour cells can be isolated; they are able to grow on a basic culture medium, whereas normal plant cells in addition require the presence of the plant hormones auxin and cytokinin for growth in such a culture medium. The ability to proliferate unlimitedly is retained, when a small amount of bacteria-free crown gall callus tissue is transplanted on a healthy plant : the graft developsin a tumour. Transplantation of normal callus tissue does not result in an overgrowth. Crown gall tissue contains higher levels of auxin than normal tissue and also produces several cytokinins [9]. It is assumed that this is the reason for the autonomous growth of crown gall cells.

Several types of crown gall tumours are known. The bacteria that induce the tumour determine which kind of tumour will arise. For instance, in tumours often unusual amino acid derivatives ('opines') are found, namely either octopine, lysopine, octopinic acid and histopine (N^2-(D-1-carboxyethyl)-L-arginine, -L-lysine, -L-ornithine, -L-histidine respectively) or nopaline and ornaline (N^2-(1,3 dicarboxypropyl)-L-arginine respectively -L-ornithine), or none of them. Tumours which do not contain any of these compounds may contain other, hitherto unknown substances. On *Kalanchoë daigremontiana* tumours can be distinguished by their morphology [10](Fig.1). Tumours that contain octopine have a rough surface and are surrounded by many adventitious roots, while tumours that contain nopaline have a smooth surface and have a limited number of root-like structures only at the bottom of the tumour. Tumours of the latter type often show leaflike structures. Finally, *Agrobacterium rhizogenes* strains induce the development of only roots from woundsites on Kalanchoë.

It is conceivable that during the period spent in the tumour the bacteria actually profit from the tumour specific compounds, since they not only induce the synthesis of them by the plant cells, but they are also able to utilize them as a source of carbon and nitrogen. The genes coding for the enzymes involved in catabolism of these tumour specific compounds are controlled by repressor molecules that are specifically inducible by the 'opines' which can be utilized [12]. Agrobacterium strains that induce octopine producing tumours, catabolize octopine, but not nopaline and are therefore called 'octopine strains'. Strains that induce nopaline producing tumours, catabolize nopaline, but not octopine and are called 'nopaline strains'. Strains that induce tumours without octopine or nopaline, and utilize neither octopine nor nopaline, are called 'null type strains'.

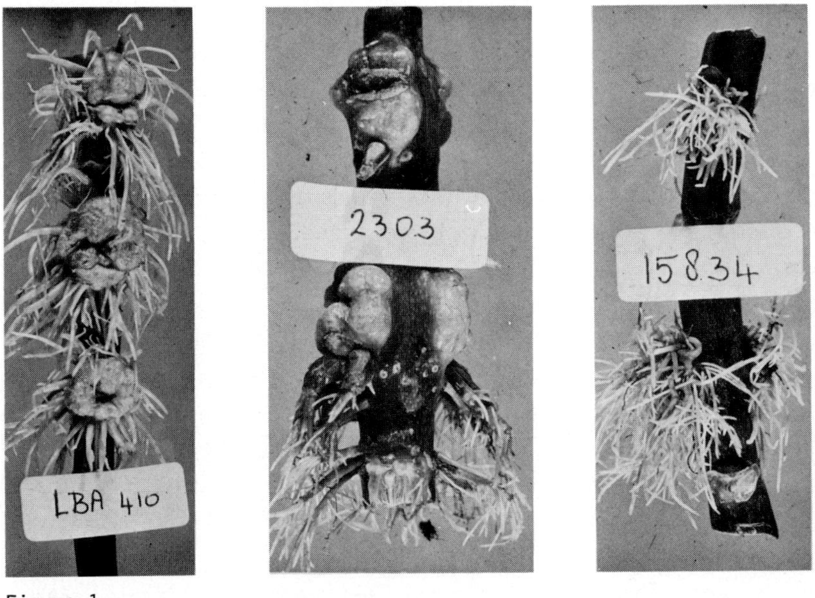

Figure 1.
At the left a rough tumour, in the middle a smooth tumour and at the right root proliferation induced by different Agrobacterium strains.

PLASMIDS

Plasmids have been detected in all Agrobacteria and Rhizobia. Most strains carry large plasmids ranging in size from 100 to > 300 megadaltons. Smaller plasmids have also been found, however, in only a limited number of strains. While the small plasmids have been isolated by the classical cleared lysate procedure, new procedures have been developed for the detection and isolation of the large covalently closed circular molecules [13,14].

Ti plasmids. When Agrobacterium strains are cultured at 37°C instead of 28°C, which is the normal growth temperature, avirulent derivatives are frequently found. These strains have lost a plasmid of a size between 90 and 150 megadaltons [15,16]. Such avirulent derivatives may acquire virulence from a virulent donor strain in mixed infections *in planta* [18]; strains that have acquired virulence, have also received a plasmid from the donor strain [16,17]. These plasmids have therefore been named Ti plasmids. The same experiments have shown, that Ti plasmids code for the ability to breakdown octopine or nopaline, for exclusion of phage AP1, sensitivity to agrocin K84, and furthermore determine, whether octopine or nopaline or neither of these will be synthesized in crown gall cells.

Transfer of Ti plasmids has also been observed *ex planta*. Recipients that have received a Ti plasmid, are selected on media with octopine or nopaline as a sole nitrogen source. Since agar contains impurities that Agrobacteria can utilize as a nitrogen source, exhaustive washing of agar is required before it can be used in the selective medium, and even then the results are not always satisfactory. Therefore a new selective medium was developed. This medium contains octopine or nopaline as a nitrogen source and bromothymolblue as a pH indicator in the presence of minimal amount of phosphate buffer. Ti plasmid carrying strains form yellow colonies on such green plates, while Ti plasmid lacking strains remain translucent [19].

Using the selective medium it has been found that the incP-1 plasmids RP4, R702, R68.45, R772, R751.pMG1≠2 and RP1.pMG1 are able to mobilize Ti plasmids [19]. Furthermore it has been shown that Ti plasmids are self-transmissible [20,21]. However, transfer only takes place after induction of the transfergenes by octopine, octopinic acid or lysopine for octopine Ti plasmids or by nopaline for nopaline Ti plasmids. The transfergenes and the genes for catabolism of octopine, octopinic acid and lysopine belong to separate operons, which ate coordinately controlled by the same repressor [19,22,23].

All these transfer experiments have shown, that octopine Ti plasmids carry genes for octopine catabolism (occ), transmissibility (tra), exclusion of phage AP1 (ape) and oncogenicity (onc), and that they also determine the rough tumour morphology and whether or not tumours will synthesize octopine (ocs). Nopaline Ti plasmids generally carry genes for nopaline catabolism (noc), transmissibility (tra), exclusion of phage AP1 (ape), sensitivity to agrocin K84 (agr) and oncogenicity (onc), and they also determine the smooth tumour morphology and whether or not tumours will synthesize nopaline (nos).

There is evidence that A.rhizogenes strains carry plasmids of a size comparable to that of Ti plasmids [24], and that these carry the genes that determine the induction of root proliferation at woundsites [25].

Rhizobium plasmids. Large plasmids have been detected in almost all Rhizobium strains studied so far. The number and size of plasmids present in strains of a given species seem to be scarcely variable; however, no exact correlation could be demonstrated as yet between plasmid presence and species [14,26]. In many Rhizobia more than one large plasmid is present and consequently 10 to almost 30% of the total genetic information in such strains is plasmid-encoded, so possibly some or all of the symbiotic properties are genetically controlled by these large DNA molecules.

One strain of *R.leguminosarum* could be 'cured' of its smallest plasmid of 110 megadaltons by thermal shock [14]. The resulting derivatives were no longer able to induce nodule formation on its own hosts. They found 'rough' colonies on yeast-mannitol agar; studies of its surface polysaccharides showed that this strain contains a lectin-agglutinable fraction, but that neither unhydrolyzed LPS nor viable bacteria were agglutinated[56].So the molecular basis for the inf^- character of these strains may well be their inability to recognize their proper host plant. Recently another group described the transfer of host-specificity from a strain of *R. leguminosarum* into other Rhizobia [27]. So it seems that in any case host-specificity in *R.leguminosarum* is plasmid controlled.

The *K.pneumoniae* structural and regulatory genes for nitrogenase have been cloned into small multicopy plasmids [28]. These were used to detect homology between the Klebsiella genes and Rhizobium plasmids by hybridization. Hybridization could not be detected between the Rhizobium plasmids and the vector nor with a vector carrying the regulatory genes for nitrogenase of Klebsiella. However, homology was found between Rhizobium plasmids from strains of *R.leguminosarum* and *R. meliloti* and the vector carrying the Klebsiella structural genes for nitrogenase[29]. In conclusion, also effectiveness is a plasmid-encoded trait in these Rhizobia.

HOSTRANGE OF TI PLASMIDS; ROLE OF CHROMOSOMAL GENES IN TUMORIGENESIS

Ti plasmids can be transferred into *E.coli* [30]; however, they are unstable in this host and the E.coli transconjugants are neither able to breakdown octopine nor do they induce tumours. However, upon transfer into different Rhizobium species, Ti plasmids are stably maintained and the Rhizobium exconjugants are able to breakdown octopine or nopaline, depending on the Ti plasmid that has been rereceived [10,14,31]. *R.trifolii*, *R.phaseoli* and some *R.leguminosarum* recipients become oncogenic upon receipt of a Ti plasmid [31]. A number of such transconjugants only induce small tumours; the tumours induced by these strains contain octopine and/or nopaline and are rough or smooth depending on the Ti plasmid received. Some *R. leguminosarum* and *R.meliloti* transconjugants, remain avirulent after introduction of a Ti plasmid; they do, however, acquire the ability to breakdown octopine or nopaline [31]. R.trifolii strains 5 and 0403 transconjugants are able to nodulate their proper hosts (*Trifolium* species) effectively; the plasmids, which these strains naturally carry are compatible with Ti plasmids [14,31].

These findings show, that chromosomal genes influence the oncogenicity of a Ti plasmid carrying strain. Attachment of Agrobacterium to the plant cell wall is an essential step in tumour induction. Agrobacteria cured of their Ti plasmid still attach to plant cells and may inhibit tumour induction by competition with virulent bacteria for attachment sites. For these strains attachment is determined by chromosomal genes. However, a few non-pathogenic Agrobacteria have been isolated from nature that do not compete with pathogenic bacteria for attachment sites. These strains can be converted into pathogens by the introduction of a Ti plasmid. LPS isolated from the converted strains does inhibit tumour induction by a pathogenic strain, but LPS of the original recipient does not. So genes which code for such LPS alterations that attachment becomes possible, are located on the Ti plasmids, but for most strains also on the chromosome [32].

To be able to study the possible role of chromosomal genes in oncogenicity a system for the transfer of chromosomal genes has to be developed and a chromosomal map has to be constructed. We recently succeeded in mobilizing chromosomal genes in A.tumefaciens strain C58 with R plasmids RP4, R702 and R68.45. R68.45 gave the highest recombination frequencies. By four factor crosses linkage between a number of auxotrophic markers could be calculated [31].

GENETIC MAP OF AN OCTOPINE TI PLASMID

Ti plasmid insertion mutants. These have been isolated in two different ways: 1) Ti plasmids can be mobilized by R plasmids of incompatibility group P-1 [19]. Recipients that have received a Ti plasmid in such an experiment behave as if the Ti plasmid they carry has acquired 'affinity' for the R plasmid [4,33]. When such recipients are used as donors in further crosses, the Ti plasmid is mobilized by the R plasmid with a higher frequency than from the original donor strain. 'Affinity' appeared to be based on transpositional and other recombinational events. Some recipients were found to carry cointegrated plasmids consisting of R plasmid and Ti plasmid. In such cointegrates often either one or both of the component plasmids are partially deleted. Other recipients have been found to carry Ti plasmids with an insertion of a transposon originating from the R plasmid [4,33].

In this way about 200 Ti plasmids have been obtained that carry an insertion of Tn1, which codes for carbenicillin resistance or an insertion of a new transposon of 10 megadaltons derived from R702, that carries genes for resistance against streptomycin and spectinomycin and has been named Tn1831. In fact, other new transposons may be discovered by using R plasmids with possible transposons as vehicles mobilizing other plasmids.

2) In *E.coli* cointegrate plasmids were constructed between RP1.pMG1, carrying the transposons Tn1, coding for carbenicillin resistance and Tn904, coding for streptomycin resistance, and bacteriophage Mu*cts*62 [34]. A cointegrate in which Mu*cts*62 was found to be integrated in the R plasmid near its site for kanamycin resistance and between its transfer regions, was selected for further use. It is known that genomes with bacteriophage Mu are prone to the formation of deletions. This is especially the case when R::Mu plasmids are transferred from *E. coli* into *Rhizobium* or *Agrobacterium* [35,36,37]. When our R::Mu cointegrate was transferred into Agrobacteria or Rhizobia, deleted R plasmids that lacked the Mu-insertion, were obtained. These R plasmids were frequently tra$^-$. One strain carrying such a tra$^-$ R plasmid and a Ti plasmid was used as a donor in conjugation experiments with a Ti plasmid cured recipient. Recipients were selected that had received streptomycin resistance, encoded by Tn904, from the donor. In this way 155 strains were isolated that had a stable insertion of Tn904 in the Ti plasmid [34].

Ti plasmid deletion mutants. In strain Ach5 Ti plasmid mutants that are constitutive for utilization of octopine, are sensitive to homooctopine, an analogue of octopine consisting of homoarginine, which is toxic for *Agrobacterium*,and pyruvate [22,38]. Non-reverting mutants, resistant to the toxic effect of homooctopine are affected in the utilization of octopine and generally carry deleted Ti plasmids. Transposons generate deletions starting at one end of the transposon with high frequency. So, when non-reverting homooctopine resistant derivatives were selected from strains having a Ti plasmid containing a transposon (either Tn1 or Tn904), these derivatives were almost always found to carry Ti plasmids with deletions starting from the transposon-insertion [39,40].

Genetic map of an octopine Ti plasmid. A physical map of an octopine Ti plasmid has been described for the restriction enzymes SmaI [41], HpaI [41], KpnI [42], and Xba [42]. Ti plasmid deletion mutants can thus easily be mapped by restriction enzyme analysis of the isolated plasmid DNA. Since both R plasmids and transposons have a fixed size their insertion position can be mapped in the same way. While deletion mutants are very suitable to produce a gross genetic map, by the use of insertion mutants a fine map can be constructed. Each insertion takes place at a different site and prevents the expression of the gene, in which the insertion has occurred. A disadvantage of insertion mutants is that the inserted DNA may exert polar effects on distal genes. Especially inserted R plasmids may influence genes, which are located far from the insertion site. For example the insertion of R702 in a site, which is located between the replicator region and the genes for octopine catabolism (Fig.2), results in loss of virulence towards *Kalanchoë*; however, deletion mutants have been isolated that lack a large region of the Ti plasmid including this R702 insertion-site, and these still induce tumours on *Kalanchoë* [39,40]. When looking for the effect of mutations on the oncogenicity of the bacterium, it is important to test the mutants for virulence on a number of different plants. We experienced that some of our mutants were avirulent on pea and/or *Kalanchoë*, but were virulent on tomato or *Petunia*.

With the aid of the physical maps mentioned above, deletions generated by transposons inserted at different positions in the Ti plasmid have been mapped. By comparing the plasmid genotype of the mutants to their phenotype, a gross genetic map has been constructed [39,40](Fig.2). Genes involved in replication and incompatibility, conjugative transfer, catabolism of octopine, virulence and AP1 exclusion have been localised.

Although there was a high correlation between loss of AP1 exclusion and loss of virulence, also mutants have been isolated that lacked AP1 exclusion, but were still virulent. So, the AP1 exclusion marker is not essential for oncogenicity, but is located close to genes involved in virulence [40]. Most mutants that lack the octopine catabolism genes are also transfer-deficient, suggesting that at least some transfer-genes are located very close to these genes [22,39]. This was also expected from the fact that octopine catabolism and transfer genes are controlled by the same repressor. However, generally many genes are involved in the conjugative transfer of plasmids, so it is quite possible that these are located elsewhere on the Ti plasmid. It has been postulated by others that the bacterial conjugation mechanism is also used in the transfer of Ti plasmid DNA into plant cells. Since many of the transfer-deficient mutants that we isolated were still virulent, this seems to be unlikely. One of the strains isolated carries a Ti plasmid from which more than 80% has been deleted, leaving a plasmid of only 28 megadaltons [39]. This plasmid has been shortened by *in vitro* manipulation, whereby a plasmid of about 13 megadaltons has been obtained that carried the replicator-genes and the genes for incompatibility [43].

Crown gall cells have been found to contain a piece of Ti plasmid DNA, which has been called T-DNA [8]. This T-DNA is contiguous to or overlaps a region, which is conserved in all Ti plasmids (hereinafter referred to as the common sequence) studied so far [44,45].

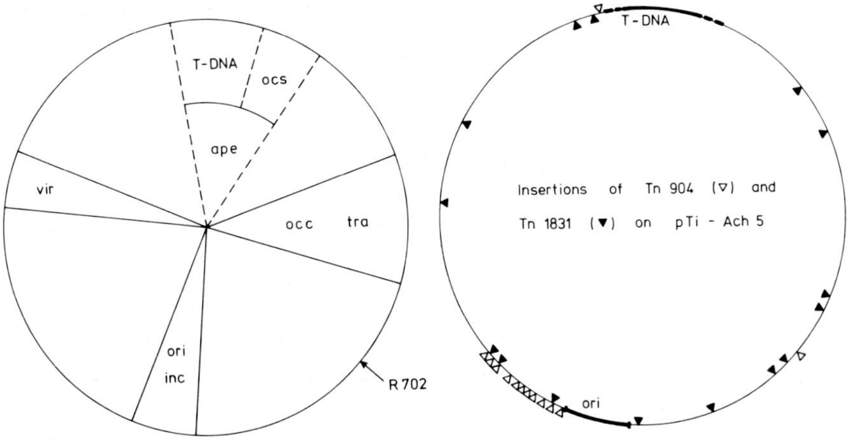

Figure 2.
At the left : a genetic map of an octopine Ti plasmid. At the right : the positions of a representative number of insertions of Tn904 respectively Tn1831 in this same plasmid.

It is not yet shown, where the T-DNA is located in the plant cell. *In situ* hybridization experiments on infected tissues with highly labelled ^{125}I-Ti plasmid DNA [26] and data obtained from fusion experiments between tobacco crown gall protoplasts and normal tobacco leaf protoplasts [47] suggest a nuclear origin. At least part of the T-DNA is transcribed into RNA in tumour cells [48,49]. Some of these RNA molecules have polyadenylic acid at their 3'end, which is characteristic of eukaryotic mRNA [49]. This suggests that T-DNA transcripts are subject to an eukaryotic control system. Whether the RNA is also translated into proteins is not yet known. One of the proteins that might be encoded by this RNA, is the enzyme lysopine dehydrogenase, which catalyzes the synthesis of octopine, octopinic acid and lysopine in crown gall cells.

Mutants that carry Ti plasmids from which the entire T-DNA region and the common sequence have been deleted, are avirulent [39,40]. However, when the deletions extend far into the T-DNA, but leave behind all or most of the common sequence, the mutants are still able to induce tumours, but the tumours grow slowly and do not synthesize octopine [39]. Apparently, besides a gene for lysopine dehydrogenase, genes have been eliminated which are needed to obtain vigorously growing tumours.

These results indicate that the common sequence contains genes that are essential for oncogenicity; they also show that the T-DNA does not behave like a transposon as proposed by others, since transposons only jump if both ends are intact.

Restriction enzyme analysis of 20 strains with Tn904 insertions in the Ti plasmid showed that Tn904 has a high preference for a certain region of the Ti plasmid [42]. Furthermore it was found that a number of strains did not carry a single copy of Tn904, but had double insertions. In most cases the single or double Tn904 insertions were deleted or partially integrated. Analysis of 25 strains with Tn1831 insertions in the Ti plasmid showed that Tn1831 inserted at random; all insertions consisted of a single copy of Tn1831 [50].

Tn904 and Tn1831 insertions that caused an altered phenotype have also been mapped. Transfer experiments demonstrated that in all cases the mutations were located on the Ti plasmid. The data obtained confirmed those from the analysis of deleted plasmids. Interestingly a new site essential for oncogenicity was found [42]. This site is located far from the T-DNA at the left side of the Ti plasmid. Insertions in this site cause loss of virulence on at least 4 different plant species. One mutant with an insertion just above this region was less virulent [42]; the tumours formed were comparable to tumours induced after infection with a low number of bacteria. This suggests that possibly due to a polar effect of the transposon on genes in this region, the bacterial cells are less effective in transforming plant cells. The mutants are still tra$^+$, but of course this region may be needed by the bacterium to accomplish transfer of its plasmid DNA into the plant cell specifically.

Several mutants with insertions in or near the common sequence have been isolated [42]. These mutants gave phenotypically changed tumours. One particular mutant induced small tumours with abundant formation of adventitious roots on *Kalanchoë*. A spontaneous mutant that gave tumours lacking adventitious root formation, but induced sprouting of axillary buds on *Kalanchoë*, was found to have a Ti plasmid with an insertion of an Agrobacterium IS element in the common sequence [51]. The results with these mutants suggest that at least two sets of genes are present in this region, which may be affected independently: one set of genes responsible for an 'auxin-like symptom', the other set for a 'cytokinin-like symptom'. This supports the view that T-DNA genes themselves are responsible for auxin- and cytokin-autotrophic growth of crown gall cells.

COMPLEMENTATION ANALYSIS OF MUTANTS

Recently we have isolated a recombination deficient (Rec$^-$) A.tumefaciens strain which will be useful for the analysis of the functions involved in tumour induction by complementation studies of avirulent mutants [52]. This Rec$^-$ strain has been employed already to isolate, by selecting for markers of two plasmids, a deletion mutant from a R702::Ti cointegrate plasmid that had lost the Ti coded incompatibility and replicator [23]. The mutant plasmid lacked a large piece of Ti DNA, but it had retained the occ genes. By means of this mutant plasmid it was possible to make cells diploid for the occ genes. The repression pattern that was found in complementation analysis using different types of regulation mutants for the occ and tra-genes demonstrated that the occ and tra operons are controlled by a common repressor system [23] as was postulated earlier [19,22]. Results with some mutants however indicated that there may be additional transcriptional relations between both operons [23].

INCOMPATIBILITY BETWEEN DIFFERENT TI PLASMIDS

Octopine, nopaline and nulltype Ti plasmids share regions of DNA homology [53,54] One specific region of about 5 megadaltons is highly conserved and is called the 'common sequence' [44,45]. This is part of the plasmid DNA that has been found in crown gall cells. However, also in the replicator and incompatibility region octopine and nopaline Ti plasmids share DNA homology [39,53,54]. Incompatibility is generally regarded as an indication for evolutionary relatedness. Therefore we studied the (in)compatibility between octopine and nopaline Ti plasmids. For these studies a number of different octopine and nopaline plasmids were introduced into a Ti plasmid cured C58 derivative; into the strains containing these different plasmids, plasmid pAL208 (pTiB6Δocc::Tn1) was introduced. Loss of the ability to utilize either octopine or nopaline was considered an indication for incompatibility. In this way it was found that the octopine Ti plasmids of strains B6, 147, NCPPB4 and the nopaline Ti plasmids of strains C58, T37, AG43, NCPPB1651, NCPPB2303, Kerr27, Kerr108, M3/73 and 1D135 belong to the same incompatibility group [4,55]. However, when pAL208 was introduced into the nopaline strain Kerr14, the exconjugants remained able to utilize nopaline [55]. Furthermore agarose gel analysis showed, that none of the 3 plasmids that are naturally present in this strain was lost upon the introduction of pAL208. Tumours induced by such exconjugants were rough and only synthesized octopine, indicating that pTiKerr14 does not influence oncogenicity at all. When pAL208 was introduced into strains carrying plasmids that code dor octopine or nopaline utilization but not for virulence, it was found that these (pAtAG60, pAtKerr14, pAtG12/73) are compatible with the Ti plasmid [4,55]. Interestingly also plasmids naturally occurring in *A.rhizogenes* were found to be compatible with Ti plasmids [24].

When an octopine Ti plasmid was introduced into strains carrying a nopaline plasmid not all recipients lost the ability to utilize nopaline [4]. These were purified and examined in more detail. It was found that they harboured plasmids of > 200 megadaltons being the result of cointegration of the two parental plasmids. The cointegrates were quite stable : cotransfer of all markers was 100% and when selecting for homooctopine resistant mutants, deleted plasmids were found that had the entire nopaline plasmid and part of the octopine plasmid. Cointegration had occurred in the common sequence, which is shared by these plasmids [55]. The tumours that were induced by these strains were smooth, but in tumour cells both octopine and nopaline synthesis activity were detected [4]. So DNA from both partners is introduced into the plant cells. However, interestingly the tumour morphology was identical to that of nopaline tumours; indicating a dominancy of this character over the rough phenotype of octopine tumours.

ACKNOWLEDGEMENTS

The authors wish to thank their collegues Drs Huisman, Krens, Molendijk, Otten, den Dulk-Ras, Wullems and Würzer-Figurelli of the "MOLBAS" research group for their help.

This work was sponsored by the Netherlands Foundation of Biological Research (BION) and the Netherlands Foundation for Chemical Research (SON) with financial aid from the Netherlands Organisation for the Advancement of Pure Scientific Research (ZWO).

REFERENCES

1. Graham, P.H.(1964) J.Gen.Microbiol.35, 511.
2. Dazzo, R.B. and Hubbell, B.H.(1975) Appl.Microbiol.30, 1017.
3. Sidloi-Lumbroso, R., Kleiman, L. and Schulman, H.M.(1978) Nature 273, 558.
4. Hooykaas, P.J.J., Schilperoort, R.A. and Rörsch, A.(1979) Genetic Engineering Volume 1, 151. Plenum Press, New York. Edited by J.K.Setlow and A.Hollaender.
5. Kado, C.I.(1978) EMBO Workshop on Plant Tumour Research, Noordwijkerhout, The Netherlands.
6. Lippincott, B.B. and Lippincott, J.A.(1969) J.Bacteriol.97, 620.
7. Schilperoort, R.A.(1969) Ph.D.Thesis, Leiden, The Netherlands.
8. Chilton, M.-D.et al.(1977) Cell 11, 263.
9. Braun, A.C.(1978) Biochim.Biophys.Acta 516, 167.
10. Hooykaas, P.J.J.et al.(1977) J.Gen.Microbiol.98, 477.
11. Petit, A.et al.(1970) Physiol.Vég.8, 205.
12. Klapwijk, P.M., Oudshoorn, M. and Schilperoort, R.A.(1977) J.Gen.Microbiol. 102, 1.
13. Ledeboer, A.M.et al.(1976) Nucleic Ac.Res.3, 449.
14. Prakash, R.K.et al.(1978) Proc.3rd Int.Symp.on N_2-Fixation, Madison,Wisconsin, USA (in the press).

15. Watson, B.et al.(1975) J.Bacteriol.123, 255.

16. Van Larebeke, N.et al.(1974) Nature 252, 169.

17. Van Larebeke, N.et al.(1974) Nature 255, 742.

18. Kerr, A.(1971) Physiol.Plant Pathol.1, 241.

19. Hooykaas, P.J.J., Roobol, C. and Schilperoort, R.A.(1979) J.Gen.Microbiol.110, 99.

20. Kerr, A., Manigault, P. and Tempé, J.(1977) Nature 265, 560.

21. Genetello, C.et al.(1977) Nature 265, 561.

22. Klapwijk, P.M., Scheulderman, T. and Schilperoort, R.A.(1978) J.Bacteriol. 136, 775.

23. Klapwijk, P.M. and Schilperoort, R.A.(1979) J.Bacteriol.(submitted).

24. Costantino, P., Hooykaas, P.J.J. and Schilperoort, R.A.(in preparation).

25. Albinger, G. and Beiderbeck, R.(1977) Phytopathol.Z.90, 306.

26. Nuti, M.P. (unpublished results).

27. Johnston, A.W.B. et al.(1978) Nature 276, 634.

28. Cannon, F.C.(1978) Proc.Int.Symp.on Genetic Engineering, Milan, Italy. Edited by Boyer and Nicosia, Elsevier North Holland.

29. Nuti, M.P., Cannon, F.C., Prakash, R.K. and Schilperoort, R.A. (in preparation).

30. Holsters, M.et al.(1978) Mol.Gen.Genet.163, 335.

31. Hooykaas, P.J.J.(unpublished results).

32. Whatley, M.H.et al.(1978) J.Gen.Microbiol.107, 395.

33. Hooykaas, P.J.J., den Dulk-Ras, A. and Schilperoort, R.A.(in preparation).

34. Klapwijk, P.M.et al.(1979) J.Bacteriol. (submitted).

35. Boucher, C.et al.(1977) J.Gen.Microbiol.98, 253.

36. Beringer, J.E.et al.(1978) Nature 276, 633.

37. Van Vliet, F.et al.(1978) Plasmid 1, 446.

38. Petit, A. and Tempé, J.(1978) Mol.Gen.Genet.167, 147.

39. Koekman, B.P.et al.(1979) Plasmid (in the press).

40. Ooms, G., Hooykaas, P.J.J. and Schilperoort, R.A.(in preparation).

41. Chilton, M.-D.et al.(1978) Plasmid 1, 254.

42. Ooms, G.et al.(1979) Cell (submitted).

43. Koekman, B.P. (unpublished results).

44. De Picker, A., van Montagu, M.and Schell, J.(1978) Nature 275, 150.

45. Chilton, M.-D.et al.(1978) Nature 275, 147.

46. Gordon, M.P.(1979) Proteins and Nucleic Acids Volume 6 (in the press) Academic Press, New York. Edited by A.Marcus.

47. Wullems, G., Molendijk, L.and Schilperoort, R.A.(1979) Mol.Gen.Genet. (submitted).

48. Drummond, M.H.et al.(1977) Nature 269, 535.

49. Ledeboer, A.M.(1978) Ph.D.Thesis, Leiden, The Netherlands.

50. Koekman, B.P., Hooykaas, P.J.J. and Schilperoort, R.A.(in preparation).

51. Ooms, G., Klapwijk, P.M. and Schilperoort, R.A. (unpublished results).

52. Klapwijk, P.M., van Beelen, P. and Schilperoort, R.A.(1979) Mol.Gen. Genet.(in the press).

53. Hepburn, A.G. and Hindley, J.(1979) Mol.Gen.Genet.169, 163.

54. Drummond, M.H. and Chilton, M.-D.(1978) J.Bacteriol.136, 1178.

55. Hooykaas, P.J.J., den Dulk-Ras, A. and Schilperoort, R.A.(in preparation).

56. Van der Schaal, I., de Vries, G. and Kijne, J. (in preparation).

© 1979 Elsevier/North-Holland Biomedical Press
Plasmids of Medical, Environmental and Commercial Importance
K.N. Timmis and A. Pühler, editors

The Role of Opines in the Ecology of the Ti-Plasmids of Agrobacterium

Jacques Tempé, Pierre Guyon, David Tepfer and Annik Petit
Station de Génétique et d'Amélioration des Plantes. C.N.R.A. 78000 Versailles, France.

Ecology is the study of relationships between organisms and environments. Therefore, plasmid ecology (if we may invent such a term) would describe the relationships between a plasmid and its host, i.e. its immediate environment, as well as interactions between the host and its own environment, which are in some way influenced by the plasmid.

In this paper we shall review the crown gall system and attempt to fit the bulk of the experimental observations into an ecological framework. We conclude that the crown gall ecosystem is most easily described from the point of view of the Ti-plasmid. Such an approach may be of value in describing other host-parasite systems.

CROWN GALL IN THE LABORATORY

Crown gall tumors are cancerous overgrowths that develop on dicotyledonous plants after they have been infected, through a wound, by the soil pathogen *Agrobacterium tumefaciens*[1]. The disease has serious economic consequences. In France, it affects primarily nurseries where current losses can be estimated at 10 to 20 % of production. The pathogenic action is of short duration, a few days of contact between the bacterium and the wounded tissues of the host suffice to induce a malignant transformation of some of the plant's cells. These cells escape the plant's normal cell division controls and start proliferating, forming a tumor. In nature, such tumors are often located at or near the crown of the plant, hence the name of the disease. Crown gall tumor tissues from most species can be cultured aseptically *in vitro*, on chemically defined media, where sustained cell division is not dependent on the presence of exogenous growth factors that are usually necessary for the proliferation of normal cells. The malignant character of tumor cells can be easily demonstrated by grafting onto a compatible host, resulting in the development of a tumor resembling a primary tumor in every respect except for the fact that it is free of the inciting bacterium.

The oncogenic properties of *Agrobacterium tumefaciens* are carried on covalently closed circular DNA molecules called Ti-plasmids[2,3,4], a segment of which, the T-DNA, is transferred during oncogenesis to the plant cell, where it is maintained and transcribed[5,6,7]. It is not known whether this T-DNA carries structural genes that are transcribed and translated in the tumor cells. However, it has been demonstrated, by insertional mutagenesis, that the tumor cell phenotype is determined by the T-DNA[6]. Two regions have thus been defined on the T-DNA : the ONC region which is responsible for the malignant properties of the plant cells, and the OPS region, which dictates the production by these cells of novel substances (that we have named the opines[8]), which are biochemical markers for crown gall[9].

Opines were first discovered in crown gall tumor tissues cultured *in vitro*, as unusual α-N substituted amino acid derivatives (Table 1).

TABLE 1

STRUCTURAL FORMULAE OF THE OPINES : 1° The Octopine family

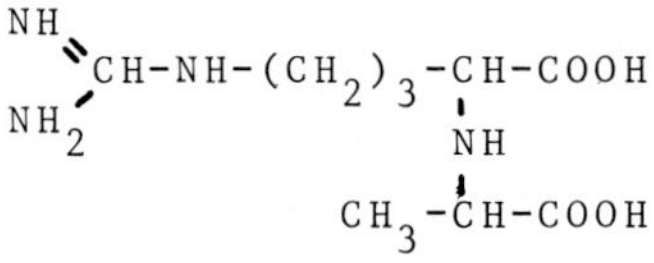

octopine

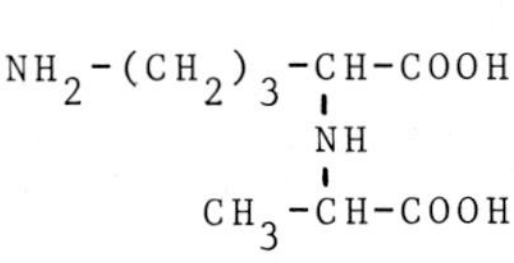

octopinic acid

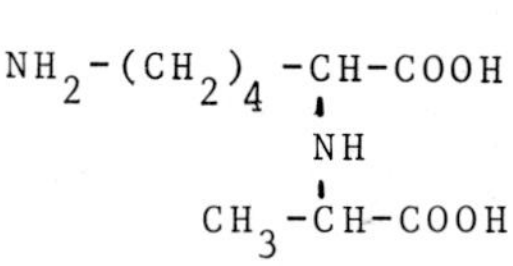

lysopine

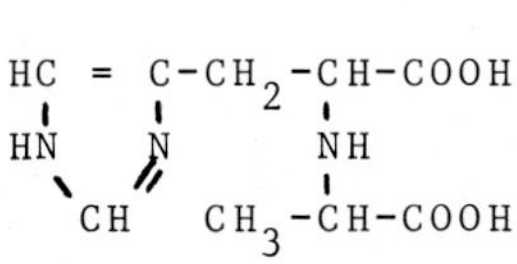

histopine

TABLE 1 (contd)

STRUCTURAL FORMULAE OF THE OPINES : 2° The Nopaline family

```
NH
  \\
   CH-NH-(CH2)3-CH-COOH          NH2-(CH2)3-CH-COOH
  /              |                          |
NH2              NH                         NH
                 |                          |
    HOOC-(CH2)2-CH-COOH          HOOC-(CH2)2-CH-COOH
```

nopaline — nopalinic acid

Lysopine[10], octopine[11], octopinic acid[12] and histopine[13] are found in "octopine tumors or tissues". Nopaline[14] was first isolated from *Opuntia vulgaris* crown gall tissue cultured *in vitro* ; its presence in other crown gall tumors was shown later. Nopaline tumors also contain nopalinic acid[15,16]. The most recently discovered opine, agropine[17], whose structure is still unknown, was found in octopine tumors.

The specificity of the opines as markers of crown gall was questionned when lysopine was found in normal tomato and tobacco tissues[18]. The subject became more controversial later, when octopine was also found in tissue cultures of various origins and in normal plants[19,20]. These observations could not be reproduced[21,22,23,24].

The significance of the production of opines in crown gall tumors remained unclear for some time. Some considered them to be normal plant products[18,19,20,25], an opinion still defended by one group[26]. On the other hand, Goldmann and Morel, studying the occurence of octopine and nopaline, proposed that opine synthesis was the result of a fault in the metabolism of nitrogenous substances in crown gall cells[27]. These authors realized that there were two types of tumorous tissues. They thought that the presence of either octopine or nopaline reflected a metabolic feature of the host. Goldmann *et al.*[28] found later that the type of opine synthesized was specified by the bacterial strain, not by the host plant. Petit *et al.*[29] then showed that most *Agrobacterium tumefaciens* strains induce the development of tumors containing either octopine or nopaline, and that a few strains induced tumors that did not contain either. We took this finding as evidence for the transfer of genetic information from the bacterium to the plant cell during tumorigenesis.

The finding that opine synthesis in crown gall tumors is specific to the bacterial strain that has induced the tumor prompted us to search for a bacterial phenotype that would be correlated with this specificity. We could not demonstrate opine synthesis by *Agrobacterium tumefaciens* cells, but we did discover that bacterial strains could catabolize octopine or nopaline with the same specificity that they displayed for the induction of opine synthesis in crown gall tumors[29]. There was a one to one correspondance between bacterial phenotypes and tumor cell phenotypes. This was the first genetic evidence supporting the idea of transfer of genetic information from *Agrobacterium tumefaciens* to the plant cell during the process of tumor transformation. These results have been generalized to the other opines[13,15,17,30,31,32].

At about the same time, another important feature of the system was discovered. As already mentioned, crown gall causes heavy losses in some nurseries. A. Kerr[33,34] found that one could achieve excellent protection of stone fruit cuttings or seedlings by dipping them in a suspension of the non-pathogenic strain *Agrobacterium radiobacter* 84. This strain had been isolated from the soil around infected plants. The reason for this protection, which is now widely used to control crown gall, was first thought to be competition, the pathogen being unable to reach specific sites already occupied by the non-pathogen. However it soon became evident that a more specific interaction was involved. Kerr and Khin Htay discovered that the strain 84, which was being used for biological control of crown gall, produces a bacteriocin that inhibits the growth of the pathogen[34]. Later it was shown that only nopaline strains were sensitive to the bacteriocin[35].

Kerr provided another clue to understanding the crown gall system. He found that the oncogenic properties of *Agrobacterium tumefaciens* could be transferred to a non-oncogenic recipient *Agrobacterium radiobacter* when both strains were simultaneously present in a developping tumor[36,37]. This genetic exchange was suggestive of conjugation, but no conditions could be found that permitted *in vitro* reproduction of the phenomenon.

THE TI-PLASMID AS THE PROGRAMMER OF THE SYSTEM

The Ti-plasmid was discovered by ZAENEN *et al.*[2] who showed that high molecular weight plasmids were present in oncogenic strains but not in avirulent isolates of *Agrobacterium*. The involvement of the Ti-plasmid in the oncogenic properties of *Agrobacterium tumefaciens* was soon confirmed[3,4]. The Ti-plasmid can now be considered to be the biological entity which carries the dominant genetic program for the crown gall system. It is clear that the earlier observations, discussed above, are manifestations of the activity of the Ti-plasmid. Besides containing the T-DNA, which is transferred to the plant cell during oncogenesis, and is responsible for the tumor phenotype and opine synthesis, the Ti-plasmid also carries genes responsible for the utilization of opines as carbon and/or nitrogen sources by *Agrobacterium*[24,32,38]. The synthesis of specific opines in tumor cells and their specific degradation by the bacteria which induced the tumor are both encoded in defined regions of the Ti-plasmid. These functions define 3 classes of Ti-plasmids : the octopine, the nopaline and a 3rd class for which no opine was previously known and which was called the null type plasmid.

We have studied opine degradation and we have shown that octopine is first degraded to arginine, which is then converted to glutamic acid[29]. We proposed a pathway for this degradation which involves at least 3 Ti-plasmid genes grouped in one operon, the occ operon[39,40] :

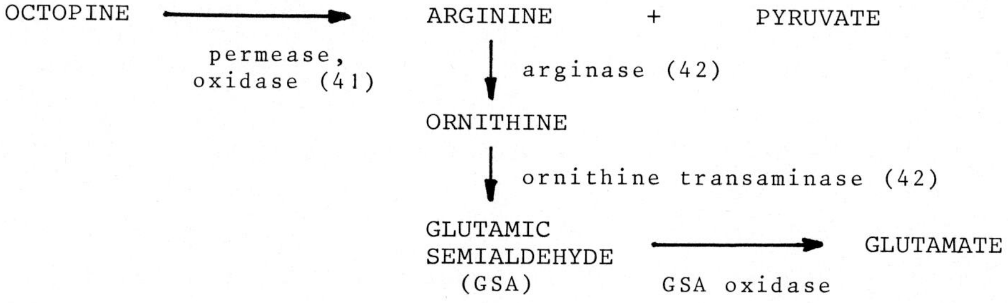

(proposed pathway for octopine degradation [39], Ti-plasmid coded steps, shown by horizontal arrows, make carbon available for growth)

Our data suggest a similar pathway for nopaline degradation, with α-ketoglutaric acid substituted for pyruvic acid.

Studies with mutants affected in octopine utilization suggest that there is only one oxidase gene on the Ti-plasmid, and that the corresponding enzyme can act on the opines derived from other amino acids. Thus synthetic compounds like meso-iminodipropionic acid (I) and D-iminopropioacetic acid (II) can be used as growth substrates by strains harbouring an octopine Ti-plasmid containing a functionnal oxidase gene.

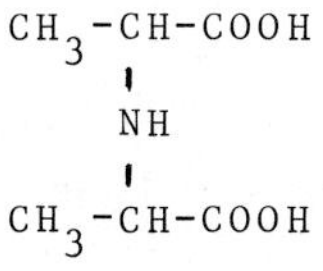

(I)

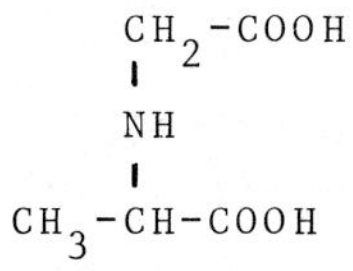

(II)

The Ti-plasmids can be considered catabolic plasmids that have acquired the capacity to trigger the production of their substrates by the host plant, the degradation of which supplies *Agrobacterium tumefaciens* cells with amino acids and an α-ketoacid. Thus the Ti-plasmids cooperate with their bacterial host to commandeer the plant cell's anabolic activity for their exclusive benefit. The conversion of plant amino acids into opines assures that only Ti-plasmid harbouring bacteria can participate, since opine catabolism is an exclusive property of the Ti-plasmids.

The transfer of the oncogenic properties described by Kerr is best explained by Ti-plasmid conjugation. All the previous attempts to demonstrate conjugation failed until it was recognized that the conjugative activity of the Ti-plasmids is inducible by opines[43,44]. Inducibility of plasmid transfer by plasmid substrates may confer a selective advantage since it would restrict conjugation to those situations where plasmid functions are desirable. Petit *et al.*[45] suggested that substrate inducibility of conjugation might occur for other degradative plasmids or for resistance transfer factors, a prediction recently confirmed for a plasmid coding for tetracycline resistance in *Bacteroides fragilis*[46]. Petit *et al.*[45] also showed that both the catabolic ans transfer function of the octopine Ti-plasmid are controlled by the same regulatory gene.

Finally, the nopaline Ti-plasmid was shown to be responsible for the sensitivity of *Agrobacterium* strains to agrocin 84, the bacteriocin produced by *Agrobacterium radiobacter* 84[34], which was shown to utilize nopaline[47]. Later studies by Ellis and Kerr (personnal communication) have demonstrated that nopaline utilization by this strain is identical to that of the oncogenic strains. Strain 84 carries a conjugative plasmid that enables its host to catabolize nopaline (Ellis, personnal communication). Whether this plasmid is a deleted nopaline Ti-plasmid, that has lost the oncogenic properties, is not yet known.

CROWN GALL IN NATURE : AN ECOSYSTEM

The laboratory observations discussed above can be easily integrated into the relatively little known about crown gall in nature, if one looks at the tumor as an ecosystem. Ecological terms can be used in this case to describe molecular biological phenomena.

Opine production has no known function in plant tumor cells. On the other hand we have seen that, *in vitro*, opines can serve as growth substrates for oncogenic bacterial strains, and that they induce the conjugative activity of the Ti-plasmids. We believe that the opines play the same role in nature as *in vitro*, and that opine production by crown gall cells serves the unique function of promoting plasmid dissemination through bacterial multiplication and conjugation. If this were the case, one would predict the presence of opines in tumors induced by *Agrobacterium* harbouring the null type Ti-plasmid. We have looked for cryptic opines in such tumors, using a test that relies on the predicted differences between catabolic capabilities of a plasmid-less strain and the same strain harbouring the null type Ti-plasmid. If cryptic opines were present in such tumors, than, an extract of these tumors, containing the opines, should allow better growth of the transconjugant than of the plasmid-less strain. We have shown that this is actually the case, and we have purified from null type tumors an extract that contains a substance absent in healthy tissues, which is a specific growth substrate for strains carrying the null type Ti-plasmid. We also found that octopine strains can utilize, by a Ti-plasmid coded pathway, the cryptic opine(s) of our extract, whereas the nopaline strains cannot (Guyon *et al.* in preparation). A purified sample was compared with agropine[17], it had the same behaviour in gas chromatography and the same mass spectrum (Firmin and Fenwick, personnal communication ; our unpublished data). We conclude that null type tumors do contain, as predicted, an opine,which is identical with agropine.

The fact that the opine concept is valid even for the so-called "null type" plasmid supports our theory that the genetic information carried by the T-DNA serves to create an ecological niche for *Agrobacterium tumefaciens*. This T-DNA is transferred from the pathogen to the plant cell, where it directs 1) differenciation of plant tissues to form a tumor (which constitutes a physical delineation of the ecosystem), and 2) the production of opines. The environment created by the expression of the T-DNA in the tumor favors the multiplication of bacteria carrying the Ti-plasmid. The location of the tumors on the roots or at the crown of the plants encourages secondary inoculation by soil microorganisms, including saprophytic *Agrobacteria* which could acquire the Ti-plasmid through opine-induced conjugation. Assuming that host range is at least partly determined by the genetic background of the bacterium, this process would result in host range diversification, which would ensure survival of the Ti-plasmid in changing conditions (such as disintegration of the ecosystem and liberation of its pathogens into the soil).

Some non-pathogenic strains of *Agrobacterium* can utilize opines for their growth[47], which, as we have seen, is the case for strain 84. The properties of this strain fit into an ecological description of crown gall if we consider it as a super-parasite. It can utilize nopaline but not direct its production, being non-oncogenic. It nevertheless manages to get a supply of opines by invading the ecosystem and killing the resident pathogen with agrocin 84.

CONCLUSION

Our present knowledge of crown gall allows us to summarize the system in simple terms. However, we recognize the complexity of the molecular interactions among the organisms involved. In casting the Ti-plasmid as protagonist and using some of the langage of ecology we have doubtless underestimated some of this complexity. On the other hand, we have shown that an ecological description is possible. We describe the tumor as an ecosystem and treat the Ti-plasmid as an organism capable of modifying its environment by genetically manipulating other organisms, namely *Agrobacterium tumefaciens* and higher plants. The expression "genetic colonization" has been used by SCHELL and coworkers[6] to describe the establishment of this parasitic relationship. From an anthropomarphic point of view the Ti-plasmid appears to engage in *opine farming*.

Opines are a key feature of crown gall. They appear to serve as the essential energy currency for the plasmid-bacterium as well as the bacterium-plant interactions. If the evolution of the Ti-plasmid were primarily constrained by the quest for nutrients, we could speculate that the Ti-plasmid evolved from a relatively simple catabolic precursor. We wonder if the opine concept can be applied to other host-parasite or host-symbiont relationships.

REFERENCES

1. Braun, A.C. & Stonier, T. (1958) Protopl. Handb. Protoplasmaforschung, 10, 1-93

2. Zaenen, I., Van Larebeke, N., Teuchy, H., Van Montagu, M. & Schell, J. (1974) J. Mol. Biol. 86, 109-127.

3. Van Larebeke, N., Engler, G., Holsters, M., Van Den Elsacker, S., Zaenen, I., Schilperoort, R.A. & Schell, J. (1974) Nature 252, 169-170.

4. Watson, B., Currier, T.C., Gordon, M.P., Chilton, M.D. & Nester, E.W. (1975) J. Bacteriol. 123, 255-264.

5. Chilton, M.D., Drummond, M.H.,Merlo, D.J., Sciaky, D., Montoya, A., Gordon, M.P. & Nester, E.W. (1977) Cell, 11, 263-271.

6. Belgian Crown Gall Research Group (1979), Proc. IV Int. Conf. Plant Pathog. Bact. Ed. M. RIDE, Angers, 1, 115-126, and these Proceedings.

7. Drummond, M.H., Gordon, M.P., Nester, E.W. & Chilton, M.D. (1977) Nature 269, 535-536.

8. Tempé, J., Petit, A., Holsters, M., Van Montagu, M. & Schell, J. (1977) Proc. Natl. Acad. Sci. USA 74, 2848-2849.

9. Goldmann-Ménagé, A. (1971) Ann. Sci. Nat. Bot. 11, 223-310

10. Biemann, K., Lioret, C., Asselineau, J., Lederer, E. & Polonski, J. (1960) Bull. Soc. Chim. 17, 979-991.

11. Ménagé, A. & Morel, G. (1964) C.R. Acad. Sci. 259, 4795-4796.

12. Ménagé, A. & Morel, G. (1965) C.R. Soc. Biol. 159, 561-562.

13. Kemp, J.D. (1977) Biochem. Biophys. Res. Comm. 74, 862-868
14. Goldmann, A., Thomas, D.W. & Morel, G. (1969) C.R. Acad. Sci. 268, 825-854.
15. Firmin, J.L. & Fenwick, R.G. (1977) Phytochem. 16, 761-762.
16. Sutton, D., Kemp, J.D. & Hack, E. (1977) Plant Physiol. 59,108
17. Firmin, J.L. & Fenwick, R.G. (1978) Nature 276, 842-844.
18. Seitz, E.W. & Hochster, R.M. (1964) Can. J. Bot. 42, 999-1004.
19. Johnson, R., Guderian, R.H., Eden, F., Chilton, M.D., Gordon,M. P.,Nester, E.W. (1974) Proc. Natl. Acad. Sci. USA 71, 536-539.
20. Wendt-Gallitelli, M.F. & Dobrigkeit, I. Zeitsch. Naturforsch. 28 c, 768-771.(1973).
21. Bomhoff, G.H. (1974). Thesis dissert. Univ. Leyden.
22. Holderbach, E. & Beiderbeck, R. (1976) Phytochem. 15, 955-956.
23. Kemp, J.D. (1976) Biochem. Biophys. Res. Comm. 69, 816-822.
24. Montoya, A., Chilton, M.D., Gordon, M.P., Sciaky, D. & Nester, E.W. (1977) J. Bacteriol. 129, 101-107.
25. Lioret, C. (1966) Physiol. Vég. 4, 89-103.
26. Lippincott, J.A., Chi-Cheng Chang, V.R., Creaser-Pence, P.R., Birnberg, P.R., Rao, S.S., Margot, J.B., Whatley, M.H. & Lippincott, B.B. (1979). Proc. IV Int. Conf. Plant Path. Bact. Ed. INRA - Angers, I, 189-197.
27. Ménagé, A. & Morel, G. (1966) C.R. Soc. Biol. 160, 52-54.
28. Goldmann, A., Tempé, J. & Morel, G. (1968) C.R. Soc. Biol. 162, 630-631.
29. Petit, A., Delhaye, S., Tempé, J. & Morel, G. (1970) Physiol. vég. 8, 205-213.
30. Lejeune, B. & Jubier, M.F. (1967) C.R. Acad. Sci. 264, 1803-05
31. Jubier-Fabre, M.F. (1975) Thesis dissert. CNRS, Paris.
32. Petit, A. & Tempé, J. (1978) Molec. gen. Genet. 167, 147-155.
33. New, P.B. & Kerr, A. (1972) J. Appl. Bact. 35, 279-287.
34. Kerr, A. & Htay, K. (1974) Physiol. Pl. Pathol. 4, 37-44.
35. Engler, G.,Holsters, M., Van Montagu, M. & Schell, J. (1975) Molec. gen. Genet. 138, 345-349.
36. Kerr, A. (1969) Nature, 223, 1175-1176.
37. Kerr, A. (1971) Physiol. Pl. Pathol. 1, 241-246.
38. Bomhoff, G.H., Klapwijk, P.M., Kester, H.C., Schilperoort, R.A. Hernalsteens, J.P. & Schell, J. Molec. gen. Genet. 145, 171-181.
39. Ellis, J.G., Kerr, A., Petit, A. & Tempé, J. (1979) Molec. gen. Genet. in press.
40. Petit, A., Dessaux, Y. & Tempé, J. (1979) Proc. IV Int. Conf. Plant Path. Bact. Ed. INRA - Angers, 1, 143-152.
41. Klapwijk, P.M., Hooykaas, P.J.J., Kester, H.C.M., Schilperoort, R.A. & Rorsch, A. (1976) J. Gen. Microbiol. 96, 155-163.

42. Wu, L. & Unger, L. Embo Workshop Pl. Tumor Res. (1978)

43. Kerr, A., Manigault, P. & Tempé, J. (1977) Nature, 263, 560-561

44 Genetello, C., Van Larebeke, N. Holsters, M. De Picker, A. Van Montagu, M. & Schell, J. (1977) Nature, 265, 561-563.

45 Petit, A., Tempé, J., Kerr, A., Holsters, M., Van Montagu, M. & Schell, J. (1978) Nature 271, 570-572.

46 Privitera, G., Sebald, M. & Fayolle, F. (1979) Nature, 278, 657-659.

47 Kerr, A. & Roberts, W.P. (1976) Physiol. Pl. Pathol. 9, 205-211.

Plasmids of Medical, Environmental and Commercial Importance
K.N. Timmis and A. Pühler, editors

SEARCH FOR PLASMID-ASSOCIATED TRAITS AND FOR A CLONING VECTOR IN *PSEUDOMONAS PHASEOLICOLA*

N. J. PANOPOULOS, B. J. STASKAWICZ AND D. SANDLIN
Department of Plant Pathology, University of California, Berkeley, CA. 94720 (U.S.A.)

INTRODUCTION

Agrobacterium tumefaciens is a prime example of a plant pathogen in which the involvement of plasmids in the control of pathogenicity is well established[1,2]. The presence of naturally occurring plasmids has been reported in many other plant pathogenic bacteria[3-14], and there is good evidence for the involvement of a plasmid in the control of phytopathogenicity in *P. syringae*. Gonzales and Vidaver[5,7] reported that production of the toxic oligopeptide syringomycin (SR) and pathogenicity on corn (holcus spot disease) is controlled by a 35 Mdal plasmid in at least one strain, HS191. This strain frequently loses these traits in culture and several non-pathogenic, non-SR producing derivatives were obtained after acridine orange treatment. These differed from their parent strain by the lack of a 35 Mdal plasmid, designated pCG131. Attempts to reintroduce the pCG131 plasmid into a pCG131⁻ (cured) derivative of HS191 by transformation were unsuccessful.

We have investigated the possibility of plasmid-controlled pathogenicity traits in *P. phaseolicola*. The possible involvement of plasmids in the control of toxigenicity in this bacterium was tentatively suggested recently by Gantotti *et al.*[10,12] This report summarizes the results of our continuing investigation into detection and characterization of plasmids of *P. phaseolicola*, their possible involvement in the control of its pathogenicity and toxigenicity and the potential utility of plasmid RSF1010 as a cloning vector in this bacterium.

Symptomatology of the bean halo blight disease and assay for phaseolotoxin

Pseudomonas phaseolicola causes the halo blight disease of bean which is characterized by localized watersoaked lesions in leaves and a chlorotic halo in the surrounding tissues as well as in distant non-inoculated leaves (systemic chlorosis[15]. The chlorotic reaction is caused by a toxin (halo blight toxin). Non-toxigenic strains or mutants isolated in the laboratory fail to produce chlorosis although they are capable of causing typical water soaked lesions. The chlorotic symptoms can be induced both by cell-free culture extracts as well as by purified preparations of toxin[16-18]. The chemical identity of halo blight toxin has been in dispute[16, 17, 19], but recent studies indicate that the tripeptide (N^{δ}-phosphosulfamyl)ornithylalanylhomoarginine (trivial name phaseolotoxin) is the major if not the sole pathologically significant toxin of *P. phaseolicola*[20,21].

Phaseolotoxin inhibits the enzyme ornithine transcarbamylase[18, 20] and can be detected in picogram quantities by the growth inhibition test using *E. coli* or *S. typhimurium* as indicators[22, 23]. The use of mutants deficient in oligopeptide permease (Opp^-), which are phaseolotoxin resistant[23], and the reversal of inhibition by supplementation of the bioassay plates with citrulline[22] makes the microbiological detection of this toxin unambiguous. Whenever reference is made here regarding the toxigenicity of *P. phaseolicola* strains a citrulline-reversible Opp-dependent toxicity towards *E. coli* or *S. typhimurium* is implied. In our experience this toxicity corresponds well with the chlorosis-inducing ability of the strains following inoculation on bean leaves, cultivar 'Red Kidney'.

Occurrence of plasmids in *P. phaseolicola*

We have conducted a survey of 18 strains from culture collections. All strains examined contained from one to three plasmids ranging in size from 5.2 to 110 Mdal (Fig. 1). For convenience the plasmids will be referred to by their size range: class L includes relatively large plasmids ranging from 78 to 110 Mdal; class M1 plasmids had a molecular weight from 28-32 Mdal; class M2 plasmids ranged from 23-27 Mdal; and class S plasmids were *ca.* 5.5 Mdal. The degree of similarity between plasmids within each size class has not been established and, therefore, the above classification may be entirely arbitrary. With this reservation, we have classified our strains in five groups (I-V) based on their plasmid patterns (Table 1).

Search for correlation between plasmids and ability to form watersoaked lesions

All but one of the strains examined were pathogenic on *Phaseolus vulgaris* cultivar 'Red Kidney'. Therefore, it was not possible to establish a correlative relationship between plasmid content and pathogenicity on this host. The universal occurrence of L plasmids in all strains suggested the possibility that they may be required for pathogenicity. However, this was ruled out by the isolation of a mutant, designated Tox-231, of strain G50 (an L, M1, M2 strain) which lacked the L plasmid but caused typical watersoaked lesions on this host. The parent strain had been genetically marked with an auxotrophic mutation (His^-) prior to mutagenesis. This, coupled with the M1-M2 plasmid

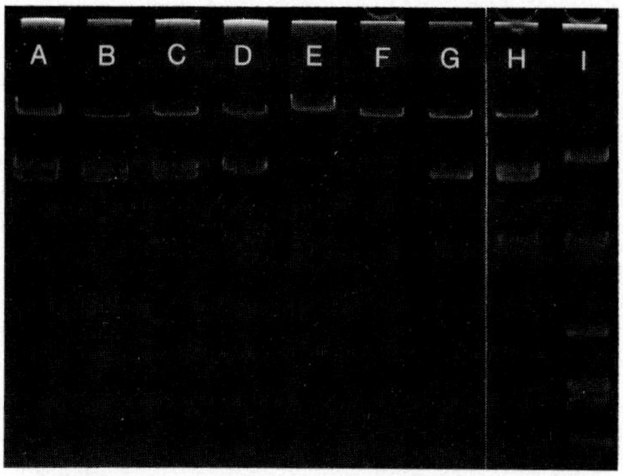

FIGURE 1

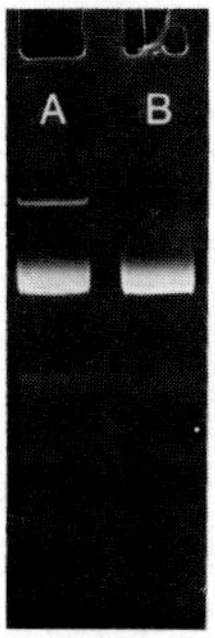

FIGURE 2

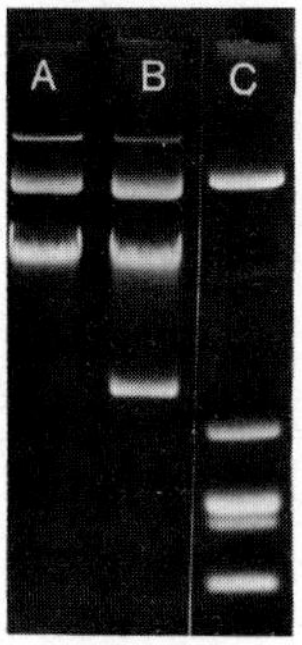

FIGURE 3

Fig. 1. Agarose gel (0.7%) electrophoresis of plasmid DNA preparations from *P. phaseolicola*. Plasmid DNA was extracted as described in ref. 14. Sample wells contained DNA from strains G50 (A), G50 Tox$^-$ (B, C; two different stocks), HB36 (D), HB31 (E), HB26 (F), HB18 (G), and HB13 (H). Well (I) contains a mixture of plasmid DNA from *E. coli* K12 (R6-5) and *E. coli* V517[24] as molecular weight standards. Molecular weights of the standards are (from top to bottom): 65 (faint band), 35.8, 4.8, 3.7, 3.4 and 2.6 Mdal. (The 2.0 and 1.8 Mdal plasmids of strain V516 are not visible in this picture.) The diffuse band present in all wells is chromosomal DNA. The large plasmid, present in all *P. phaseolicola* strains, is designated L; the medium-size plasmids, present in strains G50, G50 Tox$^-$ and HB13, are designated M1 and M2; the small plasmid of strain HB31 is designated S. Thus, G50, G50 Tox$^-$ and HB13 contain L, M1 and M2 plasmids; HB36 contains plasmids L and M1; HB31 contains plasmids L and S; HB18 has plasmids L and M2; and HB26 contains only plasmid L.

Fig. 2. Plasmid patterns of strain G50 (A) and its mutant Tox-231 (B). DNA extraction and electrophoresis were as described in Fig. 1 legend. The L plasmid of G50 is absent in Tox-231.

Fig. 3. Agarose gel electrophoresis of DNA extracted from strain G50 (A), G50 (RSF1010$^+$) (B) and plasmid standards (C) (see Fig. 1 legend). The smallest three plasmids of strain V517 are not seen in this picture; the faint bands in the middle of well C are OC or linear forms of the V517 plasmids.

pattern which was not found in any other *P. phaseolicola* strains, rules out the possibility that Tox-231 was a contaminant. Possible correlation between these plasmids patterns and race specificity, i.e. pathogenicity on other host varieties carrying specific resistance genes, is currently under investigation.

TABLE 1

CLASSIFICATION OF *P. PHASEOLICOLA* STRAINS ACCORDING TO PLASMID CONTENT[a]

Group	Plasmids Present				No. of Strains	No. of Strains that were	
	L	M1	M2	S		Tox$^+$	Tox$^-$
I	+	-	-	-	3	2	1
II	+	+	-	-	3	1	2
III	+	-	+	-	1	0	1
IV	+	+	+	-	7	4	3
V	+	-	-	+	4	4	0

[a]Plasmid DNA was prepared as described in ref. 14. Molecular weight of plasmids was determined by coelectrophoresis with a DNA standard, prepared similarly, containing RG-5 and the plasmids of *E. coli* V517[24]. Plasmids designated L had molecular weights from *ca.* 75 to 110 Mdal; M1 plasmids were from 28 to 32 Mdal; M2 plasmids from 23 to 27 Mdal; S plasmids were from 5.2 to 6.0 Mdal. Toxin production was determined as described in ref. 22 and in the text.

Search for correlation between plasmids and toxigenicity

The toxigenicity of *P. phaseolicola* strains was established by the *in vitro* assay described previously and in most cases was confirmed *in planta*[22]. Approximately 2/3 of the strains were toxigenic, although they differed in the amount of bioassayable toxin they produced. There was no obvious correlation between plasmid patterns found in our survey and toxigenicity (Table 1). Four strains, all toxigenic, contained only plasmids of the L and S class but not M1 or M2 class plasmids. Two other toxigenic strains contained only the L plasmid. Three strains, one of which was toxigenic, contained a large and one medium-sized plasmid (M1). One strain (Tox$^-$) had an L plasmid and a medium-sized M2 plasmid. Finally, three non-toxigenic and four toxigenic strains had plasmids of the L, M1 and M2 size class.

Gantotti *et al.*[10,12] recently reported that a UV-induced Tox$^-$ mutant of strain G50[24] designated G50 Tox$^-$, differed from its parent strain by the lack of the 27 Mdal plasmid (plasmid M1 in our terminology). They also reported that another toxigenic strain (HB36) contained one large and one medium-sized plasmid, the latter similar in size to the 28 Mdal plasmid of G50 whereas a non-toxigenic strain (HB20) contained a large and a medium-sized plasmid, the latter was comparatively smaller than that of the toxigenic strains G50 and HB36. These results suggested an apparent association between the 27 Mdal plasmid and toxigenicity. Our finding concerning the plasmid content of strains

G50, HB36 and HB20 are in agreement with those reported by Gantotti *et al.*[10,12] However, regarding the plasmid content of G50 Tox⁻ our data are at variance. Plasmid DNA preparations obtained by CsCl-EtBr density gradient or with the method of Gonzales and Vidaver[14] from strains G50 and G50 Tox⁻ contained both medium-sized plasmids, 23 and 28 Mdal, here referred to as M1 and M2. A large plasmid (80 Mdal) was also present in both strains when the DNA was prepared by the latter method although it was absent from our CsCl-EtBr preparations. Among several different stocks of strain G50 Tox⁻ stored in our collection all contained the above three plasmids. The apparent discrepancy between our findings concerning strain G50 Tox⁻ and those of Gantotti *et al.*[10,12] cannot be adequately explained.

In other experiments we attempted to isolate Tox⁻ mutants following treatment with plasmid curing agents such as acidine orange and ethidium bromide. Although we have made repeated attempts and have examined several thousand colonies in each of three different strains we have been unable to obtain such mutants. However, we have isolated many Tox⁻ mutants following mutagenesis with nitrosoguanidine and ethylmethanesulfonate and screening of the survivors for ability to inhibit *E. coli* on miminal medium[22]. The majority of these mutants had no apparent changes in plasmid content. However, one mutant designated Tox-231 and isolated from strain G50 which contains the L, M1 and M2 plasmids, had lost the L plasmid and another, designated Tox-246, had apparently suffered a large deletion in the same plasmid. The Tox-231 mutant, although less toxigenic than its parent strain, produced low amounts of phaseolotoxin (i.e. citrulline-reversible Opp-dependent inhibitory activity) suggesting that the structural genes for phaseolotoxin synthesis are probably not located on the L plasmid of strain G50.

<u>RSF1010 as a cloning vector in *P. phaseolicola*</u>

To aid in the genetic analysis of pathogenicity in *P. phaseolicola* we have investigated the replicative ability of plasmids with known molecular vehicle properties. Among the plasmids studied thus far is RSF1010, a 5.5 Mdal non-conjugative multicopy plasmid specifying resistance to streptomycin (Sm^r) and sulfanamides (Su^r), which has been used as a cloning vector for EcoR1-generated DNA fragments in non-phytopathogenic Pseudomonads.[26-28] *E. coli* C600 (RSF1010) (RK2) carrying RSF1010 and RK2 (Amp^r, Kan^r/Neo^r, Tet^r), the latter as a mobilizing plasmid, was mated with *P. phaseolicola* and transconjugants resistant to streptomycin were selected. Of the Sm^rSu^r transconjugants ca. 80% were Amp^s, Tet^s, Kan^s/Neo^s, i.e. presumably $RSF1010^+$ $RK2^-$. The presence of RSF1010 in these transconjugants was confirmed by agarose gel electrophoresis of DNA extracts

(Fig. 3). The Sm^r character remained stable through several subcultures in media devoid of streptomycin. Our data, therefore, show that RSF1010 can be preferentially mobilized by P-1 group plasmids into *P. phaseolicola* where it replicates autonomously in a stable manner.

There is little published information concerning the presence of cloning sites, other than the EcoR1 site, on the RSF1010 replicon. Unpublished data indicate that there are no Hind III or BamH1 sites on the RSF1010 molecule (Fred Heffron, pers. communication). We have constructed an RSF1010-*kan* derivative by inserting the *kan* fragment of plasmid pML2 [29] in the EcoR1 site of RSF1010. This fragment contains a Hind III site and should be useful for cloning Hind III-generated fragments by insertional inactivation. Furthermore, the RSF1010::Ap (Tn3) derivatives described previously by Heffron *et al.*[30] offer similar possibilities for cloning BamH1 fragments into the BamH1 site present on Tn3. These and other derivatives are currently being studied in this laboratory.

SUMMARY

Strains of *P. phaseolicola* contain one or more plasmids ranging from approximately 5.5 to 110 Mdal and were classified in five groups according to their characteristic plasmid patterns. A survey of 18 strains did not reveal an association between plasmids and toxigenicity or pathogenicity on Red Kidney bean. More detailed studies with the toxigenic strain G50 and several non-toxigenic mutants did not support an earlier suggestion that one of its plasmids (*ca.* 28 Mdal) might be involved in toxigenicity. The isolation of another mutant of this strain which had lost a 80 Mdal plasmid indicated that neither the structural genes for phaseolotoxin biosynthesis nor the gene(s) responsible for pathogenicity, at least on one host, are located on this plasmid. We have succeeded recently in isolating Tn7 insertion mutants in two strains of *P. phaseolicola* and are in the process of determining the physical location of the insertions. In one mutant the insertion appears to be on the large plasmid. This and other similar mutants should facilitate the search for other plasmid associated properties in *P. phaseolicola*

We also showed that plasmid RSF1010, which has been used as a vector in molecular cloning experiments, can replicate autonomously in *P. phaseolicola* and have constructed a RSF1010-*kan* derivative which should permit the identification of Hind III inserts by insertional inactivation. This and other derivatives currently under construction will make possible the use of cloning for the genetic analysis of host pathogen interaction in the bean halo blight disease.

ACKNOWLEDGEMENTS

Thanks are extended to Drs. Richard Meyer, Fred Heffron, Donald Helinski, Kenneth Timmis, Suresh Patil and to Basu Gantotti for helpful discussions on various aspects of these experiments and/or for supplying bacterial strains.

REFERENCES

1. Van Larebeke, N., Genetel-o, C., Schell, J., Schilperoort, R. A., Hermans, A. K., Hernalsteens, J. P. and Van Montagu, M. (1975) Nature (London) New Biol. 255, 742-743.

2. Watson, B., Currier, T., Gordon, M. P., Chilton, M.-D. and Nester, E. W. (1975) J. Bacteriol. 123, 255-264.

3. Curiale, M. S. and Mills, D. (1977) J. Bacteriol. 131,224-228.

4. Panopoulos, N. J., Guimaraes, W. V., Hua, S.-S., Sabersky-Lehman, C., Resnik, S., Lai, M. and Shaffer, S. (1978) In (D. Schlessinger, ed.), Microbiology 1978. Amer. Soc. Microbiol., Washington, D. C., pp.238-241.

5. Gonzalez, C. F. and Vidaver, A. K. (1977) Proc. Amer. Phytopathol. Soc. 4, 107 (Abstr.).

6. Gross, D. C. and Vidaver, A. K. (1977) Proc. Amer. Phytopathol. Soc. 4,138 (Abstr.).

7. Gonzalez, C. F. and Vidaver, A. K. (1978) Abstracts of papers, 4th Internat. Congr. on Plant Pathogenic Bacteria. Angers, Fr., p.3.

8. Lacy, G. H. (1977) Proc. Amer. Phytopathol. Soc. 4,198 (Abstr.).

9. Dahlbeck, D., Pring, D. R. and Stahl. R. E. (1977) Proc. Amer. Phytopathol. Soc. 4,176 (Abstr.).

10. Gantotti, B. V., Patil, S. S. and Mandel, M. M. (1978) Proc. 3rd Internat. Congr. of Plant Pathology. Munich, FRG. p. 68 (Abstr.).

11. Panopoulos, N. J. (1979) Proc. 4th Intrnat. Congr. on Plant Pathogenic Bacteria. Angers, Fr. in press.

12. Gantotti, B. V., Patil, S. S. and Mandel, M. (1978) Phytopathol. News 12, 154 (Abstr.).

13. Coplin, D. L. (1978) Phytopathology 68, 1637-1643.

14. Gonzalez, C. F. and Vidaver, A. K. (1979) J. Gen Microbiol. 110,161-170.

15. Zaumeyer, W. J. and Thomas, H. R. (1957) USDA Technical Bulletin No. 868.

16. Patil, S. S. (1974) Annu. Rev. Phytopathol. 12, 56-75.

17. Mitchell, R.E.(1976) Phytochemistry 15, 1941-1947.

18. Patil, S. S., Tam, S. Q. and Sakai, W. S. (1972) Plant Physiol. 49, 803-807.

19. Patil, S. S., Youngblood, P., Christiansen, P. and Moore, R. E. (1976) Biochem. Biophys. Res. Commun. 69, 1019-1027.

20. Mitchell, R. E. (1978) Physiol. Plant Pathol. 13, 37-49.

21. Mitchell, R. E. (1979) Physiol. Plant Pathol. 14, in press.

22. Staskawicz, B. J. and Panopoulos, N. J. (1979) Phytopathology 69, in press.

23. Panopoulos, N. J. and Staskawicz, B. J. (1979) Proc. 4th Internat. Congr. on Plant Pathogenic Bacteria. Angers, Fr. in press.

24. Macrina F. L., Kopecko, D. J., Jones, K. R., Ayers, D. J. and McCowen, S. M. (1978) Plasmid 1,417-420.

25. Patil S. S., Hayward, A. C. and Emmons, R. (1974) Phytopathology 64,590-595.

26. Nagahari, K. and Sakaghuchi, K. (1978) J. Bacteriol. 133,1527-1529.

27. Nagahari, K., Tanaka, T., Nishimura, F., Kuroda, M. and Sakaghuchi, K. (1977) Gene 1,141-152.

28. Nagahari, K. (1978) J. Bacteriol. 136,312-317.

29. Hershfield, V., Boyer, H. W., Yanofsky, C., Lovett, M. A. and Helinski, D. R. (1974) Proc. Nat. Acad. Sci. (USA) 71,3455-3459.

30. Rubens, C. F., Heffron, F. and Falkow, S. (1975) J. Bacteriol.128, 425-434.

VI – BROAD HOST RANGE PLASMIDS

Plasmids of Medical, Environmental and Commercial Importance
K.N. Timmis and A. Pühler, editors

ESSENTIAL REGIONS FOR THE REPLICATION AND CONJUGAL TRANSFER OF THE BROAD HOST RANGE PLASMID RK2

C.M. THOMAS, D. STALKER, D. GUINEY and D.R. HELINSKI
Department of Biology, B-022, University of California at San Diego, La Jolla, California 92093 (U.S.A.)

INTRODUCTION

The plasmid RK2 belongs to the incompatibility group P-1. It is thus capable of conjugal transfer between a wide range of gram-negative bacterial species[1,2] in which it is stably maintained. The properties of the plasmid replicon of RK2 that provide for such generalized maintenance are of fundamental interest and attempts have been made to define by deletion analysis the regions of the plasmid genome essential for autonomous replication in *Escherichia coli*[3,4,5]. An important finding was that the RK2 genome could be split into two functionally distinct regions, neither of which is capable of autonomous replication[6]. One of these regions contains the origin of replication, as defined by electron microscopy[6a]. An origin-containing segment of RK2 can replicate as a plasmid in *E. coli* if the rest of the RK2 replicon, cloned into a different plasmid vehicle, is present in the same cell[6]. Initial problems in obtaining mini replicons from RK2, due to the few cleavage sites for commonly used restriction enzymes, have been overcome by the use of the restriction enzyme *Hae* II[7]. In this paper we describe the identification of three non-contiguous regions of RK2 that are required for replication of this plasmid. In addition the region that contains a functional replication origin of RK2 has been located within a 380 bp *Hpa* II fragment. Sequencing of this fragment has allowed us to propose a tentative model for control of initiation at the RK2 replication origin.

Finally, we have mapped the site of the nick induced by the RK2 relaxation complex and have demonstrated that the region around this site is required for mobilization of RK2 derivatives by the RK2 conjugal transfer system. This region lies on or near the end of the larger group of tra genes mapped previously in RK2[3,7] and RP4[10,11], and we propose that it contains the origin of transfer, *ori* T, utilized during conjugal transfer.

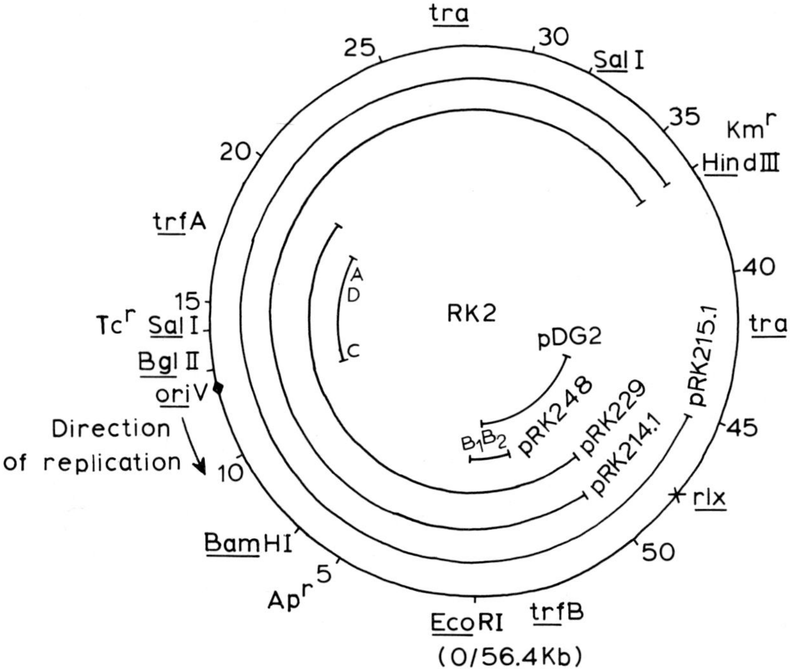

Fig. 1. Physical map of the regions of RK2 included in various deletion derivatives. Ap^r, Tc^r and Km^r refer to genes conferring resistance to ampicillin, tetracycline and kanamycin respectively. '*Ori*V' refers to the origin of vegetative replication as mapped by electron microscopy[6a] and by functional assay in the *trans* system[13]. *trf*A and *trf*B refer to *trans*-acting replication functions[13]. *rlx* refers to the relaxation complex site. *Tra* refers to genes required for transmissibility.

MAPPING OF REGIONS ESSENTIAL FOR REPLICATION

Small derivatives of RK2 (56.4 kb) that are capable of autonomous replication in *E. coli* were obtained by partial digestion of RK2 plasmid DNA with the restriction enzyme *Hae* II[7]. The first derivative obtained, pRK229 (24 kb), was the result of a single deletion, while a smaller plasmid, pRK248 (9.6 kb), was derived from pRK229 by partial digestion and ligation and, therefore, may represent two or more segments of pRK229[7]. The cleavage sites for restriction enzymes *Hae* II and *Hin*d II in pRK248 have been mapped by the analysis of fragments produced by digestion with one or both of these enzymes in addition to enzymes whose cleavage sites have previously been mapped and also by deletion analysis of hybrids consisting of pRK248 joined to pMK20 (a small, Km^r derivative of ColE1)[12,13]. pRK248 has been mapped with respect to pRK229 by comparison of restriction enzyme cleavage maps and digestion patterns of the

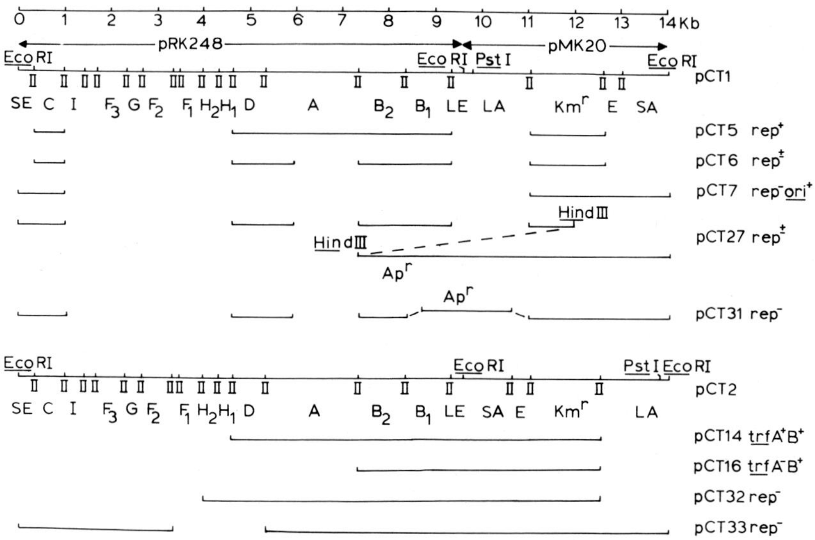

Fig. 2. Map of the cleavage sites for the restriction enzymes *Eco*RI, *Pst* I and *Hae* II in plasmids pCT1 and pCT2 (pRK248/pMK20 hybrids) and their derivatives. Cleavage sites for restriction enzyme *Hae* II are designated II. 'rep' refers to the RK2 replicon: +, functional; -, non-functional; ±, poorly functional. *trf*A and *trf*B refer to *trans*-acting replication functions.

two plasmids. The end points of the deletion that gave rise to pRK229 have been mapped with respect to RK2 by determining which of the *Hae* II fragments from the small *Sal* I-*Eco*RI fragment of pRK229 are derived from the region of the *Eco*RI site of RK2. The relationship between the various plasmid derivatives and the parent RK2 plasmid is shown in Fig. 1.

To determine the essential regions for replication of pRK248, hybrids between this plasmid and pMK20 were constructed by cleavage and ligation at the unique *Eco*RI sites present in the two plasmids. The restriction maps for both orientations of the hybrid (pCT1 and pCT2) and some of their derivatives are shown in Fig. 2. Partial digestion with restriction enzyme *Hae* II was used to delete the Tc^r region of pRK248. The hybrid was still able to replicate in a *pol*A1 mutant indicating that the RK2 replicon was still functional, since the ColE1 replicon requires high levels of *pol*I. Similar partial digestions were used to remove all of the other *Hae* II fragments of the hybrid except five fragments from pRK248, fragments A, B_1, B_2, C and D, and the Km^r fragment from pMK20[13]. The resulting plasmid, pCT5 (Fig. 2), containing 5.4 kb of RK2 DNA, is the smallest plasmid derived from RK2 that is stably maintained and has a similar copy number to RK2 (4-10 copies/chromosome equivalent). A smaller derivative, pCT6, containing 4.1 kb of RK2 (Fig. 2), was obtained by *Hha* I partial digestion; however, this derivative is present at low copy number and is not stably maintained[13].

As can be seen from Fig. 1 and 2, the essential five *Hae* II fragments are derived from three distinct regions of the RK2 genome. Fragment D is adjacent to A, fragment B_1 is next to B_2 and fragment C is separate from either pair. In the case of the pRK248-pMK20 hybrids and their derivatives, deletions that remove fragment C, D or B_1 of the essential *Hae* II fragments (pCT32, pCT33 and pCT31, respectively; see Fig. 2) will not replicate in a *pol*A1 strain indicating that the RK2 replicon is no longer functional. In addition, it has been found that the low copy number defect in pCT6 (or its Ap^r equivalent, pCT27; see Fig. 2) can be transcomplemented by a hybrid plasmid, designated pCT16 (Fig. 2), consisting of the B_1-B_2 region cloned into a ColE1 derivative. It seem reasonable therefore to conclude that these three well separated regions of the RK2 genome are essential for autonomous replication of RK2. Recently deletion mapping of the replication regions of RP4 (similar or identical to RK2) has yielded results that are consistent with these findings[5].

Replication origin and *trans*-acting regions

It has been demonstrated that the RK2 replicon can be cleaved with a restriction endonuclease into two fragments[6], one that contains the replication origin and the other providing essential *trans*-acting replication functions. From the locations of the *Hae* II fragments required for replication, and the site identified by electron microscopy as the replication origin[6a], it was proposed that the *Hae* II C fragment of pRK248 contains the replication origin[13]. Plasmid pCT7 (Fig. 2), a derivative of pCT1, containing only the *Hae* II C fragment of pRK248 in addition to a Km^r fragment and the ColE1 replicon, was found to be capable of replication in a *pol*A1 strain so long as a second plasmid, pRK2045[6], carrying the *trans*-acting replication genes of RK2 was present[13]. Deletion of the *Hae* II C fragment from pCT7 removed this capability and so it was concluded that the *Hae* II C fragment contains a functional RK2 replication origin. A derivative of pCT2, pCT14, carries *Hae* II fragments D, A, B_1 and

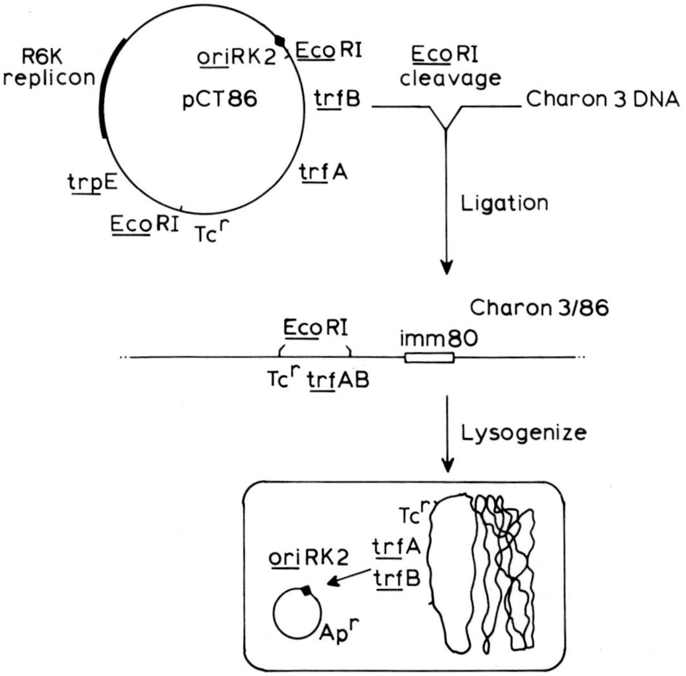

Fig. 3. Construction of a Charon 3 derivative containing the *trf*A and *trf*B regions of RK2. pCT86 is a hybrid plasmid consisting of pRK353, an R6K derivative[16], and pRK248. Tc^r lysogens have the helper regions integrated into the host chromosome. This strain is capable of supporting the replication of a derivative of RK2 consisting of the *ori*RK2 segment joined to a selective marker. (The drawing is not to scale.)

B_2 from pRK248 but not fragment C (Fig. 2). pCT14 was found to provide all of the *trans*-acting functions required for replication of the '*ori*' plasmid, pRK2067, derived from RK2[6], that is incapable of autonomous replication in the absence of a 'helper' plasmid[13]. The two regions (D/A and B_2/B_1) are designated *trf*A and *trf*B for _trans_-acting _r_eplication _f_unction.

To facilitate analysis of the region of RK2 that contains a functional replication origin *E. coli* strains were constructed that carried the *trans*-acting genes, integrated into the host chromosome. This was achieved by first cloning the *trans*-acting genes into the λ cloning vehicle Charon 3[14], and then constructing lysogens with this hybrid bacteriophage (Fig. 3)[15]. Despite the fact that these genes are present in these lysogens at a copy number of one per chromosome (RK2 normally is present at about 4-10 copies per chromosome equivalent) a plasmid containing only the replication origin of RK2 joined to a selective marker replicates in this strain with a copy number characteristic of RK2[15].

The *Hae* II C (0.75 kb) fragment, containing the replication origin of RK2, has been mapped for the location of sites of various restriction enzymes. *Hpa* II cleaves this fragment at four sites. The largest *Hpa* II fragment has been cloned into a derivative of pBR322 by partial digestion of the pBR322 derivative with *Taq* I and complete digestion of pCT7 with *Hpa* II. Since this hybrid plasmid was found to replicate in a *pol*A1 'helper' RK2 strain and also in a *pol*A^+ helper strain even when the ColE1 replicon is deleted[15], it is concluded that this *Hpa* II fragment contains a functional RK2 replication origin.

The DNA sequence of this 380 bp fragment has been determined[17] using the chemical method of Maxam and Gilbert. The general features of the sequence are outlined in Fig. 4. Several features are worth emphasizing. First, there is a set of five tandem repeats around the *Mbo* I site; the repeat unit is 22-24 bp of which a group of 15 are highly conserved. The group of repeats is preceded by a sequence, GATGATG, which is similar in sequence to a Pribnow box[18]. It is conceivable that transcription starting from this putative promoter, in the direction of replication, is controlled by binding of a regulatory protein to the direct repeats. Such a transcription event may result in synthesis of a primer for initiation of replication or may be required only for activation of the initiation process. After the direct repeats there is a region, consisting of five AT base pairs surrounded by GC rich segments, analogous in sequence to the oop RNA terminator of bacteriophage lambda[19]. This is followed by a 60 bp region with a high (67%) A+T content and then a 65 bp region that is rich in G+C (80%) and can be folded into several possible hairpin structures due to inverted repeat sequences. Although the RNA primer/DNA junction at the origin of RK replication has not been identified it is conceivable that initiation of replication occurs at the AT rich region. The adjacent GC rich region may be required for either promoting separation of the DNA strands in the AT rich region (see Sobell 1978[20]) or for the binding of an essential protein(s) for replication, or both. As mentioned above, an intriguing possibility is that initiation of DNA replication in this origin region is regulated by binding of a regulatory protein to the set of five tandem repeats which thus modulates transcription required for either priming of DNA synthesis or transcriptional activation. Analogous proposals for the role of tandem repeats in regulation

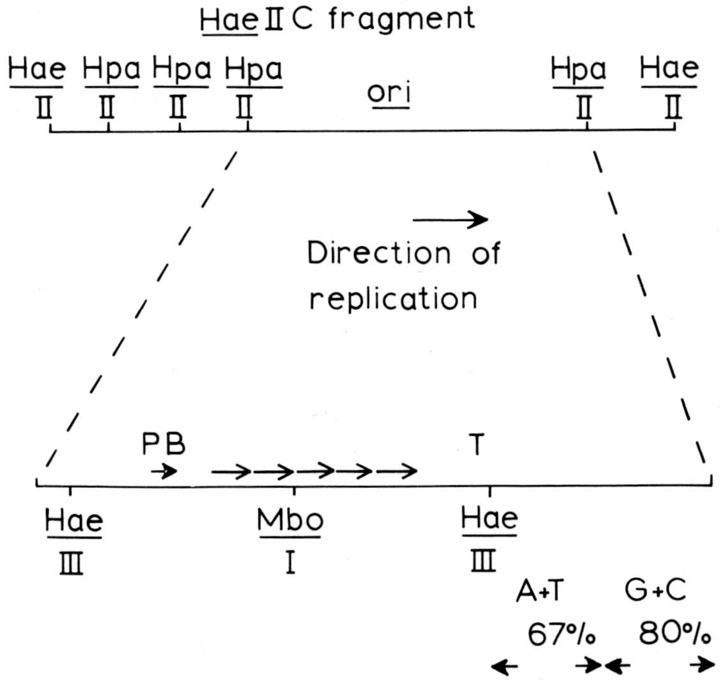

Fig. 4. Summary of nucleotide sequences in the replication origin region of *Hae* II fragment C of RK248. PB refers to a putative Pribnow box sequence[18]; T designates the position of a nucleotide sequence analogous to the oop RNA terminator of lambda[19]; → indicates the location of the tandem repeats of a nucleotide sequence.

of transcription have been made for the *cro* and C_I genes of bacteriophage λ[20a,20b]. As yet, however, there is no evidence for a negatively acting protein in RK2--both of the *trans*-acting regions appear to be required positively for replication of the origin-containing plasmid. It is possible of course that one of the *trans*-acting regions could be coding for both positive and negative regulatory proteins or one of the *trans*-acting proteins could itself be acting both positively and negatively. In any case, the independence of the origin plasmid copy number from gene dosage of *trans*-acting genes, suggests that any regulatory protein specified by these genes would be autoregulated. It should be emphasized that to date the evidence does not distinguish between a role for the *trans*-acting proteins in the initiation of replication at the RK2 origin of replication or as part of an elongation complex formed at the *ori*RK2. One of our major efforts at the present time is the further analysis of the *trans*-acting genes to determine whether they specify a negatively acting component and to determine the nature of the role of the positively acting genes.

RELAXATION COMPLEX AND PLASMID MOBILIZATION

RK2 can be isolated as a DNA-protein relaxation complex after gentle lysis of cells by lysozyme-EDTA-TritonX100 treatment[21]. Approximately 30-50% of the supercoiled DNA is converted to the open circular form on treatment of the complex with SDS, pronase or ethidium bromide. These properties of the RK2 complex are similar to other plasmid DNA-protein complexes studied in this laboratory[22,23,24]. A single nick occurs per molecule as a result of the relaxation event as shown by alkaline sucrose gradient analysis of the open circular DNA. The site of the nick was mapped by digesting of the open circular product of the nicking event in separate incubations with *Hin*d III, *Bam* HI and *Bgl* II and analyzing the products on alkaline sucrose gradients. The site of nicking (designated *rlx*) is located at 48 kb on the RK2 map, a considerable distance from the origin of replication (Fig. 1)[21].

Using deletion derivatives of RK2, a *cis*-acting function required for plasmid transfer, the presumptive origin of transfer, has been mapped in the same region as *rlx*. The structure of these RK2 derivatives, designated pRK215.1, pRK214.1 and pRK229, is shown in Fig. 1. All of them are non-self-transmissible since they lack portions of a relatively large region shown to be required for self-transfer which is located clockwise from the *Hin*d III site (Fig. 1)[3]. The derivatives were tested for mobilization from a *rec*A strain containing a hybrid plasmid, pRK2013, that consists of the RK2 transfer genes cloned into ColE1[6]. The results presented in Table 1 show that pRK215.1 is mobilized efficiently while pRK214.1 and pRK229 are not. pRK215.1 contains *rlx* and can be isolated as a relaxation complex, while pRK214.1 and pRK229 are deleted for *rlx* and cannot be purified in the form of a detectable relaxation complex. The region containing *rlx* has been cloned from pRK215.1 into the non-mobilizable plasmid pBR322. This hybrid plasmid, pDG2, is now efficiently mobilized by an RK2 derivative and can be purified in the form of a relaxation complex (Table 1).

These results define a *cis*-acting function that is required for RK2 transfer and that maps in the 5 kb region contained in pRK215.1 but deleted from pRK214.1. This *cis*-acting function most likely represents the origin of transfer, *ori* T, of RK2, and maps in the same region as the relaxation complex site, *rlx*. These results with RK2, analogous to the previous work

TABLE 1

PRESENCE OF RELAXATION COMPLEX AND MOBILIZATION OF RK2 DERIVATIVES

Plasmid	Mobilizability[a]	Relaxability (%)
pRK215.1	1.7×10^{-3}	30
pRK214.1	2.0×10^{-7}	< 5
pRK229	4.3×10^{-6}	< 5
pDG2	4.1×10^{-1}	25
pBR322	$< 10^{-5}$	< 5

[a]Transfer frequency of mobilized plasmid/transfer frequency of the self-transmissable helper plasmid. pRK2013[6] was used to mobilize pRK215.1, pRK214.1 and pRK229, while pRK231[7] was used to mobilize pDG2 and pBR322. The relaxability of each plasmid was determined in the absence of the helper plasmids.

with ColE1[25,26] support a role for the relaxation complex in plasmid DNA transfer.

CONSTRUCTION OF BROAD HOST RANGE CLONING VEHICLES

The replication and conjugal transfer systems of RK2 and similar P group plasmids provide the potential for transferring genes between bacterial species and also for the cloning and study of foreign bacterial DNA in *E. coli* with subsequent transfer back to the original species. Important for biological containment in such cloning experiments is a broad host range vehicle that is non-self-transmissible. In addition, the ability to mobilize this vehicle is very important for studies with bacteria for which no convenient transformation procedure exists. While non-self-transmissible derivatives of RK2 (and both RP4 and RP1) have been isolated[8,9,3], the mapping of both the essential replication genes and the regions required for mobilization provide the basis for a more systematic construction of broad host range cloning vehicles.

Low molecular weight of the vehicle is an important factor in facilitating analysis of the physical structure of cloned fragments. pRK248 and its derivative pRK2501[27], obtained by inserting a *Hae* II Km^r fragment into pRK248, are useful in this respect. pRK2501 is cleaved at a single site by five restriction endonucleases (*Sal* I, *Hind* III, *Xho* I, *Bgl* II and *Eco*RI); the first three of which show 'insertional inactivation' of one or the other of the selective markers when a restriction fragment is cloned into them. The broad host range properties of these two plasmids have not yet been properly tested. Transfer of pRK248 and pRK2501 relies on transformation since they lack *rlx*. Construction of similar plasmids that also contain the relaxation complex site region and are mobilizable by the RK2 conjugal transfer system is in progress in our laboratory[28]. Such vehicles which are non-self-transmissible would require for mobilization a helper system to provide the necessary *tra* genes. This can be provided in a *rec*A$^-$ host in the form of a narrow host range hybrid plasmid carrying the RK2 *tra* genes. Such a binary vehicle system potentially could provide a useful plasmid cloning vehicle that affords a high level of biological containment[29].

CONCLUSIONS

Recent progress in the study of the vegetative replication and conjugal transfer of the broad host range plasmid RK2 has been described. In the case of both phenomena interest was centered on the region in which it is thought each process is initiated, the *ori*V and putative *ori*T. Each region can be isolated by cloning of restriction fragments and their function can be assayed by providing other required genes in *trans*. Having defined these regions our major goal at present is to attempt to understand the nature of the interactions of these *trans*-acting genes with *ori*V and *ori*T.

ACKNOWLEDGEMENTS

This work was supported by grants from the National Institute of Allergy and Infectious Disease (A1-07194) and the National Science Foundation (PCM77-06533). CMT was supported by a British Medical Research Council Travelling Fellowship, D.S. by a U.S. Public Health Service postdoctoral fellowship (1-F32-GM06542) and D.G. by a U.S.P.H.S. grant (A1-07036-03).

REFERENCES

1. Datta, N. and Hedges, B. W. (1972) J. Gen. Microbiol., 70, 453-460.
2. Olsen, R. H. and Shipley, P. (1973) J. Bacteriol., 113, 772-780.
3. Figurski, D., Meyer, R., Miller, D. S. and Helinski, D. R. (1976) Gene, 1, 107-119.
4. Meyer, R., Figurski, D. and Helinski, D. R. (1977) Molec. Gen. Genet., 152, 129-135.
5. Sakanyan, V. A., Yakubov, L. Z., Alikhanian, S. I. and Stepanov, A. J. (1978) Molec. Gen. Genet., 165, 331-341.
6. Figurski, D. and Helinski, D. R. (1979) Proc. Natl. Acad. Sci., USA, in press.

6a. Meyer, R. and Helinski, D. R. (1977) Biochim. Biophys. Acta, 478, 109-113.

7. Meyer, R., Figurski, D. and Helinski, D. R., submitted for publication.

8. Shipley, P. L. and Olsen, R. H. (1975) J. Bacteriol., 123, 20-27.

9. Hedges, R. W., Cresswell, J. M. and Jacob, A. E. (1976) FEBS Letters, 61, 186-188.

10. Barth, P. T. and Grinter, N. J. (1977) J. Mol. Biol., 113, 455-474.

11. Barth, P. T., Grinter, N. J. and Bradley, D. E. (1978) J. Bacteriol., 133, 43-52.

12. Kahn, M. and Helinski, D. R. (1978) Proc. Natl. Acad. Sci., USA, 75, 2200-2204.

13. Thomas, C. M. and Helinski, D. R., submitted for publication.

14. Blattner, F. R., Williams, B. G., Blechl, A. E., Dennistron-Thompson, K., Faber, H. E., Furlong, L.-A., Grunwald, D. J., Kiefer, D. O., Moore, D. D., Schuman, J. W., Sheldon, E. L. and Smithies, O.(1977) Science, 196, 161-169.

15. Thomas, C. M., unpublished.

16. Kolter, R. and Helinski, D. R. (1978) Plasmid, 1, 571-580.

17. Stalker, D., unpublished.

18. Pribnow, D. (1975) J. Mol. Biol., 99, 419-443.

19. Schwarz, E., Scherer, G., Hobom, G. and Kössel, H. (1978) Nature, 272, 410-414.

20. Sobell, H. M. (1978) Cold Spring Harbor Symp. Quant. Biol., 43 (in press).

20a. Ptashne, M., Backman, K., Humayun, M. Z., Jeffrey, A., Manier, R., Meyer, B. and Sauer, R. T. (1976) Science, 194, 156-161.

20b. Johnson, A., Meyer, B. J. and Ptashne, M. (1978) Proc. Natl. Acad. Sci., USA, 75, 1783-1787.

21. Guiney, D. and Helinski, D. R., submitted for publication.

22. Clewell, D. and Helinski, D. R. (1969) Proc. Natl. Acad. Sci., USA, 62, 1159-1166.

23. Kline, B. C. and Helinski, D. R. (1971) Biochem., 10, 4975-4980.

24. Kupersztoch-Portnoy, Y., Miklos, G. and Helinski, D. R. (1974) J. Bacteriol., 120, 545-548.

25. Inselburg, J. (1977) J. Bacteriol., 132, 332-340.

26. Warren, G., Twigg, A. and Sherratt, D. (1978) Nature, 274, 259-261.

27. Kahn, M., Kolter, R., Thomas, C., Figurski, D., Meyer, R., Remaut, E. and Helinski, D. R. (1979) Methods in Enzymology, in press.

28. Ditta, G. and Stanfield, S., unpublished.

29. Meyer, R. J., Figurski, D. and Helinski, D. R. (1977) in DNA Insertion Elements, Plasmids and Episomes. Bukhari, A. I., Shapiro, J. A. and Adhya, S. L. (eds.) Cold Spring Harbor, pp. 559-566.

Plasmids of Medical, Environmental and Commercial Importance
K.N. Timmis and A. Pühler, editors

NATURALLY OCCURING INSERTION MUTANTS OF BROAD HOST RANGE PLASMIDS RP4 AND R68

H.J.Burkardt, U. Priefer, A. Pühler, G.Rieß and P. Spitzbarth
Lehrstuhl für Mikrobiologie, Universität Erlangen, Egerlandstr.7
D-8520 Erlangen, FRG

INTRODUCTION

In recent years broad host range plasmids of the P1 incompatibility group have been widely used in genetic and molecular biological studies. Because of their self transmissibility, chromosome mobilizing activity and abilities to be propagated in a wide range of host bacteria they are extremely useful for analysing recombination in asexual bacteria and chromosome mapping.[1-7] We have investigated the relationships of some IncP1 plasmids which are listed in table 1.

TABLE 1

Plasmid	Source
RP1	Lowbury et al. (1969)[8]
RP4	Datta et al. (1969)[9]
RP4-2	Ap-s[a] RP4 derivative which emerged spontaneously in _Rhizobium lupini_
RP4-3	Km-s[b] RP4 derivative which emerged spontaneously in _R.lupini_
RP4-5	Ap-s and Km-s RP4 derivative which emerged spontaneously in _R.lupini_
RP8	Black and Girdwood (1969)[10]
RK2	isolation from the same hospital as RP1; Ingram et al. (1973)[11]
R68	isolation from the same hospital as RP1; Stanisich and Holloway (1971)[1]
R68.45	an R68 derivative that shows chromosome mobilizing ability (cam); Haas and Holloway (1976)[4]
pMO47	cma plasmid, kindly provided by Dr. B. Holloway
pMO60	cma plasmid, isolated from a natural source in Japan
pMO61	cma plasmid, kindly provided by Dr. B. Holloway
pMO62	" "
pMO90	" "

pMO91	cma plasmid, kindly provided by Dr. B. Holloway
pMO92	" "
pMO93	" "
pMO94	" "

[a]Ap-s: ampicillin sensitive
[b]Km-s: kanamycin sensitive

RESULTS AND DISCUSSION

We have used electron microscopical and restriction enzyme cleavage analysis techniques to resolve the molecular structure of DNA molecules and have found two relationship classes:

(i) identity of the plasmids RP1, RP4, R68, RK2 and
(ii) insertion mutants of RP4 and R68, respectively, for RP4-2, RP4-3, RP4-5, RP8, R68.45 and the pMO plasmids.

Demonstration of the identity of those plasmids in class (1) is reported elsewhere[12]. Those plasmids in class (2) can be subdivided into three groups, namely those carrying

1) a large insertion (RP8),
2) small insertions, which are located near the kanamycin gene and are characterized by having one SmaI and two PstI restriction sites (e.g. R68.45 and the pMO plasmids) and
3) small insertions, which are located near or in the kanamycin gene (RP4-2) or within the Tn1 transposon (RP4-3) or in both (RP4-5). These insertions are characterized by one BamHI, one HindIII and one PstI site. In addition they can cause deletions in neighbouring genes.

RP8 insertion

Thus far, only one large insertion of the type found in RP8 has been identified. Length measurements of the RP8 plasmid have revealed that the inserted DNA measures 12 µm. It can be demonstrated in a heteroduplex experiment as large single-stranded loop (Fig. 1). In contrast to the original RP4 molecule which has single EcoRI and HindIII sites, the inserted DNA contains a lot of sites for these enzymes: 7 for EcoRI and 3 for HindIII. This could be the reason for the restricted host range of RP8 in comparison to that of RP4[13]. Nothing is known about the origin of the additional RP8 DNA but it possibly was derived from Pseudomonas aeruginosa, since RP8 was isolated from this bacterium and has the same GC ratio as RP4, namely 60%[14].

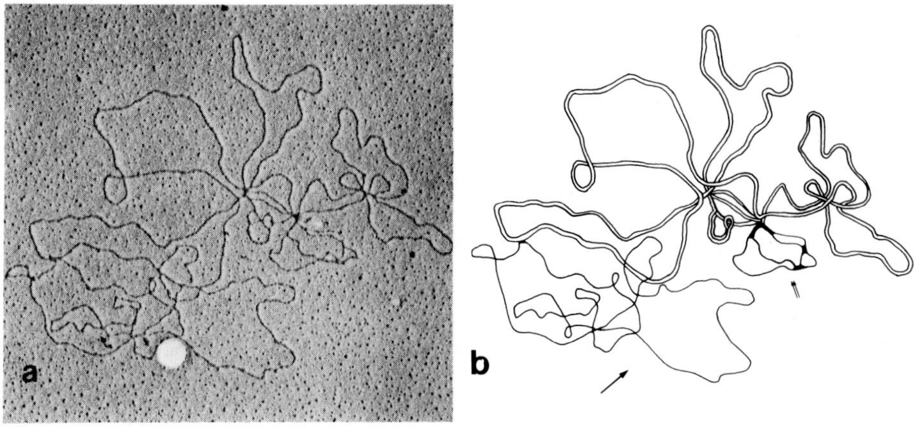

Fig. 1. RP4/RP8 heteroduplex molecule. Arrow points to single-stranded RP8 DNA, double arrow points to Tn1 transposon.
a) Micrograph b) Diagramatic representation

R68.45-like insertions (ISP1)

Type 2) insertions were first detected in R68.45[15]. This plasmid exhibits an enhanced chromosome mobilizing ability in comparison to its parental plasmid R68. Molecular studies have revealed that a small DNA insertion is involved in this change of the plasmid character. Length measurements showed that the R68.45 DNA is 0.6 µm longer than the parental R68 DNA. Heteroduplex experiments with RP8 and R68.45 (Fig. 2) have localized the insertion close to the kanamycin gene, 61.4% of the molecule length from the *Eco*RI site (in Tn1 direction). Restriction enzyme analysis revealed two new *Pst*I and one new *Sma*I site for R68.45 in comparison to R68 (Figs. 3,4). When the other cma plasmids pMO47, 60, 61, 62, 90, 91, 92, 93, 94 were examined by restriction enzyme analysis, they all showed the same characteristics as R68.45, no matter by which chromosomal marker transfer they had been detected (Figs. 5-7). We therefore assume that these plasmids have acquired an IS element of *P.aeruginosa*, their original host. We propose the designation ISP1 for this element (ISP1: insertion sequence of *Pseudomonas* 1).

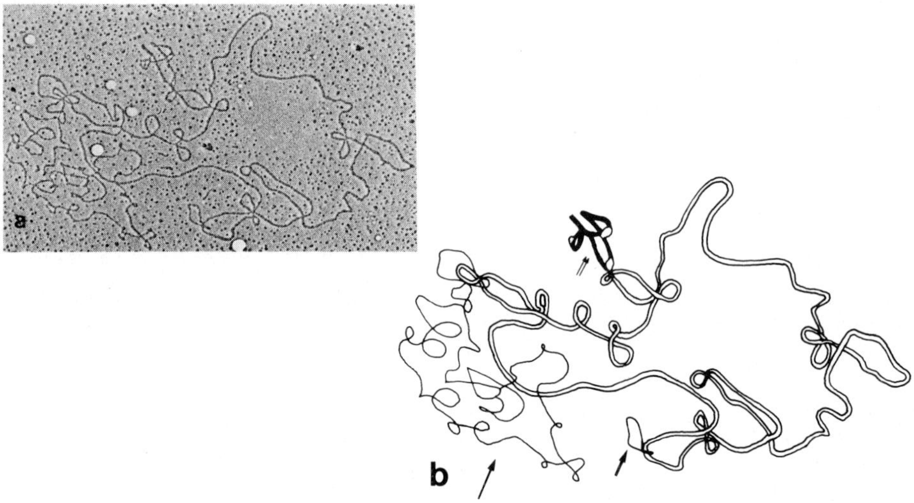

Fig. 2. RP8/R68.45 heteroduplex molecule. Thin arrow points to single-stranded RP8 DNA, thick arrow points to single-stranded R68.45 DNA, double arrow points to Tn1 transposon.
a) Micrograph b) Diagramatic representation

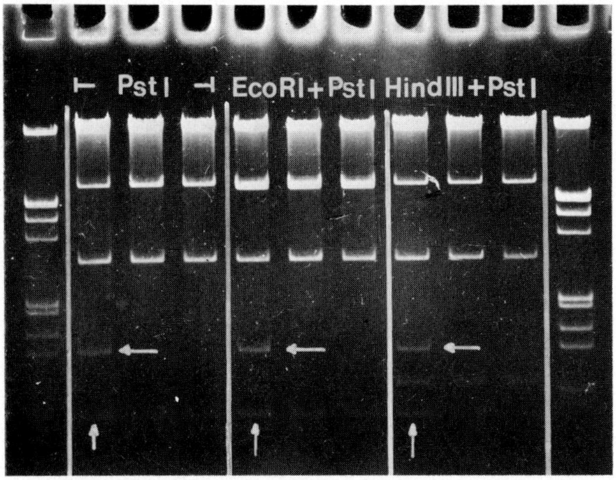

Fig. 3. Restriction endonuclease pattern of R68.45 (channels 2,5, 8), R68 (channels 3,6,9) and RP4 (channels 4,7,10). λ DNA cleaved with EcoRI and HindIII enzymes was used as reference (channels 1, 11). Arrows point to PstI bands of R68.45 (one is a double band) that are additional to those of R68 and RP4, respectively.

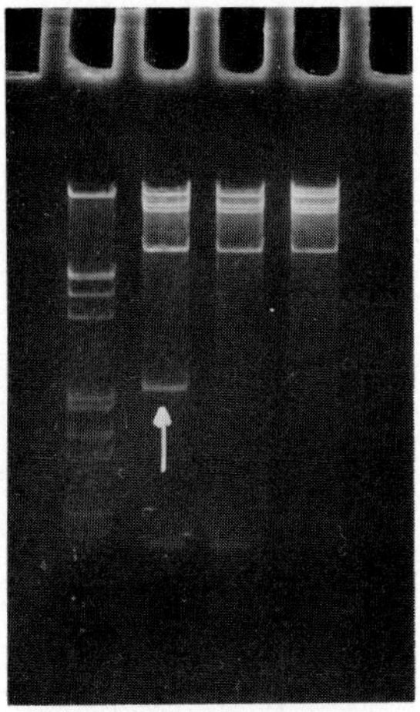

Fig. 4. SmaI restriction pattern of R68.45 (channel 2), R68 (channel 3) and RP4 (channel 4). Reference: λ DNA cut by EcoRI and HindIII enzymes. Arrow points to the additional SmaI band.

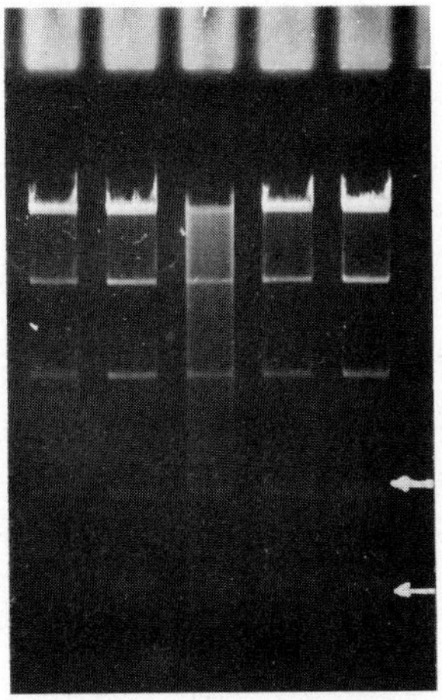

Fig. 5. PstI restriction pattern of pMO47 (channel 1), pMO60 (channel 2), pMO61 (channel 3), pMO62 (channel 4) and R68.45 as reference (channel 5). Arrow point to ISP1-specific PstI bands (see text).

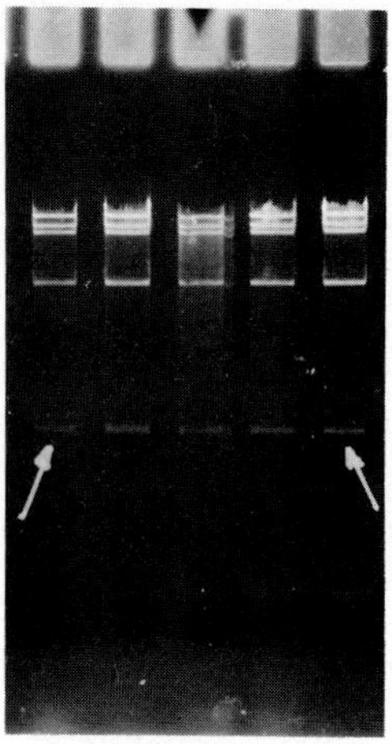

Fig. 6. *SmaI* restriction pattern of pMO47 (channel 1), pMO60 (channel 2), pMO61 (channel 3), pMO62 (channel 4) and R68.45 as reference (channel 5). Arrow points to ISP1-specific *SmaI* band (see text).

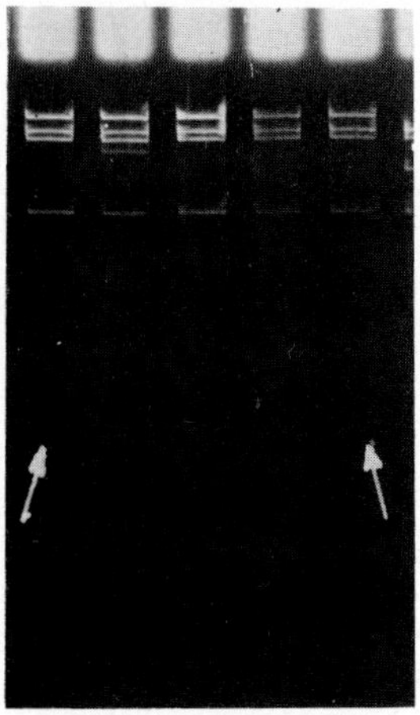

Fig. 7. *SmaI* restriction pattern of pMO90 (channel 1), pMO91 (channel 2), pMO92 (channel 3), pMO93 (channel 4) and pMO94 (channel 5). Arrow points to ISP1-specific *SmaI* band (see text).

RP4-2-like insertions (ISR1)

RP4 in R.lupini exhibits unstable behaviour[16]: Greater than 1% of plasmids in this host are mutant derivatives that have lost either their own transmissibility and the transfer inhibition of the R.lupini conjugational system[16] or their antibiotic resistances. Mutant plasmids which have lost ampicillin resistance were designated RP4-2, those which lost kanamycin resistance, RP4-3, and those which have lost both resistances, RP4-5 (loss of tetracycline resistance has never been observed). Length measurements of the mutant RP4-2 and RP4-3 plasmids indicated a DNA insertion into the RP4 molecule (Table 2).

TABLE 2

LENGTH OF RP4 PLASMID AND RP4-2 AND RP4-3 MUTANT PLASMIDS

Plasmid	Length and Standard Deviation	Molecules Measured
RP4	19.1 ± 0.3 um	21
RP4-2[a]	20.0 ± 0.13 um	20
RP4-3[a]	21.2 ± 0.11 um	19

[a]Recently, mutant plasmids which were smaller than the original RP4 plasmid could also be isolated. Their reduced size can be assumed to result from deletion events, involving a DNA segment of more than 7 μm in the case of one particular RP4-5 mutant. This deletion process is discussed later.

This interpretation has been tested by heteroduplex experiments. In the first experiment, RP8 was hybridized with RP4-2 and RP4-3 DNA. RP8 plasmid DNA was used in preference to RP4 because of its additional single-stranded hybridization marker. In the second experiment, RP4 and RP4-2 were hybridized after cleavage with the EcoRI endonuclease. Fig. 8 shows an RP8/RP4-2 heteroduplex molecule. The single-stranded RP8 DNA (thin arrow) and a substitution bubble (thick arrow) are clearly visible. This substitution bubble was localized by comparison of measurements obtained from this heteroduplex with those from an RP4/RP4-2 heteroduplex, in which both plasmids were cut by EcoRI. The heteroduplex shown in Fig. 9 provided the distances of the termini of the substitution bubble from the EcoRI site. Since the distance between the EcoRI site and the RP8 insertion is known[17], the bubble can be unambigously loca-

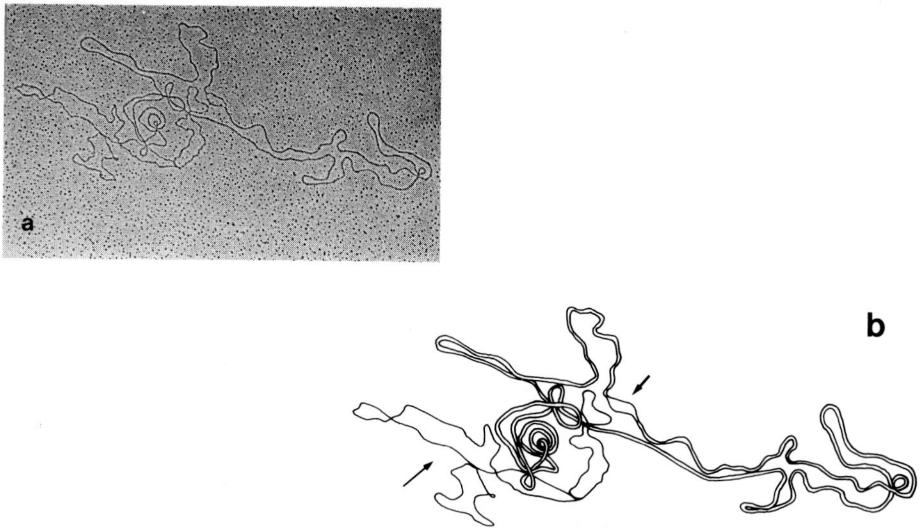

Fig. 8. RP8/RP4-2 heteroduplex molecule. Thin arrow points to single-stranded RP8 DNA, thick arrow points to heterologous RP4-2 DNA ("bubble"). a) Micrograph, b) Diagramatic representation

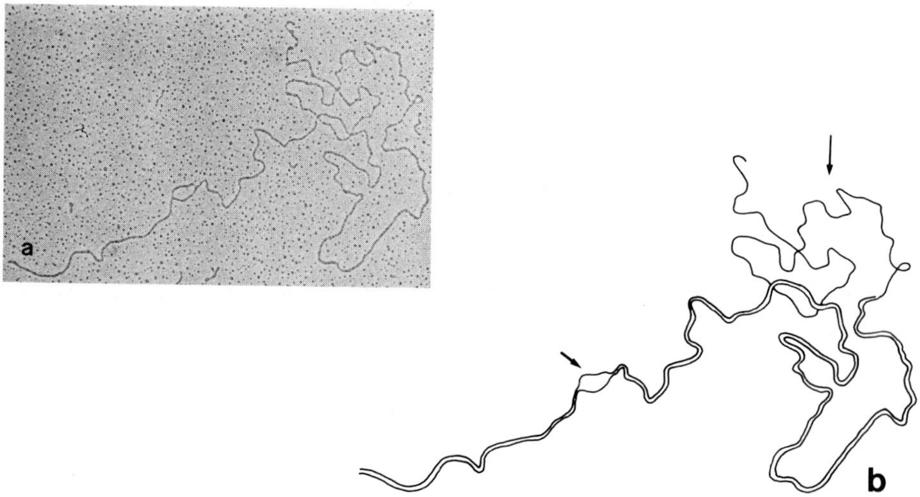

Fig. 9. RP4/RP4-2 heteroduplex molecule. Both molecules had been linearized by <u>Eco</u>RI digestion before hybridizing. Thick arrow points to heterologous DNA ("bubble"). The heteroduplex molecule has one single-stranded end, because one complete single-stranded molecule and one single-stranded molecule fragment had hybrized. a) Micrograph, b) Diagramatic representation

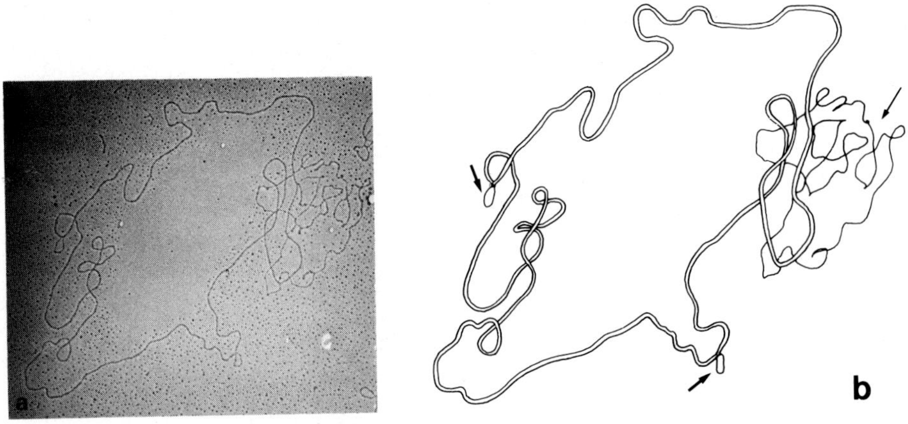

Fig. 10. RP8/RP4-3 heteroduplex molecule. Thin arrow points to single-stranded RP8 DNA, thick arrows point to single-stranded inserted DNA. a) Micrograph, b)Diagramatic representation

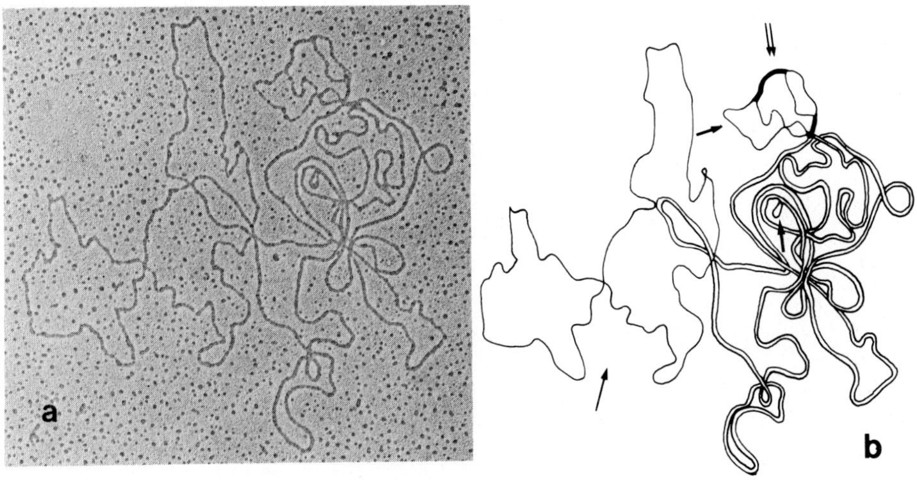

Fig. 11. RP8/RP4-3 heteroduplex molecule the Tn1 transposon had looped out before hybridization[12]. Thin arrow points to single-stranded RP8 DNA, thick arrows point to single-stranded inserted DNA, double arrow points to Tn1 transposon.
a) Micrograph, b) Diagramatic representation

ted within Tn1. From the RP4-2 measurements (table 2) one would expect an insertion loop rather than a substitution bubble in the heteroduplex molecules. On closer examination, however, the bubble branches were found to be of different length, the longer arm probably belonging to RP4-2 and the shorter to RP4. A possible interpretation of the substitution bubble is that in RP4-2, foreign DNA has integrated into the plasmid followed by deletion of orinal RP4 DNA in the neighbourhood of the inserted DNA. According to this interpretation the deleted DNA must be slightly smaller than the inserted segment. Restriction enzyme analysis demonstrated this explanation to be correct. Some new antibiotic resistance mutants of RP4 that were isolated from R.lupini appear to be deleted of a rather large DNA segment, in contrast to the small RP4-2 deletion described above. In the case of one RP4-5 mutant plasmid a molecule length of only 12 μm was measured, indicating a DNA deletion of more than 7 μm (see table 2).
In RP8/RP4-3 heteroduplices three single-stranded structures can be recognised (Fig. 10): a large loop due to the RP8 DNA and two small ones of identical size (0.95 kb). Their positions could be determined in the heteroduplex molecule shown in Fig. 11, in which the Tn1 transposon had looped out before hybridizing[12]. One insertion occurred in Tn1, the other in the kanamycin resistance determinant which inactivated its function. In this case, the insertion was not followed by deletion. Deletion formation may occasionally, however, take place producing kanamycin- and ampicillin sensitive mutant plasmids, that have been termed RP4-5. It should be mentioned that two newly isolated RP4-3 mutant plasmids carry this insertion in Tn1 without causing deletions, whereas the kanamycin sensitivity is caused by such a process. The insertions into these two resistance genes not only have the same size (0.95 kb) but are also characterized by identical restriction sites: one BamHI, one HindIII and one PstI site (figures not shown). We therefore assume that the inserted DNA is a special IS element originating from the R.lupini genome. For this element we propose the designation ISR1 (insertion sequence of R.lupini 1).

CONCLUSION

IncPI plasmids R68 and RP4 are obviously effective targets of

IS elements of soil bacteria. In P.aeruginosa the method to select for these sequences is to isolate mutants with donor ability for a chromosomal gene. ISP1 acquisition is independent of the gene selected. ISP1 has a preferential insertion site near the kanamycin gene region. In R.lupini no selection method for ISR1 is available. The frequency of the ISR1 transfer from the Rhizobium genome to the plasmid, however, is sufficiently high for the detection of RP4::ISR1 molecules without selection. No special function of ISR1 in RP4 could be detected. ISR1 preferentially integrates into DNA sequencies in the neighbourhood of the kanamycin resistance gene and into the Tn1 transposon itself. In this context it is of interest to note that both these DNA segments are characterized by an extraordinarily high A+T content in comparison to the average A+T content of the plasmid molecule (Burkardt et al., in preparation). The use of IncPI plasmids to pick up new IS elements from soil bacteria would appear to be a promising approach to clarify the role of these elements in bacterial genetics.

ACKNOWLEDGEMENTS

We thanks Mrs. K.Bauer for excellent technical assistance. This research was supported by Deutsche Forschungsgemeinschaft (Pu 28/2, Pu 28/4, Pu 28/8).

REFERENCES

1. Stanisich, V.A. and Holloway, B.W. (1971) Genet. Res., Cambridge, 17, 169-172.
2. Beringer, J.G. (1974) J. Gen. Microbiol., 84, 188-189.
3. Lacy, G.H. and Leary, J.V. (1976) Genet. Res., Cambridge, 27, 363-368
4. Haas, D. and Holloway, B.W. (1976) Molec. Gen. Genet. 144, 243-251.
5. Towner, K.J. and Vivian, A. (1976) J. Gen. Microbiol., 93, 355-360.
6. Meade, H.M. and Signer, E.R. (1977) Proc. Nat. Acad. Sci. USA, 74, 2076-2078.
7. Haas, D. and Holloway, B.W. (1978) Molec. Gen. Genet., 158, 229-237
8. Lowbury, E.J.L., Kidson, A., Lilly, H.A., Ayliffe, G.A. and Jones, R.J. (1969) Lancet, ii, 448-452.

9. Datta, N., Hedges, R.W., Shaw, E.J., Sykes, R.B. and Richmond, M.H. (1971) J. Bacteriol., 108, 1244-1249.

10. Black, W.A. and Girdwood, R.W.A. (1969) Brit. Med. J., iv,234.

11. Ingram, L.C., Richmond, M.H. and Sykes, R.B. (1973) Antimicrobial Agents Chemotherapeutical, 3, 279-288.

12. Burkardt, H.J., Pühler, A. and Rieß, G. (1979) J. Gen. Microbiol., submitted for publication.

13. Burkardt, H.J., Mattes, R., Pühler, A. and Heumann, W. (1978) J. Gen. Microbiol., 105, 51-62.

14. Holloway, B.W. and Richmond, M.H. (1973) Genet. Res., Cambridge, 21, 103-105.

15. Rieß, G., Burkardt, H.J. and Pühler, A. (1978) Hoppe-Seyler's Z. Physiol. Chem., 359, 1139.

16. Pühler, A. and Burkardt, H.J. (1978) Molec. Gen. Genet., 162, 163-171.

17. Spitzbarth, P. (1978) Thesis, FA Universität, Erlangen-Nürnberg, FRG.

© 1979 Elsevier/North-Holland Biomedical Press
Plasmids of Medical, Environmental and Commercial Importance
K.N. Timmis and A. Pühler, editors

RP4 AND R300B AS WIDE HOST-RANGE PLASMID CLONING VEHICLES

PETER T BARTH
ICI Corporate Laboratory, The Heath, Runcorn, Cheshire, England.

INTRODUCTION

The exploitation, both commercial and scientific, of the enormous potential of restriction enzyme techniques ("genetic engineering") depends upon the development of cloning vehicles that can be used not only in *Escherichia coli* but in organisms with different interesting attributes, such as *Pseudomonas aeruginosa, Rhizobium leguminosarum* or *Klebsiella pneumoniae*. The plasmids of incompatibility group P, such as RP4[1], provide an obvious and attractive source of material from which to fashion cloning vehicles for use in such strains because of their, by now legendary, wide host-range. Most IncP plasmids can be transferred to any Gram-negative bacterial species that has been tried [2,3,4]. Their conjugative ability makes their transfer simple and means that they can be used in species which have, as yet, no transformation system. Presumably, as an evolutionary adaptation to a wide host-range, IncP plasmids generally have few sites susceptible to restriction endonucleases. Those they have, tend to cluster around their antibiotic resistance genes which have probably been recently acquired. Having few restriction sites means that it is easier to achieve the ideal for cloning, that is, a plasmid with a single copy of each useful restriction site within a recognisable non-essential gene. Successfully cloned plasmids then are easily recognised by the loss of this gene function. RP4 has already achieved the ideal with respect to the unique *Hind*III site within its Km^R marker[5].

Genetic and restriction maps of suitable IncP plasmids are a prerequisite for the development of cloning vectors for particular purposes. Within the last few years maps of RP1[6,7], RP4[8,9,10], R68[11] and RK2[12,13] have been produced. These plasmids are very closely related if not identical. One advantage of mapping with a transposon, as we have done for RP4, is that the restriction sites introduced with the transposon in each clone can be used, together with the pre-existing restriction sites in the plasmid, to attempt the excision of specific segments of RP4. Some of these experiments have been described previously[8]. Where excisions failed, we concluded that the segment involved contains an essential function. Where they succeeded we have isolated smaller plasmids that have lost functions or restriction sites which may confer particular cloning advantages. This paper describes a continuation of this work.

For some cloning purposes, small, wide host-range multi-copy, non-conjugative plasmids have advantages. Plasmids of incompatibility group Q such as R300B and RSF1010 appear to be suitable. Although we described their properties, relationship to one another and wide host-range (including *P.aeruginosa*) some years ago[14,15], the latter has been recently rediscovered[16] apparently engendering new interest in this group. I have constructed a restriction map of R300B and performed a few simple cloning experiments.

MATERIALS AND METHODS

Bacterial strains. *E.coli* K12 strains used were J53, W3110T$^-$, C600 as described by Bachmann[17] and the *recA* derivative of W3110T$^-$, HH27[18]. For host-range testing the following strains were also used: *Salmonella typhimurium* TA1535 and TA1537, *Proteus mirabilis* 13, *Serratia marcescens* 10450 and 11045, *Pseudomonas aeruginosa* 280 and *Klebsiella aerogenes* K16.

Plasmids. Plasmids used were RP4, R702, RP4-δ1, RP4::Tn7 isolates and R300B as described previously[10,14].

Media. Growth media used were as previously described[10].

Isolation and screening of transposon derivatives of RP4. These were isolated by crossing RP4 out of suitable strains containing a chromosomally-located Tn7 or Tn76, selecting for trimethoprim resistant (10μg/ml) transconjugants and tested as previously described[10].

Test for surface exclusion. An overnight standing broth culture of W3110T$^-$ (R702) was diluted 100 fold and 0.1 ml samples were spread onto half the surface of minimal agar plates selective for J53 (proline and methionine) plus 1mg sulphonamide per ml. Newly isolated colonies of J53 (RP4::Tn7) were streaked in one direction across the plate into the donor. Colonies of J53 and J53 (RP4-δ1) (Sex$^-$) and J53(RP4) and J53(pRP3))Sex$^+$) were used as controls[10].

Isolation of plasmid DNA. Plasmid DNA was isolated as previously described[14] or by similar methods using 20 ml cultures and lysis with triton X100.

Restriction enzyme digests. Plasmid DNA in TNE buffer supplemented with 0.1M $MgCl_2$,0.1M mercaptoethanol was incubated with restriction enzymes at 37° for 1 hour (generally in a total volume of 20 μl), except for *Sma*I which required replacement of NaCl by KCl. Enzymes were obtained from commercial sources.

Sucrose gradient analysis. Sedimentation analyses in freeze-thaw generated sucrose gradients were as previously described[8].

Gel electrophoresis. For the electrophoresis of plasmid DNA we used 60ml 0.8%

agarose gels (tris acetate buffer pH 7.7) on horizontal glass plates (15x15cm) in Shandon-Southern electrophoresis tanks. DNA samples were mixed with 5μl 10% ficol and bromophenol blue, loaded into the wells and run at a constant 70 mA for about 2½ hours.

In vitro plasmid deletions. These were generated with restriction enzymes and ligase and isolated by transformation of *E.coli* as previously described[8].

RESULTS

An ampicillin sensitive RP4::Tn76 mutant. We have reported previously[8] our failure to isolate Ap^s mutants of RP4 by Tn7 insertion: their incidence was less than 0.1% of RP4::Tn7 clones. Recently, Datta *et al* [19] have identified further examples of transposons indistinguishable from Tn7 (by their conferred resistances, molecular weight and restriction sites). Amongst these was Tn76 from *Klebsiella aerogenes* R269A which was used to isolate RP4::Tn76 plasmids. Out of only six such plasmids (isolated by V.M.Hughes) one was found to be Ap^s (pRP764). This remarkable difference in incidence may be due to undetected differences between Tn7 and Tn76 or to the use of *K.aerogenes* rather than *E.coli* as the transposon donor.

By sucrose gradient analysis, pRP764 has a molecular weight of 45.0 Mdal (data not shown). Following restriction with *Eco*R1 or *Bam*HI, analysis showed that its transposon is inserted 3.6 Mdal to the right of the *Eco*R1 site in RP4. Gel electrophoresis analysis put the insertion just a little clockwise of the *Pst* site at the 3.6 Mdal position(Table 1 and Fig.1). This coincides with the known position of the *bla* gene on Tn1[20] which is located on RP4 in this region.

Surface exclusion mutants of RP4. Fresh isolates of J53 (RP4::Tn7) clones were screened for loss of the surface exclusion phenotype (Sex) by plate tests as described in Materials and Methods. Clones that looked Sex^- were retested in liquid matings for their exclusion of properties[10]. Out of about 500 clones tested, three had reduced surface exclusion to the entry of R702. One of the RP4::Tn76 clones, ie pRP765 was also found to be Sex^-. All four had simultaneously lost conjugal transferability. They could be classified into two groups on the basis of the degree of loss of surface exclusion compared to the parental RP4. pRP101 confers no surface exclusion (like RP4-δ1[21]) but pRP102, 103 and 765 have a surface exclusion index of about 5 (Table 1).

Restriction enzyme mapping of these mutants correlated with this grouping: pRP101 mapped at a new *tra*3 site, whereas the other three extended the previously defined *tra*2 site[10] (Table 1 and Fig.1).

TABLE 1
RP4::Tn PLASMIDS

Plasmid	Phenotypic change[a]	Molecular mass (Mdal)	Restriction enzymes used for analysis[b]	Deduced structure[c]
pRP764	Ap^s	45	*Eco*R1,*Pst*I,*Bam*HI, *Kpn*I,	Tn76 at 3.6
pRP101	Tra^- Sex 50	45	*Eco*R1,*Pst*I,*Bgl*II *Bam*HI,*Kpn*I.	Tn7 at 11.8
pRP102	Tra^- Sex 5	45	ditto	Tn7 at 15.4
pRP103	Tra^- Sex 5	45	ditto	Tn7 at 16.0
pRP765	Tra^- Sex 5	44.4	*Eco*R1,*Bam*HI (SG)	Tn76 at 15.7
pRP761	Unstable	44.7	*Eco*R1,*Pst*I,*Sma*I, *Bam*HI,*Xho*I	Tn76 at 34.6
pRP761.6	$Tra^{\pm}$ stable	30	ditto	See Figure 2
pRP301	$KmTpSm^s$	34.9	*Eco*R1 *Bam*HI *Hind*III *Sal*I	See Figure 4

[a] Ap^s = ampicillin sensitive; Tra^- = loss of conjugal transferability; $Tra^{\pm}$ = reduced transfer frequency; Sex = surface exclusion index ie ratio of R702 transconjugants in crosses of $\frac{\text{W3110T}^- \text{ (R702) X J53 (test plasmid)}}{\text{W3110T}^- \text{ (R702) X J53 (RP4)}}$;

Km = kanamycin; Tp = trimethoprim; Sm = streptomycin.

[b] Restriction enzymes were used singly or in combination, the DNA fragments being analysed by agarose gel electrophoresis or sucrose gradient sedimentation (SG).

[c] Transposons were found to be inserted into RP4 in the usual Tn7 orientation[10] at the Mdal co-ordinate given (see Figure 1).

An unstable RP4::Tn76 mutant. It was noticed that one of the RP4::Tn76 clones that had been isolated was unstable, that is, the strain W3110T⁻ (pRP761) produced antibiotic sensitive colonies when grown without selection. This stability depended on the plasmid and not the host. After transfer to C600, the plasmid's stability was quantified. A selected (kanamycin) overnight culture of C600 (pRP761) was innoculated into broth and grown aerated at 37°C, without selection for about 17 generations, keeping the cell titre between 10^6/ml and 10^8/ml by subculture. Dilutions were plated out for single colonies. Out of 48 colonies tested for kanamycin and trimethoprim resistance, only two had retained their resistances and hence the plasmid. Plasmid pRP761 therefore had a half-life of about 3.5 generations under these growth conditions.

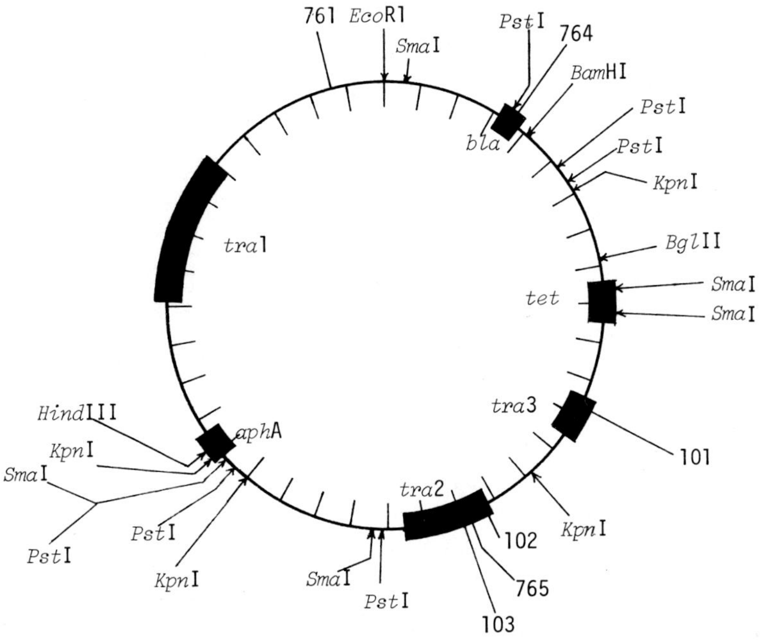

Fig.1. Map of RP4 showing the Tn7 or Tn76 insertion sites described in this paper plus a few restriction sites relevant to their mapping (Table 1).

Restriction enzyme analysis of pRP761 showed that its Tn76 was inserted (in the usual Tn7 orientation[10]) about 1.4 Mdal to the left of the *Eco*R1 site in RP4 (Fig.1). This insertion may therefore influence an essential gene (or genes) which is thought to be situated in this region[8,12].

A reduction of plasmid copy number by the Tn76 insertion is one way in which the instability could be caused. The copy numbers of pRP761 and RP4 (differentially labelled) were therefore measured by the mixed lysis technique[8]. It was found that they had 1.8 and 2.4 copies per chromosome respectively. Considering that pRP761 is about 25% larger than RP4 and therefore presumably more fragile, these figures may not be significantly different.

Host-range of pRP761. The instability of pRP761 means that it has an alteration in a plasmid maintenance function. As the wide host-range of IncP plasmids may be a consequence of their possessing a particularly sophisticated plasmid maintenance system, independent of host functions, it seemed worthwhile examining the host-range of pRP761. The donors used were W3110T$^-$(pRP761) and as controls, the same host with pRP764 or RP4. These were plate-mated with the strains of *S.typhimurium, P.mirabilis, S.marcescens, P.aeruginosa, K.aerogenes* and *E.coli* (C600) as given in Materials and Methods. The Thy$^-$ phenotype of the donors allowed their counterselection by using Isosensitest (Oxoid) medium containing kanamycin (25μg/ml) or trimethoprim (1mg/ml) as selection for the plasmids. It was found that whereas RP4 and pRP764 transferred readily to all these hosts, pRP761 would not give transconjugants with *P.aeruginosa* (but did with the other species tested).

Spontaneous deletions of pRP761. The instability of strains containing pRP761 takes more than one form. As described above, loss of the whole plasmid commonly occurs but clones that have lost the Tn76 markers (Tp^R Sm^R) but retained the RP4 markers can also be found. Plasmid DNA isolated from such clones, analysed on gels, shows that the plasmid has suffered a considerable deletion of about 12-15 Mdal. Plasmid preparations from pRP761 strains grown without Tp selection also show a minority band of deleted plasmids. These deletions therefore occur spontaneously.

Using a *rec*A strain HH27 (pRP761) it was found that Tp^sSm^s clones also appeared. These similarly contained a deletion of the plasmid. This process is therefore not dependent on the *rec*A$^+$ gene.

Plasmid DNA from one of these clones (pRP761-6) was analysed using restriction enzymes. The gel in Figure 2 shows that it has lost all the restriction sites attributable to Tn76 plus the RP4 *Eco*R1 site. It has retained the *Sma*I site at 0.6 Mdal. The conjugal transferability of this plasmid is considerably reduced in comparison to RP4 or its parent pRP761. It therefore seems likely that the deletion in pRP761-6 extends from the *Eco*RI site of RP4, through the inserted Tn76 and into the *tra*1 region around 30-31 Mdal. We have previously reported that mutants mapping in this region have a reduced rather than undetectable transfer frequency[10]. Another clone analysed, pRP761-2, had a similar gel pattern but was completely *tra*$^-$. The deletion had presumably extended further into the *tra*l region.

No two clones analysed had exactly the same restriction pattern. Even those retaining all the characteristics of the parental pRP761 had acquired small deletions or structural changes. For example the gel analysis shown in Figure 3 from a clone of C600 (pRP761), differs from the parental in having an extra, non-stochiometric, *Eco*R1 fragment of 5.2 Mdal. This is most easily explained by considering that a minority of the plasmid molecules have acquired an extra *Eco*R1 site close to the *Pst*1 site on Tn76, perhaps by a deletion in that region. Other derivative plasmids appear to have undergone complex structural changes which have not been fully analysed.

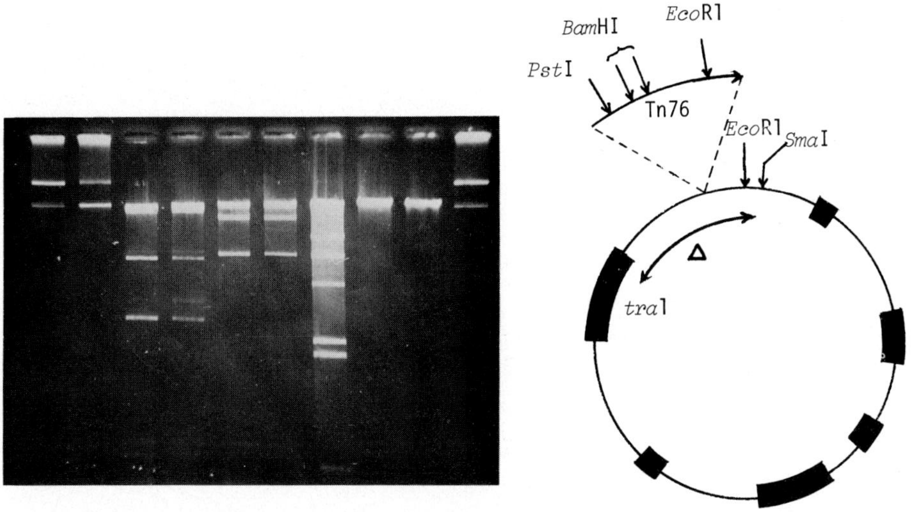

Fig.2. Analysis by agarose gel electrophoresis of pRP761-6, a spontaneous deletion of HH27 (pRP761). The tracks in left to right sequence are: 1 untreated, 2 *Eco*R1, 3 *Pst*I, 4 *Eco*R1 + *Pst*I, 5 *Sma*I, 6 *Eco*R1 + *Sma*I, 7*Hind*III fragments of λ, 8 *Bam*HI, 9 *Eco*R1 + *Bam*HI 10 untreated. The diagram shows the probable extent of the deletion.

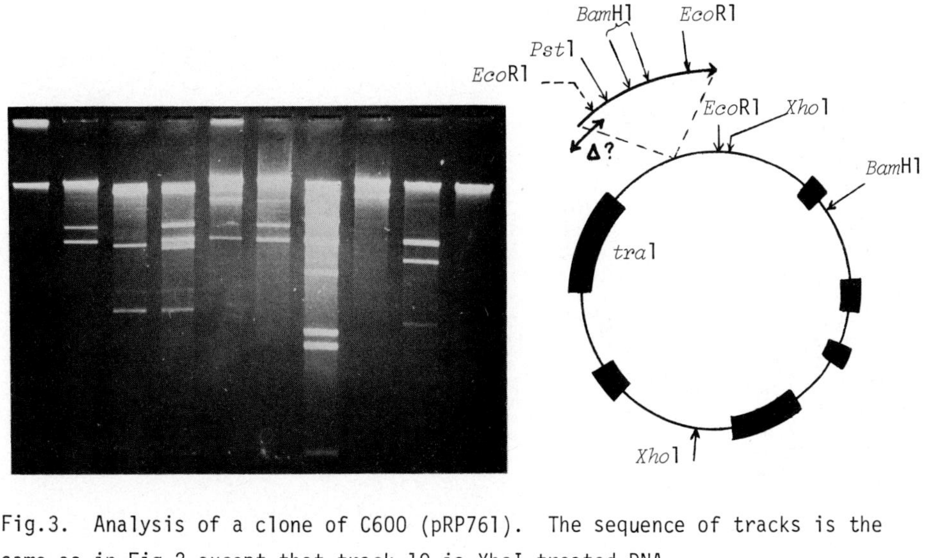

Fig.3. Analysis of a clone of C600 (pRP761). The sequence of tracks is the same as in Fig.2 except that track 10 is XhoI treated DNA.

Stability of the pRP761 derivatives. Cultures of the various spontaneous derivatives of pRP761 described above were grown for several generations without selection. Single colonies were tested for retention of the plasmid markers. It was found that the Tp^SSm^S derivatives were no longer unstable with respect to the remaining RP4 markers ($ApTcKm^R$). The original pRP761 and those having small deletions, although retaining the Tn76 markers, remained unstable.

In vitro deletions of RP4::Tn7 plasmids. As described in the Introduction, we have used our mapped RP4::Tn7 (and Tn76) plasmids to attempt to isolate deletions systematically around the RP4 genome. These results will be published elsewhere. One interesting plasmid created in these experiments will be described here as it has become a useful wide host-range cloning vector.

Cleavage of pRP3 with *Hind*III followed by ligation and transformation of resulting molecules into *E.coli* led to the isolation of pRP301 which had lost the small *Hind*III fragments, as shown in Figure 4. This structure was confirmed by sucrose gradient and agarose gel analysis (Table 1). This plasmid has lost Km^R and the $TpSm^R$ of Tn7 but remains Tra^+. The excised segment of DNA clearly has no essential functions. It has lost one of the two *Sal*I sites on RP4 (there are none on Tn7) leaving one close to or within the *tet* gene. Cloning exogenous *Sal*I fragments into this site (J. D. Windass, personal communication) generates plasmids that have become Tc^S thus confirming my previous suggestion[5] that this site permits insertional mutation.

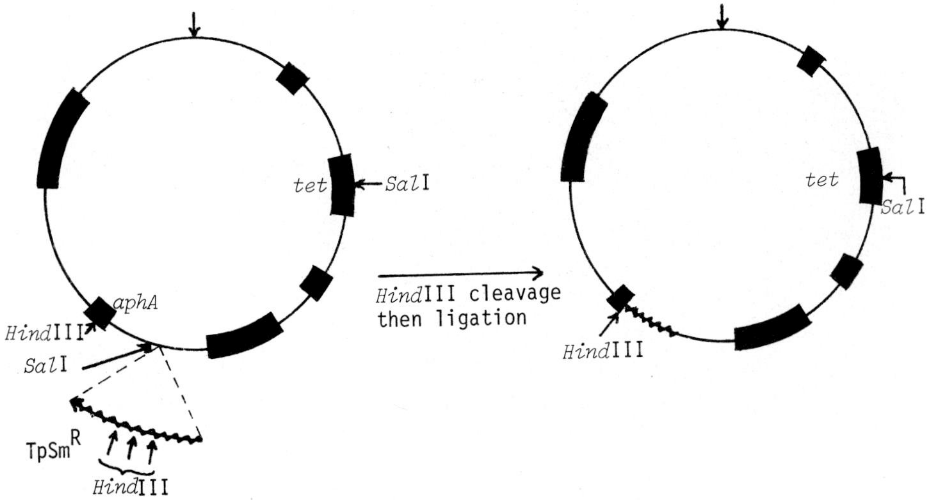

Fig.4. Construction of pRP301 from the RP4::Tn7 plasmid pRP3 by excision of the small *Hind*III fragments.

A restriction map of R300B. R300B DNA and as a control, λDNA, were treated with various restriction enzymes and analysed by gel electrophoresis. R300B was found to have no sites susceptible to *Bam*HI, *Bgl*II, *Hind*III, *Sal*I, *Xba*I, *Xho*I or *Xma*I. However, it has a single *Eco*Rl site (as we have noted before[15]), a single *Hpa*I site and two *Pst*I sites. The cleavage pattern with these enzymes is shown in Figure 5. It was not possible to distinguish the distance between the *Eco*Rl and *Hpa*I sites with these gels. Excision of the 0.45 Mdal fragment between the two *Pst*I sites generated plasmids that had lost the Su^R of R300B although they retained Sm^R. Cloning of exogenous *Pst*I fragments into *Pst*I-cut R300B also generated Su^S hybrid plasmids. One or both of the *Pst*I fragments must therefore be within the *sul* gene as shown in Figure 5. The wide host-range vector properties of R300B are not altered by cloning exogenous DNA into this gene. Also, the available *Hpa*I site permits the cloning of randomly-sheared DNA fragments by blunt-ended ligation. R300B, therefore, as a mobilizable (by eg IncP plasmids), 5.7 Mdal[14], multicopy plasmid, appears to be a potentially useful cloning vector.

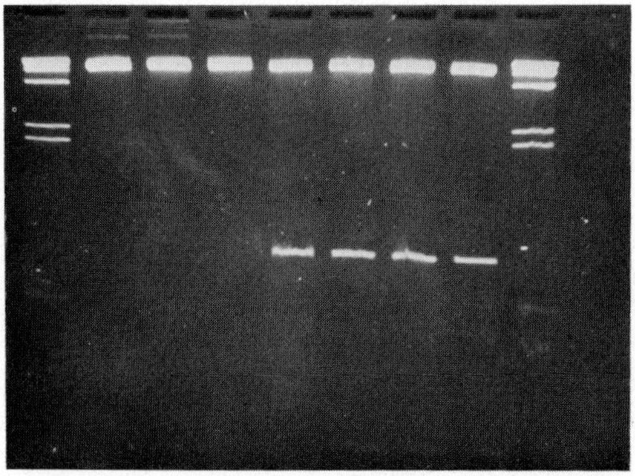

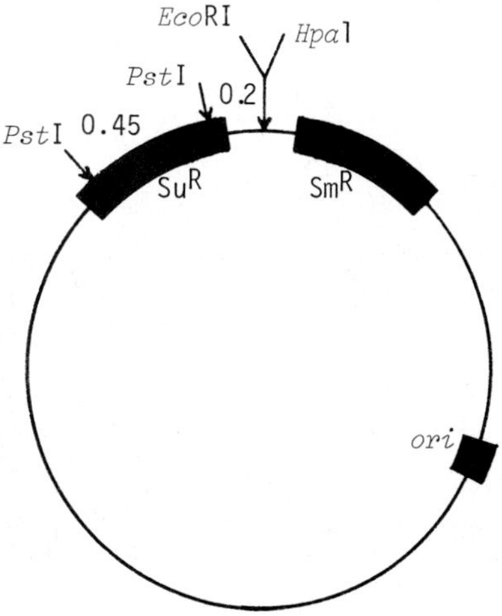

Fig.5. R300B analysed by electrophoresis through 2% agarose gels. Tracks 1 and 9 are *Hind*III fragments of λ, the others of R300B are: 2 *Hpa*I, 3 *Eco*RI, 4 *Eco*RI + *Hpa*I, 5 *Pst*I, 6 *Pst*I + *Eco*RI, 7 and 8 *Hpa*I + *Pst*I.

RSF1010 is very similar if not identical to R300B. They have the same characteristics, single *Eco*R1[22] and *Hpa*I[16] restriction sites and are mutually related by heteroduplex analysis[22] or DNA-DNA hybridization[14] to R684. I have therefore superimposed the genetic map of Heffron *et al*[22] onto our cleavage map in constructing the diagram in Figure 5.

DISCUSSION

This paper describes further work on two plasmids that are, or can be adapted to be, useful cloning vectors in a wide range of bacterial species. They are not only of inherent interest because of their wide host-range but also, by learning more about their restriction and genetic maps, we can make intelligent attempts to reconstruct them for particular cloning purposes.

In RP4, we have confirmed the position of the *bla* gene by mapping a direct "hit" with Tn76. The difference in insertion specificity between Tn7 and Tn76, which are otherwise indistinguishable ([19] and unpublished data) is puzzling. Two genes concerned with surface exclusion have been mapped. One, giving only a five fold effect, maps in the *tra*2 region around 15-16 Mdal where we previously speculated it to be [10]. The other gene, which when disrupted by Tn7 loses all surface exclusion effect, maps separately at about 12 Mdal, which we have called *tra*3. Plasmid F also has two genes concerned with surface exclusion but these are adjacent in the F *tra* operon[23]. The concomitant loss of transferability of our Sex mutants may be directly due to the *sex* genes disrupted by the transposons or to the latter's polar or deletion effects, as discussed more fully elsewhere[10]. A paradox arises with respect to RP4-δ1. This carries no residual Sex phenotype[21] like pRP101, but if it has a single deletion this cannot stretch to the *tra*3 region.

The insertion site of the unstable plasmid pRP761 (Figure 1) presumably locates a plasmid maintenance gene. The reduced host-range of pRP761 suggests that this maintenance gene is involved in plasmid promiscuity. Copy number control does not appear to be much, if at all, affected. A plasmid partition mechanism may be affected or some other aspect of the complicated mechanism of replicon initiation or replication. It seems unlikely that promiscuity is just a reflection of a wide conjugal transfer host-range. For example, Inc Iα plasmids can mobilize IncQ plasmids into *P.mirabilis* and must therefore make success ful conjugal bridges, but they cannot themselves inhabit this host[2]. Promiscuity seems more likely to be a reflection of a plasmid replication and partition system that is peculiarly independent of bacterial host functions.

The spontaneous deletions of pRP761 are presumably selected because of their greater stability. They may originate from the ends of Tn76:transposon termini are noted hot spots for deletion formation. But it seems odd that these deletions can remove the presumed maintenance gene (close to the RP4 *Eco*Rl site, Figure 2) that is being affected by the inserted Tn76. Perhaps these plasmids switch to being reliant on a host-generated function. Being Tra$^-$ these plasmids are difficult to test for a predicted drastic reduction in host-range.

R300B is no exception to the rule that appears to be emerging, namely that promiscuous plasmids have few restriction sites and those they have are situated in or near their drug resistance markers, which may have been recently acquired. It seems reasonable to speculate that the reduction of restriction sites to a minimum is a selective advantage to wide host-range plasmids.

The use of transposon insertion to delete (or clone) specific segments of a plasmid, as illustrated in Figure 4, is clearly a powerful tool in the construction of, for example, new cloning vectors from RP4. We have not been able to isolate Tn7 derivatives of R300B but both Tn1[22] and Tn5 derivatives (N. Datta and P. T Barth, unpublished data) are available to this end.

ACKNOWLEDGEMENTS

Some of this work was done in the Bacteriology Department of the Royal Postgraduate Medical School, London where it was supported by a grant to Naomi Datta and myself from the Medical Research Council of the UK. I thank Naomi Datta for constant help and encouragement, N. J. Grinter for technical assistance and V. M. Hughes for isolation of the RP4::Tn76 clones. I am grateful to S. A. Withe for current technical help and his photographic skills, to B. Symington for the diagrams and A. V. Greaves for typing this manuscript.

REFERENCES

1. Datta, N., R.W. Hedges, E.J. Shaw, R.B. Sykes and M.H. Richmond. (1971) Properties of an R factor from *Pseudomonas aeruginosa*. J.Bacteriol.108, 1244-1249.

2. Datta, N. and R.W. Hedges (1972) Host ranges of R factors. J.Gen. Microbiol.70, 453-460.

3. Olsen, R.H. and P. Shipley (1973). Host ranges and properties of the *Pseudomonas aeruginosa* R factor R1822. J.Bacteriol 113,772-780.

4. Towner, K.J. and A. Vivian (1976) RP4-mediated conjugation in *Acinetobacter calcoaceticus*. J.Gen.Microbiol.93,355-360.

5. Barth, P.T. (1979) Plasmid RP4, with *Escherichia coli* DNA inserted *in vitro*, mediates chromosomal transfer. Plasmid 2, 130-136.

6. Stanisich, V.A. and P.M. Bennett (1976). Isolation and characterization of deletion mutants involving the transfer genes of P.group plasmids in *Pseudomonas aeruginosa* Molec.Gen.Genet.149,211-216.

7. Grinsted, J., P.M. Bennett and M.H. Richmond (1977). A restriction map of R-plasmid RP1. Plasmid 1, 34-37.

8. Barth, P.T. and N.J. Grinter (1977). Map of plasmid RP4 derived by insertion of transposon C. J.Mol.Biol.113, 455-474.

9. Depicker, A., M. Van Montagu and J. Schell (1977). Physical map of RP4 in "DNA insertion elements, plasmids and episomes" (ed.Bukhari,Shapiro & Adya) 678-679. Cold Spring Harbor Laboratory.

10. Barth, P.T., N.J. Grinter & D.E. Bradley (1978). Conjugal transfer system of plasmid RP4: analysis by transposon 7 insertion. J.Bacteriol.133,43-52.

11. Holloway, B.W. (1979). Plasmids that mobilize bacterial chromosome. Plasmid 2, 1-19.

12. Figurski D., R. Meyer, D.S. Miller & D.R. Helinski (1976). Generation *in vitro* of deletions in the broad host range plasmid RK2 using phage Mu insertions and a restriction endonuclease. Gene 1, 107-119.

13. Meyer, R., D. Figurski and D.R. Helinski (1977). Restriction enzyme map of RK2 in "DNA insertion elements, plasmids and episomes" (eds.Bukhari,Shapiro and Adya) 680. Cold Spring Harbor Laboratory.

14. Barth, P.T. and N.J. Grinter (1974). Comparison of the deoxyribonucleic acid molecular weights and homologies of plasmids conferring linked resistance to streptomycin and sulphonamides. J.Bacteriol.120, 618-630.

15. Grinter, N.J. and P.T. Barth (1976). Characterization of SmSu plasmids by restriction endonuclease cleavage and compatibility testing. J.Bacteriol. 128,394-400.

16. Nagahari, K. and K. Sakaguchi (1978). RSF1010 plasmid as a potentially useful vector in *Pseudomonas* species. J.Bacteriol.133,1527-1529.

17. Bachmann, B.J. (1972). Pedigrees of some mutant strains of *Escherichia coli* K-12. Bacteriol.Rev.36,525-557.

18. Barth, P.T. and N. Datta (1977). Two naturally-occurring transposons indistinguishable from Tn7. J.Gen.Microbiol.102,129-134.

19. Datta, N., V.M. Hughes, M.E. Nugent and H. Richards (1979). Plasmids and transposons, their stability and mutability in bacteria isolated during an outbreak of hospital infection. Plasmid 2, in press.

20. Heffron, F., P. Bedinger, J.J. Champoux and S. Falkow (1977). Deletions affecting the transposition of an antibiotic resistance gene. Proc.Natl. Acad.Sci.USA 74, 702-706.

21. Hedges, R.W., J.M. Cresswell and A.E. Jacob (1976). A non-transmissible variant of RP4 suitable as cloning vehicle for genetic engineering. FEBS Letters 61, 186-188.

22. Heffron, F., C. Rubens and S. Falkow (1975). Translocation of a plasmid DNA sequence which mediates ampicillin resistance: molecular nature and specificity of insertion. Proc.Nat.Acad.Sci.USA 72,3623-3627

23. Manning, P.A. and M. Achtman (1979). Cell to cell interactions in conjugating *E.coli*: the involvement of the cell envelope. Bacterial Outer Membranes. Ed.Inouye. Wiley and Sons.

Plasmids of Medical, Environmental and Commercial Importance
K.N. Timmis and A. Pühler, editors

NEW VECTOR PLASMIDS FOR GENE CLONING IN PSEUDOMONAS

M. BAGDASARIAN, M.M. BAGDASARIAN, S. COLEMAN and K.N. TIMMIS
Max-Planck-Institut für Molekulare Genetik, Berlin-Dahlem, BRD.

INTRODUCTION

The Pseudomonads are a group of bacteria that exhibit a wealth of exotic metabolic activities[1]. The most remarkable of these include abilities to degrade a wide range of organic compounds, including xenobiotics such as pesticides, chlorinated hydrocarbons and organomercurials, and to produce disease in animals and plants[1-4]. Many activities of *Pseudomonas* are therefore of commercial, environmental or medical importance and it will be of some interest to carry out *in vivo* and *in vitro* genetic manipulations of the DNA of these organisms.

Although excellent vector plasmids such as pBR322[5] and pACYC184[6] have been developed for gene cloning in *E.coli*, no vectors of equivalent versatility exist for cloning in *Pseudomonas*. In some experiments it may be convenient to use *E.coli* as a recipient bacterium for cloned *Pseudomonas* genes. In others, however, this will not be practicable because *E.coli* is known to lack some auxiliary biochemical pathways that are essential for the phenotypic expression of certain functions (e.g. toluene degradation[7], plant pathogenicity), in particular those which can only be detected and investigated in bacteria that are able to colonise specific ecological habitats.

Two broad host range plasmids that can be stably maintained in *Pseudomonas* have been proposed as cloning vectors for this host. These are RP4/RP1/RK2[2] and RSF1010[8]. RK2 is a large, low copy, non-amplifiable plasmid that contains determinants for three selectable antibiotic resistances, two of which (ampicillin and kanamycin) contain unique endonuclease cleavage sites (*Bam*HI and *Hind*III, respectively)[9]. Smaller derivatives of RK2 that are more suitable as cloning vectors were recently generated by the Helinski group (see Helinski et al, this volume). RSF1010 is also a non-amplifiable plasmid[10] but has the advantage that it is relatively small (5.5 mD) and is present in a large number of copies

(15-40) in host bacteria. We have examined a number of additional plasmids for their suitability as vectors for gene cloning in Pseudomonas but thus far have not found any more convenient than RSF1010. This plasmid was therefore used as a basic replicon from which to construct a series of more versatile cloning vectors. In this report we describe the construction of some of these RSF1010 derivatives and present their restriction endonuclease cleavage maps.

MATERIALS AND METHODS

Bacterial strains and plasmids. SK1592 is *Escherichia coli* F^- *gal thi* $T1^r$ *endA sbcB15 hsdR4* $hsdM^{+11}$, 2003 is *Pseudomonas aeruginosa* strain AC161 *leu res* mod^+, and 2104 is *Pseudomonas putida* strain AC34 *ade*. Both were obtained from A.Chakrabarty. Plasmids used in this work are listed on Table 1.

TABLE 1

PLASMIDS USED IN THE PRESENT WORK

Plasmid	Molecular weight (mD)	Antibiotic resistances	Host	Source
pBR322	2.6	Ap,Tc	*E.coli*	5
pML21	6.7	Km	*E.coli*	12
pACYC184	2.65	Tc,Cm	*E.coli*	6
pSC101	5.8	Tc	*E.coli*	13,14
pKT007	9.8	Tc	*E.coli*	18
pKT101	3.1	Km	*E.coli*	ColD-Km (K.N.Timmis unpubl.)
S-a	25.0	Km,Sm,Cm,Su	*E.coli*	15
RSF1010	5.5	Sm,Su	*E.coli*	16
			P.putida	A.Chakrabarty
Rms149	36.0	Cb,Gm,Sm Sp,Su	*P.aeruginosa*	17
pKT209	17.0	Cb	*P.aeruginosa*	mini Rms149, (M.Bagdasarian unpubl.)

Abbreviations: Ap, ampicillin; Tc, tetracycline; Km, kanamycin; Cm, chloramphenicol; Sm, streptomycin; Su, sulfonamide; Cb, carbenicillin; Gm, gentamycin; Sp, spectinomycin.

Experimental procedures. Media and growth conditions for bacterial strains were as described previously[18] except that *Pseudomonas* strains were grown at 30°C. Plasmid DNA was purified from *E.coli* cells as previously described[18] and from *Pseudomonas* cells by the method of Hansen and Olsen[19]. Digestion of DNA with restriction endonucleases, electrophoresis of DNA fragments through agarose gels, and ligation of DNA fragments were performed as previously described[18]. Transformation of bacteria with plasmid DNA was carried out according to the method of Kushner[11] except that, for *Pseudomonas* strains, the concentration of $CaCl_2$ was raised to 0.1 M and DMSO was omitted. The concentrations of antibiotics used in selective media for *Pseudomonas* were as follows: Cb and Sm, 1 mg/ml; Tc, 0.02 mg/ml; Cm and Km, 0.1 mg/ml; Gm, 0.01 mg/ml; and Su, 0.2 mg/ml.

RESULTS

Plasmids which are particularly appropriate as cloning vectors are small, genetically and structurally well defined replicons that are maintained at high cellular copy numbers and that can be amplified in host bacteria by the inhibition of cellular protein synthesis[20]. They code for at least one selectable property and contain unique cleavage sites for the commonly used cloning enzymes. Frequently, cloning sites are located within the genetic determinants of easily scored phenotypes, e.g. antibiotic resistances, which enable the rapid screening or selection of bacterial clones that have been transformed with recombinant plasmids ("insertional inactivation"; see refs. 18,21).

It is desirable to have both broad and narrow host range vector plasmids for gene cloning in *Pseudomonas*. Broad host range vectors would be useful for the cloning and manipulating of genes (e.g. those of environmental importance) which the investigator wishes to introduce into different laboratory or naturally-occurring bacteria. On the other hand, narrow range plasmid vectors would be useful for the cloning of genes whose products are biohazardous and for which it will be necessary to employ appropriate levels of biological containment. In order to identify suitable replicons from which to generate cloning vectors for *Pseudomonas*, we have examined a number of small plasmids for their ability to be transformed into and be stably maintained in Pseudomonas bacteria. The results are presented in Table 2.

TABLE 2
ABILITY OF DIFFERENT PLASMIDS TO BE TRANSFORMED INTO P.aeruginosa AND P.putida BACTERIA

Plasmid DNA[1]	Transformation frequency relative to E. coli		
	E.coli SK1592 $r_K^- m_K^+$	P.aeruginosa 2003 $r^- m^+$	P.putida 2104 $r^+ m^+$
$pBR322_c$	1	$< 8 \times 10^{-7}$	$< 8 \times 10^{-7}$
$pML21_c$	1	n.t.	$<1.1 \times 10^{-5}$
$pKT007_c$	1	$< 3 \times 10^{-5}$	$< 3 \times 10^{-5}$
$pACYC184_c$	1	$< 8 \times 10^{-5}$	$< 8 \times 10^{-5}$
$pKT101_c$	1	n.t.	$<1.7 \times 10^{-5}$
$pSC101_c$	1	$<1.0 \times 10^{-5}$	$<1.0 \times 10^{-5}$
$RSF1010_p$	1	0.14	0.13
$Rms149_a$	1	0.04	0.01
$pKT209_a$	1	0.10	0.02
$pKT210_c$	1	0.46	$<6.6 \times 10^{-5}$

[1] Subscripts refer to the origin of the plasmid DNA: c, a and p correspond to E.coli, P.aeruginosa and P.putida, respectively.

Cells of P.putida and P.aeruginosa can be rendered competent by treatment with $CaCl_2$ and RbCl. We have used a modification of the method of Kushner[11] in which the concentration of $CaCl_2$ is increased to 0.1 M[22] and from which DMSO is omitted. The frequency of transformation of P.aeruginosa or P.putida obtained with this method is about 10^5 transformants per μg of RSF1010 DNA prepared from P.putida (Table 3). As shown in Table 2, ColE1-type plasmids such as pBR322 and pACYC184 (the most widely used cloning vectors

TABLE 3

ABILITY OF RSF1010 DNA PREPARED FROM DIFFERENT HOST BACTERIA TO TRANSFORM *E.coli*, *P.aeruginosa* AND *P.putida*

Source of plasmid DNA	Transformation frequency[1]		
	E.coli SK1592 r_K^- m_K^+	*P.aeruginosa* 2003 r^- m^+	*P.putida* 2104 r^+ m^+
E.coli SK1592	3.6×10^5	3.0×10^3	$< 3 \times 10^{-6}$
P.putida	3.0×10^5	6.7×10^4	6.2×10^4

[1] Transformation frequency is expressed as the number of transformants per μg of DNA.

for *E.coli*) could not be introduced into and stably maintained in *Pseudomonas* strains. In contrast, RSF1010 and Rms149, a plasmid that was originally isolated from *Pseudomonas*, could be readily introduced into this host. Of these two plasmids, RSF1010 gave the highest transformation frequencies with both *E.coli* and *Pseudomonas* hosts. It should be noted that *P.putida* possesses a powerful restriction system; we have not been able to introduce into this strain plasmid DNAs prepared from host bacteria other than *P.putida* or *P.aeruginosa* (Table 3).

Restriction endonuclease cleavage map of RSF1010. As shown by the results of the preceding section, RSF1010 is the only small plasmid tested that could be efficiently transformed into *Pseudomonas* bacteria. This plasmid is stably maintained in both *E.coli* and *Pseudomonas* cells and can be purified easily from either host. We therefore decided to use RSF1010 as a starting point for the construction of a series of more useful broad range cloning vectors. The first step in this procedure was to generate a restriction endonuclease cleavage map of the plasmid using a variety of enzymes. The RSF1010 molecule contains single cleavage sites for *Eco*RI, *Bst*EII, *Hpa*I, *Sst*I, and *Pvu*II, and two *Pst*I sites that are separated by a DNA segment only 0.46 mD in size. No cleavage sites on the RSF1010 molecule were detected for the *Bam*HI, *Bgl*II, *Hin*dIII, *Kpn*I, *Sma*I and *Xba*I endonucleases. The map positions of

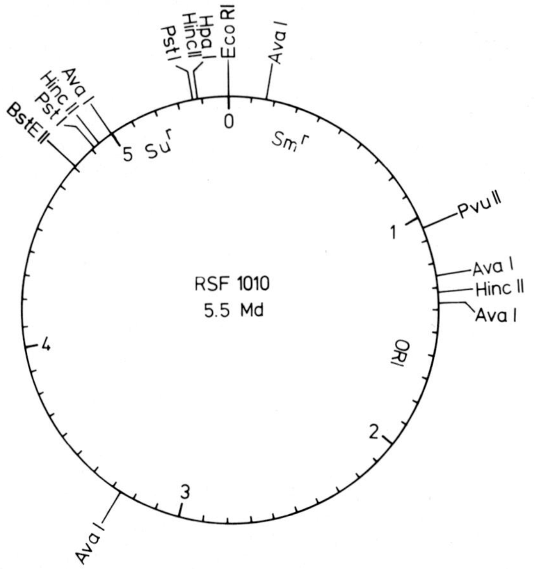

Fig. 1. Restriction endonuclease cleavage map of RSF1010.

the origin of replication and the Sm resistance determinant, relative to the EcoRI cleavage site, were taken from the data of Rubens et al.[23]. The Su resistance determinant was located also by Rubens et al.[23] and by *in vitro* insertion of DNA fragments between the two PstI cleavage sites of RSF1010 (see below).

Construction of new derivatives of RSF1010. The usefulness of RSF1010 as a cloning vector is limited by the fact that it contains single cleavage sites for only few endonucleases that are ordinarily used in cloning experiments. New derivatives of RSF1010 that contain additional cleavage sites and antibiotic resistance determinants were therefore generated. We had previously cloned into the PstI site of pBR322 a DNA fragment containing the Cm resistance determinant of the broad host range plasmid S-a. This 2.4 mD DNA fragment contains a single HindIII cleavage site. DNAs

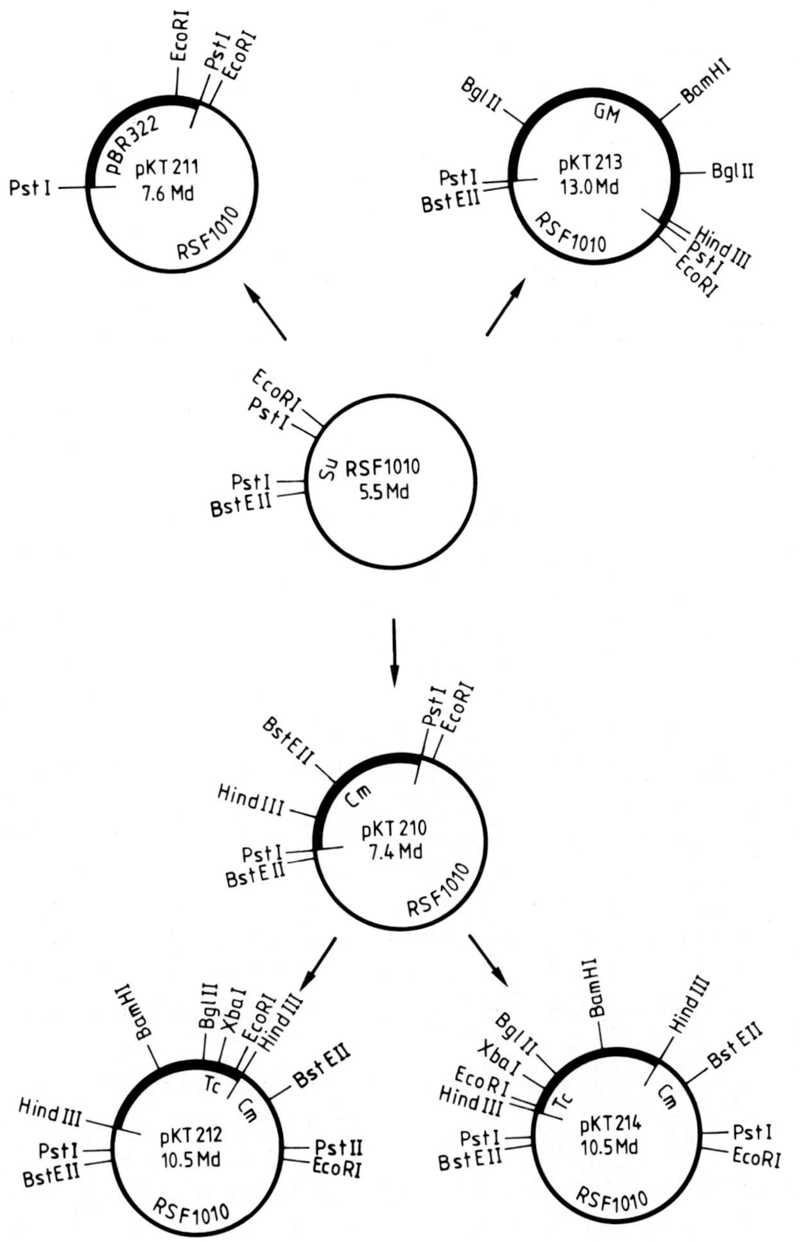

Fig. 2. Schematic diagram of genealogy of new derivatives of RSF1010.

of RSF1010 and the pBR322-Cm hybrid plasmid, pKT205, were digested with PstI endonuclease, treated with T4 DNA ligase, and transformed into E. coli SK1592. Transformant clones that were resistant to Cm and Sm, but sensitive to Tc, contained plasmids which upon digestion with PstI gave two fragments, a 2.4 mD Cm DNA fragment and a 5.0 mD DNA fragment which corresponds to the large PstI fragment of RSF1010. RSF1010-Cm derivatives, namely pKT210 and pKT215, that contain the S-a Cm fragment in both possible orientations were identified (Fig.2). These new plasmids do not specify resistance to Su, indicating that part or all of the Su resistance determinant is located on the smaller PstI fragment of RSF1010. The results also indicate that this fragment is not essential for the replication of RSF1010 because the recombinant plasmid pKT210 can be readily transformed and stably maintained in both E. coli and P.aeruginosa bacteria (Table 2). The pKT210 and pKT215 plasmids contain Sm and Cm resistance determinants and can be used for the cloning of HindIII-generated DNA fragments.

Plasmid pKT007 is a pML21 hybrid containing a 3.1 mD HindIII-generated DNA fragment that carries part of the Tn10 element and the complete tetracycline resistance determinant from plasmid R6[18]. This Tc resistance fragment was inserted into pKT210 in both possible orientations by in vitro recombination of HindIII-digested pKT210 and pKT007 DNAs. The resulting plasmids, designated pKT212 and pKT214, confer resistance to Sm, Tc and Cm, indicating that the HindIII cleavage site in the PstI Cm fragment of S-a lies outside the Cm resistance determinant. The structures of pKT212 and pKT214 are shown in Fig.2. It can be seen that these plasmids contain unique cleavage sites for BamHI, BglII and XbaI, the latter two sites apparently being located within the Tc resistance determinant[24].

Another RSF1010 derivative, designated pKT213, was constructed by the insertion of a PstI-generated 8.0 mD gentamycin-resistance DNA fragment of Pseudomonas plasmid Rms149 between the two PstI sites of RSF1010 (Fig.2). This PstI Gm fragment had previously been cloned into the pBR322 vector to form plasmid pKT208. The structure of pKT213 is shown in Fig.2.

Our attempts to introduce plasmid pBR322 into Pseudomonas strains by transformation were unsuccessful, even though the restriction-deficient *P.aeruginosa* strain 2003 could be transformed efficiently with RSF1010 DNA prepared from *E. coli* (Table 3). Because pBR322 contains unique cleavage sites for the endonucleases commonly used for cloning, we decided to contruct a recombinant between RSF1010 and pBR322. To do this a mixture of *PstI*-digested RSF1010 and pBR322 DNAs were ligated *in vitro* and transformed into *E. coli* SK1592. Plasmid DNA purified from 4 transformant clones that were resistant to Sm and Tc, but sensitive to Ap were analysed by cleavage with *PstI* endonuclease followed by gel electrophoresis. All 4 plasmids contained the two expected *PstI* fragments of molecular weights 2.6 mD and 5.0 mD but digestion with *EcoRI* revealed that only one of the two possible relative orientations of the two fragments was obtained (Fig.2). Although the RSF1010-pBR322 hybrid plasmids are stable in *E.coli*, we have not thus far been able to introduce them into *Pseudomonas*. This interesting observation is currently being investigated further.

DISCUSSION

An efficient and reproducible transformation system is one of the most essential requirements for the establishment of a cloning system in a new host. The results of the present work confirm the previous findings of Chakrabarty et al.[22] and Nagahari and Sakaguchi[8] which demonstrated that *Pseudomonas* cells may be rendered competent for the uptake of plasmid DNA by treatment with $CaCl_2$. The efficiencies of transformation obtained with this host are, however, considerably lower than those obtained with *E.coli* (Table 2). It is possible that isolation of endonuclease-deficient strains of *Pseudomonas*, similar to the *endA sbcB* strain of *E.coli* used by Kushner[11], may lead to an increase in the efficiency of the *Pseudomonas* transformation system. On the other hand, the lower transformation proficiency of *Pseudomonas* may reflect differences in the cell envelope structures of *E.coli* and *Pseudomonas* bacteria.

Our results indicate that the *Pseudomonas* restriction system is a highly effective barrier to the transfer of DNA from *E.coli* to *Pseudomonas* by transformation (Table 3) and confirm the previous results of Nagahari and Sakaguchi[8]. The existence of this barrier makes the interpretation of results concerning the host range specificity of plasmids somewhat difficult. In these studies we found that several small E.coli plasmids could not be introduced into P.putida by transformation while, on the other hand, the Pseudomonas plasmid Rms149 (and its mini-derivative pKT209), which was previously designated a narrow host range plasmid that could not be transferred to E.coli by conjugation[17], was readily introduced into the restriction-negative SK1592 strain by transformation (Table 2).

The unique EcoRI site of RSF1010 has been used previously for the construction of recombinant DNA molecules in vitro[8,10,25]. Results presented here show that RSF1010 can also be used for cloning PstI-generated DNA fragments. Several different DNA fragments were inserted into RSF1010 and the hybrid plasmids thus generated were introduced into E.coli where they were analysed. After characterization, appropriate hybrid plasmids were transferred to a restriction deficient P.aeruginosa. These RSF1010-based plasmids are therefore broad host range cloning vectors[26].

The introduction of DNA fragments carrying new antibiotic resistances into the RSF1010 molecule has increased its usefulness. Insertion of the PstI-generated Cm fragment of the S-a plasmid resulted in the gain of a unique HindIII site. It should be noted that, in contrast to the Cm resistance determinant of R6-5[18], the Cm determinant of the S-a plasmid does not contain an EcoRI cleavage site. Subsequent introduction of a Tc resistance HindIII fragment from plasmid R6 resulted in the gain of unique BamHI, BglII and XbaI sites and a second EcoRI site. In vitro deletion of the small EcoRI fragment of plasmid pKT212 (Fig.2) produced a derivative plasmid that no longer expresses resistance either to Cm or to Tc[26]. This finding demonstrates that the EcoRI cleavage site in the cloned Tc resistance fragment is located within the Tc resistance determinant itself and is consistent with our previous inability to clone an EcoRI-generated Tc resistance fragment from the R6 plasmid[18]. Furthermore, we have inserted DNA fragments into the unique BglII site of pKT212 and have shown that hybrid plasmids thus created are Tc sensitive[26]. The Tc resistance determinant of pKT212 therefore contains cleavage sites for BglII, XbaI and EcoRI (Fig.2). This will permit the ready detection and enrichment of recombinant plasmids containing DNA fragments generated by the XbaI and BglII (and EcoRI, after removal of the EcoRI site in the RSF1010 part of pKT212) endonucleases by insertional inactivation of tetracycline resistance.

The new RSF1010 derivative plasmids described here clearly require substantial reduction in size and extensive characterization in order to serve as optimal cloning vectors. We have recently generated a variety of deletion derivatives of pKT212 and pKT215 and these are now being analysed. We anticipate that a detailed investigation of the structure and replication properties of these small RSF1010 derivatives will increase their usefulness as cloning vectors and provide some insight into the genetic and biochemical basis of the broad host range property of this type of plasmid.

ACKNOWLEDGEMENTS

We are grateful to H.Mayer for generous gifts of PstI endonuclease and to H.Boyer, A.Chakrabarty, S.Cohen, S.Falkow, G. Jacoby and F.Schöffl for providing us with strains.

REFERENCES

1. Clarke, P.H. and Richmond, M.H. (1975) Genetics and Biochemistry of Pseudomonas, John Wiley and Sons, London, New York, Sydney, Toronto.
2. Chakrabarty, A.M. (1976) in: Microbiology-1976, D.Schlessinger ed., ASM Publications, Wash. pp. 579-582.
3. Chakrabarty, A.M. (1976) Ann. Rev. Genet. 10, 7-30.
4. Farrell, R. and Chakrabarty, A.M. (1979). This volume.
5. Bolivar, F., Rodriguez, R.L., Greene, P.J., Betlach, M.C., Heyneker, H.L., Boyer, H.W., Crosa, J. and Falkow, S. (1977) Gene 2, 95-113.
6. Chang, A.C.Y. and Cohen, S.N. (1978) J. Bacteriol. 134, 1141-1156.
7. Benson, S. and Shapiro, J. (1978) J. Bacteriol. 135, 278-280.
8. Nagahari, K. and Sakaguchi, K. (1978) J. Bacteriol. 133, 1527-1529.
9. Barth, P.T. and Grinter, N.J. (1977) J. Mol. Biol. 113, 455-474.
10. Tanaka, T. and Weisblum, B. (1975) J. Bacteriol. 121, 354-362.
11. Kushner, S.R. (1978) in: Genetic Engineering, H.W.Boyer and S.Nicosia eds., Elsevier/North Holland, Amsterdam, pp. 17-23.

12. Lovett, M.A. and Helinski, D.R. (1976) J. Bacteriol. 127, 982-987.

13. Cohen, S. and Chang, A.C.Y. (1973) Proc. Natl. Acad. Sci. USA 70, 293-297.

14. Cohen, S.N. and Chang, A.C.Y. (1977) J. Bacteriol. 132, 734-737.

15. Watanabe, T., Furuse, C. and Sakaizumi, S. (1968) J. Bacteriol. 96, 1791-1795.

16. Guerry, P., Embden, J.V. and Falkow, S. (1974) J. Bacteriol. 117, 619-630.

17. Shapiro, J.A. (1977) in: DNA Insertion Elements, Plasmids and Episomes. A.I.Bukhari, J.A.Shapiro and S.L.Adhya, eds., Cold Spring Harbor Laboratory, pp. 601-670.

18. Timmis, K.N., Cabello, F. and Cohen, S.N. (1978) Molec. Gen. Genet. 162, 121-137.

19. Hansen, J.B. and Olsen, R.H. (1978) J. Bacteriol. 135, 227-238.

20. Timmis, K.N., Cohen, S.N. and Cabello, F. (1978) Progr. Molec. Subcell. Biol. 6, 1-58.

21. Timmis, K.N., Cabello, F. and Cohen, S. (1974) Proc. Natl. Acad. Sci. USA 71, 4556-4560.

22. Chakrabarty, A.M., Mylroie, D.E., Friello, D.A. and Vacca, J.G. (1975) Proc. Natl. Acad. Sci. USA 72, 3647-3651.

23. Rubens, C., Heffron, F. and Falkow, S. (1976) J. Bacteriol. 128, 425-434.

24. Jorgensen, R.A., Berg, D.E., Allet, B. and Reznikoff, W.S. (1979) J. Bacteriol. 137, 681-685.

25. Nagahari, K., Tanaka, T., Hishinuma, F., Kuroda, M. and Sakaguchi, K. (1977) Gene 1, 141-152.

26. Bagdasarian, M., Bagdasarian, M.M., Coleman, S. and Timmis, K.N. (1979). To be submitted.

Plasmids of Medical, Environmental and Commercial Importance
K.N. Timmis and A. Pühler, editors

SCREENING OF SERRATIA MARCESCENS STRAINS FOR EXTRACHROMOSOMAL DNA: DETECTION OF A SMALL PLASMID USEFUL AS A CLONING VECTOR IN GRAM-NEGATIVE BACTERIA

RUDOLF EICHENLAUB AND CHRISTIANE STEINBACH
Lehrstuhl Biologie der Mikroorganismen, Ruhr-Universität Bochum, D-4630 Bochum, FRG

INTRODUCTION

Serratia marcescens is a gram-negative, red-pigmented bacterium. The increasing clinical interest in this bacterium (see review by Grimont and Grimont; 1978) is based on the rising frequency of infections caused by *S. marcescens* strains. Multiple antibiotic resistance of most clinical isolates complicates the therapy of these infections. In some cases R factors were demonstrated in *Serratia* strains which could be transferred to *Escherichia coli*[15]. Many strains of *S. marcescens* produce exoenzymes[12,30] and bacteriocins[9,14], which have become useful epidemiological markers for classification of strains[11,27].

We wanted to determine whether there was a correlation between multiple antibiotic resistance, bacteriocin production and presence of plasmid DNA in *S. marcescens*. In the course of the experiments 16 laboratory strains were investigated. It was found that only two of the strains were harboring plasmid DNA. The plasmids were designated SMP-1 and SMP-2 and were characterized. Plasmid SMP-2 was studied in detail since it has properties allowing the use as a broad host range vector.

MATERIALS AND METHODS

Bacterial strains. *S. marcescens* strains used in this study are listed in Table 1. *E. coli* C600 (*thr-1 leu-6 thi-1 supE44 lac Y1 ton A21* λ^-); *E. coli* W3110 Sm^r; *E. coli* SF8 (*recB21 recC22 lop-11 ton A1 thr-1 leu-6 thi-1 lac Y1 supE44* r_K^- m_K^-); *Salmonella typhimurium* LT2; *Acinetobacter calcoaceticus* BD413 (from D.R. Helinski).

Media. L-broth medium consisting of 10 g Bacto-tryptone, 5 g Bacto-yeast extract, 5 g NaCl per liter. L-broth agar contained

15 g Agar per liter. Selective media contained 15 μg tetracycline per ml, 200 μg penicillin per ml, and 50 μg kanamycin per ml respectively.

Test for antibiotic sensitivity. An assay kit obtained from Bayer-Leverkusen, Germany was used. Paper discs containing kanamycin (30 μg), chloramphenicol (30 μg), penicillin G (10 I.E.), ampicillin (25 μg), oxacillin (5 μg), propicillin (1 E), tetracycline (30 μg), streptomycin (10 μg), and sulfonamides (300 μg) were placed on agar plates layered with soft agar seeded with 10^8 cells of the test bacteria. After incubation the zones of inhibition were evaluated.

Assay for bacteriocin production. Bacteriocin production was induced with mitomycin C and bacteriocin was assayed as previously described[9].

Screening for plasmid DNA. Presence of plasmid DNA in strains of *S. marcescens* was determined as described by Hughes and Meynell (1977). The SDS-lysates obtained by this procedure were analyzed by sedimentation in 5-20 % sucrose gradients and by electrophoresis on 0.8 % agarose gels.

Transformation. Transformation procedures for *E. coli* and *S. typhimurium* were as described[5,18]. Transformation of *A. calcoaceticus* was as reported by Juni (1972) with some modifications: 1 to 2 μg of plasmid DNA were added to 200 μl of L-broth, containing 8 x 10^8 log-phase cells, incubated for 60 min in an ice-bath followed by heat treatment at 42°C for 1 min. The mixture was then incubated with shaking for 90 min and aliquots of 0.1 ml were plated on selective L-broth agar plates.

Isolation of plasmid DNA. For the preparation of cleared lysates the lysozyme-dodecyl sulfate-salt procedure was used[13]. Cleared lysates of *E. coli* were also prepared with lysozyme and Triton X-100[4]. Plasmid DNA was isolated from the lysates by equilibrium centrifugation in CsCl gradients containing ethidium bromide.

Electron microscopy. Plasmid DNA was spread on Parlodion-coated copper grids (400 mesh) using the aqueous technique as described by Davis et al. (1971) and examined in an Siemens Elmiskop 101. Micrographs were taken on 6 by 9 cm sheet film. Open circular ColE1 DNA (2.15 μm) was used as internal length standard.

Restriction reactions: Restrictions with *HindIII* and *BamH1* were carried out in 10 mM Tris-HCl, pH 7.5, 20 mM KCl, 6 mM $MgCl_2$. Digestions with EcoRI and PstI were carried out in 100 mM Tris-HCl, pH 7.5, 50 mM NaCl, and 10 mM $MgCl_2$.

TABLE 1

STRAINS OF SERRATIA MARCESCENS

Strain	Origin	Ref.	Strain	Origin	Ref.
HY	R. Eichenlaub	9,14	CCEB417	O. Lysenko	
CN	U. Winkler	29	Sr20	H. Matsumoto	19
EQ	U. Winkler	29	Nima	R.P. Williams	28
Sm42/S	H.E. Prinsloo	21	CH30	R. Takata	24
Hb	H.E. Prinsloo		ATCC25179	C. Tabor	23
E2	H.E. Prinsloo		Sb	C. Mulder	
FR	R.W. Kaplan		SM-6	S. Falkow	10
CCEB213	O. Lysenko		SM-6F'lac	S. Falkow	10

RESULTS

Antibiotic resistance, bacteriocin production and presence of plasmid DNA. Sixteen strains of Serratia marcescens (Table 1) were tested for their sensitivity for antibiotics, bacteriocin production and presence of plasmid DNA (Table 2). Each of the strains showed a different resistance pattern. With one exception (strain E2) all strains had resistance against sulfonamides, penicillin, propicillin, oxacillin and tetracycline in common. It was observed that most of the strains were sensitive to the aminoglycoside antibiotic kanamycin.

Bacteriocin production was tested with and without induction by mitomycin C. Eigth out of sixteen strains were producing bacteriocins which killed the test strain E. coli W3110.

Plasmid DNA was observed only in three strains, S. marcescens HY, EQ, and SM-6 F'lac, the latter strain acting only as a control. All other strains, although they did not contain plasmid DNA had multiple antibiotic resistance and in some cases were producing bacteriocins. The procedure employed should have allowed the detection of plasmids ranging in size between 2 and 150 Md. The results indicate that in many strains of S. marcescens drug resistance and bacteriocin production are not encoded by extrachromosomal elements.

TABLE 2

ANTIBIOTIC RESISTANCE, BACTERIOCIN PRODUCTION AND PLASMIDS IN STRAINS OF *SERRATIA MARCESCENS*

Strain	Antibiotic resistance[a]									Bacteriocin production	Plasmid DNA
	Tc	Cm	Pn	Pr	Ox	Ap	Sm	Km	Su		
HY	●	●	●	●	●	●	○	○	●	+	+
CN	●	○	●	●	●	○	○	○	●	+	−
EQ	○	●	●	●	●	○	○	○	●	−	+
Sm42/S	●	●	●	●	●	●	●	●	●	+	−
Hb	○	●	●	●	●	○	●	●	●	−	−
E2	○	○	●	●	○	○	○	○	○	−	−
FR	○	○	●	●	●	●	●	○	●	+	−
CCEB213	○	●	●	●	●	○	●	○	●	−	−
CCEB417	●	●	●	●	●	●	●	○	●	+	−
Sr2O	●	●	●	●	●	○	○	●	●	+	−
Nima	○	○	●	●	●	○	○	○	○	−	−
CH3O	●	●	●	●	●	○	●	●	●	−	−
ATCC25179	●	●	●	●	●	●	●	○	●	+	−
Sb	●	●	●	●	●	○	●	●	●	+	−
SM-6	●	●	●	●	●	●	●	●	●	−	−
SM-6 F'*lac*	●	●	●	●	●	●	●	●	●	−	+[b]

[a] Abreviations used: Tc - tetracycline; Cm - chloramphenicol; Pn - penicillin; Pr - propicillin; Ox - oxacillin; Ap - ampicillin; Sm - streptomycin; Km - kanamycin; Su - sulfonamides.
[b] Only plasmid detected F'*lac*. ●, resistant; o, sensitive.

Characterization of S. marcescens plasmids. The plasmids detected in *S. marcescens* strains HY and EQ were designated SMP-1 and SMP-2 respectively.

In the electron microscope plasmid SMP-1 has a contour length of 17.9 $\pm$ 0.6 μm (Fig. 1A), corresponding to a molecular weight of 37 Md. It is present in the cells of *S. marcescens* with 3-4 copies per chromosome. SMP-1 is most probably identical with a plasmid from HY strains described by Timmis and Winkler (1973a). They showed that this plasmid was bacteriocinogenic, since curing of strain HY from the plasmid by acridine orange resulted in loss of bacteriocin production (marcescin A). This plasmid can neither be transfered into *E. coli* nor co-transfered by the derepressed R factor R1-drd19[26].

Plasmid SMP-2 has a contour length of 1.25 ± 0.1 μm (Fig. 1B), corresponding to a molecular weight of 2.6 Md. The copy number per *Serratia* chromosome is 8-10. Plasmid SMP-2 has single restriction sites for the restriction endonucleases *EcoRI*, *HindIII*, and *BamH1* (Fig. 2).

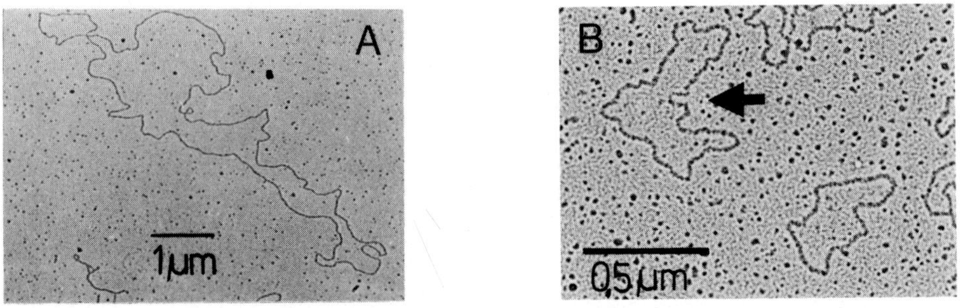

Fig. 1. Electron micrographs of plasmids SMP-1 (A) and SMP-2 (B). Arrow in (B) indicates ColE1 marker.

Construction of hybrid plasmids derived from SMP-2. In order to be able to select for the presence of plasmid SMP-2 in a bacterial host and to facilitate the study of its properties, it was necessary to link a selective marker to the plasmid. An *EcoRI* generated fragment, carrying resistance to kanamycin (Km^r) originally derived from pSC105[6], was integrated into the single *EcoRI* site of SMP-2. The resulting plasmid was designated pCS21. In a similar experiment an *EcoRI* fragment from the *Staphylococcus aureus* plasmid pI258, coding for resistance against penicillin-ampicillin (Ap^r)[20], was linked to SMP-2 resulting in plasmid pCS31 (Fig. 2). Both plasmids were cloned in the restriction and modification deficient *E. coli* SF8 to overcome the restriction barrier, which otherwise drastically reduced transformation frequencies. Besides confering Km^r or Ap^r on the host the only phenotypic change observed in *E. coli* cells harboring SMP-2 hybrids was a reduced sensitivity to colicin E1. Plasmids pCS21 and pCS31 are stably maintained in *E. coli*, when grown for 20 generations no segregation was observed. Plasmid pCS21 replicates in E. coli polAts214 at 42°C indicating no dependency for DNA-polymerase I. The copy number is 5-7 per chromosome and SMP-2 derivatives are not amplified by chloramphenicol (250 μg/ml).

When SMP-2 is linked to plasmid ColE1 Ap (RSF2124) the hybrid replicon plasmid pCS22 seems to be under replication control of SMP-2. In E. coli C600 the copy number of pCS22 is only 3-5 per chromosome and chloramphenicol treatment (250 μg/ml) does not result in an amplification.

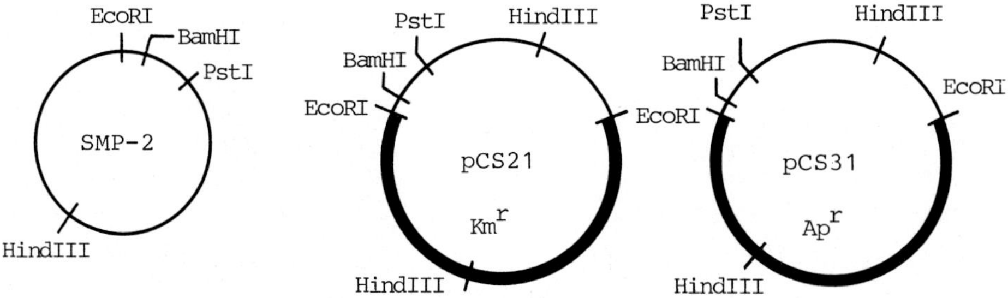

Fig. 2. Restriction map of plasmids SMP-2, pCS21, and pCS31. Heavy line indicates Km^r and Ap^r fragments respectively.

Bacterial hosts for derivatives of plasmid SMP-2. As shown above E. coli can be used as a host bacterium for derivatives of plasmid SMP-2, in which it is stably maintained.

Repeated experiments to reintroduce the hybrid plasmids derived from SMP-2 into S. marcescens using $CaCl_2$ treatment to achieve competence for DNA uptake were unsuccessful. However, in other gram-negative bacteria such as Salmonella typhimurium and Acinetobacter calcoaceticus derivatives of the Serratia plasmid could be established. Transformation of these strains gave rise to antibiotic resistant transformants with frequencies of 0.1 - 1 x 10^7. transformants per viable cell per μg of plasmid DNA. Reisolation of plasmid DNA from transformed clones of S. typhimurium and A. calcoaceticus and subsequent transformation improved transformation frequencies to 1 - 4 x 10^{-3} transformants per viable cell per μg of plasmid DNA. Low transformation frequencies obtained with plasmid DNA isolated from E. coli can be explained by restriction phenomena. Similar observations have been reported when S. typhimurium was transformed with plasmid pSC101 DNA isolated from E. coli[18].

The stability of pCS21 and pCS31 in S. typhimurium and A. calcoaceticus is not as high as in E. coli. We observed that in S. typhymurium 5 % of the cells lost the plasmid per generation. In A. calcoaceticus the segregation rate was 15 % per generation. However, under selective pressure both plasmids can be maintained in these strains without difficulty.

In control experiments with plasmid pBR322[22], widely used as a cloning vector in E. coli, no viable transformants of Salmonella and Acinetobacter were obtained.

Construction of a cloning vector derived from SMP-2. The properties of SMP-2 derivatives described above, may allow use of these plasmids as cloning vectors. To facilitate cloning with the Serratia plasmid we have constructed plasmid pCS41 by a series of manipulations which included (i) deletion of one HindIII and one BamH1 site in the SMP-2 part of pCS21 and (ii) insertion of transposon TnA into the EcoRI fragment carrying Km^r (Fig. 3). Details of the construction of pCS41 will be described elsewhere (Eichenlaub, in preparation). The molecular weight of pCS41 is 9.9 Md. The plasmid does not self-transfer and when mobilized by a derepressed R100 only 0.03 % of the recipient cells show co-transfer of pCS41 with R100 drd. The single HindIII site which is retained in pCS41 allows cloning of HindIII generated fragments, resulting in insertional inactivation of Km^r [3]. Bacterial clones harboring hybrid plasmids are easily detected by selection for Ap^r and screening for kanamycin sensitivity.

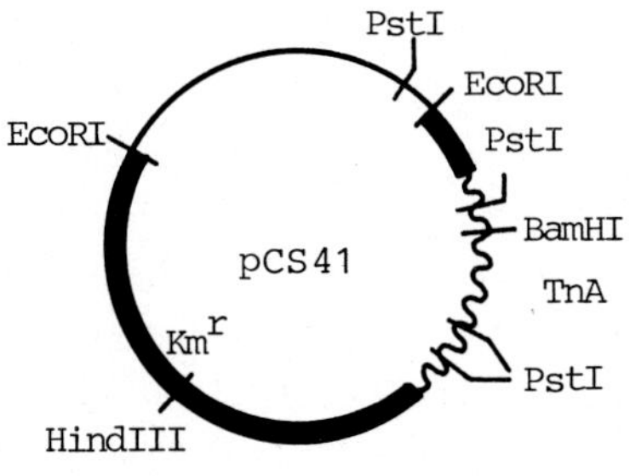

Fig. 3. Restriction map of pCS41. Heavy line indicates Km^r fragment, waved line indicates position of TnA.

DISCUSSION

The screening of 16 strains of *S. marcescens* showed that multiple antibiotic resistance is not necessarily plasmid encoded, although clinical isolates often carry R factors[12,15]. Other possibilities are that in *Serratia* the antibiotics do not interact normally with their target sites or that the membrane is not permeable for the drugs. The latter possibility seems to be more likely as an explanation for multiple antibiotic resistance in *S. marcescens*. This is supported by the observation that oxacillin sensitive mutants derived from *S. marcescens* SM-6 have an altered membrane composition and have become simultaneously sensitive to tetracycline and chloramphenicol[30].

The extensive use of antibiotics, especially in intensive care units will suppress competing bacteria, thus allowing naturally resistant *S. marcescens* to colonize new areas. During contact with other bacteria *Serratia* can acquire R factors which may further extend its spectrum of resistance. Consequently, on top of causing infections resistant to therapy by most antibiotics *S. marcescens* may become a dangerous reservoir for R factors.

The bacteriocin production parallels the resistance against antibiotics in being not related to the presence of plasmid DNA in six out of sixteen strains. Strain HY is the only one where there is evidence for a bacteriocinogenic plasmid[8,26]. Bacteriocins of *S. marcescens* also include high molecular weight bacteriocins[9] which may originate from defective phages[1]. Surprisingly SMP-2 derivatives reduce the sensitivity of an *E. coli* host to colicin E1. However, we observed that resistance to the killing effect of colicin E1 on SMP-2 containing cells is much lower as compared to colicin E1 immune *E. coli* cells. At present the question remains open whether this phenomenon is related to colicin immunity or not.

Plasmids derived from SMP-2 have been introduced into *E. coli*, *S. typhimurium* and *A. calcoaceticus*. However, all experiments to transform these plasmids into *S. marcescens* failed. A possible explanation is that no competence for DNA uptake was achieved when using the conventional $CaCl_2$ treatment[5]. It is also conceivable that restriction prevented successful transformation, since the plasmid DNA used was isolated from *E. coli*. All SMP-2 derivatives have recognition sites for the restriction endonuclease *SmaI*.

In the hybrid replicon pCS22, constructed by linking of EcoRI cleaved SMP-2 and RSF2124, replication is under the control of SMP-2. This observation which is in contrast to the dominance of ColE1 replication in hybrid replicons such as pSC134 (pSC101-ColE1)[2] deserves closer examination.

The hybrid plasmid pCS41 which we have described in this report may be useful for DNA cloning by the following reasons. (i) It has a moderate copy number which still allows to isolate plasmid DNA in reasonable yields. This is also of advantage when gene products of hybrid plasmids excert negative effects on the viability of the host[16]. (ii) Plasmid pCS41 facilitates detection of hybrid clones by screening for inactivation of Km^r. (iii) pCS41 does not transfer and shows low mobilization frequencies which are important factors in biological containment. (iiii) pCS41 posses an extended host range which is advantageous for cloning in other gram-negative bacteria. Plasmid pCS41 has a size of 9.9 Md. We are presently reducing the size of the plasmid by deleting PstI and Sau3A fragments. This strategy should also delete parts of TnA needed for translocation. In addition it is possible to reduce TnA to the size of the β-lactamase gene, containing a single PstI site. The plasmid may then also be used for cloning of PstI fragments allowing screening for hybrids by insertional inactivation of the bla gene[22].

ACKNOWLEDGEMENTS

We thank all colleagues who have supplied us with bacterial strains. The valuable technical assistance of U. Raeder, M. Hilgemann, and R. Bortlisz is gratefully acknowledged.

REFERENCES

1. Bradley, D.E. (1967) Bacteriol. Rev., 31, 230-314.
2. Cabello, F., Timmis, K. and Cohen, S.N. (1976) Nature, 259, 285-290.
3. Cabello, F., Timmis, K. and Cohen, S.N. (1978) Microbiology-1978, American Society for Microbiology, Washington, D.C., pp. 42-44.
4. Clewell, D.B. and Helinski, D.R. (1969) Proc. Natl. Acad. Sci. U.S.A., 62, 1159-1166.
5. Cohen, S.N., Chang, A.C.Y. and Hsu, L. (1972) Proc. Natl. Acad. Sci. U.S.A., 69, 2110-2114.

6. Cohen, S.N., Chang, A.C.Y., Boyer, H.W. and Helling, R.B. (1973) Proc. Natl. Acad. Sci. U.S.A., 70, 3240-3244.

7. Davis, R.W., Simon, M. and Davidson, N. (1971) Methods Enzymol., 21, 413-428.

8. Eichenlaub, R. and Winkler, U. (1974) Biochem. Biophys. Res. Comm., 59, 133-139.

9. Eichenlaub, R. and Winkler, U. (1974) J. Gen. Microbiol., 83, 83-94.

10. Falkow, S., Marmur, J., Carey, W.F., Spilman, W.M. and Baron L.S. (1961) Genetics, 46, 703-706.

11. Grimont, P.A.D., Grimont, F. and Dulong de Rosnay, H.L.C. (1977) J. Gen. Microbiol., 99, 301-310.

12. Grimont, P.A.D. and Grimont, F. (1978) Ann. Rev. Microbiol., 32, 221-248.

13. Guerry, P., Le Blanc, D.J. and Falkow, S. (1973) J. Bacteriol., 116, 1064-1066.

14. Hamon, Y. and Péron, Y. (1966) Ann. Inst. Pasteur (Paris), 110, 556-561.

15. Hedges, R.W., Rodriguez-Lemoine, V. and Datta, N. (1975) J. Gen. Microbiol., 86, 88-92.

16. Hershfield, V., Boyer, H.W., Yanofsky, C., Lovett, M. and Helinski, D.R. (1974) Proc. Natl. Acad. Sci. U.S.A., 71, 3455-3459.

17. Juni, E. (1972) J. Bacteriol., 112, 917-931.

18. Lederberg, E.M. and Cohen, S.N. (1974) J. Bacteriol., 119, 1072-1074.

19. Matsumoto, H., Tazaki, T. and Hosogaya, S. (1973) Japan. J. Microbiol., 17, 473-479.

20. Novick, R.P. and Bouanchaud, D. (1971) Ann. N.Y. Acad. Sci., 182, 279-294.

21. Prinsloo, H.E. (1966) J. Gen. Microbiol., 45, 205-212.

22. Rodriguez, R.L., Tait, R., Shine, J., Bolivar, F., Heyneker, H., Betlach, M. and Boyer, H.W. (1977) Molecular cloning of recombinant DNA, Academic Press, Inc. N.Y., pp. 73-84.

23. Tabor, C.W. and Kellogg, P.D. (1970) J. Biol. Chem., 245, 5424-5433.

24. Takata, R. and Kobata, K. (1976) Molec. gen. Genet., 149, 159-165.

25. Timmis, K. and Winkler, U. (1973,a) J. Bacteriol., 113, 508-509.

26. Timmis, K. and Winkler, U. (1973,b) Mol. gen. Genet., 124, 207-217.

27. Traub, W.H., Raymond, E.A. and Startsman, T.S. (1971) Appl. Microbiol., 21, 837-840.

28. Williams, R.P., Green, J.A. and Rappoport, D.A. (1956) J. Bacteriol., 71, 115-120.

29. Winkler, U. (1968) Z. Naturforsch., 18 b, 118-123.

30. Winkler, U., Heller, K.B. and Folle, B. (1978) Arch. Microbiol., 116, 259-268.

VII – GENE CLONING WITH PLASMIDS

Plasmids of Medical, Environmental and Commercial Importance
K.N. Timmis and A. Pühler, editors

CLONING IN ESCHERICHIA COLI THE GENOMIC REGION OF KLEBSIELLA PNEUMONIAE WHICH ENCODES GENES RESPONSIBLE FOR NITROGEN FIXATION

A.Pühler, H.J.Burkardt and W.Klipp
Institut für Mikrobiologie und Biochemie, Lehrstuhl Mikrobiologie der Universität Erlangen,
852o Erlangen, FRG

INTRODUCTION

The genetics of nitrogen fixation (nif) is best known from investigations of Klebsiella pneumoniae which is a free living N_2-fixing species. This organism is closely related to Escherichia coli and the his-nif-region of K.pneumoniae could be transferred to E.coli. N_2-fixing E.coli strains were obtained[1]. Various plasmids have been constructed in E.coli which carry the his-nif-region e.g. the F-prime plasmids FN39 and FN68[2]. The most important plasmid pRD1, formerly called RP41, was derived from RP4 by fusion with FN68[3]. Besides the RP4 markers Ap, Km, Tc and Tra^+, pRD1 carries the nif-region and additional genes which can complement gnd, his and shiA mutations in E.coli. Because RP4 has a broad host range[4] plasmid pRD1 can be transferred to a large variety of gram negative bacteria. Some of them e.g. Salmonella typhimurium[5] were able to fix nitrogen after acquisition of pRD1.

Plasmid pRD1 has a molecular weight of 101 Mdals and is unstable in $recA^+$ hosts[6]. It is therefore not ideal for studying the nif-genes of K.pneumoniae. Cloning of the nif-region onto appropriate plasmid vehicles should facilitate these studies. To construct such plasmids we first analyzed the genetic and physical organization of pRD1 and isolated deletion mutants which enabled us to localize the nif-region on the restriction map of pRD1.

EcoRI AND HindIII RESTRICTION MAPS OF pRD1

In order to construct a physical map of pRD1 we first digested pRD1 DNA using the endonucleases EcoRI and HindIII. The resulting fragments, separated by agarose gel electrophoresis, are shown in

Fig. 1. 19 EcoRI and 15 HindIII fragments could be identified. However there is at least 1 additional EcoRI and 1 additional HindIII fragment with molecular weights less than 0.4 Mdals.

A minimum of 36 fragments would be expected following an EcoRI/HindIII double digestion of pRD1. 31 of these fragments can be seen on the gel in Fig. 1. The molecular weights of the fragments are given in the legend to this figure. Summing the molecular weights of EcoRI and HindIII fragments provides a molecular weight estimation for pRD1 between 101 and 103 Mdals which is in good agreement with electron microscopic length measurements[6].

To construct the pRD1 restriction map we cloned the EcoRI and HindIII fragments separately in λ and plasmid vehicles and determined the EcoRI and HindIII sites on the cloned fragments. Using this information a fairly complete restriction map could be established. In Fig. 2 the EcoRI and HindIII restriction maps of pRD1 are shown. The fragment designations are according to their numbering in Fig. 1. 20 EcoRI fragments, 16 HindIII fragments and 36 EcoRI/HindIII fragments are arranged in an unequivocal sequence.

ISOLATION AND CHARACTERIZATION OF DELETION MUTANTS OF PLASMID pRD1

The aim of our study was to clone the nif-region on a multicopy plasmid vehicle. We therefore decided to map first the nif-region on the restriction map of pRD1 by isolation of deletion mutants of pRD1 using the P1 transduction method. Phage P1, capable of generalized transduction, is not able to package the whole pRD1 plasmid whereas deletion mutants of pRD1, more or less in the size of the P1 genome, should be packaged. Since we did not use deletion inducing methods the transduction rate was low. Two transductants were further characterized. They carry plasmids with the following markers: Ap, Km, Tc, Tra^+, His^+ and Nif^+. We isolated the plasmid DNAs and by electron microscopy we found that one plasmid was 41 μm and the other 32 μm long. The longer plasmid was designated pAP1, the other pAP2. Evidently

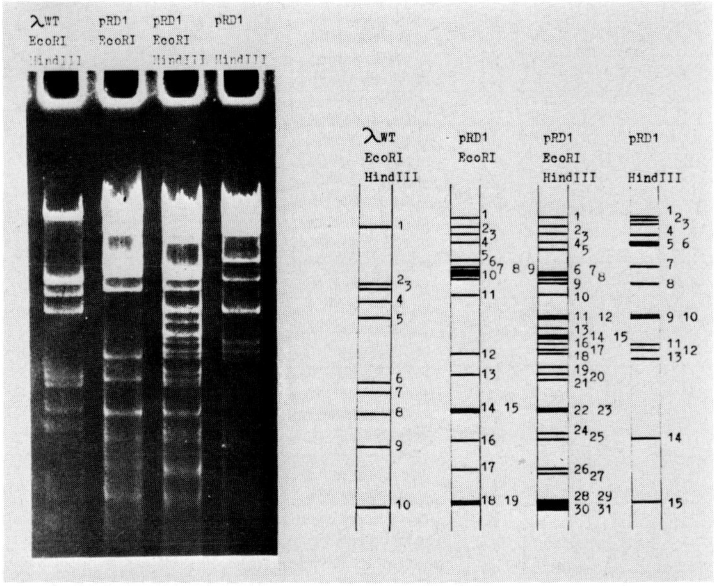

Fig. 1. Restriction fragments of plasmid pRD1 separated by agarose gel electrophoresis.
pRD1 DNA was digested by EcoRI, by HindIII and by EcoRI/HindIII. The fragments were separated on an agarose gel. In addition the EcoRI/HindIII fragments of λ -DNA were used as molecular weight reference markers. For the different fragments of pRD1 the following molecular weights (in Mdals) were obtained:

EcoRI digestion		HindIII digestion		EcoRI/HindIII double digestion			
E1:	34	H1:	24	EH1:	22	EH16:	1.60
E2:	14	H2:	20	EH2:	13.6	EH17:	1.54
E3:	10	H3:	15	EH3:	10	EH18:	1.50
E4:	8.8	H4:	10	EH4:	8.2	EH19:	1.30
E5:	5.0	H5:	8.2	EH5:	7.0	EH20:	1.20
E6:	4.2	H6:	8.2	EH6:	3.7	EH21:	1.10
E7:	3.9	H7:	4.4	EH7:	3.7	EH22:	0.90
E8:	3.5	H8:	3.1	EH8:	3.4	EH23:	0.90
E9:	3.5	H9:	2.0	EH9:	3.1	EH24:	0.84
E10:	3.4	H10:	2.0	EH10:	3.0	EH25:	0.80
E11:	3.0	H11:	1.6	EH11:	2.0	EH26:	0.70
E12:	1.5	H12:	1.54	EH12:	2.0	EH27:	0.66
E13:	1.2	H13:	1.5	EH13:	1.8	EH28:	0.60
E14:	0.9	H14:	0.8	EH14:	1.7	EH29:	0.60
E15:	0.9	H15:	0.6	EH15:	1.7	EH30:	0.60
E16:	0.8					EH31:	0.60
E17:	0.7						
E18:	0.6						
E19:	0.6						

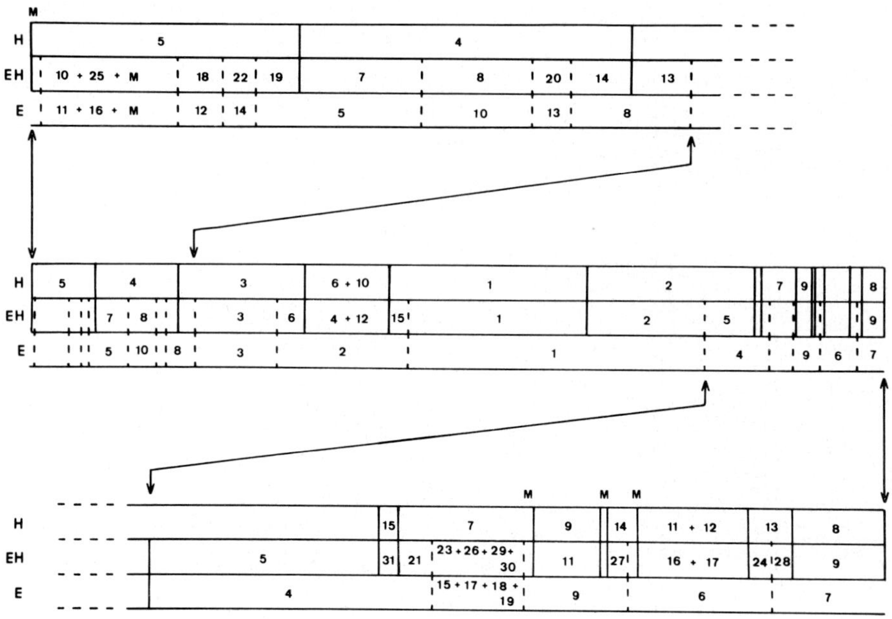

Fig. 2. EcoRI and HindIII restriction map of plasmid pRD1. The complete restriction maps of pRD1 for the enzymes EcoRI and HindIII are given in the middle of the figure. The ends of the linearized maps are drawn separately at a larger scale (4x). The EcoRI fragments (E), the HindIII fragments (H) and the EcoRI/HindIII fragments (EH) are numbered according to their position on the agarose gel of Fig. 1. Fragments which could not be detected on that gel were marked with "M" (minifragments).

pAP1 and pAP2 are deletion mutants of pRD1, which still carry the RP4 markers and the his-nif-region of K.pneumoniae. The deleted segments in the plasmids pAP1 and pAP2 were mapped by HindIII restriction analysis. The HindIII fragments of the plasmids pRD1, pAP1 and pAP2 separated by agarose gel electrophoresis are shown in Fig. 3. Comparing the restriction patterns of the different plasmids the following results are obtained:

- The first 6 HindIII fragments of pRD1 and pAP1 appear identical.
- The smaller HindIII fragments of pRD1, with the exception of

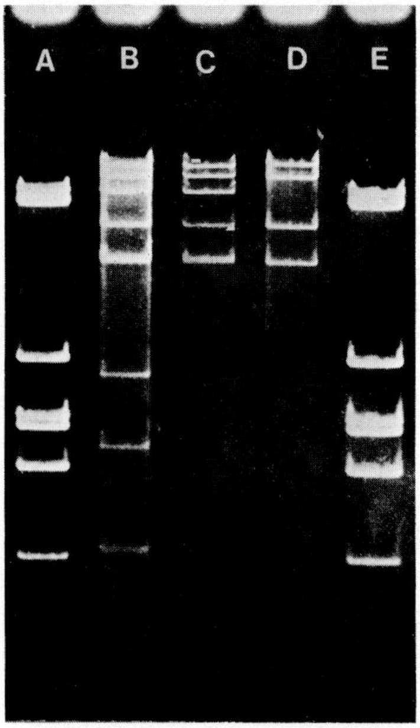

Fig. 3. The HindIII fragments of the plasmids pRD1, pAP1 and pAP2 separated by agarose gel electrophoresis.
DNAs of plasmids pRD1, pAP1 and pAP2 were digested by the endonuclease HindIII and the fragments were separated by agarose gel electrophoresis: (B) HindIII fragments of pRD1, (C) HindIII fragments of pAP1, (D) HindIII fragments of pAP2, (A) and (E) EcoRI fragments of λ-DNA.

fragment H10, are absent in the HindIII digest of pAP1.

- pAP2 consists of only 4 large HindIII fragments.

With this information the deleted segments in pAP1 and pAP2 can easily be located.

Plasmid pAP1 can be derived from pRD1 by deleting a DNA segment with 8 HindIII sites. The deletion designated A, is depicted in Fig. 4. The remaining DNA still possesses 7 HindIII sites whose locations are consistent with the HindIII pattern of the plasmid pAP1.

Plasmid pAP2 can be derived from plasmid pAP1 by an additional deletion, designated B, which is also shown in Fig. 4. The 4 HindIII sites remaining after removal of DNA by deletions A and B generate the HindIII fragment pattern of plasmid pAP2.

The length of the deletions A and B can be calculated from the contour lengths of the plasmids pRD1, pAP1 and pAP2. Deletion A is 8 μm and deletion B 9 μm long.

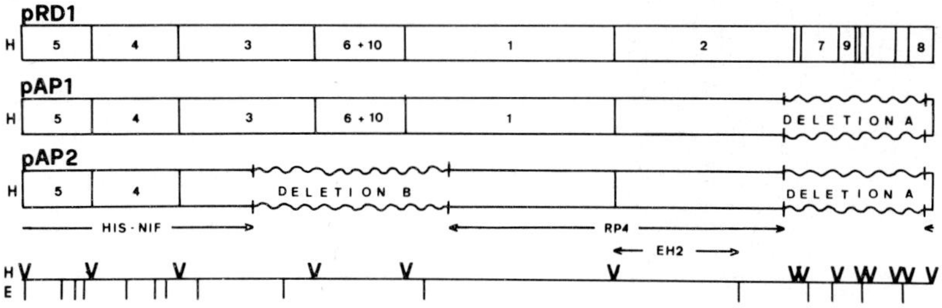

Fig. 4. The HindIII restriction maps of the plasmids pRD1, pAP1 and pAP2 and the location of the integrated RP4 plasmid and the his-nif-region.
The complete EcoRI and HindIII restriction map of pRD1 is drawn. The structure of the plasmids pAP1 and pAP2 is derived from this map using the results described in Fig. 3.
(H,**V**: HindIII sites, E, **|** : EcoRI sites. The HindIII fragments are numbered according to Fig. 1)

CORRELATION OF GENETIC MARKERS WITH THE PHYSICAL MAP OF pRD1: THE LOCATION OF THE NIF REGION

The location of the RP4 molecule within the physical map of pRD1 can be determined because plasmid RP4 is cut only once by the enzymes EcoRI and HindIII. An EcoRI/HindIII double digestion of RP4 DNA (38.2 Mdals[7]) provides two fragments of 24.4 Mdals and 13.8 Mdals[8]. The 13.8 Mdals fragment can also be found amongst the EcoRI/HindIII double digest fragments of pRD1. Fragment EH2, with a molecular weight of 13.6 Mdals, indicates the location of RP4. In Fig. 4 the RP4 part of pRD1 is drawn so that the deletions A and B flank the RP4 DNA. One can speculate that the ends of the RP4 DNA are IS-like and may therefore have the ability to induce deletions in the adjacent DNA. As plasmid pAP2 is His^+, Nif^+ and only contains the fragments H5, H4 and parts of H3 and H8 in addition to RP4 DNA it must be assumed that the nif encoding genes are located wihtin the remaining fragments.

CLONING THE NIF-REGION of pRD1 ONTO THE MULTICOPY PLASMID VEHICLE pWL625

To clone the entire nif-region of K.Pneumoniae we used the following strategy. A combination of fragments H8, H5, H4 and H3 of the plasmid pRD1 should give the functional nif-region and since there were some hints that the fragments H8 and H5 of pRD1 did not carry nif-genes, we first designed an experiment to connect fragments H4 and H3 in the same alignment as in pRD1 and to clone them in a multicopy plasmid vehicle. The cloning scheme was as follows. We first cloned H4 and H3 separately in the plasmid vehicle pWL625[9], which carries genes for ampicillin and kanamycin restistance. pWL625 possesses 1 HindIII site which is located directly in the kan-r gene. Cloning HindIII fragments into this site destroys the gene and therefore hybrid plasmids are recognizable by ampicillin resistance and kanamycin sensitivity of transformed cells. The hybrid plasmids pWK1 (pWL625 + H4) and pWK2 (pWL625 + H3) constructed in this way were not able to provide nitrogen fixation ability to E.coli cells (Table 1). Incomplete HindIII digestion was used to linearize pWK1 and fragment H3, obtained from a complete digestion of pWK2, was inserted - resulting in the hybrid plasmid pWK120 (pWL625 + H4 + H3). The fragments H4 and H3 were in the same alignment as in pRD1. To ensure the correct orientation an E.coli C strain was transformed by the ligation mixture and transformants were slected for ability to fix nitrogen in nitrogen free medium. A significant increase in optical density indicated E.coli C cells, which were Nif^+. Single colonies of this culture were ampicillin resistant, kanamycin sensitive and could reduce acetylene, indicating an active nitrogenase (Table 1).

TABLE I

PROPERTIES OF THE PLASMIDS pWL625, pWK1, pWK2 AND pWK120

Plasmid	Ap	Km	Nif^+
pWL625	+	+	-
pWK1	+	-	-
pWK2	+	-	-
pWK120	+	-	+

pWK1 = pWL625 + H4
pWK2 = pWL625 + H3
pWK120 = pWL625 + H4 + H3

Plasmid DNA purified from such a colony was HindIII digested and analyzed by agarose gel electrophoresis. A comparison of the Hind III fragments of plasmids pWL625, pWK1, pWK2 and pAP1 with the HindIII fragments of the newly isolated plasmid shows that pWK120 has the desired structure (Fig. 5).

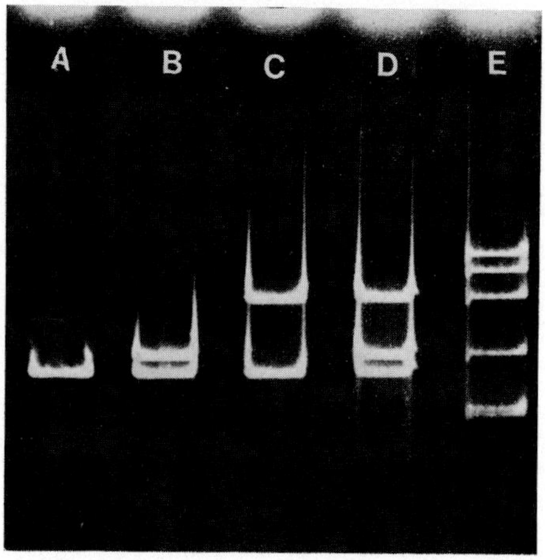

Fig. 5. Agarose gel electrophoresis of HindIII fragments of the plasmids pWL625, pWK1, pWK2, pWK120 and pAP1.
The HindIII fragments of pWL625 (A), pWK1 (B), pWK2 (C), pWK120 (D) and pAP1 (E) are shown on an agarose gel.

Since E.coli C (pWK120) is able to fix nitrogen it may be assumed that pWK120 carries all essential genes for nitrogen fixation. In addition, EcoRI digestion of pWK120 demonstrated the presence of fragment E8 of pRD1 in pWK120, indicating that H4 and H3 are combined in pWK120 in the same alignment as in pRD1.

MOLECULAR PROPERTIES OF THE MULTICOPY NIF-PLASMID pWK120

We have shown by electron microscopy that plasmid pWK120 has a contour length of 17 μm which corresponds to a molecular weight of 34 Mdals. This measurement is in good agreement with the molecular weight determination of pWK120 derived from its HindIII fragments (pWL625: 10 Mdals, H4: 10 Mdals and H3: 15 Mdals).

The plasmid vehicle pWL625 is a multicopy plasmid. In order to determine the copy number of pWK120, E.coli C (pWK120) was grown in nitrogen free medium, harvested and lysed by the sarkosyl method. The crude lysate was then analyzed in a CsCl-gradient by analytical ultracentrifugation. The resulting DNA-profile is shown in Fig. 6. In addition to the chromosomal DNA of E.coli C (1.71 g/cm^3) there is a second peak of 1.715 g/cm^3 density. The denser material is pWK120 DNA. Since the plasmid peak, is higher than the chromosomal peak a high copy number of the plasmid pWK120 is indicated. By comparing these two peaks, and taking into account the lengths of the E.coli chromosome (1100 µm) and the plasmid pWK120 (17 µm), a copy number of nearly 65 can be deduced.

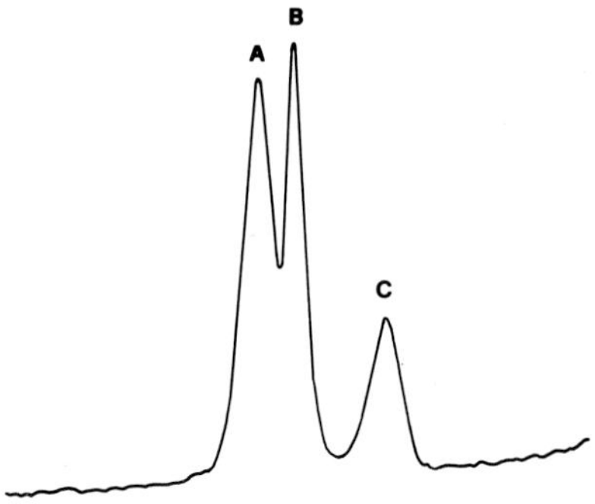

Fig. 6. Copy number determination of plasmid pWK120
A crude lysate of E.coli C (pWK120) cells was centrifuges in a CsCl gradient of an analytical ultracentrifuge. The resultant DNA profile is shown.
A: chromosomal DNA of E.coli C (1.710 g/cm^3)
B: pWK120 DNA (1.715 g/cm^3)
C: reference DNA of Micrococcus lysodeicticus (1.7308 g/cm^3)

REMARKS ON THE EVOLUTION OF THE NIF-REGION FROM K.PNEUMONIAE

Since we have cloned most of the EcoRI and HindIII fragments of pRD1 we were able to measure the density of these fragments by analysis of CsCl-gradients using analytical ultracentrifugation. Surprisingly the K.pneumoniae part of pRD1 has a very heterogeneous density. The results for a larger DNA-segment of pRD1 are compiled in Fig. 7. The highest density was found for fragment H4 which is believed to contain exclusively nif-genes. Its density, 1.72 g/cm^3, is very different from the normal chromosomal density of Klebsiella (1.715 g/cm^3). Since a similar high density is found for Rhizobium DNA we postulate that the nif-region of Klebsiella was obtained from soil bacteria. This hypothesis can be tested by hybridization experiments of nif-DNA from Klebsiella with nif-DNA from Rhizobium species.

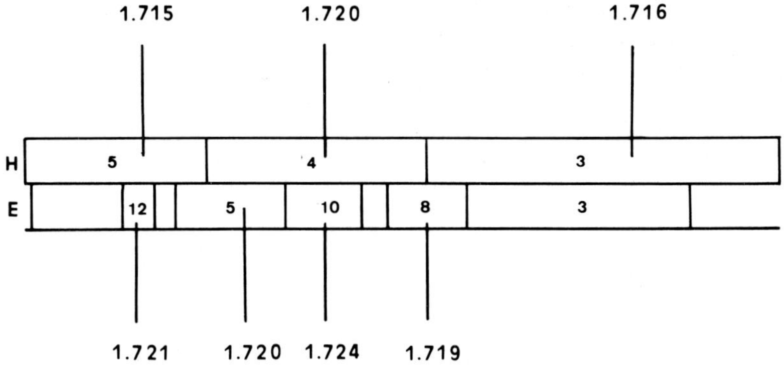

Fig. 7. Part of the density map of pRD1.
The buoyant density of different EcoRI and HindIII fragments of pRD1 was determined by CsCl-gradient centrifugation in an analytical ultracentrifuge. Most of the nif-genes are localized on fragment H4.

In addition to this very high density we found that the nif-region is flanked by a series of inverted repeats. These inverted repeats are located in the middle of fragment H5 and at the beginning of fragment H3. Figure 8 demonstrates how these inverted repeats were found. We started with the hybrid plasmid pWK120 which consists of the vector pWL625 and fragment H3 and H4. In a homoduplex experiment we looked for "snap-back" structures in single stranded pWK120 DNA. Fig. 8 shows such a typical single stranded pWK120 molecule. Several snap back structures, indicating inverted repeats are shown. Two of them belong to the plasmid vector pWL625, which contains Tn3 and the kan-fragment of pML21[9]. In addition an interesting structure, called "cat's head with two ears", can be detected which we located in fragment H3. A similar structure has been found in the fragment H5, which borders the nif-region on the left side. This structure resembles a "cat's head with one ear". These snap back structures bordering the nif-region of K.pneumoniae are summarized in Fig. 9. We interpretate these two sets of inverted repeats at the left and at the right end of the nif-region as follows: The nif-region of pRD1 flanked by two sets of inverted repeats represents a transposon like

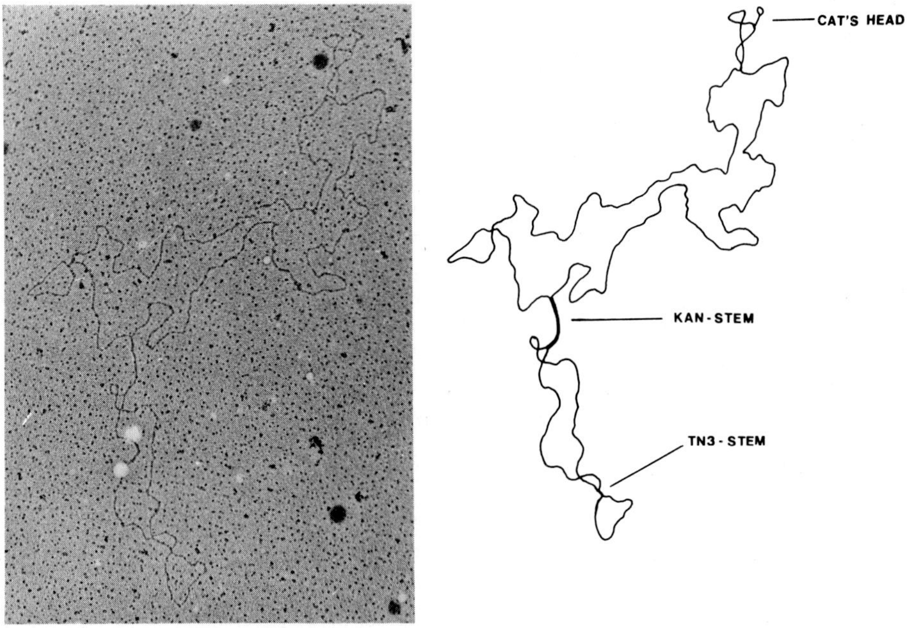

Fig. 8. A circular single stranded pWK120 molecule with snap back structures
The hybrid plasmid pWK120 contains the plasmid vector pWL625 and the cloned fragments H4 and H3 of pRD1. pWL625 carries the ampicillin transposon Tn3. In Tn3 the kan-fragment of pML21 was inserted by genetic engineering. This pWL625 contains the inverted repeats of Tn3 and of the kan-fragment forming the Tn3-stem and the kan-stem in homoduplex experiments. In the cloned H3 + H4 segment the typical "cat's head with two ears" can be seen.

structure. This structure is perhaps a precondition for the postulated genetic exchange process by which the nif-genes from soil bacteria were integrated into the Klebsiella chromosome. The "cat's head" structures border very closely the segment of high density DNA of the pRD1 plasmid. It remains to be shown whether parts of fragment H5 carry genes encoding functions which are involved in the nitrogen fixation process. Such possible nif-genes are not necessary for the nif^{+} phenotype in E.coli C, as this fragment is not present in plasmid pWK120.

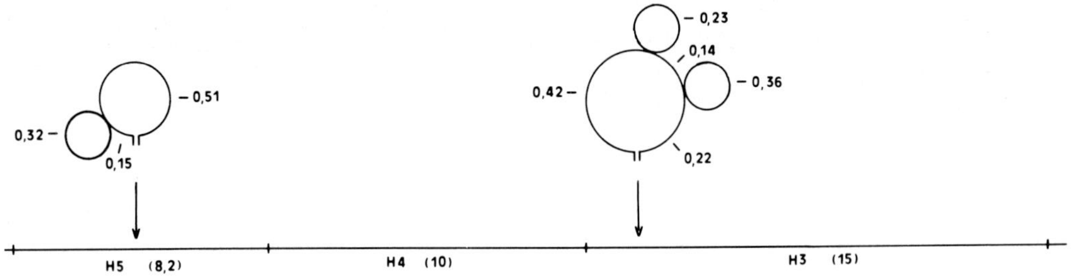

Fig. 9. Snap back structures bordering the nif-region of K.pneumoniae
The location of snap back structures (cat's heads) found in homoduplex experiments with the fragments H5, H4 and H3 of pRD1 cloned in pWL625 are indicated. They flank the genomic region encoding nitrogen fixation. Lengths are given in µm and were calculated from several molecules.

SUMMARY

The nif-region of K.pneumoniae present on the plasmid pRD1 was cloned on a multicopy plasmid vehicle in E.coli C. EcoRI and HindIII restriction maps of pRD1 have been established. Deletion mutants were isolated and used to localize the his-nif-region on the restriction map. The reconstruction of the nif-region on a hybrid plasmid requires the combination of two HindIII fragments in the same alignment as found in pRD1. The hybrid plasmid with the Nif^+ phenotype in E.coli C has been designated pWK120. It has a molecular weight of 34 Mdals and a copy number of 65. A density study of pRD1 showed that the nif-region DNA possesses a very high density compared with the average density of the Klebsiella chromosome. In addition, this high density region is flanked by two sets of inverted repeats. We therefore postulate that the nif genes of K.pneumoniae were not evolved in Klebsiella but were received by a recombination process from soil bacteria.

ACKNOWLEDGEMENTS

We would like to thank Mrs. Angelika Simon for excellent technical assistance, Mrs. Kathrin Bauer for photographic aid, Mrs. Anneliese Hofmann for typing, Dr. W.Heumann and Dr. J.Reeve for critical reading the manuscript.

This work was supported by a research grant of the Deutsche Forschungsgemeinschaft (Pu 28/7).

REFERENCES

1. Cannon, F.C. et al. (1974) J. Gen. Microbiol., 80, 227 - 239.
2. Cannon, F.C. et al. (1976) J. Gen. Microbiol., 93, 111 - 125.
3. Dixon, R. et al. (1976) Nature, 260, 268 - 271.
4. Datta, N. et al. (1971) J. Bact., 108, 1244 - 1249.
5. Postgate, J.R. and Krishnapillai, V. (1977) J. Gen. Microbiol., 98, 379 - 385.
6. Pühler, A. et al. (1979) Molec. Gen. Genet., 171, 1 - 6.
7. Burkardt, H.J. et al. (1978) J. Gen. Microbiol., 105, 51 - 62.
8. De Picker, A. et al. (1977) DNA-Insertions, Elements, Plasmids and Episomes, Cold Spring Harbor Laboratory, USA, pp. 678 - 679.
9. Goebel, W. et al. (1977) Molec. Gen. Genet., 157, 119 - 129.

Plasmids of Medical, Environmental and Commercial Importance
K.N. Timmis and A. Pühler, editors

MOLECULAR AND GENETIC ANALYSIS OF KLEBSIELLA NIF

FRANK CANNON
ARC Unit of Nitrogen Fixation, University of Sussex, Brighton, Sussex, BN1 9RQ, England

INTRODUCTION

The initial transductional[1] and transconjugational[2] transfer of nif showed that his and nif were linked in Klebsiella pneumoniae. Large numbers of K.pneumoniae nif mutants were subsequently isolated and characterised in several laboratories[3-11] and all of these have been mapped in one region close to the operator end of his. The isolation of Nif^+ hybrids[12] by the intergeneric transfer of his and nif from K.pneumoniae to a non-nitrogen fixing Escherichia coli indicated that all the genes specific to nitrogenase synthesis were located near his.

The close linkage of his and nif facilitated the construction of F1 and P1 incompatibility group plasmids which carry the his-nif region of the K.pneumoniae genome.[13,14] pRD1, a P1-type plasmid, and several derivatives have been used for complementation analyses and fine structure mapping of nif mutations in K.pneumoniae. Dixon et al. (1977)[7] identified nine nif genes initially and later Merrick et al. (1978)[8] and Elmerich et al. (1978)[9] reported eleven and twelve nif genes respectively, arranged six operons. More recently, MacNeil et al. (1978)[10] reported fourteen nif genes arranged in seven operons. The identification of yet another gene, nifU, brings the total number of nif genes now recognised to fifteen (Merrick et al. 1979).[11]

Apart from the products of the three structural genes for nitrogenase, relatively little is known about the products of the other genes. K.pneumoniae nitrogenase is composed of two redox proteins (see ref. 15 for review). The Fe-protein also termed Kp2, (M.W., 68,000 dals) is a dimer comprising two identical sub-units specified by nifH. It contains a single Fe_4S_4 cluster and acts as an electron donor to the Mo-Fe-protein (Kp1) (M.W., 218,000 dals) which contains 2 Mo atoms and approximately 32 Fe atoms, some of which are present as Fe_4S_4 clusters. Kp1 is a tetramer of two non-identical subunits of molecular weights 56,000 and 60,000 dals which are the products of nifD and K respectively. A cofactor containing Fe and Mo (FeMoCo) has recently been isolated from Kp1, and Kp1 activity which is impaired in cell extracts of nifB, nifE and nifN mutants is partially restored by the addition of FeMoCo.[16] Polypeptides of 46,000 and 50,000 dals have been assigned to nifE and nifN respectively.[16] Kp2 activity, which is impaired in nifM and nifS mutants is partially restored in cell extracts of these mutants by the addition of purified Kp2.[16] Preliminary evidence suggests that the products of nifJ (120,000 dals) and nifF (17,000 dals) are involved in the electron transport system for nitrogenase.[16] A regulatory role has tentatively been assigned to nifA and nifL products which have so far not been identified.[7,16] Moreover the status of nifL as a separate gene from nifA has yet to be confirmed. The three remaining genes, nifQ, nifV and nifU have 'leaky' Nif phenotypes and have not been assigned functions.[11,16]

Genetic studies of nif which would be greatly facilitated by small amplifiable plasmids carrying individual or groups of nif genes are: investigation in vitro and in vivo of the transcriptional control of nif genes, the identification of their products, the sequencing of nif DNA and the use of Klebsiella nif DNA as a probe for locating nif genes in other Nif^+ bacteria on which little or no nif genetic studies have been done. This report describes the cloning, characterisation and some uses of nif DNA sequences which together encode 14 of the 15 Klebsiella nif genes.

CLONING OF K.PNEUMONIAE NIF GENES

The source of DNA for the primary cloning of nif genes was pRD1, a 10^8 daltons plasmid, carrying the complete cluster of Klebsiella his-linked nif genes. In general, this phase of cloning involved the following steps, (i) Digestion of pRD1 and cloning vector DNAs with the restriction endonucleases EcoR1 or HindIII. (ii) Ligation of the resulting fragments with T4 polynucleotide ligase. (iii) Transformation of His^-Nif^- derivatives of K.pneumoniae with the resulting mixture of recombinant molecules. (iv) Selection of transformants carrying genetic determinants on the cloning vectors followed by screening for a Nif^{+} phenotype. In some cases, transformants carrying a genetic determinant (*his*) on the 'passenger' DNA were selected and subsequently screened for Nif^{+}.

Kennedy (1977) estimated from cotransduction frequencies with bacteriophage P1 that the length of DNA between *his*D and the most distal *nif* gene was approximately 30 kb. Because of its size the *his*-*nif* region probably contains several sites for most class II restriction endonucleases with six base specifities. Therefore distamycin A which preferentially protected some EcoR1 sites was used to generate a series of plasmids, pCRA10, pCRA13 and pCRA37, each containing one or more contiguous EcoR1 restriction fragments of *his*-*nif* DNA which extended in the direction of *his* through *nif*.[17] The cloning vehicle used, pMB9, is a 3.5×10^{6} dals plasmid, carrying tetracycline resistance genes (Tc), the replicator region for ColE1 and a single recognition site for EcoR1. Recombinant plasmids were isolated by selecting $His^{+}Tc^{r}$ transformants of KP5058, a *his*D *nif*B derivative of *K.pneumoniae*. The EcoR1 restriction fragments cloned in pCRA10, pCRA13 and pCRA37 are listed in Table 1 in the order of their occurrence in the plasmids and in the *K.pneumoniae* chromosome.

TABLE 1
PROPERTIES OF pCRA10, pCRA13 and pCRA37

Plasmid	Phenotype conferred on KP5058*	MW x 10^{6} dals of EcoR1 digestion fragments
pMB9	Tc^{r}	3.5
pCRA10	Tc^{r} His^{+}	3.5, 1.6
pCRA13	Tc^{r} His^{+}	3.5, 1.6, 1.1
pCRA37	Tc^{r} His^{+} Nif^{+}	3.5, 1.6, 1.1, 5.3, 3.6

* KP5058 is *his*D2 *nif*B *hsd*R1 *rsp*L4

Since pCRA10 complements *his*D mutations in *K.pneumoniae* and *E.coli* the 1.6×10^{6} dals DNA fragment present in all three plasmids must contain the promoter-operator region and the *his*GD genes of the *his* operon assuming the operon structure and direction of transcription is the same in *K.pneumoniae* as in *E.coli*. Complementation studies established that *nif*B and *nif*F but not the structural genes for nitrogenase[17] were present on pCRA37.

The presence of a single HindIII cleavage site in the *nif* DNA of pCRA37 suggested that pRD1 carried a HindIII restriction fragment which partially overlapped the *nif* DNA of pCRA37 and extended beyond its *his*-distal end[18] (figure 1). A recombinant plasmid, pCM1, carrying this fragment inserted into the HindIII site of pCRA10 was isolated in CK317, a *his*D *nif*K derivative of *K.pneumoniae*.[18,19] The size of the fragment was 10.6 x 10^6 dals and its insertion resulted in the loss of tetracycline resistance. pCM1 restored CK317 to Nif^+ by recombination (marker rescue) at a frequency which was significantly above the reversion rate of the *nif*K mutation in CK317. However, marker rescue in *nif*D and *nif*H mutants carrying pCM1 was not observed indicating that *nif*H and at least part of *nif*D were not on the plasmid.

In an attempt to clone the remaining *nif* genes a 4.3 x 10^6 dals EcoR1 restriction fragment of DNA which partially overlapped the *his* distal end of *nif* DNA in pCM1 was cloned in pSA30.[18,19] The cloning vector used, pACYC184, is a 2.65 x 10^6 dals amplifiable plasmid carrying chloramphenicol (Cm) and Tc resistance genes and a single recognition site for EcoR1 which maps within Cm. pSA30 was isolated in a *nif*H *K.pneumoniae* strain, CK260, which gave rise to Nif^+ colonies at a high frequency when pSA30 was present.

NIF GENES ASSIGNED TO DNA RESTRICTION FRAGMENTS BY GENETIC AND PHYSICAL MAPPING OF *NIF* MUTANTS

The plasmids used to determine the distribution of *nif* genes on DNA restriction fragments were, pCRA37, pCM1, pSA30 and pGR112. pGR112[20] contains a 5.3 x 10^6 dals EcoR1 restriction fragment from pCRA37 inserted into the EcoR1 site of pACYC184.

To map the end points of the *nif* DNA cloned in the above plasmids by genetic fine structure mapping, (for details see Cannon *et al*. 1979[19]) derivatives of pRD1 or pMF100 with *nif* point or insertion mutations were transferred to a *his*-*nif* deletion strain carrying one of the plasmids. Transconjugants were then tested for marker rescue or complementation.

Physical mapping of *nif* insertion mutations involved the following steps. Details are given by Riedel *et al*. (1979).[20] 1) Partial purification of total DNA from strains carrying Mu, Tn5 or Tn10 insertions in *nif* genes. 2) Digestion of the DNA with restriction endonucleases. 3) Separation of the fragments by agarose gel electrophoresis. 4) Denaturation of the fragments in the gels. 5) Transfer of the denatured fragments to nitrocellulose sheets using the Southern technique. 6) Hybridization with ^{32}P-labelled probe DNAs obtained by 'nick-translation' of small amplifiable plasmids carrying *nif* genes. Restriction fragments which had been altered in size by a Mu, Tn5 or Tn10 insertion were identified by comparing the pattern of radioactive bands for DNA from insertion mutants with that for DNA from a wild type *nif* control strain.

The assignment of *nif* insertion mutations to EcoRl and HindIII restriction fragments and the *nif* genes genetically mapped on pGR112, pCRA37, pCM1 and pSA30 are shown in figure 1.

pCRA37. All of the *nif* genes between *his* and *nif*N are on pCRA37 and the *his* distal end of the *nif* DNA cloned in pCRA37 maps in *nif*N.

pGR112. The *his*-distal end of the 5.3 x 10^6 dals *nif* DNA fragment cloned in pGR112 maps in *nif*F and the *nif* genes on the *his* proximal side of *nif*F, *nif*Q,B,A,L, are on this fragment.

pCM1. The *nif* genes between *nif*Q and *nif*K are on the HindIII restriction fragment of *nif* DNA in pCM1 and part if not all of *nif*Q and *nif*K are carried at the *his* proximal and *his* distal ends respectively of this fragment.

pSA30. The operon, pHDK, for the three structural genes for nitrogenase are on pSA30. The *his*-proximal end of the pSA30 *nif* DNA mapped within *nif*E and the *his*-distal end mapped beyond the promoter adjacent to *nif*H but there is no evidence that any part of *nif*J was on pSA30.

A PHYSICAL MAP OF *NIF* GENES

The precise locations (±250 basepairs) of *nif* insertion

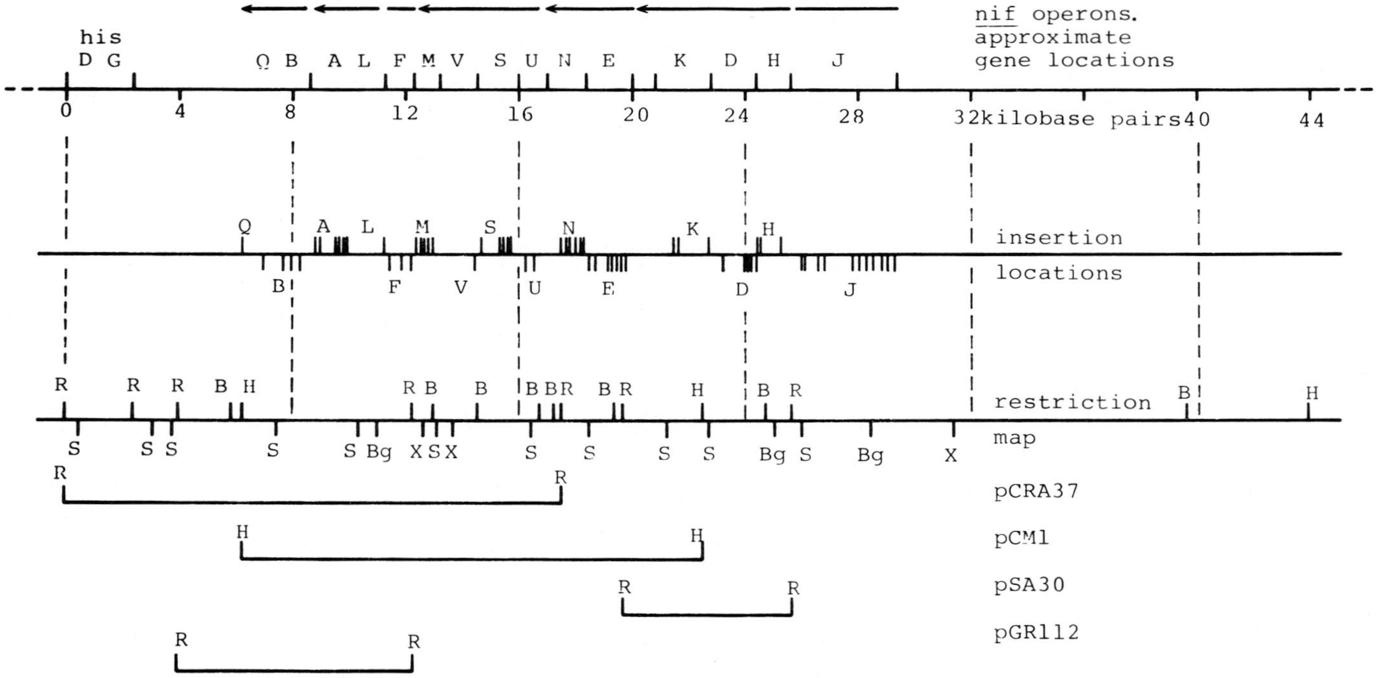

Fig. 1. Maps of his and nif genes, nif insertion mutations (±250 basepairs) and restriction sites in the his-nif region of the K.pneumoniae chromosome. The arrows indicate nif operons and their direction of transcription - nifpF and nifpJ are monocistronic operons. Symbols for restriction endonucleases are: EcoRl, R; HindIII, H; BamHl, B; Sall, S; Xhol, X; and BglII, Bg. pCRA37, pCM1, pSA30 and pGR112 are small amplifiable plasmids carrying the nif DNA restriction fragments indicated.

mutations were obtained by determining the molecular distances between the points of insertion and restriction endonuclease sites which had previously been mapped on the DNA of the his-nif region.[20] This entailed a rapid and simple procedure for the partial purification of DNA from nif::Mu[9] and nif::Tn[8] insertion mutants, digestion of the DNA with restriction endonucleases and the preparation of Southern transfers for hydridization with ^{12}P-labelled plasmid probes which contained cloned nif DNA sequences. The mapping of mutations in all of the known nif genes required the strategic use of restriction enzymes for which sites had been mapped on the DNA of the nif region, bacteriophage Mu and transposons Tn5 and Tn10. It also required the use of probes appropriate for different parts of the nif region. The end points of deletions ending within the nif region were also determined by this method. Details of the procedure which was similar in outline to that described above for the assignment of nif genes to restriction fragments and the various strategies adopted were given by Riedel et al. 1979.[20]

The mapping results for 43 nif::Tn5, 9 nif::Tn10 and 17 nif::Mu mutations are summarised in figure 1. The map of nif genes derived from these results and the order of the mutations examined are the same as those obtained from genetic fine structure mapping. Four of the nif::Tn10 mutations which were originally isolated on the plasmid pMF100 and subsequently transferred to the K.pneumoniae chromosome by P1 transduction were mapped on both the plasmid and chromosome. The same allele on two different genomes mapped at locations between 200 and 250 basepairs of each other in all four cases, confirming the reliability of the method. The minimum size of the nif gene cluster derived from these results is 23kb which is in good agreement with that obtained from P1 transduction studies (20 kb) by Kennedy (1977).[6] The maximum separation between insertions in two adjacent genes (nif K-E) is 1.6kb - this and smaller gaps between other genes could accommodate small nif genes which have not yet been identified. The boundaries between nifL and F, nifN and E and nifD and H were accurately determined, because in these cases, insertions in the two adjacent genes mapped very close to each other. From 13 insertions mapped in nifJ an expected size of 124,000 dals was determined for the gene product which is in good agreement with the size of the *nif*J product (120,000 dals)[8,9,16] obtained from SDS polyacrylamide gel electrophoresis. However, useful estimates of the expected size of gene products could not be determined for other *nif* genes because the number of insertions mapped were insufficient.

NIF TRANSCRIPTION IN *K.PNEUMONIAE*

Earlier physiological studies of *nif* regulation in *K.pneumoniae* which showed that nitrogenase was repressed in the presence of ammonia have been substantiated by results which show that nitrogenase polypeptides are not synthesised in cultures grown on ammonia.[21] Furthermore, the synthesis of polypeptides recently assigned to other *nif* genes is also similarly repressed.[16] Tubb and Postgate (1973)[22] demonstrated that ammonia represses nitrogenase synthesis at the transcriptional level using inhibitors of initiation of mRNA synthesis (rifampicin) and of protein synthesis (chloramphenicol) during derepression of samples taken from fully repressed, sulphate-limited continuous cultures.

The construction of small amplifiable plasmids carrying *nif* DNA sequences has made it possible to detect *nif* specific mRNA directly using the Southern hybridization technique. Janssen *et al*. (1979)[23] detected mRNA which hybridized to DNA restriction fragments representing 6 of the 7 known *nif* operons in cultures derepressed for *nif*. The *nif*pJ operon was not examined since it had not been cloned at the time of the experiment. In contrast *nif* specific mRNA was not detected in ammonia grown cultures which were repressed for *nif*. The reliability of the technique was confirmed using hybridization of *his* DNA with RNA from cultures repressed and derepressed for *his*. The conclusion therefore from these results is that the *nif* operons for which mRNA was detectable under the experimental conditions used are not transcribed *in vivo* in ammonia grown cultures. Since we now know that most of the DNA restriction fragments used carried more than one *nif* operon the results do not exclude the possibility that mRNAs for some *nif* operons were present in undetectably low levels in derepressed cultures. However each operon can now be investigated separately since the physical map of nif genes provides us with the possibility of representing each operon exclusively with a DNA restriction fragment.

Genetic studies of polar mutations in the 5 multicistronic nif operons have shown that these operons are transcribed in the same direction as the his operon of *K.pneumoniae*. RNA-DNA hybridization using separated strands of DNA which represented the his and 6 of the 7 nif operons showed that his specific and nif specific mRNA synthesised *in vivo* hybridized exclusively to one strand. The conclusions from these results is that those nif operons for which detectable levels of mRNA were present under the conditions used are transcribed from the same DNA strand as the his operon.

ACKNOWLEDGEMENTS

I thank M.C. Cannon for constructive criticism of the manuscript.

REFERENCES

1. Streicher, S.L., Gurney, E.G. and Valentine, R.C. (1971) Proc. nat.Acad.Sci.(Wash.), 69, 1174-1177.

2. Dixon, R.A. and Postgate, J.R. (1971) Nature (London) 234, 47-48.

3. Streicher, S.L., Gurney, E.G. and Valentine, R.C. (1972) Nature (London) 239, 495-499.

4. Shanmugam, K.T., Loo, A.S. and Valentine, R.C. (1974) Biochemica et biophysica acta 338, 545-553.

5. St. John, R.T., Johnston, M.H., Seidman, C., Garfinkel, D., Gordon, J.K., Shah, V.K. and Brill, W.J. (1975) J.Bact., 121, 759-765.

6. Kennedy, C. (1977) Molec.gen.Genet., 157, 199-204.

7. Dixon, R.A., Kennedy, C., Kondorosi, A., Krishnapillai, V. and Merrick, M. (1977) Molec.gen.Genet., 157, 189-198.

8. Merrick, M., Filser, M., Kennedy, C. and Dixon, R.A. (1978) Molec.gen.Genet., 165, 103-111.

9. Elmerich, C., Houmard, J., Sibold, L., Monheimer, I. and Charpin, N. (1978) Molec.gen.Genet., 165, 181-189.

10. MacNeil, T., MacNeil, D., Roberts, G.P., Supiano, M.A. and Brill, W.J. (1978) J.Bact., 136, 253-266.

11. Merrick, M., Filser, M., Dixon, R., Elmerich, C., Sibold, L. and Houmard, J. (1979) In preparation.

12. Dixon, R.A. and Postgate, J.R. (1972) Nature (London) 237, 102-103.

13.Cannon, F.C., Dixon, R.A. and Postgate, J.R. (1976) J.gen. Microbiol., 93, 111-125.

14.Dixon, R.A., Cannon, F.C. and Kondorosi, A. (1976) Nature (London) 260, 268-271.

15.Mortenson, L.E. and Thorneley, R.N.F. (1979) Ann.Rev.Biochem. (In press).

16.Roberts, G.P., MacNeil, T., MacNeil, D. and Brill, W.J. (1978) J.Bact. 136, 267-294.

17.Cannon, F.C., Riedel, G.E. and Ausubel, F.M. (1977) Proc.Natl. Acad.Sci.(Wash.) 74, 2963-2967.

18.Cannon, F.C. (1978) In: Genetic Engineering (H.W. Boyer and S. Nicosia eds.), pp.181-188, Amsterdam, Elservier/North - Holland, Biochemical Press.

19.Cannon, F.C., Riedel, G.E. and Ausubel, F.M. (1979) Molec.gen. Genet. (In press).

20.Riedel, G.E., Ausubel, F.M. and Cannon, F.C. (1979) Proc.Natl. Acad.Sci.(Wash.) (In press).

21.Eady, R.E., Issack, R., Kennedy, C., Postgate, J.R. and Ratcliffe, H.D. (1978) J.gen.Microbiol., 104, 277-285.

22.Tubb, R.S., and Postgate, J.R. (1973) J.gen.Microbiol., 79, 103-117.

23.Janssen, K., Riedel, G.E., Ausubel, F.M. and Cannon, F.C. (1979) Proceedings of the 3rd International Symposium on Nitrogen Fixation (W.H. Orme-Johnson and W. Newton, eds) (In press).

Plasmids of Medical, Environmental and Commercial Importance
K.N. Timmis and A. Pühler, editors

CLONING OF THE PENICILLIN G ACYLASE GENE OF ESCHERICHIA COLI ATCC 11105 ON MULTICOPY PLASMIDS

HUBERT MAYER, JOHN COLLINS and FRITZ WAGNER*
Gesellschaft für Biotechnologische Forschung mbH, Mascheroder Weg 1, D-3300 Braunschweig
*Lehrstuhl für Biotechnologie und Biochemie, Technische Universität, D-3300 Braunschweig, W.-Germany

INTRODUCTION

The advances in modern genetic methodology, so called "genetic engineering" has so far had little impact in the development of improved strains directly for application in biotechnological processes. We present here a step in this direction in which a newly developed highly efficient method of gene cloning, cosmid packaging[1,2] was used initially to isolate a gene directly from a bacterial total genome without positive selection pressure. Subsequent cloning of fragments of this region has led to the production of bacterial strains with high constitutive levels of enzyme, whereas the gene product in the initial strain was dependent on induction.

The gene chosen for investigation codes for penicillin G acylase, an enzyme which converts penicillin G to 6-amino-penicillanic acid (6-APA). 6-APA has singular importance as substrate for the production of novel synthetic or semisynthetic penicillins which exhibit a broader antibacterial spectrum, a greater resistance to acids and resistance to β-lactamase. The penicillin G acylase is important not only for the production of 6-APA from natural penicillins but can also be used to catalyse, at low pH, the synthesis of semisynthetic penicillin derivatives[3]. Two distinct classes of acylases (also called amidases) are encountered in nature which exhibit different specific hydrolytic activities when either penicillin V (phenoxymethyl penicillin) or penicillin G (benzyl penicillin) are presented as substrate. The penicillin G acylase (EC.3.5.1.11) from *Escherichia coli* has a much higher activity with penicillin G than penicillin V. This enzyme in *E.coli* ATCC 11105 hydrolyses penicillin G to yield phenylacetic acid and 6-APA, has a molecular weight of about 70 K daltons, and can be dissociated partially, in 1% SDS to yield a subunit of 20.5 K daltons[3,4].

MATERIALS AND METHODS

Plasmids and bacteria. The plasmids pJC720[1] and pBR322 have been described[7]. *E.coli* strain N205 (r_k^+ m_k^+ *recA*$^-$ *su*$^-$) was obtained from N. Sternberg, 5K (r_k^- m_k^+ *thr*$^-$ *thi*$^-$) from S. Glover, HB101 (r^- m^- *leu*$^-$ *pro*$^-$ *recA*$^-$) from H.W. Boyer. pOP203-3 was obtained from F. Fuller, to whom we are grateful for communication of details of the construction of this plasmid before publication. *Serratia marcescens* ATCC 27117 was used for assay of 6-APA by an overlay technique.

Media. *E.coli* ATCC 11105 was grown in a medium containing 5 g beef extract, 10 g yeast extract, 2 - 5 g NaCl per litre. Induction of penicillin acylase was accomplished by the addition of 1 g of sodium phenylacetate per litre of medium. *S.marcescens* was grown in medium containing 6 g Bacto-peptone, 2 g yeast extract, 50 mM sodium phosphate, pH 7.5, per litre.

Assays for 6-APA

a) Plate overlay test[8]

Colonies were overlayed on 80 mm plates with 5 ml of soft agar overlay containing 0.5 ml of an overnight culture of *S.marcescens* per 100 ml overlay medium (Difco antibiotic medium Nr. 4 in 0.05 M $NaPO_4$, pH 7.5, 20 mg/l penicillin G). Zones of inhibition were scoved after 14 hrs growth at 37°C.

b) Quantitative test

This test is based on the condensation of 6-APA with p-dimethylaminobenzylaldehyde (Schiff's base) and was carried out as described on liquid grown cultures[9]. The unit of specific activity is given as the condensation of 1 µmole 6-APA per minute per milligram bacterial dry weight at 37°C and pH 7.5.

Cloning with cosmids and transformation. The use of pJC720 to clone chromosomal genes has been described[1,2].

Cell-free protein synthesis. The method of plasmid coded cell-free protein synthesis has been described in detail elsewhere[10]. Plasmid DNA concentrations were 100 µg/ml of synthesis mix. Protein labelling was made with ^{35}S-methionine in the presence of cyclic AMP.

RESULTS

Initial cloning of the penicillin G acylase gene

The cosmid cloning system was used for the cloning of the gene for penicillin G acylase from the total DNA of *E.coli* ATCC 11105. Cosmids are plasmids carrying the Lambda bacteriophage *cos* site. This site allows the plasmid, including large foreign DNA fragments which have been ligated into suitable restriction enzyme sites, to be taken up into bacteriophage particles. These particles containing large hybrid molecules are then adsorbed to *E.coli* recipients and the hybrid plasmids are introduced with a high efficiency into the recipient bacteria. Since the original cosmids are very small they are not carried through this process and a direct physical selection for hybrids is imposed. Furthermore the average size of the hybrid DNA is very large and very few colonies need be screened in order to find a particular gene, even in the absence of a selective marker.

The *E.coli* gene library was established with *Hind*III-fragments from *E.coli* ATCC 11105 total chromosome and *Hind*III cleaved pJC720. After ligation in the ratio 1 to 3 at 300 μg/ml and in vitro packaging[1,2] some 30,000 rifampicin resistant HB101 clones were obtained from 100 ng of the ligated DNA. The library was replicated onto minimal plates to test the frequency of *pro*$^+$ and *leu*$^+$ plasmids amongst the colonies. Both *pro*$^+$ and *leu*$^+$ clones were obtained at a frequency of about 0.2%. An investigation of some 5,000 hybrid clones showed that the hybrids were on average 26 Md. Since the original vector was 16 Md, the average size of the foreign DNA per hybrid is estimated as 10 Md. The physical size of the hybrids and the frequency of the complementation of auxotrophic markers, taken together, indicated that a fairly statistical representation of the entire genome of ATCC 11105 was present in this library and that any particular gene should be found in a screening of some 500 clones. In fact, however, in the screening some 10,000 clones by the *S.marcescens* overlay technique only one positive clone (pHM5) was obtained.

Subcloning onto pBR322

The plasmid pBR322 codes for a β-lactamase acitivity which would interfere with the penicillin G acylase assay. The β-lactamase gene was therefore destroyed by replacement of the small *Eco*RI-*Pst*I fragment with *Eco*RI-*Pst*I fragments of pHM5. Clones were then selected by virtue of the remaining tetracycline resistance region. One of the acylase positive clones, pHM6, contained only the 4.9 Md *Eco*RI-*Pst*I carried by pHM5. Initial restriction endonuclease mapping of pHM6 allowed further deletions to be made. Thus pHM8 (5.55 Md) was derived from pHM6 by *Hind*III-induced deletion of 1.8 Md of the foreign DNA. This led to slight reduction in penicillin G acylase activity but the plasmid still caused relatively high yields of enzyme and still gave resistance to 5 μg/ml of tetracycline. Further deletions were induced by partially digesting pHM6 with either *Taq*I (pHM9) or *Sau*3A (pHM10) followed by ring closure-ligation and selection for resistance to 10 μg/ml tetracycline. In these two plasmids the hybrid fragment was reduced in size to about 3.4 Md and still contained three *Hind*III sites.

Subcloning onto pOP203-3

pOP203-3 is a plasmid made by F. Fuller (personal communication) in which the operator-promoter region of the *lac* operon has been built onto the tetracycline resistance plasmid pMB9, such that fragments inserted at the *Eco*RI site (in the N-terminal region of the β-galactosidase gene) should be transcribed with high efficiency from this very active promoter. It was attempted to clone *Taq*I fragments from pHM6 into the *Eco*RI site of pOP203-3 by flush-end joining of single strand endonuclease treated ends. One acylase-plus hybrid obtained in this manner (pHM11) still has one of the *Eco*RI site of the pOP203-3 vector and it is therefore unclear exactly how the fusion of the hybrid fragments arose, since apparently the S1-nuclease treatment was incomplete. This plasmid contains only 1.7 Md of chromosomal DNA and represents the most accurate localisation of the penicillin G acylase gene up to now. As will be seen from the data on enzyme levels below, it does not yet seem that the acylase gene is under control of

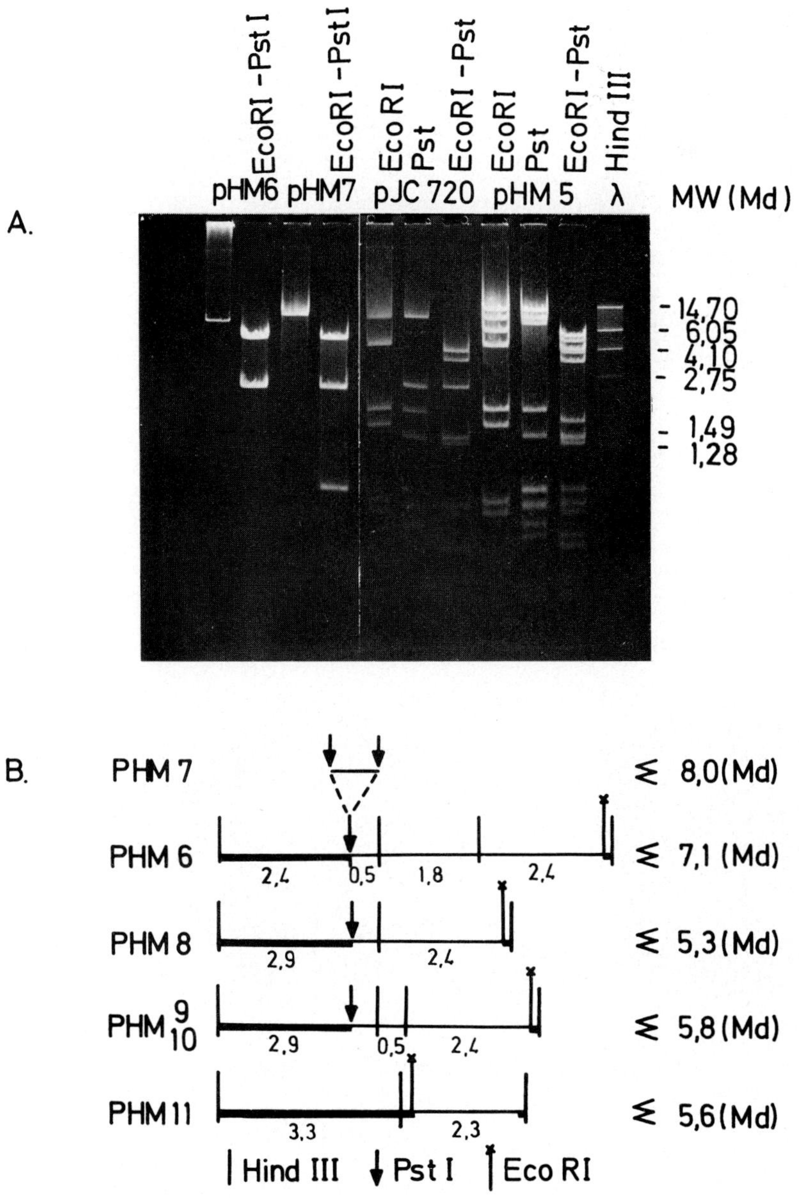

Fig. 1. A) 1% Agarose gel electrophoresis of restriction endonuclease cleaved DNAs.
B) Restriction endonuclease maps.

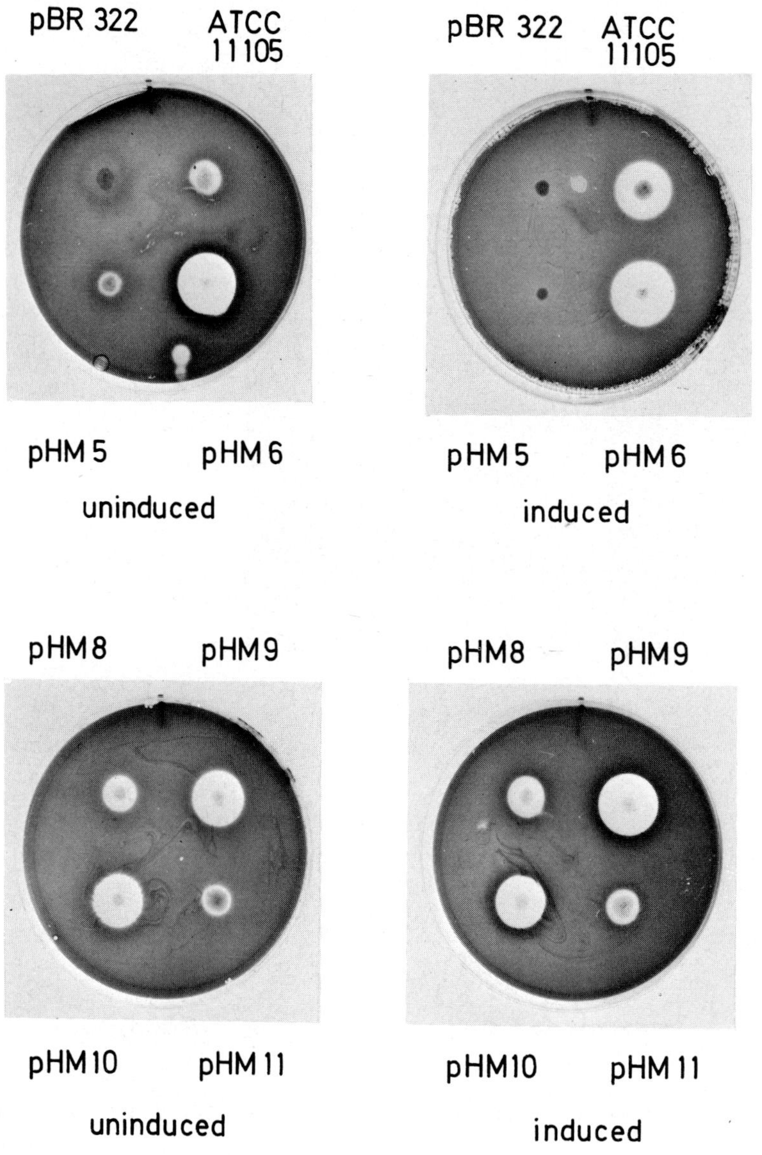

Fig. 2. Plate test for penicillin G acylase activity.

the *lac* promoter in pOP203-3 and these experiments are being continued with the expectation of increasing levels of enzyme by a further order of magnitude.

Increased levels of penicillin G acylase in the plasmid bearing strains

It was to be expected that by cloning the penicillin G acylase gene on plasmids such as pOP203-3 and pBR322 which are present at about fifty copies per cell the level of penicillin G acylase in the cell would increase in accordance with the increased gene dosage. Furthermore, if the structural gene could be separated from the regulatory gene regions (e.g. repressor gene, CAP binding site, operater) then high constitutive levels of gene expression are to be expected.

Initial studies of the penicillin G acylase activity, using the colony-overlay method, showed rather surprisingly that the initial cosmid-hybrid pHM5 strain had equivalent or even lower acylase levels than ATCC 11105. Whether or not this observation is due to altered repressor levels or to post-translational modification of the penicillin G acylase is presently under investigation.

In Table 1 the relative acylase activities of the hybrid strains are presented, firstly as inhibition zones as measured on the plate test and secondly using the quantitative test on broth grown cells with penicillin G as substrate.

The first hybrid formed during the subcloning shows very much higher levels of penicillin G acylase and is essentially constitutive for this enzyme in the absence of the inducer, phenylacetate. This property of constitutive production is common to the series pHM6 through-11 inclusive. The lower activity obtained on the further cloning of smaller fragments has not been explained. In particular the low activity obtained with pHM11 did not fulfill hopes that the penicillin G acylase gene had been brought under control of the *lac* promoter in pOP203-3. Attempts to increase the specific activity of penicillin G acylase in these hybrid strains by allowing plasmid amplification in chloramphenicol were without success.

The hybrid plasmid pHM6 still shows a strong (ten fold) cata-

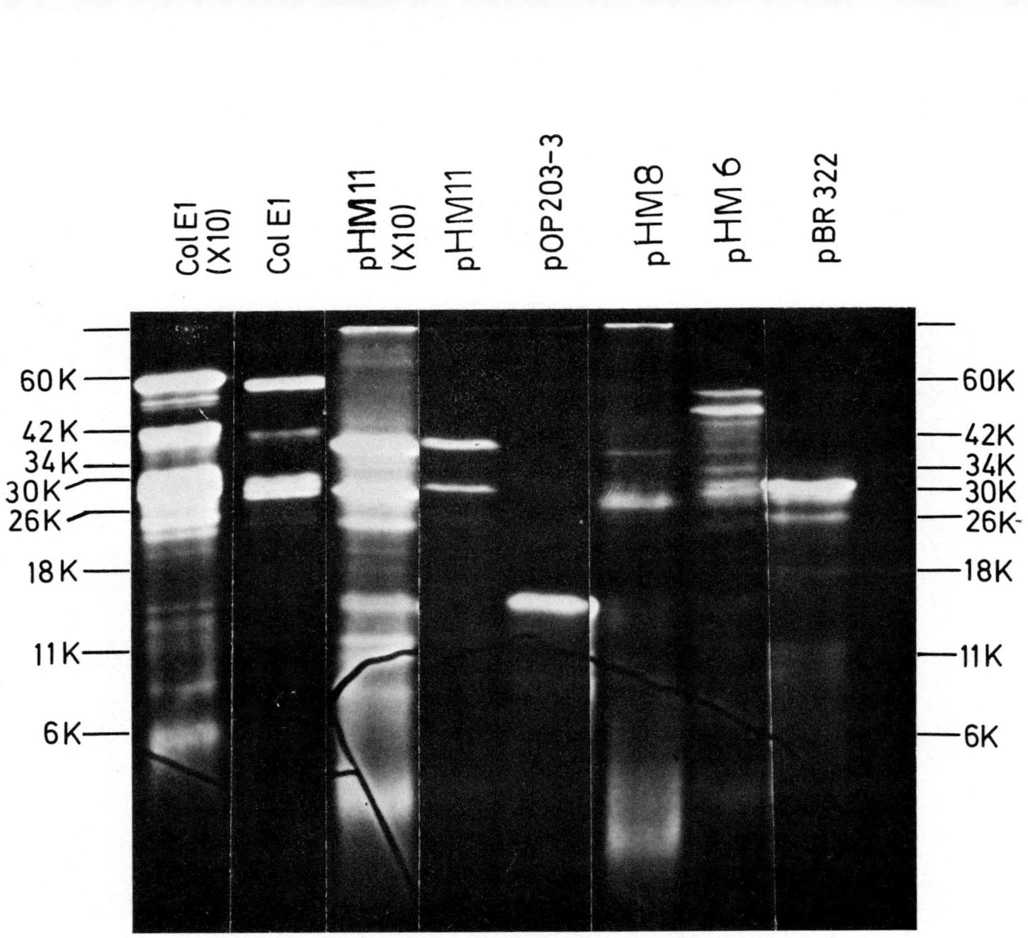

Fig. 3. ^{35}S-methionine labelled proteins produced in a cell-free protein synthesis system, by the indicated plasmids as visualised in an autoradiogram of a 20% acrylamide gel electrophoresis. Where indicated the exposure time of the autoradiogram was increased ten fold (x10).

bolite repression effect in the presence of glucose (data not shown).

TABLE 1

PENICILLIN G ACYLASE LEVELS IN HYBRID STRAINS

Strain	*Relative plaque area uninduc./induced		Spec. activity uninduc./induced		*Relative spec. activity uninduc./induced	
ATCC 11105	1	3.6	0.02	0.12	1	6
E.coli 5K pHM6	10	15.7	0.56	0.70	28	35
E.coli 5K pHM7	9	12.3	NT	0.44	NT	22
E.coli 5K pHM8	4	6	NT	0.36	NT	18
E.coli 5K pHM11	3	4	NT	0.28	NT	14

* Values are normalised to the uninduced ATCC 11105 level. Induction was carried out by the addition of phenylacetate and in the absence of glucose.

Stability of the plasmid pHM6

E.coli 5K pHM6 was grown in LB broth to stationary phase. 10^3 colonies were then plated on each of the following media, a) LB broth plus 0.1% phenylacetate, b) LB broth plus 1% glucose, c) LB broth, 1% glucose and 20 µg/ml tetracycline. On plate "a" 2% of the colonies had lost the ability to produce acylase. On the other plates all colonies were found to be acylase positive.

Cell-free protein synthesis

The plasmid hybrids and original vector plasmids were used to direct cell-free DNA dependent protein synthesis. The proteins made (labelled with ^{35}S-methionine were analysed on a 10% acrylamide gel (Fig.3).

Comparing pBR322 with pHM6 it can be seen that in the cloning of the acylase the β-lactamase (30K) gene has been removed and a number of new bands are prominent particularly at 55 K, 50 K, 48 K, 37 K, 30 K, 28 K and a pair at 15 K. The deletion induced in pHM6 which gave rise to pHM8 leads to loss of the three largest protein bands (55 to 48 K). Further subcloning led to the production of pHM11 in which only 1.7 Md of ATCC 11105 chromosomal DNA is present.

Proteins of 34 K, 26 K, 18 K and 11 K are associated with tetracycline resistance in both pBR322 and pOP2O3-3 and in the hybrid plasmids although in some cases they are difficult to detect, for example in pOP2O3-3 and pHM11 in which the *lac* promoter appears to be predominant. The following proteins can be considered common to all the acylase plus plasmids studied: 37 K, 28 K and the 15 K pair, as well as those proteins associated with tetracycline resistance. A very weak band is also seen in all acylase plus clones at 20.5 Kd.

DISCUSSION

We demonstrate here the successful application of some of the most recent methodology in the area of gene cloning to the improvement of penicillin G acylase output from a strain of singular biotechnological importance. The key step in this operation was the production of a very large number of hybrids containing DNA fragments cloned directly from the prokaryotic genome. This then allowed the identification of the required clone by a simple screening procedure, without a direct selection.

For some reason, still unexplained, the initial clone had low specific activity for penicillin G acylase but on further subcloning constitutive levels of enzyme were obtained considerably in excess of that obtainable with the fully induced parent strain. This is most easily explained on the basis of a model in which the structural gene for penicillin G acylase has become separated from either the controlling repressor gene or its operater during the later cloning steps.

Through further subcloning the acylase G gene will be brought under the control of a stronger promoter. This should lead to enzyme levels far in excess of those obtained here, which are already higher than uninduced levels of strains in commercial use.

The difference between the present study and classical strain improvement programmes based on random mutagenesis, or more recently cell fusion, lies basically in the fact that during the present programme we gain more knowledge about the gene concerned, are able to localise it fairly precisely and to isolate it essentially one thousand fold purified. This knowledge is then directly applicable to further manipulation of the gene either in terms of higher output or by *site-directed* mutagenesis to alter the specificity of the enzyme. In our present example this could lead directly to the synthesis of new antibiotics. One can be optimistic in this direction in that one can saturate the isolated gene with mutations *in vitro* and screen thousands of such mutations in a bacterium which is otherwise unmutated.

The study of the isolated plasmids in the cell-free protein synthesis system showed the presence of a few proteins that were present in all acylase plus hybrids. Proteins of the expected molecular weight for penicillin G acylase (70 K, 20.5 K) were not seen amongst the major products. This could be interpreted to mean either that the proteins are produced in very low amounts, or that the penicillin G acylase is normally subject to a post-translational proteolysis and is in fact represented *in vitro* by one of the 37 K, 28 K, or 26 K proteins.

ACKNOWLEDGEMENTS

We thank C. Martin for helpful discussion and B. Gosch, H. Stephan und W. Westphal for their expert technical assistance.

REFERENCES

1. Collins, J. and Hohn, B. (1978) Proc. Natl. Acad. Sci. USA, 9, 4242-4246.

2. Collins, J. and Brüning, H.J. (1978) Gene, 4, 85-107.

3. Rolinson, G.N., Batchelor, F.R., Butterworth, D., Cameron-Wood, J., Cole, M., Eustache, G.C., Hart, M.V., Richards, M. and Chain, E.B. (1960) Nature, 187, 236-237.

4. Kutzbach, C. and Rauenbach, E. (1974) Hoppe-Seyler's Z. Physiol. Chem., 354, 45-53.

5. Price, K.E. (1969) Adv. Appl. Microbiol., 11, 17-75.

6. Collins, J., Johnsen, M., Jørgensen, P., Valentin-Hansen, P., Karlström, H.O., Gautier, F., Lindenmaier, W., Mayer, H. and Sjöberg, B.M. (1978) in Microbiol. 1978, eds. David, J. and Novik, R. (Am. Soc. of Microbiol., Wash. D.C.), 150-153.

7. Bolivar, F., Rodriguez, R.L., Greene, P.J., Betlach, M.C.,

Heyneker, H.L., Boyer, H.W., Crosa, J.H. and Falkow, S. (1977) Gene, 2, 95-113.

8. Oostendorp, I.G. (1972) Antonie van Leeuwenhoek, 38, 201.

9. Bomstein, J. and Evans, W.G. (1965) Anal. Chem., 37, 576.

10. Collins, J. (1979) Gene, 5, in press.

Plasmids of Medical, Environmental and Commercial Importance
K.N. Timmis and A. Pühler, editors

MOLECULAR CLONING IN BACILLUS SUBTILIS

W. GOEBEL, J. KREFT AND K.J. BURGER
Institut für Genetik und Mikrobiologie der Universität Würzburg,
Röntgenring 11, D-8700 Würzburg

ABSTRACT

The construction of hybrid plasmids is described, which consist of the tetracycline resistance plasmid pBS161-1 of *B. subtilis* and the *E. coli* plasmids pBR322 and pACYC184, respectively. Most of these composite replicons undergo extensive deletions preferentially in *B. subtilis*. The final products are relatively stable plasmids, which are able to replicate in *E. coli* and in *B. subtilis* and can be used as cloning vectors in both bacteria.

It is shown that tetracycline resistance from *E. coli* can be expressed in *B. subtilis* and vice versa. However, *B. subtilis* is unable to express the ß-lactamase gene of *E. coli*. The block of gene expression is located on the posttranslational level, i. e. the precursor protein of ß-lactamase can not be processed by *B. subtilis*. Chloramphenicol resistance from *E. coli* is also not expressed in *B. subtilis*. It is shown by hybridization that the *E. coli* gene determining the chloramphenicol acetyltransferase is not transcribed in *B. subtilis*.

INTRODUCTION

The technique of gene cloning has opened new possibilities in basic as well as applied genetics[1-3]. Most experiments exploiting this new technique were performed in *Escherichia coli*. Recently, progress has been also made in developing *B. subtilis* as an alternative host[4-7], which may have some decisive advantages compared to *E. coli*[8-9]. However, previous results have already signaled new problems which arise, when foreign genes are introduced in *B. subtilis*. Out of several genes of *E. coli* which were transferred into *B. subtilis* none was expressed in this new host[4,6,8,9].

This paper describes the construction of new recombinant plasmids, which are useful for gene cloning in *B. subtilis* and *E. coli*. The expression of foreign genes, which were transferred into *B. subtilis* with these plasmids is analysed.

MATERIAL AND METHODS

The applied material and methods will be published in detail elsewhere (Kreft et al., manuscript in preparation). For strains see ref. 6.

RESULTS

1. New recombinant plasmids capable of replication in *B. subtilis* and *E. coli*

We have recently described the construction of the recombinant plasmid pJK3 by in vitro ligation of the *Bacillus* plasmid pBS161-1 with the *E. coli* plasmid pBR322[6], which is stably maintained in both hosts. Whereas no structural changes have ever been observed in pJK3 when it is propagated in *E. coli*, a deletion in the pBR322 part of about 900 bp occurs when it is propagated in *B. subtilis* (Fig. 1). The remaining plasmid pJK33 is stable in both hosts, carries genes for tetracycline and ampicillin resistance from *B. subtilis* and *E. coli*, respectively and has a single recognition site for *Bam*HI and two for *Eco*RI, *Hind*III and *Pst*I.

The recombinant plasmid pJK201 was obtained by transformation of *E. coli* 5K with the ligated mixture of *Hind*III cleaved *B. subtilis* plasmid pBS161-1 and *E. coli* plasmid pACYC184 as previously described[6]. It is derived from the expected recombinant plasmid by in vivo deletion affecting both parental plasmids. Subsequent propagation of this plasmid in *B. subtilis* leads to pJK202, which has suffered a further substantial deletion *in vivo* as indicated in Fig. 2. Plasmid pJK201 carries genes for resistance to tetracycline and

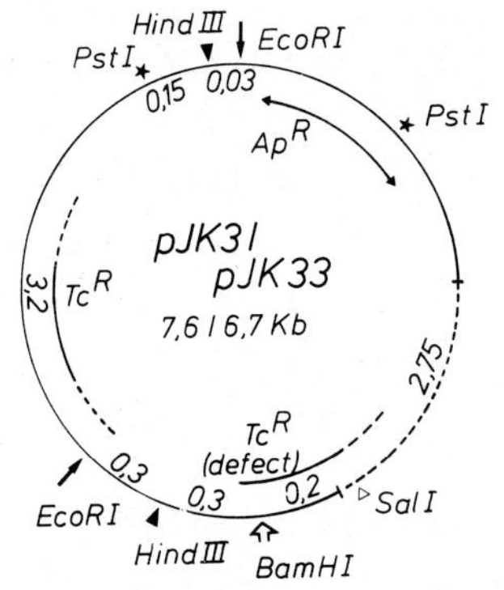

Fig. 1 Restriction map of pJK3/pJK33. The dashed part of pJK3 is deleted in *B. subtilis*, yielding pJK33.

chloramphenicol and replicates in E. coli as well as in B. subtilis, whereas plasmid pJK202 has lost the capability of replication in E. coli and can only replicate in B. subtilis.

It has been reported that several plasmids from S. aureus determining chloramphenicol or tetracycline resistance can be readily transformed into B. subtilis[10]. Two of these plasmids, pUB112 and pC221, were combined in vitro with plasmids of the type described above. HindIII-linearized plasmids pJK302 and pJK3-1 (Kreft, unpublished) were ligated with HindIII digested pUB112 and pC221 DNA, respectively. B. subtilis protoplasts were transformed with the ligation mixtures and colonies resistant to chloramphenicol and tetracycline were selected. The recombinant plasmids obtained from these clones were designated pJK312 and pJK321. Both plasmids were transformed into E. coli where they are also able to replicate and express chloramphenicol and tetracycline resistance as well. Until now no restriction enzyme has been found which inactivates one of the two resistance genes.

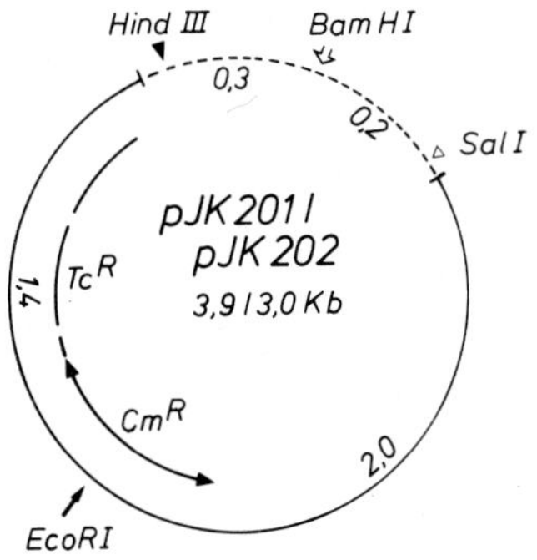

Fig. 2 Restriction map of pJK201/pJK202. The dashed part of pJK201 is deleted in B. subtilis, yielding pJK202.

2. Expression of tetracycline resistance (Tc^R) of E. coli and B. subtilis.

We have previously shown that the tetracycline resistance carried by the Bacillus plasmid pBS161-1 is expressed in E. coli although at a reduced level[6]. To test whether the tetracycline resistance from the E. coli plasmid pBR322 is expressed in B. subtilis a recombinant plasmid was constructed consisting of the cryptic B. subtilis plasmid pBS1[11] and pBR322. Both plasmids which carry single EcoRI sites were digested with this restriction enzyme and ligated in vitro with T4 ligase. E. coli 5K was transformed with the ligation mixture and the obtained Tc^R clones were screened for recombinant plasmids. One of the clones contained the complete hybrid consisting of pBS1 and pBR322, this plasmid was designated pJK501. After transformation of B. subtilis with this plasmid tetracycline resistant colonies were obtained. The plasmid which could be isolated from these colonies apparently had undergone a large deletion in vivo, which surprisingly affected only the pBS1 part of pJK501. The deleted plasmid was designated pJK502 (Fig. 3). It replicates

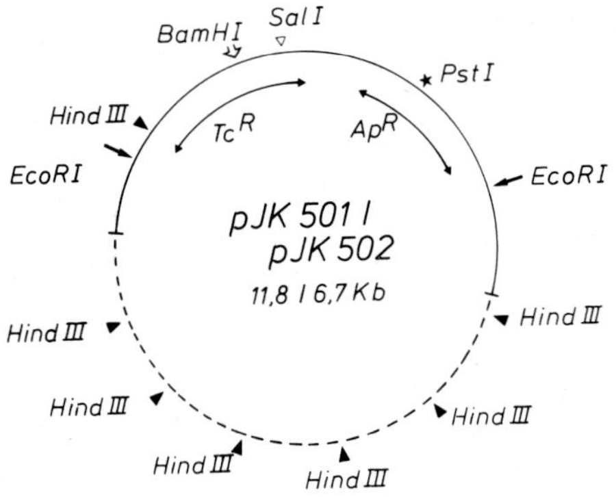

Fig. 3 Restriction map of pJK501/pJK502. The dashed part of pJK501 is deleted in B. subtilis, yielding pJK502.

in E. coli and B. subtilis. The level of resistance (>20 µg/ml) is the same in both hosts.

3. Lack of expression of ampicillin resistance (Ap^R) gene of E. coli in B. subtilis.

The ampicillin resistance deriving from E. coli in pJK3 and several other plasmids[4,6] is not expressed in B. subtilis. Previous investigations have already indicated that no mature β-lactamase (bla) is synthesized in B. subtilis carrying plasmid pJK3[6].

To test whether the inhibition of Ap^R gene expression is on the transcriptional, translational or posttranslational level, pJK3 was transformed in the minicell producing B. subtilis strain Cu403[12]. RNA produced in minicells containing pJK3 was labeled with (^{3}H)-uridine, purified and hybridized with the recombinant plasmid, the two parental plasmids, pBR322 and pBS161-1, and the miniplasmid Rsc11 deriving from the antibiotic resistance factor R1. The latter plasmid shares only the Ap^R gene with pBR322 and pJK3, but has no other homologies with these plasmids and pBS161-1. The hybridization data shown in Fig. 4 indicate that pBR322 is quite efficiently transcribed in B. subtilis.

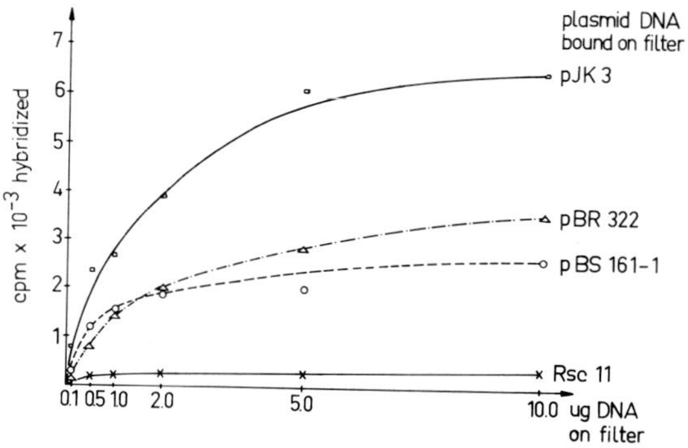

Fig. 4 Filter hybridization of in vivo labeled ^{3}H-RNA from B. subtilis minicells containing pJK33 with plasmid DNAs. Constant amounts of ^{3}H-labeled RNA were hybridized to increasing amounts of plasmid DNA reaching saturation of higher DNA concentrations.

These transcripts appear to include that of the Ap^R-gene since the RNA hybridizes also with Rsc11 to the expected level. The higher amount of hybridization of the ^{3}H-labeled RNA with pBR322 compared to Rsc11 indicates that more than the Ap^R gene of pBR322 is transcribed in B. subtilis.

The gene products synthesized in B. subtilis minicells harbouring pJK3 were determined by labeling them with ^{35}S-methionine. The labeled proteins were separated by electrophoresis on polyacrylamide gels and compared to ^{35}S-labeled proteins produced in minicells of E. coli harbouring the same recombinant plasmid (Fig. 5). Two proteins seem to be indistinguishable in both systems, designated L^+ and T. Whereas the nature of protein T is unknown, protein L^+ can be readily identified as the precursor protein of ß-lactamase carrying a 23 amino acid leader sequence at the N-terminal end[13]. To demonstrate that the molecular nature of protein L^+ in both systems is indistinguishable proteolytic cleavage of the isolated proteins from both systems was performed according to the method described by Cleveland[14] using S. aureus V8 protease and papain. As shown in Fig. 6 the three peptids (A-C) obtained after proteolysis of both proteins with papain have the same size. Digestion of the

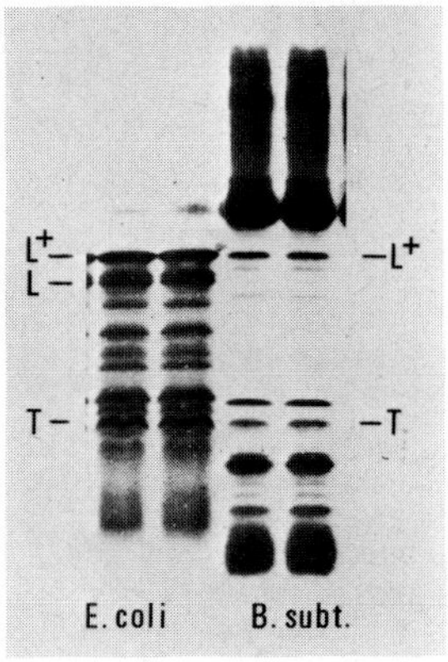

Fig. 5 SDS-acrylamide gel electrophoresis of proteins from E. coli and B. subtilis minicells containing pJK3 or pJK33, respectively.

L: β-lactamase
L^+: β-lactamase precursor

proteins with V8 protease leads to two major peptides (A', B') in the case of the E. coli preparation. The intensity of band A' suggests that it may contain two fragments of similar size. These two peptides (A', B') are also observed in a V8 digest obtained with the B. subtilis preparation. The additional band

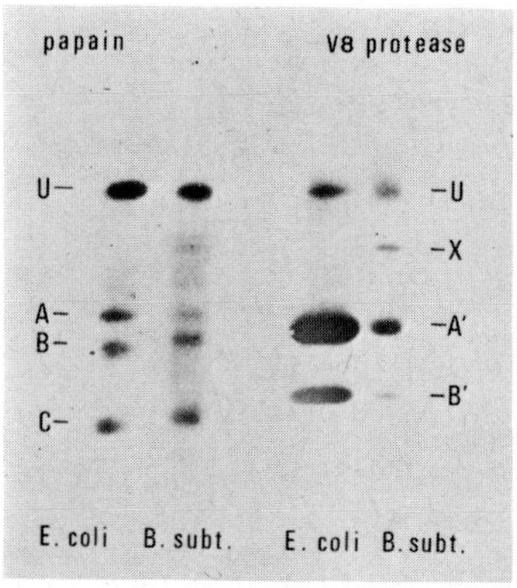

Fig. 6 Proteolysis of β-lactamase precursor protein from E. coli and B. subtilis minicells according to Cleveland et al.[14]

U: Undigested protein.

(X) visible in this digest appears to be a product of incomplete digestion, as calculated from its molecular weight.

Protein L (Fig. 5) is the biologically active mature ß-lactamase[13]. It is evident that this form of the ß-lactamase is present in the protein extract from E. coli but absent in the extract from B. subtilis minicells, indicating that B. subtilis is unable to convert the precursor ß-lactamase to ß-lactamase.

That the Ap^R gene from pJK3 and pJK33 remains structurally intact in B. subtilis was shown by retransformation of E. coli to ampicillin resistance with plasmid DNA (pJK3 and pJK33) isolated from B. subtilis.

4. Lack of expression of chloramphenicol resistance from E. coli in B. subtilis.

The recombinant plasmid pJK201 expresses Tc resistance in E. coli and in B. subtilis. Again, the level of Tc resistance caused by this plasmid is higher in B. subtilis (>5 µg/ml) than in E. coli (<5 µg/ml). The chloramphenicol (Cm) resistance deriving from E. coli, however, is expressed only in E. coli (level of resistance ∿ 100 µg/ml), but not in B. subtilis. There are no indications that a precursor form of the enzyme chloramphenicol acetyl transferase (cat), which is responsible for the Cm resistance, exists. The cat gene of E. coli is under the control of catabolic repression[15]. B. subtilis cells were therefore grown in medium containing glycerol with and without addition of cAMP,

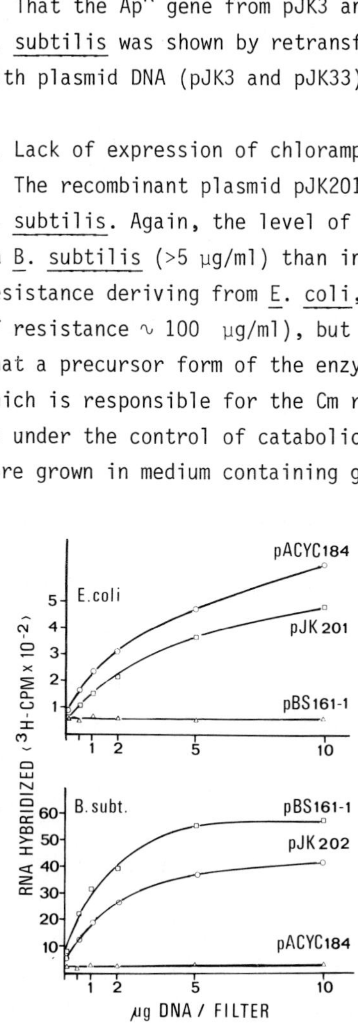

Fig. 7 Filter hybridization of in vivo-labeled ^{3}H-RNA from E. coli and B. subtilis minicells containing pJK201 and pJK202, respectively.
Hybridization was performed as described in Fig. 4.

before testing the cells for Cm resistance. B. subtilis harbouring pJK201 failed to express Cm resistance also under these growth conditions.

Transcription of the cat gene was determined in E. coli and B. subtilis minicells harbouring the recombinant plasmid pJK201. ^{3}H-labeled RNA isolated from the E. coli minicells hybridized with pJK201 and with pACYC184 but not with pBS161-1 (Fig. 7). ^{3}H-labeled RNA from B. subtilis minicells harbouring pJK201 hybridized to about equal amounts with pJK201 and pBS161-1 but not with pACYC184 (Fig. 7). This indicates that the cat gene carried by pACYC184 which is transcribed starting from a strong promotor in E. coli is not transcribed in B. subtilis.

In this case the structural integrity of the Cm^R gene in plasmid pJK201 was again demonstrated by retransformation of pJK201 into E. coli which leads to Cm resistant colonies.

DISCUSSION

The application of the gene cloning technique to Bacillus subtilis is presently hampered by two problems. a) The enhanced instability of recombinant plasmids containing foreign genes[5-9] and b) the frequent failure of B. subtilis to express foreign genes[4,6,8,9]. As shown by Canosi et al.[16] competent cells of B. subtilis are preferentially transformed by oligomeric forms of circular DNA. This may implicate that deletions occur in plasmid DNA taken up by the cell in a monomeric form. In this study we have used, however, the transformation procedure of Chang et al.[17]. Under these conditions transformation with monomeric plasmid DNAs is as efficient as with oligomeric forms (Kreft, unpublished).

The generation of the deletions within the two recombinant plasmid pJK3 and pJK201 occurs several generations after the uptake of the DNA. In the case of pJK3 the deletion affects only the E. coli part of the recombinant plasmid, but in pJK501 part of the B. subtilis plasmid is deleted and in pJK201 parts of both parental plasmids are affected. Preferential sites for these deletions can not yet be assigned, but it is evident that they occur more frequently in B. subtilis than in E. coli.

The expression of foreign genes introduced into a new host via a recombinant plasmid depends on the accurate transcription of the foreign genes, the translation of the transcripts on the ribosomes and eventually on posttranslational events, when the first translational product represents a biologically inactive precursor form. The three E. coli genes, Tc^R, Ap^R and Cm^R, have been chosen since they are representative examples for different types of gene expression: Tc^R is transcribed from a promotor which is under negative control[18]; elimination of the control element leads to a constitutive transcription of Tc^R, which seems to be the case in pBR322[19].

Ap^R is transcribed from a strong constitutive promotor. The primary gene product carries a leader sequence of 23 amino acids[13], which is removed during the passage through the cytoplasmic membrane. The final product, ß-lactamase, is located in the periplasmic space[13].

Transcription of Cm^R is under catabolic repression[15], i.e. it depends on the presence of cAMP and the cAMP-binding protein (CAP). The gene product, chloramphenicol acetyltransferase (cat) is a cytoplasmic protein, which does not seem to be subject to posttranslational processes.

The data suggest that the RNA polymerase of *B. subtilis* is able to recognize the constitutive promotors of the *E. coli* genes, Tc^R and Ap^R, although it can not be excluded presently that the transcription of these *E. coli* genes starts from a *B. subtilis* promotor located in the vicinity of these genes. However, *B. subtilis* is definitely unable to start at the promotor for the Cm^R gene, the recognition of which requires positive control elements.

The observation that minicells of *B. subtilis* are only capable of synthesizing ß-lactamase up to its precursor form suggests that *B. subtilis* can not remove the leader sequence from the N-terminus. This process which has been shown to be performed by membrane-bound proteases in several instances[20,21] and appears to be required for the transport of the protein through the cytoplasmic membrane, thus seems to be host-specific. The data further suggest that the precursor form of ß-lactamase is biologically inactive.

The result indicate that expression of foreign genes in *B. subtilis* is more restricted than in *E. coli*. It may be therefore advantageous to perform the molecular cloning of DNA first in *E. coli* and transfer subsequently the obtained recombinant DNAs into *B. subtilis* by using the bifunctional plasmids described.

Although none of the restriction enzymes tested so far inactivates one of the two antibiotic resistance genes carried by pJK312 and pJK231 and which are expressed both in *E. coli* and *B. subtilis* it should be possible to further modify these plasmids or to find other restriction enzymes, which will allow the detection of in vitro inserted DNA by marker inactivation.

ACKNOWLEDGEMENTS

We would like to thank Miss H. Brand and Miss M. Vogel for excellent technical assistance. This work was supported by a grant from the Deutsche Forschungsgemeinschaft (SFB 105 - A11).

REFERENCES

1. Cohen, S.N. (1977) Science 195, 654-657.
2. Vosberg, H.P. (1977) Hum. Genet. 40, 1-40.
3. Goebel, W. (1979) Naturw. Rdsch. (in press).
4. Ehrlich, S.D. (1978) Proc. Natl. Acad. Sci. USA 75, 1433-1436.
5. Gryczan, T.J., Dubnau, D. (1978) Proc. Natl. Acad. Sci. USA 75, 1428-1432.
6. Kreft, J., Bernhard, K., Goebel, W. (1978) Molec. gen. Genet. 162, 59-67.
7. Keggins, K.M., Lovett, P.S., Duvall, E.J. (1978) Proc. Natl. Acad. Sci. USA 75, 1423-1427.
8. Ehrlich, S.D. (1978) in "Genetic Engineering" (eds. Boyer, H.W. and Nicosia, S.) p. 36-46. Elsevier/North Holland Biomedical Press.
9. Goebel, W., Kreft, J., Bernhard, K., Schrempf, H., Weidinger, G. (1978) in "Genetic Engineering" (eds. Boyer, H.W. and Nicosia, S.) p. 47-58. Elsevier/North Holland Biomedical Press.
10. Ehrlich, S.D. (1977) Proc. Natl. Acat. Sci. USA 74, 1680-1682.
11. Bernhard, K., Schrempf, H., Goebel, W. (1978) J. Bacteriol. 133, 897-903.
12. Mertens, G., Reeve, J.N. (1977) J. Bacteriol. 129, 1198-1207.
13. Sutcliffe, J.G. (1978) Proc. Natl. Acad. Sci USA 75, 3737-3741
14. Cleveland, D.W., Fischer, S.G., Kirschner, M.W., Laemmli, W.K. (1977) J. Biol. Chem. 252, 1102-1106.
15. Smith, D.H., Harwood, J.H., Rubin, F.A. (1972) in "Bacterial plasmids and Antibiotic Resistance" (eds. Krcmery, V., Rosival, L., Watanabe, T.) p. 319-335. Avicenum Prague.
16. Canosi, U., Morelli, G., Trautner, Th. A. (1978) Molec. gen. Genet. 166, 259-267.
17. Chang, S., Cohen, S.N. (1979) Molec. gen. Genet. 168, 111-115.
18. Tait, R.C., Boyer, H.W. (1978) Cell 13, 73-81.
19. Bolivar, F., Rodriquez, R.L., Greene, P.J., Betlach, M.C., Heyneker, H.L., Boyer, H.W., Crosa, J.H., Falkow, S. (1977) Gene 2, 95-113.
20. Blobel, G., Dobberstein, B. (1975) J. Cell Biol. 67, 835-851.
21. Chang, C.N., Blobel, G., Model, P. (1978) Proc. Natl. Acad. Sci. USA 75, 361-365.

© 1979 Elsevier/North-Holland Biomedical Press
Plasmids of Medical, Environmental and Commercial Importance
K.N. Timmis and A. Pühler, editors

THE EXPRESSION OF BACTERIAL ANTIBIOTIC RESISTANCE GENES IN THE YEAST *Saccharomyces cerevisiae**

CORNELIS P. HOLLENBERG
Max-Planck-Institut für Biologie, Abt. Beermann, Spemannstr. 34, 7400 Tübingen, (Federal Republic of Germany)

INTRODUCTION

Yeast 2-μm DNA is a closed circular extra-chromosomal DNA element of which 50-100 copies are normally found in several strains of Saccharomyces cerevisiae[1-8]. The molecule contains a non-tandem inverted duplication, approximately 600 base pairs long, that is involved in intramolecular recombination leading to the inversion of the enclosed unique DNA segments[2-8] (Fig.1). No gene products of 2-μm DNA are known and no function has yet been

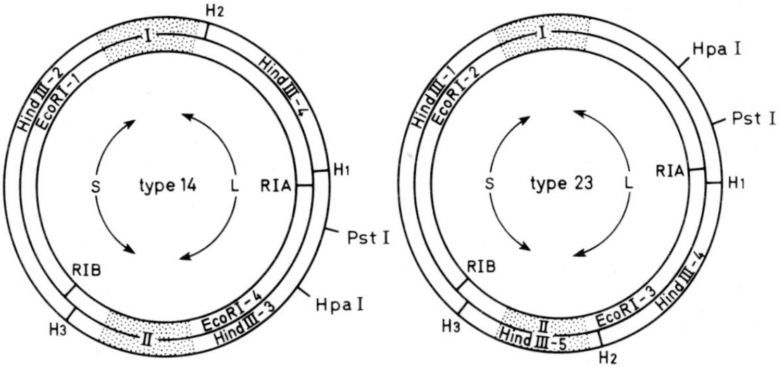

Fig. 1. Structure of the two types of yeast 2-μm DNA monomers. The monomers, type 14 and 23, are differentiated by the relative inversion of the S and L segments enclosed by the inverted duplication (id) sequences. Monomer type 14 consists of EcoRI fragments 1 and 4 and the type 23 monomer consists of EcoRI fragments 2 and 3. The id sequences are dotted. RIA and RIB refer to the EcoRI sites; H1, H2 and H3 to HindIII sites. Sizes are as follows: S segment, 2,260 base pairs; L segment, 2,700 base pairs; id sequence, 620 base pairs; EcoRI fragments 1-4 are 4.00, 3.80, 2.35, and 2.18 kilobases respectively; HindIII fragments 1-5 are 4.00, 2.75, 2.12, 1.30 and 0.93 kilobases respectively (Hollenberg, 1978).

*These results have in part been presented at the ICN-UCLA Symposium on Extrachromosomal DNA, Keystone, Colorado, 1979.

established. When inserted in different bacterial plasmids and introduced into E.coli, the 2-μm DNA promotes the synthesis of at least 4 discrete polypeptides[9,10]. The DNA regions coding for these polypeptides have been mapped[10]. In view of the functional expression of several yeast genes in E.coli[11-15], it is likely that some of these polypeptides represent products of genes located on the 2-μm DNA.

Recently, efficient transformation of yeast has been achieved[16-18] by using 2-μm DNA as a vector and the cloned yeast leu2 or his3 gene as a selective marker. We have used this technique to analyse the expression of some bacterial antibiotic resistance genes in the yeast cell in order to study the following questions:

(1) Are prokaryotic genes functionally expressed in the yeast cell and can such genes be used as selective markers for yeast transformation?

(2) Can bacterial genes on the 2-μm DNA be exploited to study the normal function of 2-μm DNA in the yeast cell?

The advantages of bacterial resistance genes as selective markers on a 2-μm DNA vector are manifold. They are present on many bacterial cloning vectors, are provided with suitable restriction sites for integration and they can be used for most yeast strains as they are not dependent on the presence of auxotrophic markers. In addition, the absence of homology with nuclear DNA sequences precludes chromosomal integration, which has been observed with homologous markers[21,17].

In this paper, I present data to show that a bacterial ampicillin resistance gene is expressed in S.cerevisiae and discuss evidence indicating that bacterial chloramphenicol and kanamycin resistance genes confer their resistance on a yeast cell.

RESULTS

Construction of recombinant plasmids

Beggs[16] has developed an efficient transformation system for yeast by using composite plasmids that consist of pMB9, 2-μm DNA and a yeast fragment bearing the leu2 gene. In this study, pJDB219 (kindly donated by Dr. J. Beggs) was digested with HindIII and ligated to pBR325[19] that had been linearized with HindIII. The ligation mixture was used to transform E.coli JA221 leuB6, in order to isolate clones that contain pBR325 carrying the yeast leu2 gene. Transformants were selected on medium minus leucine and subsequently replica plated on L-broth plates containing tetracycline, ampicillin or chloramphenicol. The plasmid DNA from clones that were sensitive to tetracycline but resistant to chloramphenicol and ampicillin was isolated and analyzed.

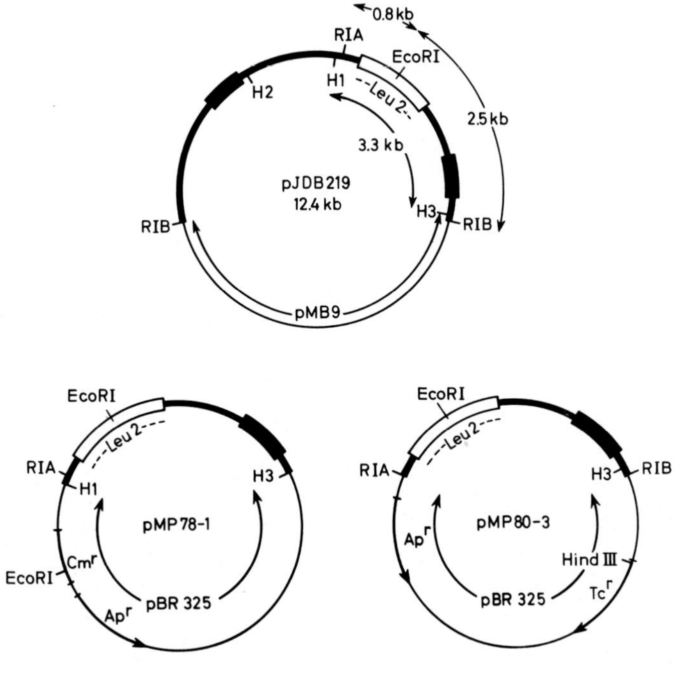

Fig. 2. Schemes of recombinant plasmids. To construct pMP78-1, the HindIII fragment of 3.3 kb from pJDB219 (16) was inserted into the HindIII site of pBR325. The 3.3 kb HindIII fragment carries the yeast leu2 gene integrated at the PstI site of HindIII fragment 3 present on 2-μm DNA type 14. pMP80-3 consists of pBR325 that has integrated at its EcoRI site two EcoRI fragments of pJDB together covering about the same region as the 3.3 kb HindIII fragment. The 2-μm DNA part is drawn thick and the inverted duplication sequences extra thick. RIA and RIB are the EcoRI sites and H1, H2, H3, are the HindIII sites of 2-μm DNA. Restriction enzyme digestions, ligations and E.coli JA221 (C600 recA1 leuB6 trp E5 hsdR$^-$ hsdM$^+$ lacY) transformation were described previously[10]. The Ap^r (ampicillin resistance), Cm^r (chloramphenicol resistance) and Tc^r (tetracycline resistance) genes are indicated with arrows.

Plasmid pMP78-1 (Fig.2) from clone T78-1 consists of pBR325 and a HindIII fragment of about 3.3 kb also present in the HindIII digest of pJDB219. This fragment carries the yeast leu2 gene in a segment of 2-μm DNA as indicated. pMP80-3 (Fig.2) was constructed in a similar way and consists of pBR325 with two EcoRI fragments of pJDB219 integrated at the EcoRI site. The two EcoRI fragments cover about the same region of pJDB219 as the HindIII segment. The integration at the EcoRI site of pBR325 inactivates the chloramphenicol resistance gene. A third plasmid, pMP81, was constructed by the integration of the two EcoRI fragments of pJDB219 in the EcoRI site of plasmid pCRI[20]. pMP81 thus carries a kanamycin resistance gene.

The intact bacterial resistance genes present on the three constructed plasmids, pMP78-1, pMP80-3 and pMP81, are listed in Table 1.

TABLE 1
RESISTANCE GENES ON RECOMBINANT PLASMIDS

Recombinant Plasmid	Bacterial vector	Integration site	Intact bacterial resistance genes	Yeast transformants
pMP78-1	pBR325	HindIII	Amp^R Cam^R	YT6
pMP80-3	pBR325	EcoRI	Amp^R Tet^R	YT8
pMP81-2	pCRI	HindIII	Kan^R	YT10

Yeast transformants

The purified pMP plasmids were used to transform S.cerevisiae AH22, a double leu2 mutant[21], following the procedure described by Beggs[16]. Colonies prototrophic for leucine were obtained with frequencies of 1-2 per 1000 regenerating cells. The same transformation frequency was obtained with plasmid pJDB219, in agreement with the results of Beggs[16]. This means that the 2-μm DNA sequences required for the stable replication of the recombinant plasmids in yeast must reside on the 2.12 kb HindIII fragment 3, present on 2-μm DNA type 14 (Compare Fig.1).

The stability of the pMP plasmids in the yeast transformants was studied by plasmid DNA and marker analysis.

Supercoiled DNA was isolated from the transformants after 50-100 generations growth on selective medium and analyzed on agarose gels. The main supercoiled band of the transformants YT6 and YT8 migrated at the same rate as the plasmids pMP78-1 and pMP80-3 that had been used for their transformation. In both cases 2-μm DNA molecules comprised only 10-20% of the circular DNA

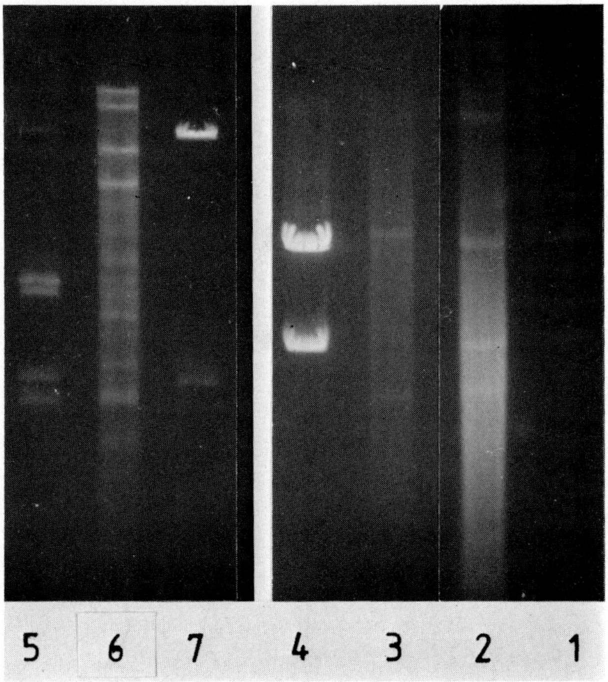

Fig. 3. Restriction digests of the supercoiled and linear DNA fractions of some yeast transformants. Yeast DNA was isolated as described before[10] and fractionated in a CsCl-ethidium gradient. The supercoiled DNA band and the pooled mitochondrial plus nuclear DNA band were isolated, prepared for digestion and analysed by agarose electrophoresis. Lanes 1-4 contain HindIII digests of supercoiled YT6-2 DNA (lane1), linear YT6-2 DNA (lane2), linear DNA of strain AH22 (lane3), and pMP78-1 DNA (lane4). Lanes 5-8 contain EcoRI digests of supercoiled YT10-2 DNA (lane5), linear YT10-2 DNA (lane6), and pMP81 DNA (lane7).

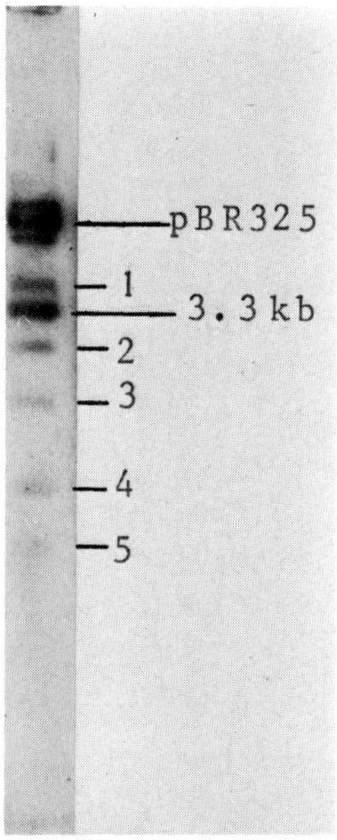

Fig. 4. Autoradiogram of a Southern blot of a HindIII digest of the supercoiled DNA fraction of YT6-2. See text for further details.

fraction. A different situation was found for YT10-2. Here the pMP81-2 plasmid made up only 10% of the total supercoiled DNA fraction, the rest being 2-µm DNA molecules.

The arrangement of the 2-µm DNA sequences in the supercoiled and the linear DNA fractions from the transformed yeast strains was examined by restriction analysis and Southern hybridization with labelled BTYP-1 DNA (BTYP-1 consists of pBR322 and total 2-µm DNA linked over the PstI site, ref 10).

The HindIII digest of the supercoiled fraction of YT6-2 (Fig.3) comprises mostly the two fragments of plasmid pMP78-1 (lane4) used for its transformation. That 2-µm DNA was still present in this transformant was shown by the hybridization assay in Fig.4. In addition to the five HindIII fragments of 2-µm DNA, two bands of about 5 kb and 6 kb have hybridized. The two HindIII fragments of pMP78-1 (lane4) are also well visible in the linear DNA digest of YT6-2 (lane2).

The amount of supercoiled DNA that could be extracted from the transformants was much higher than that from the recipient strain AH22. This was also true for YT10-2, although the majority of the supercoiled DNA of this transformant consisted of 2-µm DNA. Lane 5 in Fig.3 shows that the four EcoRI fragments of 2-µm are very prominent and only a rather faint band at the position of pCR1 indicates the presence of pMP81.

A preliminary comparision of restriction fragments from the supercoiled and the linear DNA fractions gave no indication for the integration of the plasmids in the chromosomal DNA. On Southern blots, no difference could be observed between the two DNA fractions.

<u>Expression of the ß-lactamase gene in yeast</u>

Transformants YT6 and YT8 contain a relatively stable population of the recombinant plasmids pMP78-1 and pMP80-3, which contain the bacterial ampicillin resistance gene from pBR325. Plasmid pBR325 [19,23] is derived from pBR322 and the ampicillin resistance gene originates from a R plasmid from Salmonella paratyphi B [24-25]. The gene codes for a ß-lactamase, an enzyme that hydrolyses the ß-lactam bond in penicillin antibiotics.

The expression of the ß-lactamase gene in yeast could be shown in several ways. Cell-free extracts of YT6 transformants were tested for ß-lactamase activity. Benzylpenicillin-sensitve E.coli cells were plated in soft agar containing 150 µg of benzylpenicillin per ml. 25 µl of extracts of YT6 and AH22 were spotted on these plates, which were subsequently incubated at 37°C. Fig.5 A shows strong growth of E.coli cells at the spots of the YT6 extracts and absence of growth at the AH22 control spot. Clearly, the extracts of the transformed yeast cells, carrying the ampicillin resistance gene, contain a substance that degrades benzylpenicillin and thus allows growth of the sensitive E.coli cells. That the growing cells were still penicillin-sensitive and had not been transformed by the yeast extracts, could be shown by replating on benzylpenicillin plates.

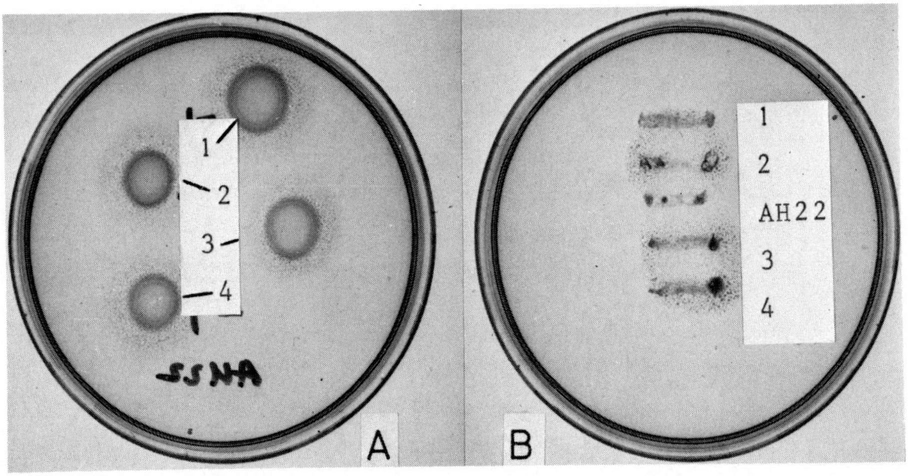

Fig. 5 A. Demonstration of ß-lactamase activity in YT6 cell-free extracts. 1 g of YT6 cells was suspended in 0.05 M Tris-HCl (pH 7.4), 5 mM EDTA and disrupted by shaking with glass beads. The homogenate was centrifuged for 15 min at 16000 xg. The Supernatants (25 µl) of four transformants were spotted on soft agar plates containing benzylpenicillin (150 µg/ml) and sensitive E.coli cells. The control spot of the AH22 extract is invisible.

B. Demonstration of ß-lactamase activity in intact YT6 cells. The transformants and control AH22 were streaked on the soft agar plates described under A and grown at 37 °C.

The synthesis of active ß-lactamase could also be demonstrated in intact YT6 and YT8 cells. The transformants were streaked on a plate of soft agar containing benzylpenicillin and sensitive E.coli cells. After 1-2 days incubation at 37°C or 30°C a clear growth of E.coli cells around the grown yeast streaks was visible (Fig.5 B). The control streak of strain AH22 did not induce growth of E.coli cells. On the contrary, a clear inhibitory effect on the background growth of E.coli cells after longer incubation times could be observed around the cell streak of AH22. This experiment shows that ß-lactamase activity is synthesized and can be demonstrated in intact yeast cells carrying the bacterial gene. Under certain conditions this assay can also be used to screen single yeast colonies for the presence of ß-lactamase activity.

O'Callaghan and Morris[26] developed a color reaction to assay ß-lactamase, in which a chromogenic derivative of cephalosporin (87/312), called nitrocefin, is used. ß-lactamase converts the yellow color of the intact molecule into the red cleavage product. When nitrocefin was applied to outgrown cell streaks, the color change could be observed after 15-60 min depending on the age of the cells. This reaction allows rapid detection of ß-lactamase-producing yeast cells.

To quantitate the relative amounts of ß-lactamase activity present in transformed yeast strains, 1 g of stationary cells was disrupted and the homogenate was centrifuged for 15 min at 16000xg. The ß-lactamase activity of the supernatant was determined by measuring the cleavage of nitrocefin at 390 nm. Fig.6 shows the activity curves of extracts from YT6-2 and YT8-3 and a sonicate of E.coli cells carrying pBR325. The E.coli (pBR325) sonicate contained about 25-100 times more ß-lactamase activity per g cells than the YT6 extract. How this value is influenced by extraction and assay conditions is not yet clear. Extracts of strain AH22 or YT10 did not show any activity in this assay.

In E.coli, the ß-lactamase of plasmid RI has been shown to be a periplasmatic enzyme of the type RTEM[27]. The enzyme is synthesized as a preprotein with a 23 amino acid leader sequence[28,29], which presumably serves as a signal to direct the transport of the protein through the cell membrane. The sizes of the ß-lactamase synthesized in E.coli and in yeast transformants were compared by

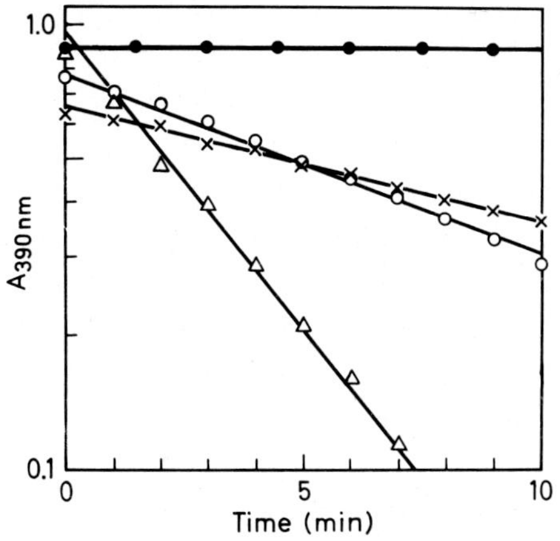

Fig. 6. ß-lactamase activity curves of cell-free extracts from two yeast transformants, YT6-2 and YT8-3. As a control an E.coli strain carrying plasmid pBR325 was extracted by sonication. ß-lactamase activity was determined in 1 ml at 37 °C with cephalosporin 87/312 (nitrocefin) as substrate as described (26). 1 g of YT6-2, YT8-3 or YT10-1 was disrupted and suspended in a final volume of 5 ml of which 25 µl were used per assay. 0.5 g E.coli (pBR325) grown overnight in the presence of benzyl-penicillin, was sonicated in 3 ml 50 mM Tris-HCl (pH 7.8), 50 µM 2-mercaptoethanol and 2 µl were used per assay. Δ——Δ, E.coli (pBR325) sonicate; o——o, YT6-2; x——x, YT8-3; •——•, YT10-1.

electrophoresis of an E.coli sonicate and yeast extracts in non-denaturing polyacrylamide gels. The ß-lactamase bands in the gel were detected with the nitrocefin reaction. No difference in mobility between the enzyme activity synthesized in the two organisms could be detected. In previously reported experiments[30], we observed a slower migration rate for the activity in the yeast extracts. This difference in mobility was not found in later extracts and we therefore have to refrain from any comment on the molecular weight of the yeast enzyme at the moment.

Evidence for the expression of a chloramphenicol and a kanamycin resistance gene in yeast

YT6 and YT8 transformants were plated on minimal plates plus 1% ethanol and 1, 1.2, 1.4, 1.6, 1.8, or 2 mg chloramphenicol per ml. Both YT6 and YT8 grow on plates containing less than 1.8 mg chloramphenicol per ml. On the higher concentrations YT8 does not grow, but YT6 shows many growing colonies next to cells that are inhibited. The only difference between the plasmids of YT6 and YT8 (see Table 1) is the integration site in pBR325. YT8 contains plasmid pMP80-3 with an interrupted chloramphenicol resistance gene, whereas this gene is intact in YT6. The observed resistance of YT6, therefore, is provisionally interpreted as a result of the expression of the bacterial chloramphenicol resistance gene, which codes for a chloramphenicol transacetylase. Experiments to directly assay the enzyme activity in cell-free extracts are under way.

YT10, which contains pMP81 bearing a kanamycin resistance gene, shows a similar mode of resistance to neomycin. Kanamycin and neomycin are aminoglycoside antibiotics and can both be inactivated by a number of aminoglycoside modifying enzymes[31,32]. After one week of growth at 35°C plates containing 0.2% glucose and 3 mg neomycin B sulphate per ml, many YT10 cells had formed colonies, whereas most of the controle YT8 cells did not grow. On replating, the YT10 cells showed neomycin resistance.

DISCUSSION

The basic question whether prokaryotic structural genes can be expressed in an eukaryotic cell has been answered in the affirmative by an extended analysis of the expression of a bacterial ß-lactamase gene in S.cerevisiae. In addition, evidence has been presented for the functional expression in yeast of two other bacterial genes, coding for resistances to chloramphenicol and kanamycin.

We have chosen to study resistance genes because they are present on many bacterial cloning vectors and are provided with suitable restriction sites for integration. Functional expression of these genes in S.cerevisiae could enable them to assume a role in yeast transformation analogous to their function as selective markers in bacterial cloning systems.

Yeast is not sensitive to penicillin antibiotics and the resistance conferred by the ß-lactamase gene, therefore, cannot be shown directly. The presence of the ß-lactamase activity, however, can be readily demonstrated. The growth of indicator bacteria in the soft agar assay is easily detectable and with replica plating we were able to screen single colonies for segregation provided that the colony density was not too high. The nitrocefin assay is very rapid and works well for larger cell streaks. On single colonies we have not yet obtained satisfactory results.

Under the proper conditions, yeast is sensitive to chloramphenicol and neomycin and the presence of active bacterial genes in the cell coding for resistances to these compounds can be directly tested. Yeast transformant YT6-2 contains the intact chloramphenicol resistance gene as shown by bacterial transformation with cell-free extracts or with the supercoiled DNA fraction. Nutrient plates with 1% ethanol and high concentrations of chloramphenicol allow a high percentage of cells to grow. For YT8-3, which contains an interrupted chloramphenicol gene, very few growing cells were seen when plated under identical conditions. Some of these colonies were not resistant upon retesting. The same phenomenon was observed on neomycin.

The presence of foreign genes on 2-μm DNA opens a new approach to the study of its function in the yeast cell. For this purpose the bacterial genes offer an alternative to the use of cloned yeast nuclear genes.

ACKNOWLEDGEMENTS

A gift of nitrocefin from Glaxo-Allenburys Research Ltd. by courtesy of Dr. C.H. O'Callaghan is gratefully acknowledged. I wish to thank Dr. J. Beggs for providing plasmid pJDB219, P. Hardy for critical reading of the manuscript and Professor W. Beermann for his interest and support.

REFERENCES

1. Hollenberg, C.P., Borst, P., and van Bruggen, E.F.J. (1970) Biochim. Biophys. Acta 209, 1.
2. Hollenberg, C.P., Degelmann, A., Kustermann-Kuhn, B., and Royer, H.-D. (1976) Proc. Natl. Acad. Sci. USA 73, 2072.
3. Guerineau, M., Grandchamp, C., and Slonimski, P.P. (1976) Proc. Natl. Acad. Sci. USA 73, 3030.
4. Beggs, J.D., Guerineau, M., and Atkins, J.F. (1976) Molec. Gen. Genet. 148, 287.
5. Livingston, D.M., and Klein, H.L. (1977) J. Bact. 129, 472.
6. Royer, H.-D., and Hollenberg, C.P. (1977) Molec. Gen. Genet. 150, 271.

7. Cameron, J.R., Philippsen, P., and Davis, R.W., (1977) Nucl. Acids. Res. 4, 1429.

8. Gubbins, E.J., Newlon, C.S., Kann, M.D., and Donelson, J.E. (1977) Gene 1, 185.

9. Hollenberg, C.P., Kustermann-Kuhn, B., and Royer, H.-D. (1976) Gene 1, 33.

10. Hollenberg, C.P. (1978) Molec. Gen. Genet. 162, 23.

11. Struhl, K., Cameron, J.R., and Davis, R.W. (1976) Proc. Natl. Acad. Sci. USA 73, 1471.

12. Ratzkin, B., and Carbon, J. (1977) Proc. Natl. Acad. Sci. USA 74, 487.

13. Struhl, K., and Davis, R.W. (1977) Proc. Natl. Acad. Sci. USA 74, 5255.

14. Clarke, L., and Carbon, J. (1978) J. Molec. Biol. 120, 517.

15. Walz, A., Ratzkin, B., and Carbon, J. (1978) Proc. Natl. Acad. Sci. USA 75, 6172.

16. Beggs, J.D. (1978) Nature 275, 104.

17. Hicks, J.B., Hinnen, A., and Fink, G.R. (1978) Cold Spring Harbor Symp. Quant. Biol. in press.

18. Struhl, K., Stinchcomb, D.T., Scherer, S., and Davis, R.W., (1979) Proc. Natl. Acad. Sci. USA 76, 1035.

19. Bolivar, F. (1978) Gene 4, 121.

20. Covey, C., Richardson, D., and Carbon, J. (1976) Molec. Gen. Genet. 145, 155.

21. Hinnen, A., Hicks, J.B., and Fink, G.R. (1978) Proc. Natl. Acad. Sci. USA 75, 1929.

22. Guerineau, M., Grandchamp, C., Paoletti, J., and Slonimski, P. (1971) Biochim. Biophys. Res. Commun. 42, 550.

23. Bolivar, F., Rodriguez, R.,Betlach, M., and Boyer, H.W. (1977) Gene 2, 75.

24. Meynell, E., and Datta, N. (1967) Nature 214, 885.

25. So, M., Gill, R., and Falkow, S. (1975) Molec. Gen. Genet. 142, 239.

26. O'Callaghan, C.H., and Morris, A. (1972) Antimicrobiol. Agents and Chemotherapy 2, 442.

27. Matthew, M., and Hedges, R.W. (1976) J. Bacteriol. 125, 713.

28. Ambler, R.P., and Scott, G.K. (1978) Proc. Natl. Acad. Sci. USA 75, 3732.

29. Suttcliffe, J.G. (1978) Proc. Natl. Acad. Sci. USA 75, 3737.

30. Hollenberg, C.P. (1979) in Extrachromosomal DNA (D. Cummings, P. Borst, I. Dawid, S. Weissman, C.F. Fox, eds.) ICN-UCLA Symp. Vol. 15, Academic Press, New York, in press.

31. Benveniste, R., and Davies, J.(1973) Ann.Rev.Biochem. 42, 471.

32. Haas, M.J., and Dowding, J.E. (1975) Methods Enzymol. 43, 611.

AUTHOR INDEX